W9-ANK-167

Rings and Geometry

NATO ASI Series

Advanced Science Institutes Series

A series presenting the results of activities sponsored by the NATO Science Committee, which aims at the dissemination of advanced scientific and technological knowledge, with a view to strengthening links between scientific communities.

The series is published by an international board of publishers in conjunction with the NATO Scientific Affairs Division

A	Life Sciences	Plenum Publishing Corporation
B	Physics	London and New York
C	Mathematical and Physical Sciences	D. Reidel Publishing Company Dordrecht, Boston and Lancaster
D	Behavioural and Social Sciences	Martinus Nijhoff Publishers
E	Engineering and Materials Sciences	The Hague, Boston and Lancaster
F	Computer and Systems Sciences	Springer-Verlag
G	Ecological Sciences	Berlin, Heidelberg, New York and Tokyo

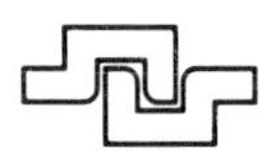

Series C: Mathematical and Physical Sciences Vol. 160

Rings and Geometry

edited by

Rüstem Kaya

University of Anadolu, Graduate School of Sciences,
Eskisehir, Turkey

Peter Plaumann

Mathematical Institute, University Erlangen-Nürnberg,
Erlangen, F.R.G.

Karl Strambach

Mathematical Institute, University Erlangen-Nürnberg,
Erlangen, F.R.G.

D. Reidel Publishing Company

Dordrecht / Boston / Lancaster / Tokyo

Published in cooperation with NATO Scientific Affairs Division

Proceedings of the NATO Advanced Study Institute on
Rings and Geometry
Istanbul, Turkey
September 2-14, 1984

Library of Congress Cataloging in Publication Data

NATO Advanced Study Institute on Rings and Geometry (1984: Istanbul, Turkey)
Rings and geometry.

(NATO ASI series. Series C, Mathematical and physical sciences; vol. 160)
"Proceedings of the NATO Advanced Study Institute on Rings and Geometry, Istanbul, Turkey, September 2–14, 1984"—T.p. verso.
Includes index.
1. Geometry—Congresses. 2. Rings (Algebra)—Congresses. I. Kaya, Rüstem. II. Plaumann, Peter. III. Strambach, Karl. IV. Title. V. Series: NATO ASI series. Series C, Mathematical and physical sciences; vol. 160.
QA447.N37 1984 516'.186 85-18284
ISBN 90-277-2112-2

Published by D. Reidel Publishing Company
P.O. Box 17, 3300 AA Dordrecht, Holland

Sold and distributed in the U.S.A. and Canada
by Kluwer Academic Publishers,
190 Old Derby Street, Hingham, MA 02043, U.S.A.

In all other countries, sold and distributed
by Kluwer Academic Publishers Group,
P.O. Box 322, 3300 AH Dordrecht, Holland

D. Reidel Publishing Company is a member of the Kluwer Academic Publishers Group

Printed in The Netherlands.

TABLE OF CONTENTS

EPILOG

PREFACE

When looking for applications of ring theory in geometry, one first thinks of algebraic geometry, which sometimes may even be interpreted as the concrete side of commutative algebra. However, this highly developed branch of mathematics has been dealt with in a variety of monographs, so that - in spite of its technical complexity - it can be regarded as relatively well accessible. While in the last 120 years algebraic geometry has again and again attracted concentrated interest - which right now has reached a peak once more - , the numerous other applications of ring theory in geometry have not been assembled in a textbook and are scattered in many papers throughout the literature, which makes it hard for them to emerge from the shadow of the brilliant theory of algebraic geometry.

It is the aim of these proceedings to give a unifying presentation of those geometrical applications of ring theory outside of algebraic geometry, and to show that they offer a considerable wealth of beautiful ideas, too. Furthermore it becomes apparent that there are natural connections to many branches of modern mathematics, e.g. to the theory of (algebraic) groups and of Jordan algebras, and to combinatorics. To make these remarks more precise, we will now give a description of the contents.

In the first chapter, an approach towards a theory of non-commutative algebraic geometry is attempted from two different points of view. Firstly through generalization of the algebraic tools, which involves in particular a study of the non-commutative analog of the function field of an algebraic variety; secondly through the use of synthetic geometry, many of whose definitions for classical objects of algebraic geometry (such as rational normal curves) carry over to the case of a non-commutative coordinate field.

The second chapter is dedicated to the more traditional theory of projective Hjelmslev planes, where homomorphisms rather than single planes lie in the center of interest. First the emphasis is put on topological and particularly locally compact Hjelmslev planes and their topological properties; also the connection between the algebraic properties of topological coordinate rings and the incidence-theoretic behaviour (closedness of configurations) of the associated affine Hjelmslev planes is investigated. Then for finite Hjelmslev planes the factorization problem is studied in detail, and a number of existence- resp. nonexistence theorems are derived from this. Some fruitful construction principles for finite Hjelmslev planes are presented, e.g.

the method of auxiliary matrices, and influences on the theory of designs are pursued.

The third chapter has its historical origins in the investigations of H. Freudenthal about Moufang planes and in the papers of T.A. Springer and F.D. Veldkamp about Hjelmslev-Moufang planes. The aim of these papers was to establish a bridge between geometry on one side and the theory of Jordan algebras and algebraic groups on the other side. In the first part of this chapter, the classical construction of a Hjelmslev-Moufang plane over any octonion algebra (by means of a reduced exceptional simple Jordan algebra) is modified in order to yield a method of constructing so-called "Moufang-Veldkamp planes" whose coordinate ring can be any alternative ring of stable rank 2; these "Moufang-Veldkamp planes" are defined abstractly by a set of axioms on a plane incidence structure which is additionally endowed with a neighborhood relation. Furthermore, a proof is indicated for the fact that every Moufang-Veldkamp plane arises in the above way. The second part of the chapter summarizes the axiomatic theory of projective planes over associative rings of stable rank 2; apart from general assumptions on the incidence- and neighborhood-relation, the existence of sufficiently many transvections, dilatations and their duals must be required in order to characterize these planes. Finally, the homomorphisms between two such ring planes are classified.

The fourth chapter begins with an account of the possibilities of extending the structural identity between classical projective geometry and linear algebra over division rings to arbitrary rings; over large classes of rings the structure of special linear, orthogonal, symplectic and unitary groups is investigated and the isomorphisms between such groups are studied. When classifying the linear maps of matrix rings (over commutative rings) that preserve certain invariants (e.g. the determinant), unexpected connections with the theory of affine group schemes become apparent.

One contribution of this chapter is dedicated to the geometrical applications of quadratic algebras (which are called kinematic algebras there); in particular, the geometrical structure of the units of alternative quadratic algebras is put in contrasting perspective to the third chapter.

The fourth chapter ends with a discussion of how to find natural coordinatizations of general geometries. When taking a lattice as a prototype of such a geometry, then coordinatization in a very general sense means a construction of an algebraic structure (mostly a module) whose lattice of substructures or congruence-relations is isomorphic to the given lattice. By simple, natural and independent axioms of a geometric flavor lattices are characterized which are isomorphic to submodule lattices of torsion free modules of Goldie dimension at least three over left Ore domains.

An appendix to the book contains an encouragement to the geometers by C. Arf, where he describes the concepts of geometry as indispensable for mathematics.

We are very grateful to the NATO Science Division for providing the funds for this Advanced Study Institute and to the participants in this summer school who were a receptive audience for the lectures. But primarily we wish to thank the authors who strove to keep a balance between their present preferences and the prescribed overall theme.
We hope that thereby a book has come into being which -in spite of the personal preferences of the authors which have not been suppressed- reveals a unifying approach. Thus it can well serve as a substitute for a lacking monograph in that area of mathematics where the ring theory occuring outside of algebraic geometry is harmonically unified with geometry.

Erlangen and Istanbul,

May 31, 1985

Rüstem Kaya, Peter Plaumann and Karl Strambach

Part I

Non-Commutative Algebraic Geometry

PRINCIPLES OF NON-COMMUTATIVE ALGEBRAIC GEOMETRY (*)

P. M. Cohn
University College London
Gower Street
London WC1E 6BT

Introduction

Algebraic geometry (of the usual kind) may be defined as the study of the solutions of sets of polynomial equations in $x_1,\ldots,x_n$ over a commutative field k. A great deal of insight into this problem is obtained by taking the geometric picture into account: The solutions form a subset of affine n-space, $\mathbb{A}^n(k)$, and if one admits points at infinity, by taking homogeneous coordinates, one has projective space $\mathbb{P}^n(k)$.

If one now allows the field to be non-commutative, the problem changes radically; it is a bit like comparing abelian and non-abelian groups, - many of the most interesting problems of the latter do not

(*) The author accepts no responsibility for this title.

R. Kaya et al. (eds.), Rings and Geometry, 3–37.

even arise in the former. So far one has very little insight into non-commutative algebraic geometry, in particular, the geometric picture seems to have got lost. In these lectures I want to explain the problems one is facing; perhaps in solving them one can regain some kind of geometric picture.

I. Free algebras and free fields

All our rings are associative, with a unit element 1 which is preserved by homomorphisms, inherited by subrings and acts unitally on modules. It may of course happen that $1 = 0$, but then we have the zero ring, consisting only of 0, and this is excluded when we are dealing with a field or a subring of a field. By a 'field' we understand a not necessarily commutative division ring; Sometimes we add 'skew' for emphasis. If A is any ring, by an <u>A-algebra</u> one understands a ring R with a homomorphism from A to the centre of R; here A may as well be taken to be commutative. By an <u>A-ring</u> we mean a ring R with a homomorphism $A \to R$.

Let k be a commutative field and $X = \{x_1, \ldots, x_n\}$ a set of indeterminates. We consider $k[X]$, the polynomial ring in the indeterminates X over k, and its field of fractions $k(X)$. The elements of the latter have the form $\phi = f/g$, $f, g \in k[X]$, so there is little difference between considering zeros of polynomials or of rational functions: $\phi = 0$ if and only if $g \neq 0$ and $f = 0$.

Our first task is to find non-commutative analogues of $k[X]$ and $k(X)$. As non-commutative polynomial ring we might take $k\langle X\rangle$, the ring of all polynomials in the non-commuting indeterminates x_i over k;

it consists of all expressions

$$\sum a_{i_1 \ldots i_r} x_{i_1} \ldots x_{i_r} \qquad a_{i_1 \ldots i_r} \in k \, .$$

with the obvious multiplication; this is the free k-algebra on X. It turns out that this is not quite general enough, in that we want to allow a skew field of coefficients. Moreover, we do not want all the field elements to commute with the variables. This suggests the following

Definition. Let D be any field, E a subfield and X a set. The free D-ring on X centralizing E is the D-ring $D_E\langle X\rangle$ generated by D and X with defining relations

(1) $$\alpha x = x\alpha \qquad \alpha \in E \, , \quad x \in X \, .$$

Usually E will be the centre of D (or a subfield of the centre). This has the following reason. If we want to replace $x \in X$ by an element a of D we must make sure that $\alpha a = a\alpha$, if the resulting map $D_E\langle X\rangle \to D$ is to be a homomorphism (by (1)). For general E let us write E' for the centralizer of E in D, then it is clear that X can always be specialized to values in E'. In particular,

(i) if C is the centre of D, X can be specialized to any values of D in $D_C\langle X\rangle$,

(ii) in $D_D\langle X\rangle$, X can be specialized to any values in C (instead of $D_D\langle X\rangle$ one usually writes $D\langle X\rangle$).

Together with $D_E\langle X\rangle$ we want to consider the field of rational functions (in non-commuting indeterminates), but we first have to define what this means. A commutative ring R has a field of fractions iff it is an integral domain, i.e. $1 \neq 0$ and R has no zero-divisors.

For a general ring R these conditions are clearly still necessary, but no longer sufficient. Moreover, even when there is a field of fractions, it need not be unique. What we shall look for is a field of fractions which is universal in a sense which need not concern us too closely here. To obtain it we have to see how fields are constructed in the non-commutative case. Taking first the commutative case again, if R is a commutative integral domain, we form its field of fractions by formally adjoining inverses for all the non-zero elements. Thus we form the set of all expressions a/b $(a,b \in R,\ b \neq 0)$ whith the rule $a/b = a'/b'$ iff $ab' = ba'$, and define addition and multiplication

$$a/b + c/d = (ad + bc)/bd, \qquad a/b \cdot c/d = ac/bd .$$

The corresponding construction in the non-commutative case is to invert certain matrices. This would give nothing new for a commutative ring, where

$$A^{-1} = \mathrm{adj}(A) \cdot \det(A)^{-1} .$$

But over a non-commutative ring we do not have a determinant at our disposal (even the Dieudonné determinant, [15] is really a rational function rather than a polynomial in the entries of A), so forming A^{-1} will give something genuinely new.

Thus for an nxn matrix A we formally adjoin n^2 indeterminates a'_{ij} forming an nxn matrix A', with defining relations (in matrix form)

$$AA' = A'A = I . \tag{2}$$

More generally we can invert any set Σ of matrices over R and so obtain the universal Σ-inverting ring or localization R_Σ . It has the property that there is a Σ-inverting homomorphism $\lambda: R \to R_\Sigma$

(i.e. a homomorphism mapping every matrix of Σ to an invertible matrix) and λ is universal in the sense that every Σ-inverting homomorphism $f:R\to S$ can be uniquely factored by λ to give a homomorphism $f':R_\Sigma \to S$ satisfying $f=\lambda f'$. For suitable sets Σ this ring can also be described as the set of all components of solutions of

$$(3) \qquad P\,u = b \qquad\qquad P \in \Sigma .$$

For we have

$$\begin{pmatrix} P & b \\ 0 & 1 \end{pmatrix}^{-1} = \begin{pmatrix} P^{-1} & -P^{1}b \\ 0 & 1 \end{pmatrix} .$$

which shows that R_Σ contains the solution of (2); conversely the jth column of P^{-1} may be obtained as solution of $Pu = e_j$.

Instead of (3) it will be convenient to put $A = (P, -b)$ and write our system as

$$(4) \qquad A\,\nu = 0 ,$$

where $A = (A_o, A_1, \ldots, A_n)$ is an nxn+1 matrix and $\nu = (1, \nu_1, \ldots, \nu_n)^T$ is a normalized solution vector (i.e. the first component is 1; T stands for 'transpose', thus ν is to be read as a column). We shall regard (4) as an admissible system for ν_n provided that $(A_1, \ldots, A_n) \in \Sigma$. We shall also use the following terminology:

$A_* = (A_1, \ldots, A_{n-1})$ is the core of A or of the system (4), (A_o, A_*) is the numerator and (A_*, A_n) the denominator. We also put $\nu_* = (\nu_1, \ldots, \nu_{n-1})^T$. In this notation (4) may be written as

$$(5) \qquad (A_*, -A_o) = (A_*, A_n) \begin{pmatrix} I & \nu_* \\ 0 & \nu_n \end{pmatrix} .$$

Essentially this expresses ν_n as a fraction $\frac{\text{numerator}}{\text{denumerator}}$. To make this precise let us call two matrices A , B associated if $B = PAQ$

where P, Q are invertible; if $\begin{pmatrix} A & 0 \\ 0 & I \end{pmatrix}$ is associated to $\begin{pmatrix} B & 0 \\ 0 & I \end{pmatrix}$, where the I's may be of different orders, then A, B are said to be stably associated. Now (5) shows that when the denominator (A_*, A_n) is a unit, then ν_n is stably associated to its numerator. We shall describe (5) as Cramer's rule. To give an example, if we are in $D_E\langle X\rangle$ and we specialize X so that both numerator and denominator in (5) become non-singular (i.e. invertible matrices over D), then we obtain a non-zero value for ν_n, i.e. the rational function ν_n is defined and non-zero under this specialization.

To get a 'universal' field of fractions we want to invert as many matrices as possible, so let us decide first which matrices cannot be inverted. If A, B are non-zero $n \times n$ matrices and $AB = 0$, it may still be possible to invert A, because B may be mapped to 0. But if

(6) $\qquad A = PQ \qquad P \; n \times r, \quad Q \; r \times n, \quad r < n,$

then under any homomorphism to a field, A must become singular (= non-invertible). Let us call A non-full if it admits a factorization (6) and full otherwise. Then the most we can hope to invert are all the full matrices; so if there is a field K with a homomorphism $R \to K$ inverting all the full matrices over R, and no proper subfield of K has this property, then K is the universal field of fractions of R (in a sense that should be intuitively clear and that can be made precise, see [3] Ch. 7). Now one has

Theorem 1. Let D be any field and E a subfield. Then the free D-ring on any set X centralizing E, $D_E\langle X\rangle$ has a field $D_E ⦓X⦔$ inverting all full matrices.

$D_E \langle\!\langle X \rangle\!\rangle$ will be called the <u>free field</u> on X over D (centralizing E).

We shall not prove this result here in its general form (it follows because $D_E \langle X \rangle$ is a semifir, cf. [3]), but we shall prove a special case which will be enough for our needs. Our main tool is the specialization lemma, also useful elsewhere, which we shall now describe.

Every infinite commutative field k has the following property, known as the <u>density property</u>: Any non-zero polynomial f in $X = \{x_1, \ldots, x_n\}$ over k takes on a non-zero value $f(\alpha)$ for some $\alpha \in k^X$.

There is a similar result for rational functions $\phi = f/g$ proved in the same way; we need only choose α so that $f(\alpha)g(\alpha) \neq 0$.

The corresponding property in the non-commutative case is <u>Amitsur's theorem on generalized polynomial identities</u> [1]. Given a field D with centre C such that $[D:C] = \infty$, if $0 \neq f \in D_C \langle X \rangle$, then $f(a) \neq 0$ for some $a \in D^X$.

We really want the generalization to the free field $D_C \langle\!\langle X \rangle\!\rangle$. In fact Amitsur [2] uses his result to prove the existence of $k \langle\!\langle X \rangle\!\rangle$. But we shall not need this further result. In view of what was said earlier, we need the following

<u>Specialization Lemma</u>. Let D be a field with centre C and suppose that A.1: $|C| = \infty$, A.2: $[D:C] = \infty$. Then for every full matrix A over $D_C \langle X \rangle$ there exists $a \in D^X$ such that A(a) is non-singular ([4, 10]).

Here $a \in D^X$ is a function from X to D and A(a) is the result of applying this function to the arguments X where they occur in A.

The condition $[D:C] = \infty$ cannot be omitted; whether $|C| = \infty$ is also necessary is not known (Amitsur needs it too in his proof). It could be omitted if we can show that for every square matrix A over an infinite field D there exists $\alpha \in D$ such that $A - \alpha I$ is non-singular.

Proof of Theorem 1 (for the case $E = C$, when A.1-2 hold cf. [13]). We have a natural mapping

$$(7) \qquad D_C\langle X\rangle \to D^{D^X}$$

given by $p \mapsto (p_f)$ where $p_f = p(Xf)$, for any $f : X \to D$. Now with each square matrix A over $D_C\langle X\rangle$ we associate a subset of D^X, its <u>singularity support</u>:

$$\mathcal{S}(A) = \{f \in D^X \mid A(Xf) \text{ non-singular}\} .$$

Of course $\mathcal{S}(A) = \emptyset$ when A is non-full, but by the specialization lemma, $\mathcal{S}(A) \neq \emptyset$ if A is full. Let us abbreviate the diagonal sum $\begin{pmatrix} A & 0 \\ 0 & B \end{pmatrix}$ as $A \oplus B$. If A, B are non-singular matrices, then so is $A \oplus B$, hence

$$\mathcal{S}(A \oplus B) = \mathcal{S}(A) \cap \mathcal{S}(B) , \qquad \mathcal{S}(I) = D^X .$$

Hence the family of supports $\mathcal{S}(A)$ (A full) is closed under finite intersections. We can therefore find an ultrafilter F on D^X containing all the sets $\mathcal{S}(A)$. By the ultraproduct theorem (cf. [12] p. 210), the natural mapping

$$D_C\langle X\rangle \to D^{D^X}/F$$

is an embedding in a field, and if A is a full matrix, then $A(\alpha)$ is non-singular for all $\alpha \in \mathcal{S}(A)$, hence A is then non-singular in the image. Therefore if U is the field generated by the image, we have a homomorphism $D_C\langle X\rangle \to U$ inverting all full matrices; by uniqueness if follows that $U = D_C \langle X \rangle$.

Although we cannot prove Theorem 1 in this way for all cases (the actual proof is constructive, but is longer), we can extend it somewhat. In the first place we must extend Amitsur's theorem. Let us write E' for the centralizer of E in D . Then we have the

<u>Generalized GPI theorem</u>. Let D be a field with infinite centre and let E be a subfield of D with centralizer E' such that $E'' = E$ and $[EaE' : E] = \infty$ for all $a \in D^X$. Then for any full matrix A over $D_E \langle X \rangle$, $A(\alpha)$ is non-singular for some choice of values α of X in E' (cf. [14] Ch. 5).

With this result we can prove Theorem 1 for $D_E \langle X \rangle$, where D, E satisfy the hypotheses listed here.

Let us note two problems at this point; given a square matrix A over a field, an element α is called a <u>singular eigenvalue</u> of A if $A - \alpha I$ is singular. Now we have

<u>Problem 1</u>. Let D be a skew field with finite centre C . Given $A \in D_n$ find an element of D which is <u>not</u> singular eigenvalue of A .

When C is infinite, such elements are easy to find, while none may exist when D is finite. This problem immediately suggests a counter-part:

<u>Problem 2</u>. Does every square matrix over a field D have a singular eigenvalue in some extension of D ?

Clearly we cannot expect a singular eigenvalue in D itself (this does not even hold in the commutative case). But it is clear that over an algebraically closed commutative field every square matrix has an eigenvalue, and it has been shown by R.M.W. Wood (unpublished) that every matrix over the real quaternions has a singular eigenvalue.

Let us mention another application of the specialization lemma. The free D-ring $D_E\langle X\rangle$ has a power series completion, denoted by $D_E\langle\langle X\rangle\rangle$ (it will simplify matters if we assume that X is finite; - for infinite X it makes a difference whether the degrees of the $x \in X$ are bounded). We have a natural embedding

$$D_E\langle X\rangle \rightarrow D_E\langle\langle X\rangle\rangle$$

and this is an <u>honest</u> homomorphism, i.e. it keeps full matrices full (by the inertia theorem, see [3] p. 103). We note that a matrix A over $D_E\langle\langle X\rangle\rangle$ is invertible iff $A(0)$ is invertible over D. For we can write $A = A_0 + A_1 + \ldots$ where A_i is homogeneous of degree i, in particular $A_0 = A(0)$. When this is invertible, we can write

$$A = A_0(I - C)$$

where C lies in the augmentation ideal, hence $A^{-1} = \left(\sum_0^\infty C^r\right) A_0^{-1}$. The converse is clear. This means that any element of $D_E\langle X\rangle$ which is defined at $X = 0$ has a power series expansion. Now if p is any element of $D_E\langle X\rangle$, with denominator A in some representation, then A is full, and for each $f \in \mathcal{S}(A)$ we can write p as a power series in $y = x - xf$.

When D, E satisfy the hypotheses of the specialization lemma, any full matrix is defined for some values of X, so if p is defined at $x = a$, we can write p as a power series in $y = x - a$ $(x \in X)$. Thus at each point of $D_E\langle X\rangle$ we have a completion; the advantage over the representation obtained from Cramer's rule is that we have a convenient normal form in the power series.

II. Specializations and the rational topology.

Let K be a field with centre C and E a field containing K. We shall consider zeros of $f \in K_C\langle X\rangle$ in E^n, or more generally, singularities of a square matrix A over $K_C\langle X\rangle$, i.e. points of E^n at which A becomes singular (cf. [11], [10] ch. 8). Not only is this more general, it also allows an important simplification to be made. It is clear that the set of points where A becomes singular is unchanged when we pass from A to a matrix stably associated to A. We shall use this freedom to simplify A as follows.

<u>Proposition 1</u> (Linearization process). Every matrix A over $K_C\langle X\rangle$ is stably associated to a matrix of the form

$$B_o + \sum B_i x_i \qquad B_o \in {}^mK^n, \quad B_i \in {}^mC^n \tag{1}$$

(Here ${}^mK^n$ denotes the set of all $m\times n$ matrices over K).

Proof. It will be enough if we show how to resolve a product ab. Consider an entry $c+ab$ in A, in the south-east corner, say. Writing only this corner, we have

$$\begin{bmatrix} c+ab \end{bmatrix} \to \begin{bmatrix} c+ab & 0 \\ 0 & 1 \end{bmatrix} \to \begin{bmatrix} c+ab & a \\ 0 & 1 \end{bmatrix} \to \begin{bmatrix} c & a \\ -b & 1 \end{bmatrix}$$

By repeated application of this process we reach the form (1).

The points in affine n-space over E, $\mathbb{A}^n(E)$ or E^n, are given by n-tuples over E. We denote them by $\alpha, \beta, \ldots$ and define: If $\alpha, \beta \in E^n$, then β is a <u>specialization</u> of α over K, in symbols $\alpha \underset{K}{\to} \beta$, if for any square matrix A over $K_C\langle X\rangle$,

$$A(\alpha) \text{ singular} \Longrightarrow A(\beta) \text{ singular.}$$

Since singularity is preserved by stable association, we need only require this for linear matrices. If the map $X \mapsto \alpha$ defines an isomorphism, we shall call α free over K. Thus $\alpha \in E^n$ is free over K iff

$$A_o + \sum A_i\alpha_i \text{ is singular} \Longleftrightarrow A_o + \sum A_i x_i \text{ is non-full.}$$

Besides affine space $\mathbb{A}^n(E)$ we have projective space $\mathbb{P}^n(E)$, whose elements are $(n+1)$-tuples $\xi = (\xi_o,\ldots,\xi_n)$ not all 0, with the understanding that ξ,η represent the same point iff $\xi_i = \eta_i\lambda$ for some $\lambda \in E^x$. To define specialization in $\mathbb{P}^n$ we use the linear expression for matrices. Thus we say $\xi \underset{K}{\rightarrow} \eta$ iff

$$\sum_o^n A_i \xi_i \quad \text{singular} \Longrightarrow \sum_o^n A_i \eta_i \quad \text{singular, for any } A_i$$

over K. For specialization to a point in K^n we can simplify this criterion:

Theorem 2. Let E/K be any skew field extension $\alpha \in E^n$, $\lambda \in K^n$. Then $\alpha \underset{K}{\rightarrow} \lambda$ iff for any $A_1,\ldots,A_n$ over K,

$$I - \sum A_i(\alpha_i - \lambda_i) \text{ is non singular.} \tag{2}$$

Proof. Assume that $\alpha \underset{K}{\rightarrow} \lambda$. If $I - \sum A_i(x_i-\lambda_i)$ becomes singular for $x_i=\alpha_i$, then it must become singular for $x_i=\lambda_i$. But then we have I, which is non-singular, a contradiction. So (2) holds.

Conversely, if (2) holds, consider the homomorphism ϕ : $K_C\langle X\rangle \rightarrow K(\alpha)$ given by $A \mapsto A(\alpha)$. The set of all square matrices becoming singular under ϕ is called its singular kernel and is denoted by $\text{Ker}\,\phi$. We have to show that under the map $X \mapsto \lambda$ every matrix in $\text{Ker}\,\phi$ becomes singular, and here it is enough to take linear matrices: $A_o + \sum A_i x_i$. So let $A_o + \sum A_i\alpha_i$ be singular, we have to show that $C = A_o + \sum A_i\lambda_i$ is singular. We note that C has entries in K; suppose that C is non-singular.

We have

$$\begin{aligned} A_o + \sum A_i\alpha_i &= A_o + \sum A_i\lambda_i + \sum A_i(\alpha_i - \lambda_i) \\ &= C + \sum A_i(\alpha_i - \lambda_i) \\ &= C\,(I - \sum B_i(\alpha_i - \lambda_i)), \quad \text{where} \quad B_i = -C^{-1}A_i \,. \end{aligned}$$

Here the left-hand side is singular by hypothesis and the right-hand side is non-singular by (2), a contradiction, which establishes the result.

Corollary. Let E/K be a field extension and $\alpha \in E^n$. Then every point of K^n is a specialization of α iff $I - \sum A_i(\alpha_i - \lambda_i)$ is non singular for all A_i over K and all $\lambda_i \in K$.

Let us call a point α satisfying the conditions of this Cor. quasi-free. It is clear that every free point is quasi-free. The converse holds when the specialization lemma can be applied, i.e. when the centre C of K is infinite and $[K:C] = \infty$. For then, if $A_o + \sum A_i x_i$ is full, there exists $\lambda \in K^n$ such that $A_o + \sum A_i\lambda_i$ is non singular, hence $A_o + \sum A_i\alpha_i$ is also non-singular.

To restate the Cor. in more intuitive form we make a couple of definitions. A point $\alpha \in E^n$ is called an inverse eigenvalue of the sequence $A_1, \ldots, A_n$ if $I - \sum A_i\alpha_i$ is singular. Given an extension E/K, we can regard K^n as a subgroup of E^n; its cosets are called the levels in E over K. Thus $\alpha, \beta \in E^n$ are on the same level iff $\alpha - \beta \in K^n$. Now the Cor. states that α is quasi-free, i.e. every point of K^n is a specialization of α iff its level contains no inverse eigenvalue of any sequence of matrices over K. Put the other way, we can say: if the level of α contains an inverse eigenvalue (of some sequence of matrices over K), then there is a $\lambda \in K^n$ which is not a specialization of α.

Let $K \subseteq E \subseteq L$ be fields. Given $\alpha \in L^n$, we define the <u>locus</u> of α in E over K as the set of all specializations of α in E over K.

Examples:

1. If $\alpha \in K^n$, the locus of α in K is just the point α.
2. The locus of α in K is all of K^n precisely when α is quasi-free/K.
3. If a square matrix A over E has an inverse eigenvalue α in L but none in E, then the locus of α in E is empty (when E is existentially closed, cf. IV below, this cannot happen).

As an illustration let us take the commutative case. Given commutative fields $E \supseteq K$, if $\alpha \in E$ is algebraic over K but not in K, then α satisfies an equation

$$f(x) = 0 \tag{3}$$

over K. If (3) also has a solution λ in K, we can replace f by a polynomial of lower degree which still has α as zero but not λ. In the general case it may not be possible to separate out the rational solutions in this way; those that always accompany α represent the locus. Here is an example of a point α which has a specialization in K without being free or itself in K.

In our example the locus of α consists of precisely one point, which may be taken to be ∞. This condition means that $I - A\alpha^{-1}$ is non-singular for all A, i.e. $\alpha I - A$ is non-singular while the condition at $\lambda \in K$ means that $I - A(\alpha - \lambda)$ is singular for some A. Let k be a commutative field of characteristic 0 and form the rational function field $k(t)$. Take $K = k(t) \langle u_1, u_2 \rangle$; then K is infinite-dimensional over its centre $k(t)$ (that $k(t)$ is the centre is intuitively clear; it may be proved using the power series representation at the end of I, or see [14] Ch. 7). Let α be a root of the equation

$$(4) \qquad (x+1)t - tx = 0$$

(one can show that $tx-xt=t$ has a root cf. [6]) in some extension field of K. We cannot specialize $\alpha \to \lambda \in K$, for this would give $t = 0$. So it only remains to show that $\alpha I - A$ is non-singular.

By (4) we have $t\alpha = (\alpha+1)t$ and $At = tA$ for any matrix A over K, hence

$$t(\alpha I - A) = ((\alpha+1)I - A)t .$$

Therefore if $A - \alpha I$ is singular, so is $A - (\alpha+1)I$. By induction $A - (\alpha+\nu)I$ is singular for all $\nu \in \mathbb{Z}$. This means that $A - \alpha I$ has infinitely many central eigenvalues (i.e. eigenvalues of A in the centre of K), whereas it cannot have more than N, if A is $N \times N$ (cf. [3] Ch. 8; essentially this follows by the PAQ-reduction over $K[t]$). So $A - \alpha I$ is non-singular and this shows that $\alpha \underset{K}{\to} \infty$. If we extend K to a field E containing a root $x = s$ of (4), then the solution α of (4) has infinitely many specializations in E, viz. ∞ and s, hence the whole level of s over K, but no point in K itself.

We have already defined for each matrix A over $K_C\langle X\rangle$ its singularity support

$$\mathcal{S}(A) = \{\alpha \in E^n \mid A(\alpha) \text{ non-singular}\}$$

and we saw that

$$(5) \qquad \mathcal{S}(I) = E^n \qquad \mathcal{S}(A \oplus B) = \mathcal{S}(A) \cap \mathcal{S}(B) .$$

It follows that we get a topology on E^n by taking the $\mathcal{S}(A)$ as a base for the open sets. This is the <u>rational</u> <u>K-topology</u> on E^n ('rational' because singularities of matrices give us more sets than zeros of polynomials). Like the Zariski topology it is not Hausdorff, but it is a T_o-topology. We also remark that E^n is irreducible: every non-empty open set is dense. For if $\mathcal{S}(A)$, $\mathcal{S}(B) \neq \emptyset$, then $\mathcal{S}(A \oplus B) = \mathcal{S}(A) \cap \mathcal{S}(B) \neq \emptyset$, by (5).

We note that $\alpha \in E^n$ is free over K precisely when

(6) $$A \text{ full} \Longrightarrow A(\alpha) \text{ non-singular.}$$

This means that $\alpha \in \mathcal{S}(A)$ unless $\mathcal{S}(A) = \emptyset$, so all the non-empty open sets have a non-empty intersection, consisting of all the free points. In K^n there are of course no free points, because $X_1 - \alpha_1$ fails to satisfy (6). Put differently, we can say that in the E-topology the non-empty open sets have empty intersection. This shows that the E-topology on E^n is in general finer than the K-topology. We also note

Theorem 3. Let E/K be a skew field extension and $\alpha \in E^n$. Then the locus of α in K^n is the closure of α in the rational K-topology.

Proof. Let $\lambda \in K^n$; we have $\lambda \in \overline{\{\alpha\}}$ iff α lies in every neighbourhood of λ, i.e. $\lambda \in \mathcal{S}(A) \Longrightarrow \alpha \in \mathcal{S}(A)$ for all matrices A over K. But this just means $A(\alpha)$ singular $\Longrightarrow A(\lambda)$ singular, i.e. $\alpha \underset{K}{\to} \lambda$.

In conclusion one may raise a rather general problem:

Problem 3. Classify the possible loci.

III. Singularities of matrices over a free ring.

Let $R = D_k\langle X\rangle$ be the free D-ring; algebraic varieties will be obtained by taking the subset of E^n $(E \supseteq D)$ where a finite collection of matrices becomes singular. Taking only one matrix A and $X = \{x_1, x_2\}$, we see that for each value a_1 of x_1 we need to find the values where

$A(a_1, x_2)$ becomes singular. So our first problem will be to find the values of x for which a matrix A over $D_k\langle x\rangle$ becomes singular. The set of all such values is called the <u>spectrum</u> of A. To avoid trivialities we shall take A to be a full matrix; this will ensure that its spectrum is a proper subset of E (assuming k to be the centre of D and the conditions of the specialization lemma to be satisfied).

The first point to notice is that singularities are of two kinds. We either have an equation

(1) $$x - a = 0$$

satisfied by a single value, or an equation

(2) $$ax - xb = c \qquad \text{(metro-equation)}$$

which has a solution of the form $x = x_o + \lambda x_1$ where x_o is a particular solution of (2), x_1 is a particular non-zero solution of the associated homogeneous equation

(3) $$ax - xb = 0 ,$$

and λ ranges over $C(a)$, the centralizer of a (cf. [6, 10]). Let us call the solutions of (1) and (2) <u>point singularities</u> and <u>ray singularities</u> respectively. Without attempting a precise definition at this stage we can say that the spectrum of a matrix consists of point and ray singularities and this immediately raises the following question:

<u>Problem 4</u>. Can the spectrum of a full matrix over $D_k\langle x\rangle$ always be written as a finite union of point and ray singularities?

To give an answer one will need to have a much more precise knowledge of the spectrum of a matrix. Here our main handicap is that the knowledge of a point in the spectrum does not permit a reduction of the matrix to one of lower order (the way that knowledge of a zero of a polynomial allows us to reduce its degree). Of course once one has a good control of the spectrum of a matrix in one variable, one will have a better chance of describing the spectrum of a matrix in two variables, giving an idea of 'algebraic sets'.

We have already observed that the spectrum of a matrix is unchanged by stable association, so we may use Theorem 2 to reduce our matrix to linear form

$$C = A + Bx . \tag{4}$$

At $x = \alpha$ we obtain $C(\alpha) = A + B\alpha$ and α lies in the spectrum iff $C(\alpha)$ is singular. Let us denote the rank and nullity of $C(\alpha)$ by $r(\alpha)$, $n(\alpha)$; thus if C is $N \times N$, then $n(\alpha) + r(\alpha) = N$. Clearly both rank and nullity are unchanged on passing to an associated matrix but this no longer holds for stable association. If C becomes $C \oplus I_s$, $r(\alpha)$ increases by s while $n(\alpha)$ remains unchanged. At ∞, for C as in (4), we have $r(\infty) = \text{rk}B$, $n(\infty) = N - \text{rk}B$, so at ∞ the rank is unchanged by stable association.

Let us examine how the spectrum changes when x is transformed. Fix $C = A + Bx$ and write n,r for nullity and rank in terms of x, and n',r' for the same in terms of y, where y is obtained from x by a linear fractional transformation. It will be enough to consider 3 cases (from which we know every fractional linear transformation can be built up):

i) $y = x + \lambda$ $(\lambda \in D)$, $C = A + Bx = A - B\lambda + By$.

$$r'(\infty) = r(\infty) , \qquad n'(\alpha) = n(\alpha-\lambda)$$

for $n'(\alpha)$ is the nullity of $A - B\lambda + B\alpha = A + B(\alpha-\lambda)$.

ii) $y = \lambda x$ $(\lambda \in D^{\times})$. $C = A + B\lambda^{-1}.y$, hence

$$r'(\infty) = r(\infty), \qquad n'(\alpha) = n(\lambda^{-1}\alpha).$$

iii) $y = x^{-1}$ $C' = Ay + B$. Here we have

$$n'(\alpha) = n(\alpha^{-1}) \text{ if } \alpha \neq 0, \quad n'(0) = N - r(\infty), \quad r'(\infty) = N - n(0)$$

if the matrices are $N \times N$. Thus we have

$$n'(0) - r'(\infty) = n(0) - r(\infty).$$

Let us define the <u>defect</u> of C as

$$d(C) = \sum n(\alpha) - r(\infty).$$

Of course this only makes sense if C has no ray singularity; but we can modify the definition either (i) by summing only over point singularities or (ii) by allowing only one point from each ray singularity. What we have shown can be summed up in

<u>Proposition 2</u>. For any matrix C over $D_k\langle x\rangle$ with a pure point spectrum, $d(C)$ is unchanged by fractional linear transformation of the variable.

We shall take k to be infinite. By a suitable transformation we may then assume C to be non-singular at infinity. So it will have the form

$$C = Ix - A,$$

and here A can be shown to be unique up to conjugacy over k ([9]). In this case (with a pure point spectrum),

$$d(C) = \sum n(\alpha) - N, \tag{5}$$

where C is $N \times N$.

<u>Problem 5</u>. Let C be a matrix with pure point spectrum. Then over an existentially closed field (see below), $d(C) = 0$.

We consider some special cases.

1. $r(\infty) = 1$. This means that $C = A + Bx$, where B has rank 1 . By elementary row transformations we can reduce all rows of B after the first row to 0 . We now apply the reduction to echelon form to A , by elementary row transformations and column permutations. The echelon form in this case is $\begin{pmatrix} I & F \\ 0 & 0 \end{pmatrix}$. Bearing in mind that C is full, we thus obtain

$$C = \begin{pmatrix} a_1 - b_1 x & b_2 x & b_3 x & \dots & b_N x \\ a_2 & 1 & 0 & \dots & 0 \\ a_3 & \dots\dots 0 \dots & 1 & \dots & 0 \\ a_N & 0 & 0 & \dots & 1 \end{pmatrix}$$

This is easily seen to be stably associated to

$$a_1 - b_1 x - b_2 x a_2 - \dots - b_N x a_N .$$

We thus have the equation

$$(6) \qquad b_1 x + b_2 x a_2 + \dots + b_N x a_N = a_1 .$$

Its solutions are ray singularities of C (for $N > 1$), since with any x_o, $x_o + \lambda x_1$ for any $\lambda \in C(b_1, \dots, b_N)$ is also a solution, where x_1 satisfies the homogeneous equation

$$b_1 x + b_2 x a_2 + \dots + b_N x a_N = 0 .$$

For $N = 1$, (6) reduces to a point singularity, so we usually assume $N > 1$, and of course take N minimal. This will be ensured by taking $b_1, b_2, \dots, b_N$ right linearly independent over k and $1, a_2, \dots, a_N$ left linearly independent over k .

2. The 2 x 2 case. Here it is possible to prove that singular eigenvalues exist. Given any matrix C , a right eigenvalue of C is an $\alpha \in D$ such that $Cv = v\alpha$ for some non-zero column vector v ; left eigenvalues are defined similarly. These eigenvalues are quite distinct

from the singular eigenvalues, e.g. unlike the latter, left and right eigenvalues are similarity invariants of C, and it can be shown that over a suitable field every square matrix has left and right eigenvalues (cf. [5, 10]). We shall use this fact to prove the

Lemma. If $a \neq 0$, then the equation

$$xax + bx + xc + d = 0 \tag{7}$$

has a solution in some extension field.

Proof. Put $C = \begin{pmatrix} -b & -d \\ a & c \end{pmatrix}$ and consider $Cv = v\alpha$. This has a non-zero solution $v = \begin{pmatrix} x \\ y \end{pmatrix}$ say:

$$-bx - dy = x\alpha ,$$

$$ax + cy = y\alpha .$$

If $y = 0$, then the second equation gives $x = 0$, a contradiction. So we may take $y = 1$ and eliminating α, we get

$$-bx - d = x(ax+c) , \qquad \text{which is just (7).}$$

Now take $A = \begin{pmatrix} a & b \\ c & d \end{pmatrix}$; we have to find x such that $A - xI$ is singular. If $c = 0$, we can take $x = a$. Otherwise on replacing x by cx, we may take $c = 1$. Then

$$\begin{pmatrix} a-x & b \\ 1 & c-x \end{pmatrix} \rightarrow \begin{pmatrix} a-x & b - (a-x)(d-x) \\ 1 & 0 \end{pmatrix}$$

and the condition on x becomes

$$x^2 - xd - ax + (ad-b) = 0 .$$

But this always has a solution, by the lemma.

We can now determine the spectrum of any 2 x 2 matrix (and verify that Problem 5 has a positive answer in this case).

Let $C = A + Bx$ be 2 x 2. If $\mathrm{rk}B = 2$, this amounts to finding a singular eigenvalue of $-B^{-1}A$ and we have seen that this always exists (in some extension). Taking the singularity at infinity, we may thus take C in the form $A + Bx$, where $\mathrm{rk}B = 1$. By case 1, we thus obtain

$$\begin{pmatrix} a_1 - b_1x & b_2x \\ a_2 & 1 \end{pmatrix}$$

and this matrix is singular iff

$$b_1x + b_2xa_2 = a_1 .$$

<u>Either</u> $b_2 = 0$. Then we have another point singularity $x = b_1^{-1}a_1$. Note that the defect is zero in this case.

<u>Or</u> $b_2 \neq 0$. Then we can take $b_2 = -1$ and get the metro-equation

$$b_1x - xa_2 = a_1$$

giving a ray singularity.

3. $N = 3$. We shall assume that $C = A + Bx$ has at least one singularity which may be taken at infinity. If $\mathrm{rk}B = 1$, we get an equation (6), so assume $\mathrm{rk}B = 2$. By row operations and column permutations we can reduce B to the form

$$B = \begin{pmatrix} 1 & 0 & p \\ 0 & 1 & q \\ 0 & 0 & 0 \end{pmatrix}$$

Now consider A. If an element in the last row is non-zero, it may be reduced to 1 and the rest of the column can be reduced to 0; e.g. if $a_{31} \neq 0$, we get the form

$$A = \begin{pmatrix} 0 & a_{12} & a_{13} \\ 0 & a_{22} & a_{23} \\ 1 & a_{32} & a_{33} \end{pmatrix}$$

If $a_{32} \neq 0$ but $a_{31} = 0$, we can interchange rows 1 and 2 and columns 1 and 2 and then reduce as before. This only leaves the case $a_{31} = a_{32} = 0$, $a_{33} \neq 0$; but then C is stably associated to a linear 2×2 matrix, so we can restrict ourselves to the above form. We have

$$A + Bx \to \begin{pmatrix} 0 & a_{12}-xa_{32} & a_{13}+px-xa_{33} \\ 0 & a_{22}+x & a_{23}+qx \\ 1 & a_{32} & a_{33} \end{pmatrix} \to \begin{pmatrix} a_{12}-xa_{32} & a_{13}+px-xa_{33} \\ a_{22}+x & a_{23}+qx \end{pmatrix} .$$

If one of the last four entries is 0, we have a product of two linear terms, a reducible case, so assume that all terms are non-zero.

We may assume that C has a further singularity which may be taken at $x = 0$. Then we have $\mathrm{rk}C(0) = 1$, i.e.

$$\begin{pmatrix} a_{12} & a_{13} \\ a_{22} & a_{23} \end{pmatrix} = \begin{pmatrix} a_1 \\ a_2 \end{pmatrix} \begin{pmatrix} b_1 & b_2 \end{pmatrix} .$$ Write $a_{32} = c_1$, $a_{33} = c_2$.

Then we get

$$C = \begin{pmatrix} a_1b_1 - xc_1 & a_1b_2 + px - xc_2 \\ a_2b_1 + x & a_2b_2 + qx \end{pmatrix} . \tag{8}$$

If $c_1 = 0$ or $q = 0$, this reduces to a quadratic equation

$$a_2b_2+qx = (a_2+xb_1^{-1})(b_2+a_1^{-1}px - a_1^{-1}xc_2) ,$$

or $$a_1b_1-xc_1 = (a_1+pxb_2^{-1} - xc_2b_2^{-1})(b_1+a_2^{-1}x) .$$

So we may assume $c_1 \neq 0$, $q \neq 0$. Similarly if $a_2 = 0$ or $b_2 = 0$, we have

$$\begin{pmatrix} a_1b_1-xc_1 & a_1b_2+px-xc_2 \\ x & qx \end{pmatrix} \quad \text{or} \quad \begin{pmatrix} a_1b_1-xc_1 & px-xc_2 \\ a_2b_1+x & qx \end{pmatrix}$$

if we put $y = x^{-1}$, these matrices become

$$\begin{pmatrix} ya_1b_1y-c_1y & ya_1b_2y+yp-c_2y \\ 1 & q \end{pmatrix} \quad \text{or} \quad \begin{pmatrix} ya_1b_1y-c_1y & yp-c_2y \\ a_2b_1y+1 & q \end{pmatrix}$$

leading to quadratic equations

$$ya_1b_2y + yp - c_2y = (ya_1b_1y - c_1y)q$$

or $$ya_1b_1y - c_1y = (yp - c_2y)q^{-1}(a_2b_1y + 1) .$$

In the general case we have (8), which presumably cannot be reduced to a polynomial equation.

4. Let $C = A - xI$, where $A = (a_{ij})$ and the a_{ij} are independent indeterminates over k. In this case there is always a singular eigenvalue as is shown in [8]. It seems plausible that one can obtain a singular eigenvalue for any matrix in this way by specialization. This would follow from a positive answer to a conjecture, to be discussed in the next lecture.

IV. Existentially closed fields and the Nullstellensatz.

The Hilbert Nullstellensatz is usually stated in two parts. The first part, in Zariski's formulation ([20], p. 165) is

Null 1. If k is a commutative field, $L \supseteq k$ a field extension and

$$f : k[x_1,\ldots,x_n] \to L$$

a surjective homomorphism, then L is algebraic over k.

The proof depends essentially on the Noether normalization lemma. We note that when k is algebraically closed, the conclusion is that $L = k$ (the usual form of the theorem). From Null 1 one can deduce (by the Rabinowitsch trick).

Null 2. If k is algebraically closed and $\mathfrak{a}$ is an ideal in $k[x_1,\ldots,x_n]$, then $g \in k[x_1,\ldots,x_n]$ vanishes at all the common zeros of $\mathfrak{a}$ iff $g^N \in \mathfrak{a}$ for some $N \geq 1$.

To find a non-commutative analogue we shall need to define a notion that corresponds to an algebraically closed field: By an existentially closed field (EC-field for short) we understand a field K (over a fixed commutative groundfield k) such that any existential sentence holding in some extension of K already holds in K itself. Any such sentence can be written as a finite conjunction of disjunctions of basic formulae $f = g$ and negations of such formulae $\neg(f=g)$, where $f, g \in K_k\langle X\rangle$. We note that $f = g$ can be written as $f - g = 0$, $\neg(f=g)$ or $f \neq g$ can be written as $\exists y . (f-g)y = 1$ and any disjunction $f_1 = 0 \vee f_2 = 0 \vee \ldots \vee f_r = 0$ can be written as $f_1 f_2 \ldots f_r = 0$. In this way any such sentence can be reduced to a finite set of equations.

Instead of equations, the vanishing of elements, we can also talk about the singularity of matrices. Given a square matrix $A = (a_{ij})$ we define the sentences

sing (A) as $\exists u_1,\ldots,u_n, v_1,\ldots,v_n \ (\sum a_{1j}u_j=0 \wedge \ldots \wedge \sum a_{nj}u_j=0 \wedge (1-u_1v_1) \ldots (1-u_nv_n) = 0)$

non-sing (A) as $\exists b_{ij}(ij = 1,\ldots,n) \ \sum a_{i\nu}b_{\nu j} = \delta_{ij}$.

The construction of an EC-field containing a given field K (over k) is fairly straightforward, by transfinite induction [12] p. 327 or [16] p. 198. But there is no uniqueness,and we cannot usually find an EC-field that is algebraic over K . On the other hand, there are certain new features not encountered in the commutative case. Thus we can describe being transcendental over k by an elementary sentence:

$$\text{transc}(x) = \exists y,z \ (xy=yx^2 \wedge x^2z=zx^2 \wedge xz\neq zx \wedge y\neq 0)$$

This tells us that $k(x) \cong k(x^2)$, because an inner automorphism of K takes x to x^2 . But there exists z centralizing x^2 but not x , hence $k(x^2) \subset k(x)$. It follows that x is transcendental over k .

This observation is due to W. H. Wheeler (cf. [16]) who more generally has an elementary sentence $\text{transc}_n(x_1,\dots,x_n)$ to say that $x_1,\dots,x_n$ commute pairwise and are algebraically independent over k. It follows that every EC-field over k contains a commutative algebraically closed subfield of infinite transcendence degree over k. One can also show that every EC-field is <u>finitely homogeneous</u>: Given $a_1,\dots,a_n$, $b_1,\dots,b_n \in K$ if there is an isomorphism $f:k(a_1,\dots,a_n) \to k(b_1,\dots,b_n)$ such that $a_i \mapsto b_i$, then there exists $t \in K^\times$ such that $t^{-1}a_i t = b_i$. In a similar vein we note

<u>Proposition 3</u>. Let K be an EC-field over K. Then for any $a_1,\dots,a_r$, $b \in K$, $b \in k(a_1,\dots,a_r)$ iff $\mathcal{C}(b) \supseteq \mathcal{C}(a_1,\dots,a_r)$.

Here $\mathcal{C}$ denotes the centralizer as before.

Proof. $\Longrightarrow$ is clear. To prove $\Longleftarrow$, let $C = k(a_1,\dots,a_r)$. If $b \notin C$, then the conclusion (which can be put as an elementary sentence) is false in $K_C\langle x\rangle$, hence it is also false in K.

In order to formulate an analogue of the Nullstellensatz, let A be a set of square matrices over $K_k\langle X\rangle$ and for any $E \supseteq K$ define (for $|X| = n$)

$$V_E(A) = \{\alpha \in E^n \mid \text{sing}(A(\alpha)) \text{ for all } A \in A\}. \tag{2}$$

We can think of this as the <u>variety</u> defined by A in E^n. Now the analogue of Null 1 (cf. [7]) is

<u>Theorem 4</u>. Let K be an EC-field over k and write $F = K_k\langle X\rangle$. If A is a finite set of matrices such that $V_E(A) \neq \emptyset$ for some $E \supseteq K$, then $V_K(A) \neq \emptyset$. This can fail if A is infinite.

The proof is almost immediate: If A is a finite set which becomes singular in E, then it does so in K, by the definition of EC-field.

To prove the last part, we take K countable (clearly any countable field is contained in a countable EC-field). We assert that K has an outer automorphism. For if K is generated by $a_1, a_2, \ldots,$ chosen so that $a_i \notin k(a_1, \ldots, a_{i-1})$, then by Prop. 3 there exists b_i commuting with $a_1, \ldots, a_{i-1}$ but not with a_i. Denote by β_i the inner automorphism induced by b_i and consider the infinite product

$$\alpha = \beta_1^{\varepsilon_1} \beta_2^{\varepsilon_2} \ldots \qquad \text{where } \varepsilon_i = 0 \text{ or } 1 . \tag{3}$$

This defines an automorphism on K; for every element of K lies in $k(a_1, \ldots, a_r)$ for some r, and here the effect of α is $\beta_1^{\varepsilon_1} \ldots \beta_r^{\varepsilon_r}$. It is clear that each choice of exponents in (3) defines a different automorphism, so we get $2^{\aleph_0}$ distinct automorphisms of K. But K only has $\aleph_0$ elements, so at least one automorphism must be outer.

If σ is outer, consider the skew polynomial ring $K[x ; \sigma]$ consisting of polynomials $\Sigma x^i a_i$ with commutation rule $ax = xa^\sigma$. It is an integral domain, with field of fractions $K(x ; \sigma)$ and here the expressions $ax - xa^\sigma$ $(a \in K)$, $1 - xy$ all have a common zero $(y = x^{-1})$. But they have no common zero in K itself, as we have seen.

We remark that in the commutative case any variety defined by a subset S of $k[x_1, \ldots, x_n]$ can also be defined by a finite set; for S can be replaced by the ideal $\mathfrak{a}$ it generates and $\mathfrak{a}$ has a finite generating set, by the Hilbert basis theorem. By contrast, in the general case not every set A can be replaced by a finite set, as the example given in the proof shows. However, we can from A form a <u>matrix ideal</u>, defined in analogy with ordinary ideals, but using matrices. We shall need an operation to take the place of addition; the usual addition of $N \times N$ matrices will not do because $A + B$ need not be singular even if A and B are. But if A, B agree in all except possibly the first column: $A = (A_1, A_2, \ldots, A_N)$, $B = (B_1, A_2, \ldots, A_N)$, then we can form the determinantal sum

$$(4) \qquad A \nabla B = (A_1 + B_1, A_2, \ldots, A_N)$$

and this will be singular whenever A, B are. We really have $2N$ such operations (4) one for each row and one for each column, and it can only be performed for special pairs of matrices (as above). We shall use ∇ for all these operations, and in using it have to remember that it is non-associative (and generally very difficult to handle). Now we can define a matrix ideal in a ring R as a set $\mathcal{A}$ of square matrices such that

M.1 $\mathcal{A}$ contains all non-full matrices,

M.2 If $A \in \mathcal{A}$ then $A \oplus B \in \mathcal{A}$ for every square matrix B,

M.3 If $A, B \in \mathcal{A}$ and $C = A \nabla B$ is defined, then $C \in \mathcal{A}$,

M.4 If $A \oplus I \in \mathcal{A}$ then $A \in \mathcal{A}$.

If moreover,

M.5 $1 \notin \mathcal{A}$,

M.6 $A, B \notin \mathcal{A} \Longrightarrow A \oplus B \notin \mathcal{A}$,

then $\mathcal{A}$ is called a prime matrix ideal. We shall not develop the theory of matrix ideals here (see [3] Ch. 7), but merely quote two results without proof. With every matrix ideal $\mathcal{A}$ we associate another one, its radical, defined as

$$\sqrt{\mathcal{A}} = \{A \mid \underbrace{A \oplus A \oplus \ldots \oplus A}_{r} \in \mathcal{A} \quad \text{for some } r\}.$$

With this definition we have (cf. [3] Ch. 7).

A. For any matrix ideal $\mathcal{A}$ over a ring R,

$$\sqrt{A} = \bigcap \{ P \mid P \supseteq A \ , P \text{ prime matrix ideal } \} .$$

B. The singular kernel of a homomorphism to a field, $R \to K$ is a prime matrix ideal and conversely, every prime matrix ideal occurs as singular kernel of some homomorphism $R \to K$.

We can now state the general form of the Nullstellensatz [7]:

Theorem 5. Let K be an EC-field over k, B a matrix and A any matrix ideal over $F = K_k\langle X\rangle$. Then $B \in \sqrt{A}$ iff $V_L(A) \subseteq V_L(B)$ for some extension L of K; if A is finitely generated, it is enough to take $L = K$.

Proof. If $B \in \sqrt{A}$ and $\phi : F \to L$ is any homomorphism in which A maps to singular matrices, then $A \subseteq \mathrm{Ker}\phi$, hence $\sqrt{A} \subseteq \mathrm{Ker}\phi$ and so $B \in \mathrm{Ker}\phi$ as claimed.

Conversely, assume $B \notin \sqrt{A}$, then by statement A above there is a prime P such that $P \supseteq A$, $B \notin P$. By statement B, P defines a field L and by definition A is singular over L but not B. Hence B is not singular at all points of $V_L(A)$.

If A is finitely generated, by $A_1, \ldots, A_r$ say, then

$$\mathrm{sing}(A_1) \wedge \ldots \wedge \mathrm{sing}(A_r) \wedge \text{non-sing}(B)$$

holds in L, hence also in K, and the proof is complete. For another version of the non-commutative Nullstellensatz see [16] p. 223.

Sometimes a refinement of the Nullstellensatz is used, in which prime ideals are replaced by maximal ideals. This can also be obtained in the non-commutative case. Write again $F = K_k\langle X\rangle$ where $X = \{x_1, \ldots, x_n\}$. For any $a \in K^n$ we have a homomorphism $\phi_a : F \to K$ where $x_i \mapsto a_i$; the singular kernel is M_a, the matrix ideal generated by $x_1 - a_1, \ldots, x_n - a_n$, and it is clear that M_a is a maximal matrix ideal. Taking K to be an EC-field and A a finitely generated maximal matrix ideal of F, we have $V_K(A) \neq \emptyset$ by Th. 4, say $a = (a_1,\ldots,a_n) \in V_K(A)$. Hence $x_i - a_i \in A$, so $M_a \subseteq A$ and by maximality we have $M_a = A$. Thus every finitely generated maximal matrix ideal has the form M_a for some $a \in K^n$. We deduce the

Corollary. Let K be an EC-field over k and A a finitely generated matrix ideal in $K_k\langle X\rangle$. Then

$$\sqrt{A} = \bigcap \{M \mid M \text{ fin. gen. maximal matrix ideal } \supseteq A\}.$$

Proof. If $\{P_\lambda\}$ is the family on the right, then clearly $\sqrt{A} \subseteq P_\lambda$. Now let $B \notin \sqrt{A}$; then by Th. 5 there exists $a \in K^n$ such that $a \in V_K(A)$ but $B(a)$ is non-singular. Hence $B(a) \notin M_a$ and so $B(a) \notin \bigcap P_\lambda$, and the result follows.

Later we shall also need the homogeneous form. Let us write $X^h = \{x_0, x_1,\ldots,x_n\}$ and consider $K_k\langle X^h\rangle$ as a graded algebra. The only matrices allowed are homogeneous in the x's; such a matrix is stably associated to one of the form $\sum_0^n A_i x_i$.

Theorem 6. Let K be an EC-field over k, $F = K_k\langle X^h\rangle$ and denote by J the singular kernel of the homomorphism $\phi : F \to K$ given by $x_i \mapsto 0$ $(i = 0,1,\ldots,n)$. If A is a homogeneous finitely generated matrix ideal, then $V_K(A) = \emptyset$ iff $\sqrt{A} = J$.

Proof. ϕ maps A to singular matrices (unless A contains a matrix of degree 0, a case that is tacitly excluded), so in any case $\sqrt{A} \subseteq J$. If $V_K(A)$ contains a point $\alpha \neq 0$ then $\sqrt{A} \subseteq M_\alpha$, so $\sqrt{A} \neq J$. Now $\sqrt{A} = \bigcap\{M \mid M \text{ fin. gen. max.} \supseteq A\}$, so if this is $\neq J$, then $V_K(A)$ contains some point $\neq 0$.

So far we have seen that while one can in many cases define the analogues of the classical notions, the corresponding theorems frequently are not true, or can only be established in very special cases. To conclude these lectures I want to discuss a result which is fundamental in algebraic geometry and whose analogue, if true, would form a cornerstone of non-commutative algebraic geometry. This is what is usually called the main theorem of elimination theory:

$\mathbb{P}^n$ is complete (cf. [17] p. 33, [18] p. 45).

We recall that a variety X is <u>complete</u> if for every variety Y the projection $p:X\times Y \to Y$ carries closed sets into closed sets.

Outline proof (for the commutative case). We have to show that

(5) $$p : \mathbb{P}^n \times Y \to Y$$

maps closed sets to closed sets. Being closed is a local property, so we can cover Y by affine open sets and it will be enough to prove the result for affine Y. If Y is closed in $\mathbb{A}^m$, then $\mathbb{P}^n \times Y$ is closed in $\mathbb{P}^n \times \mathbb{A}^m$, so we may take $Y = \mathbb{A}^m$.

Every closed subset Z of $\mathbb{P}^n \times \mathbb{A}^m$ is given by equations

(6) $$f_i(u;y) = 0 \qquad (i = 1,\dots,t) ,$$

where u are homogeneous coordinates in $\mathbb{P}^n$ and y coordinates in $\mathbb{A}^m$. If $y_o \in \mathbb{A}^m$, $p^{-1}(y_o)$ consists of all non-zero solutions in u of $f_i(u;y_o) = 0$. Therefore

(7) $$y_o \in p(Z) \Longleftrightarrow f_i(u;y_o) = 0 \text{ has a non-zero solution } u .$$

Denote by T the set of points $y_o \in \mathbb{A}^m$ satisfying the two sides of (7); we have to show that for any system (6), T is closed.

In the graded ring $k[u]$ write I for the ideal of all polynomials vanishing at 0, and recall that a homogeneous ideal $\mathfrak{a}$ defines the empty set in $\mathbb{P}^n$ iff $\mathfrak{a} \supseteq I^r$ for some $r \geq 1$. Thus T is the set of all $y_o \in \mathbb{A}^m$ such that

$$(f_1(u;y_o),\dots, f_t(u;y_o)) \not\supseteq I^r \tag{8}$$

for all r. Write T_r for the set of all y_o satisfying (8), then $\cap T_r = T$, and to show that T is closed we need only show that each T_r is closed.

Let f_i be of degree d_i in the u's and for each $y \in \mathbb{A}^m$ consider the following map, where V_k denotes the space of homogeneous polynomials of degree k in the u's :

$$\begin{cases} \phi^{(d)}(y) : V_{d-d_1} \oplus \dots \oplus V_{d-d_t} \to V_d \\ (g_1,\dots,g_t) \mapsto \sum f_i(u;y)g_i(u) . \end{cases} \tag{9}$$

In terms of a fixed basis $\phi^{(d)}(y)$ can be written as an $n_d \times m_d$ matrix $(\phi^{(d)}_{ij}(y))$ whose entries are polynomials in the y's. It follows that (8) holds iff $\phi^{(d)}(y)$ is <u>not</u> surjective, which is the case iff all $m_d \times m_d$ minors of $\phi^{(d)}(y_o)$ are singular. The set of these y_o is therefore closed.

In our case a closed subset Z of $\mathbb{P}^n \times \mathbb{A}^m$ is given by the set of points where certain matrices become singular:

$$A_i(u;y) \qquad (i = 1,\dots,t) .$$

Now $y_o \in p(Z) \Longleftrightarrow$ the matrices $A_i(u;y_o)$ have a common singularity in $\mathbb{P}^n$. Let T be the set of these y_o and write $\mathcal{A}(y_o)$ for the matrix ideal generated by the $A_i(u;y_o)$. Then $y_o \in T$ iff $V_K(\mathcal{A}(y_o)) \neq \{0\}$. By Th. 6 we find that

$$y_o \in T \Longleftrightarrow \sqrt{\mathcal{A}(y_o)} \neq \mathcal{J} \tag{10}$$

where $\mathcal{J} = \mathrm{Ker}\,\{\phi | u_i \mapsto 0\}$. Write $\mathcal{J}^{(s)}$ for the set of all $s \times s$ diagonal matrices with u's on the main diagonal; we claim that (10) holds iff

(11) $\qquad J^{(s)} \not\subseteq A(y_o)$ for all s.

For if $J^{(s)} \subseteq A(y_o)$ for some s, then each $u_i \in \sqrt{A(y_o)}$ and so we have $J = \sqrt{A(y_o)}$. Conversely, if $J = \sqrt{A(y_o)}$, then for some s_i, $u_o I_{s_o} \oplus \ldots \oplus u_n I_{s_n} \in A(y_o)$, so if $s = \sum (s_i - 1) + 1$, then $J^{(s)} \subseteq A(y_o)$. Fix s, then $J^{(s)} \subseteq A(y_o)$ means: for each $D \in J^{(s)}$

(12) $\qquad I \oplus D = B_1 \nabla \ldots \nabla B_r$

where each B_j is non-full or of the form $A_i(u;y_o) \oplus C$. Thus $J^{(s)} \not\subseteq A(y_o)$ means that for some $D \in J^{(s)}$ no expression (12) exists, and we have to show that the set of y_o for which this is the case can be described by some matrices being singular.

Repeated determinantal sums as in (12) are very difficult to handle, and so far this has not been proved. There is also no evidence against it and so we list our final problem:

Problem 6. Show that projective space over a skew field is complete.

There are different ways of stating the result, e.g. in one form it expresses the fact that specializations can be extended ([19] p. 31). This connexion also exists in the general case; it would allow us to conclude, from the fact that a matrix with indeterminate entries has a singular eigenvalue (mentioned above in III, cf.[8]), that every square matrix has a singular eigenvalue, thus solving Problem 2.

References.

1. S.A. Amitsur, Generalized polynomial identities and pivotal monomials, Trans. Amer. Math. Soc. 114 (1965) 210 - 226.

2. S.A. Amitsur, Rational identities and applications to algebra and geometry, J. Algebra 3 (1966) 304 - 359.

3. P.M. Cohn, Free rings and their relations, LMS monographs No. 2, Academic Press (London, New York 1971).

4. P.M. Cohn, Generalized rational identities, Proc. Park City Conference 1971, Ring Theory (ed. R. Gordon) Academic Press (New York 1972) 107 - 115.

5. P.M. Cohn, The similarity reduction of matrices over a skew field, Math. Zeits. 132 (1973) 151 - 163.

6. P.M. Cohn, The range of derivations on a skew field and the equation ax - xb = c, J. Indian Math. Soc. 37 (1973) 1 - 9.

7. P.M. Cohn, Presentations of skew fields I. Existentially closed skew fields and the Nullstellensatz, Math. Proc. Camb. Phil. Soc. 77 (1975) 7 - 19.

8. P.M. Cohn, Equations dans les corps gauches, Bull. Soc. Math. Belg. 27 (1975) 29 - 39.

9 P.M. Cohn, The Cayley-Hamilton theorem in skew fields, Houston J. Math. 2 (1976) 49 - 55.

10. P.M. Cohn, Skew field constructions, LMS Lecture Notes Nr. 27, Cambridge University Press (Cambridge 1977).

11. P.M. Cohn, Zum Begriff der Spezialisierung über Schiefkörpern, Beiträge zur geometrischen Algebra, Proc. Symp. über geometrische Algebra, Birkhäuser (Basel 1977) 73 - 82.

12. P.M. Cohn, Universal Algebra, revised Ed. Reidel (Dordrecht 1981).

13. P.M. Cohn, Fractions, Bull. London Math. Soc. 16 (1984)

14. P.M. Cohn, revised edition of [3], in preparation.

15. J. Dieudonné, Les déterminants sur un corps non-commutatif, Bull. Soc. Math. France 71 (1943) 27 - 45.

16. J. Hirschfeld and W.H. Wheeler, Forcing, Arithmetic, Division rings, Lecture Notes in Math. No. 454 Springer (Berlin 1975).

17. D. Mumford, Algebraic Geometry I. Complex Projective Varieties, Grundl. math. Wiss. 221, Springer (Berlin 1976).

18. I.R. Shafarevich, Basic Algebraic Geometry, Grundl. math. Wiss. 213, Springer (Berlin 1974).

19. A. Weil, Foundations of Algebraic Geometry, AMS Colloquium Publs. vol. 29, Amer. Math. Soc. (New York 1946).

20. O. Zariski and P. Samuel, Commutative Algebra II, van Nostrand (Princeton 1960).

APPLICATIONS OF RESULTS ON GENERALIZED POLYNOMIAL IDENTITIES IN DESARGUESIAN PROJECTIVE SPACES

Hans Havlicek
Institut für Geometrie
Technische Universität Wien
Wiedner Hauptstraße 8-10
A-1040 Wien
Austria

ABSTRACT. By following ideas of synthetic real projective geometry rather than classical algebraic geometry, maps in a finite-dimensional desarguesian projective space are used to generate normal curves. We aim at solving the problems of classification, automorphic collineations and generating maps of arbitrary non-degenerate normal curves and degenerate normal curves in desarguesian projective planes (also called degenerate conics). Properties of normal curves are shown, on one hand, by using methods of projective geometry as well as linear algebra and, on the other hand, by applying results on the non-existence of certain types of ordinary and generalized polynomial identities with coefficients in a not necessarily commutative field.

1. INTRODUCTION

1.1. Preface

It is well known that special curves and surfaces in a real (or complex) projective space of finite dimension permit definitions in a purely geometrical way by using *generating maps* such as projectivities, polarities, etc. Some of these definitions may be taken over word for word to more general classes of projective spaces. However, most of the classical results will no longer hold in the general case.

Normal (rational) curves in an n-dimensional real (or complex) projective space allow geometrical definitions including those of a *conic* (n=2) and a *twisted cubic* (n=3) given by J. Steiner (1832) and F. Seydewitz (1847), respectively. First results for arbitrary finite dimension are due to W.K. Clifford [11] and G. Veronese [37]. Cf. also [5,270], [6,318], [10,166], [21], [34,894] for further details and historical remarks.

By transferring one of these definitions in an even

R. Kaya et al. (eds.), Rings and Geometry, 39–77.

generalized way, *normal curves* have been introduced in finite dimensional desarguesian projective spaces by R. Riesinger [28]. Previous publications by E. Berz [7], B. Segre [33, 325], L.A. Rosati [30], [31], R. Artzy [3], W. Krüger [23] and R. Riesinger [27] are concerned with examples of these normal curves, especially conics. Recent articles [29], [15], [16], [17], [18], all but one by the author, are also dealing with normal curves. Finally, we mention some papers which contain results on (certain) normal curves in pappian projective spaces and on *conic-like figures* in non-pappian projective planes [8,55], [9,195], [20], [25], [35], [36].

A theorem by S.A. Amitsur [1] states that a field[1] satisfies a *generalized polynomial identity* if and only if it is of finite degree over its centre. Cf. also [13;141, 162]. This result is frequently applied in Riesinger's papers, because some geometrical problems, for example the coincidence of two degenerate conics, are equivalent to the fact that a generalized polynomial in a non-commutative indeterminant x with coefficients in a field K vanishes for all values of x in K. Hence the given polynomial either is the zero-polynomial or it yields a generalized polynomial identity. By Amitsur's theorem, the latter possibility can be excluded if K is of infinite degree over its centre. In this way geometrical problems have been solved completely under the assumption that the field K has infinite degree over its centre, but not necessarily complete solutions have been given for finite degree [27], [28]. However, as will be shown in this paper by a different approach, complete solutions of these problems can be found irrespective of whether the field K has infinite degree over its centre or not.

Interpretations of generalized polynomial identities in terms of geometry are included in [2].

1.2. Basic concepts

1.2.1. Suppose that Π is an n-dimensional desarguesian projective space ($2 \leq n < \infty$) which is regarded as a set of points $\mathcal{P}$, say, and a collection of subsets of $\mathcal{P}$ which are called lines (cf. e.g. [8]). The subspaces of Π form the lattice $(u\Pi, \vee, \cap)$ with $\vee$ and $\cap$ denoting the operation signs for "join" and "intersection", respectively. Any $M \in u\Pi$ determines projective spaces $\Pi(M)$, Π/M, with lattices of subspaces

---

[1] See 1.2.2.

$$u(\Pi(M)) = u\Pi(M) = \{X\in u\Pi \mid X\subset M\},$$

$$u(\Pi/M) = u\Pi/M = \{X\in u\Pi \mid X\supset M\},$$

respectively. We shall not distinguish a point $M\in P$ from the subspace $\{M\}\in u\Pi$ and $u\Pi/M$ will be called a *bundle* (of subspaces). The same symbol will denote a *collineation* (being a point-to-point map) and the associated *isomorphism* (which maps subspaces to subspaces).

1.2.2. The term *field* will be used for a not necessarily commuatative field, but *skewfield* always means a non-commutative field. We shall assume throughout this paper that Π is a projective space $\Pi(\mathbf{V})$ on a right vector space $\mathbf{V}$ over a field K. The set of points of $\Pi(\mathbf{V})$ is the set $P(\mathbf{V})$ of all one-dimensional subspaces of $\mathbf{V}$. If $\mathbf{U}\subset\mathbf{V}$ is a subspace of $\mathbf{V}$, then $P(\mathbf{U})$ denotes the subspace of the projective space $\Pi(\mathbf{V})$ given as $\{X\in P(\mathbf{V}) \mid X=\mathbf{x}K \text{ and } \mathbf{x}\in\mathbf{U}\}$. The *centre* of K will be written as Z, and we set $L^{\times}:=L\setminus\{0\}$ for any subfield L of K.

Let $\mathbf{U}$ be an $(m+1)$-dimensional subspace of $\mathbf{V}$ with $m\geq 1$. A vector $\mathbf{u}\in\mathbf{U}$ is called *central* with respect to a given basis $\{\mathbf{p}_0,\ldots,\mathbf{p}_m\}$ of $\mathbf{U}$ if $\mathbf{u}=\sum_{j=0}^{m}\mathbf{p}_j z_j$ with $z_j\in Z$, and a subspace of $\mathbf{U}$ is named *central* if it can be spanned by central vectors. The central subspaces of $\mathbf{U}$ determine *central subspaces* of $\Pi(\mathbf{U})$ with respect to the frame $F=\{P_0=\mathbf{p}_0K,\ldots,P_m=\mathbf{p}_mK, E=\mathbf{e}K\}$ of $\Pi(\mathbf{U})$ where $\mathbf{e}=\sum_{j=0}^{m}\mathbf{p}_j$. Those projective collineations of $\Pi(\mathbf{U})$ which are fixing the frame F pointwise are induced by linear automorphisms of $\mathbf{U}$ such that

$$\mathbf{p}_j \mapsto \mathbf{p}_j c \quad (j=0,\dots,m) \text{ and } c \in K^{\times}.$$

A subspace $\mathcal{P}(\mathbf{S})$ of $\Pi(\mathbf{U})$ is central with respect to $\mathcal{F}$ if and only if it is invariant under all projective collineations fixing the frame $\mathcal{F}$ pointwise. The "if"-part of this assertion is streightforward by using induction on the dimension of $\mathbf{S}$, the "only if"-part is trivial. Assume, finally, that $\mathcal{P}(\mathbf{U})$ is a projective line. Then the set of all points which are central with respect to the frame $\{P_0, P_1, E\}$ is called a Z-*chain* and will be denoted by $[P_0, P_1, E]_Z$. See [4,326]. In terms of cross-ratios (CR) a Z-chain $[P_0, P_1, E]_Z$ is the set of all points X in the line $P_0P_1 := P_0 \vee P_1$ satisfying

$$CR(X, E, P_1, P_0) \in Z \cup \{\infty\}.$$

The dual vector space of $\mathbf{V}$ is written as $\mathbf{V}^*$ and $\langle \mathbf{h}^*, \mathbf{v} \rangle$ stands for the image of $\mathbf{v} \in \mathbf{V}$ under the linear form $\mathbf{h}^* \in \mathbf{V}^*$. In order to simplify notation, we shall frequently write Π, $u\Pi$, $\mathcal{P}$ instead of $\Pi(\mathbf{V})$, $u\Pi(\mathbf{V})$, $\mathcal{P}(\mathbf{V})$, respectively.

1.2.3. We shall need the following

LEMMA 1.1. *Let* $s_0, \dots, s_m \in K$ *be linearly independent over the centre* Z *of* K. *Given elements* $r_0, \dots, r_m \in K$ *then*

$$r_0 t s_0 + \dots + r_m t s_m = 0 \text{ for all } t \in K \tag{1.1}$$

if and only if all r_i*'s vanish simultaneously.*

Proof. The dual vector space of the right vector space K^{m+1} will be identified (as usual) with the left vector space ${}^{m+1}K$. If (1.1) holds with $r_0 \neq 0$, say, then $[r_0, \dots, r_m] \in {}^{m+1}K$ is a non-trivial linear form such that

$$<[r_0,\ldots,r_m],(ts_0,\ldots,ts_m)> = 0 \text{ for all } t\in K.$$

In terms of the projective space $\Pi(K^{m+1})$ this means that the hyperplane $P(\ker[r_0,\ldots,r_m])$ contains the subspace spanned by the points $(ts_0,\ldots,ts_m)K = (ts_0t^{-1},\ldots,ts_mt^{-1})K$ with $t\in K^\times$. But this subspace is central with respect to the standard frame of $\Pi(K^{m+1})$ according to 1.2.2. This implies the existence of a non trivial linear form $[z_0,\ldots,z_m]\in{}^{m+1}Z\subset{}^{m+1}K$ with

$$<[z_0,\ldots,z_m],(ts_0,\ldots,ts_m)> = 0 \text{ for all } t\in K$$

and yields the contradiction $z_0s_0+\ldots+z_ms_m=0$.

The converse is trivial.□

In view of this, all remarks [27;247,249], [28,445] concerning skewfields which satisfy a generalized polynomial identity (1.1) are false.

1.2.4. In $\Pi=\Pi(\mathbf{V})$ we choose two different points P and Q. Let

(1.2) $\quad \zeta : u\Pi/P \to u\Pi/Q$

be a *projective isomorphism*. Then we refer to

(1.3) $\quad \Gamma(\zeta) := \{X\in P \mid X\in l \text{ and } X\in l^\zeta \text{ for some line } l\ni P\}$

as being the *point set generated by* ζ. We shall also say that ζ is a *generating map of* $\Gamma(\zeta)$. Obviously $P,Q\in\Gamma(\zeta)$, because $P,Q\in PQ$ and $P\in(PQ)^{\zeta^{-1}}$, $Q\in(PQ)^\zeta$. The *fundamental subspace of* ζ is defined as the intersection of all ζ-invariant subspaces and is denoted by $G(\zeta)$.

If $\zeta':u\Pi/P'\to u\Pi/Q'$ is any generating map of $\Gamma(\zeta)$, then $\{P',Q'\}$ is called a *fundamental pair of the point set* $\Gamma(\zeta)$ and we shall use the term *fundamental point of* $\Gamma(\zeta)$ for P' as well as Q'. The intersection of $\Gamma(\zeta)=\Gamma(\zeta')$ with the fundamental subspace of ζ' is called the *improper part of* $\Gamma(\zeta)$ *with respect to* ζ'. All other points of $\Gamma(\zeta)$ form the *proper part of* $\Gamma(\zeta)$ *with respect to* ζ'. These two point sets are denoted by $\Gamma^{\times}(\zeta')$ and $\Gamma^{o}(\zeta')$, respectively.

Suppose that H is a hyperplane of Π which contains neither P nor Q. Then, by (1.2),

(1.4) $\quad \zeta_H : X\ (\in H) \mapsto (PX)^{\zeta} \cap H$

is a projective collineation $H \to H$ which is called the *trace map of* ζ *in* H. A point is invariant under ζ_H if and only if it is an element of $\Gamma(\zeta) \cap H$.

The next result is trivial, but important.

PROPOSITION 1.1. *If* $\zeta : u\Pi/P \to u\Pi/Q$ *is a projective isomorphism, then any line which is different from* PQ *and passing through* P *contains at most one point of* $\Gamma(\zeta)$ *other than* P.

A simple example of a projective isomorphism (1.2) is a *perspectivity*. Here the generated point set $\Gamma(\zeta)$ is the union of the line PQ and a hyperplane H not passing through P as well as Q. The trace map ζ_H is the identity map in H. The fundamental subspace of ζ equals the line PQ.

The following defintions are subject to the assumption that ζ is no perspectivity: The point set $\Gamma(\zeta)$ given by (1.3) is called a *normal curve*. The map ζ as well as the normal curve $\Gamma(\zeta)$ are named *degenerate* if P equals the fundamental subspace $G(\zeta)$ and *non-degenerate* otherwise. If Π is a projective plane, then a normal curve is also called a *conic* and the map (1.2) is a *projectivity*.

To illustrate these definitions, we recall the situation in a 3-dimensional real projective space [24,135]:

(1) If ζ is non-degenerate, then $\Gamma(\zeta)$ is a twisted cubic.

(2) Let $G(\zeta)$ be a plane. Now $\Gamma^{\times}(\zeta)\subset G(\zeta)$ is a conic (in the usual sense) and $\Gamma^{o}(\zeta)$ equals an affine line $RS\setminus\{R\}$, where $R\in\Gamma^{\times}(\zeta)$, $R\neq P,Q$ and $S\in P\setminus G(\zeta)$.

(3) Assume that $G(\zeta)=PQ$. Then, since ζ is no perspectivity, there are two ζ-invariant planes at most. In every ζ-invariant plane one line other than PQ belongs to $\Gamma(\zeta)$. Any two such lines are skew and do not pass through P or Q.

We deduce from (3) that $\Gamma(\zeta)=\Gamma(\zeta')$ does not imply $\Gamma^{\times}(\zeta)=\Gamma^{\times}(\zeta')$, as is shown by a normal curve which is the union of two different lines with a common point. Hence we are not always able to speak unambigously of the proper and the improper part of a degenerate normal curve. We shall see in 4.2.1 that there are even normal curves which are degenerate as well as non-degenerate.

The preceding definitions are very close to the ones given in [28] and they still will make sense if we drop the convention about the finite dimensionality of Π.

2. NON-DEGENERATE NORMAL CURVES

2.1. Classification

2.1.1. In order to show that non-degenerate normal curves do exist in $\Pi=\Pi(\mathbf{V})$, we have to give an example of a non-degenerate projective isomorphism [15], [23].

Let $(\mathbf{p}_0,\ldots,\mathbf{p}_n)$ be an ordered basis of $\mathbf{V}$ and write[2] $\mathbf{e}=\sum_j \mathbf{p}_j$. Then

$$F = (P_0=\mathbf{p}_0K,\ldots,P_n=\mathbf{p}_nK,E=\mathbf{e}K)$$

is an ordered frame of Π. For any two consecutive points $P_{j-1},P_j\in F$ $(j=1,\ldots,n)$ there is a unique involutory perspective collineation $\gamma_j:P\rightarrow P$ whose axis is spanned by all points of F except P_{j-1},P_j with $P_{j-1}^{\gamma_j}=P_j$. The product map

$$\gamma := \gamma_1\cdots\gamma_n$$

is a projective collineation satisfying

$$P_0 \overset{\gamma_1}{\longmapsto} P_1 \overset{\gamma_2}{\longmapsto} \cdots \overset{\gamma_n}{\longmapsto} P_n,$$

[2]We shall use $\sum_j$ as a shorthand for $\sum_{j=0}^{n}$.

$$P_1 \overset{\gamma_1}{\longmapsto} P_0 \overset{\gamma_2}{\longmapsto} \cdots \overset{\gamma_n}{\longmapsto} P_0,$$

$$\cdots$$

$$P_j \overset{\gamma_1}{\longmapsto} \cdots \overset{\gamma_{j-1}}{\longmapsto} P_j \overset{\gamma_j}{\longmapsto} P_{j-1} \overset{\gamma_{j+1}}{\longmapsto} \cdots \overset{\gamma_n}{\longmapsto} P_{j-1},$$

$$\cdots$$

$$P_n \overset{\gamma_1}{\longmapsto} \cdots \overset{\gamma_{n-1}}{\longmapsto} P_n \overset{\gamma_n}{\longmapsto} P_{n-1},$$

$$E \overset{\gamma_1}{\longmapsto} E \overset{\gamma_2}{\longmapsto} \cdots \overset{\gamma_n}{\longmapsto} E.$$

Any of the perspective collineations γ_j can be induced by a linear automorphism of $\mathbf{V}$, and it is easy to see that γ is induced by the automorphism $g \in GL(\mathbf{V})$ with

(2.1) $\quad \mathbf{p}_0{}^g = \mathbf{p}_n, \ \mathbf{p}_j{}^g = \mathbf{p}_{j-1} \ (j=1,\ldots,n).$

We consider the restricted map

(2.2) $\quad \varphi := \gamma|(u\Pi/P_0) : u\Pi/P_0 \rightarrow u\Pi/P_n.$

(Figure 1 illustrates the case $\dim\Pi = n = 2$.) Our next task is

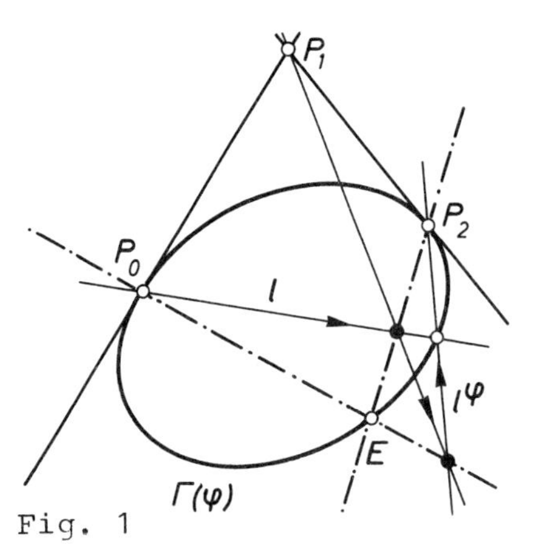

Fig. 1

to show that φ is non-degenerate. The fundamental subspace $G(\varphi)$ is invariant under φ and γ as well. But $P_n \in G(\varphi)$, and so $P_{n-j} \in G(\varphi)^{\gamma^j} = G(\varphi)$ for all $j=1,\ldots,n$, whence $G(\varphi) = \mathcal{P}$. We shall refer to this non-degenerate projective isomorphism φ

as the *normal isomorphism* determined by the ordered basis $(\mathbf{p}_0, \ldots, \mathbf{p}_n)$ of $\mathbf{V}$, and we shall also say that φ is determined by the ordered frame F of Π.

Consequently, the point set $\Gamma(\varphi)$ is a non-degenerate normal curve.

2.1.2. We need some more definitions [15]. Assume that

$$\zeta : u\Pi/P \to u\Pi/Q \quad (P \neq Q)$$

is a non-degenerate projective isomorphism. A non-empty subspace M of Π is called a *chordal subspace of* ζ if, either M is a hyperplane, or $M = L \cap L^{\zeta}$ with $L \in u\Pi/P$ and $\dim\Pi(M) = \dim\Pi(L) - -1$. Frequently we shall use the term *chord* instead of chordal subspace and a k-dimensional chord will also be called a *k-chord*.

If M is a k-chord of ζ, where $k \geq 1$ and $P, Q \notin M$, then the *trace map of* ζ *in* M is defined by formula (1.4), with H to be replaced by M, and this trace map will be denoted by ζ_M. A non-empty subspace of $\Pi(M)$ is invariant under ζ_M if and only if it is a chord of ζ.

The 0-chords of ζ are the points of the normal curve $\Gamma(\zeta)$. If we draw a line m joining two different points of $\Gamma(\zeta)$, then m is a chord in the usual sense and a 1-chord according to the definition given above. However, a 1-chord m, say, of the map ζ will not contain any point of $\Gamma(\zeta)$ at all if the trace map ζ_m has no invariant points.

The subspaces

$$(2.3) \quad S_P^{(0)}(\zeta) := P, \; S_P^{(k)}(\zeta) := (S_P^{(k-1)}(\zeta) \vee Q)^{\zeta^{-1}}$$

$$(k=1, \ldots, n-1)$$

are called *osculating subspaces of* ζ *in* P. The osculating subspaces of ζ in Q are given, by (2.3), as the osculating subspaces of ζ^{-1} in Q. We shall use the term *osculating k-subspace* for any k-dimensional osculating subspace. In addition the words *tangent* and *osculating hyperplane* will be used if k=1,k=n-1, respectively.

Clearly, we would prefer to speak of osculating subspaces of a non-degenerate normal curve rather than of osculating subspaces of a non-degenerate projective isomorphism. However, as will be shown in 2.4.2, this is not always possible if we want osculating subspaces to be *uniquely* determined by a non-degenerate normal curve. Recall that, by (1.3), a normal curve is merely a set of points and not, for example, a pair formed by the map ζ and the set $\Gamma(\zeta)$.

In the special case of a conic $\Gamma(\zeta)$, say, the tangents of ζ in P and Q can as well be defined in terms of the set $\Gamma(\zeta)$ as follows from

PROPOSITION 2.1. *If* $\zeta:u\Pi/P \to u\Pi/Q$ *is a non-degenerate projective isomorphism, then* $S_P^{(k)}(\zeta)$ *is the only* k-*chord of* ζ *which passes through* $S_P^{(k-1)}(\zeta)$ *(*k=1,...,n-1*) and meets the non-degenerate normal curve* $\Gamma(\zeta)$ *in* P *only.*

Proof. Clearly, we have $S_P^{(1)}(\zeta)\cap\Gamma(\zeta)=P$. If Π is a projective plane (n=2), then every line m, say, passing through P is a 1-chord of ζ, and $m\neq S_P^{(1)}(\zeta)$ implies $P\neq m\cap m\in\Gamma(\zeta)$.

If $n\geq 3$, then

$$S_P^{(1)}(\zeta) = (PQ)^{\zeta^{-1}} = ((PQ)^{\zeta^{-1}}\vee Q)\cap((PQ)^{\zeta^{-1}}\vee Q)^{\zeta^{-1}} = \\ = S_P^{(2)}(\zeta)^{\zeta^{-1}}\cap S_P^{(2)}(\zeta),$$

from which it follows that $S_P^{(1)}(\zeta)$ is a 1-chord of ζ. On the other hand, assume that $m=L\cap L^{\zeta}$ is any 1-chord of ζ, where L denotes a plane passing through P. Then $m,m^{\zeta}\subset L$ implies, firstly, $m\cap m^{\zeta}\in\Gamma(\zeta)$ and, secondly, $m\cap m^{\zeta}\neq P$ if and only if $m\neq$ $\neq S_P^{(1)}(\zeta)$.

Still assuming $n \geq 3$, the restricted map

(2.4) $\quad \bar{\zeta} := \zeta | (u\Pi / S_P^{(1)}(\zeta)) \; : \; u\Pi / S_P^{(1)}(\zeta) \rightarrow u\Pi / PQ$

is a non-degenerate projective isomorphism in the (n-1)-dimensional quotient space Π/P. The set of k-chords of $\bar{\zeta}$ (k=0,...,n-2 as viewed from Π/P) coincides with the set of (k+1)-chords of ζ which pass through P, and so the osculating k-subspace of $\bar{\zeta}$ in $S_P^{(1)}(\zeta)$ is identical with $S_P^{(k+1)}(\zeta)$. The proof is completed by induction on n.□

We remark that a normal isomorphism φ, as given by formula (2.2), has osculating subspaces (k=0,...,n-1)

(2.5) $\quad S_{P_0}^{(k)}(\varphi) = P_0 \vee \dots \vee P_k,$

$\qquad S_{P_n}^{(k)}(\varphi) = P_n \vee \dots \vee P_{n-k}.$

For $n \geq 3$ the restricted map $\varphi | (u\Pi / P_0P_1)$ (cf. (2.4)) is a normal isomorphism in Π/P determined by $(P_0P_1, \dots, P_0P_n, P_0E)$.

2.1.3. We are now in a position to solve the classification problem [15], [23], [28].

THEOREM 2.1. *Every non-degenerate projective isomorphism is normal.*

Proof. (1) If an ordered frame $(P_0, \dots, P_n, E)$ is to determine a given non-degenerate projective isomorphism $\zeta : u\Pi/P \rightarrow u\Pi/Q$, then necessarily $P_0 = P$ and $P_n = Q$. We shall use induction on $\dim \Pi = n$ to prove the assertion.

(2) Suppose n=2. Now ζ is a non-degenerate projectivity. We shall adopt the notations $A \overset{c}{\barwedge} B$, $a \overset{C}{\barwedge} b$, say, for perspectivities $u\Pi/A \rightarrow u\Pi/B$ with axis c $(A, B \notin c)$, $u\Pi(a) \rightarrow u\Pi(b)$ with centre

C ($C \notin a,b$), respectively, where A,B,C are points and a,b,c are lines. By a well known result on the decomposition of projectivities into a product of perspectivities [8,31], we have a factorization

$$\zeta = P \overset{q'}{\overline{\overline{\wedge}}} P_1'' \overset{p'}{\overline{\overline{\wedge}}} Q$$

with $P_1'' \notin PQ$, because ζ is non-degenerate (Fig. 2). If E:=

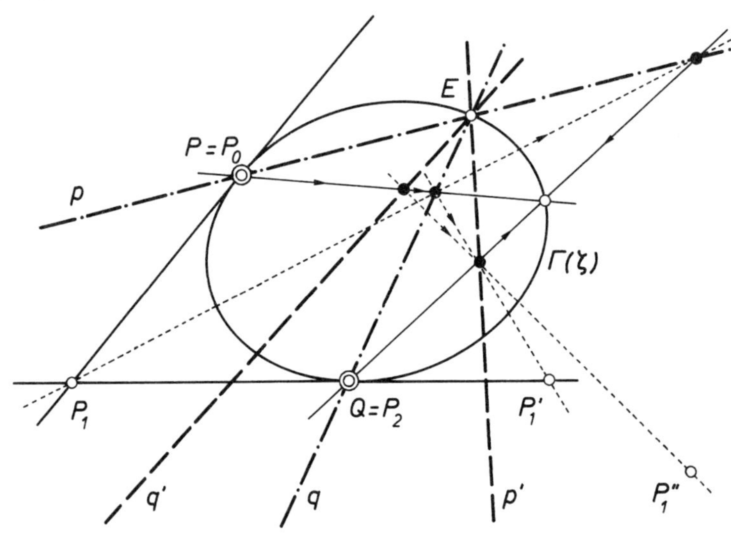

Fig. 2

$= p' \cap q'$ and q:=EQ, then $q \overset{P}{\overline{\overline{\wedge}}} q' \overset{P_1''}{\overline{\overline{\wedge}}} p'$ is a perspectivity $q \overset{P_1'}{\overline{\overline{\wedge}}} p'$, say, since q,p',q' are concurrent [8,31]. Hence

$$P \overset{q'}{\overline{\overline{\wedge}}} P_1'' \overset{p'}{\overline{\overline{\wedge}}} P_1' = P \overset{q}{\overline{\overline{\wedge}}} P_1'$$

and

$$\zeta = P \overset{q'}{\overline{\overline{\wedge}}} P_1'' \overset{p'}{\overline{\overline{\wedge}}} Q = P \overset{q'}{\overline{\overline{\wedge}}} P_1'' \overset{p'}{\overline{\overline{\wedge}}} P_1' \overset{p'}{\overline{\overline{\wedge}}} Q = P \overset{q}{\overline{\overline{\wedge}}} P_1' \overset{p'}{\overline{\overline{\wedge}}} Q.$$

Similarly, we can replace p' by the line p:=EP and P_1' by P_1. Write $P_0=P$, $P_2=Q$; then (P_0,P_1,P_2,E) determines ζ.

(3) For $n \geq 3$ we may assume, by induction hypothesis, that the restricted map $\overline{\zeta}$, as given by formula (2.4), is a

normal isomorphism in Π/P determined by an ordered frame $(PP_1,\ldots,PP_n,P\bar{E})$, say, with $P_1,\ldots,P_{n-1},P_n=Q\in S_Q^{(n-1)}(\zeta)$. Since $P\bar{E}$ is a 1-chord of ζ, there exists $(P\bar{E})\cap(P\bar{E})^{\zeta}=:E\in\Gamma(\zeta)$, and $(P=:P_0,\ldots,P_n,E)$ is an ordered frame of Π, because $E\notin$ $\notin S_Q^{(n-1)}(\zeta)$. Let φ denote the normal isomorphism determined by this ordered frame. The restrictions of ζ and φ, respectively, on $u\Pi/S_P^{(1)}(\zeta)$ coincide. Furthermore $(PE)^{\zeta}=(PE)^{\varphi}$ and

$$\begin{aligned}(PQ)^{\zeta} &= ((S_P^{(1)}(\zeta)\vee Q)\cap(S_Q^{(n-1)}(\zeta))^{\zeta^{-1}})^{\zeta} = \\ &= (S_P^{(1)}(\zeta)\vee Q)^{\varphi}\cap S_Q^{(n-1)}(\zeta) = \\ &= (P_0\vee P_n\vee P_{n-1})\cap S_Q^{(n-1)}(\zeta) = P_nP_{n-1} = (PQ)^{\varphi}\end{aligned}$$

which is sufficient for $\zeta=\varphi$ [8,126].□

THEOREM 2.2. *There is a unique non-degenerate projective isomorphism and a unique non-degenerate normal curve to within transformation under projective collineations of* Π.

Proof. By Theorem 2.1, we may restrict ourselves to normal isomorphisms. The group of projective collineations of Π, i.e. $PGL(\Pi)$, operates transitively on the set of ordered frames of Π. This completes the proof.□

2.2. Conjugate points

2.2.1 It follows from Theorems 2.1 and 2.2 that we can restrict ourselves to non-degenerate normal curves generated by normal isomorphisms. We shall investigate a normal isomorphism φ as defined by formula (2.2) and we shall use all the other notations introduced in 2.1.1.

For any point $X\in\Gamma:=\Gamma(\varphi)$ we set

$$(2.6) \quad \mathcal{H}_X := P_0 \vee \ldots \vee P_{n-2} \vee X \text{ if } X \neq P_0,$$
$$\mathcal{H}_X := P_0 \vee \ldots \vee P_{n-1} \quad \text{if } X = P_0.$$

The subspace $\mathcal{H}_X$ is a hyperplane by Proposition 2.1 and formula (2.5). Conversely, let $\mathcal{H} \supset P_0 \vee \ldots \vee P_{n-2}$ be a hyperplane. Since $\{P_i\}$ is a basis of Π, there is a point X such that

$$(2.7) \quad X = \mathcal{H} \cap \mathcal{H}^{\gamma} \cap \ldots \cap \mathcal{H}^{\gamma^{n-1}}.$$

If $\mathcal{H}=P_0 \vee \ldots \vee P_{n-1}$, then $X=P_0 \in \Gamma$. Otherwise $X \neq P_0$, $(P_0X)=\mathcal{H} \cap \ldots \cap \mathcal{H}^{\gamma^{n-2}}$ and $(P_nX)=\mathcal{H}^{\gamma} \cap \ldots \cap \mathcal{H}^{\gamma^{n-1}}$ force $X \in \Gamma$. Now, suppose $\mathcal{H}= \mathcal{P}(\ker(t\mathbf{p}^*_{n-1}-\mathbf{p}^*_n))$ with $t \in K$, where $\{\mathbf{p}^*_i\} \subset \mathbf{V}^*$ denotes the dual basis of $\{\mathbf{p}_i\}$. Then X equals the one-dimensional subspace $\mathbf{x}_t K$ of $\mathbf{V}$ with[3] $\mathbf{x}_t = \sum_j \mathbf{p}_j t^j$, because $\langle t\mathbf{p}^*_j - \mathbf{p}^*_{j+1}, \mathbf{x}_t \rangle = 0$ for $j=0, \ldots, n-1$. As t varies in K, we obtain all points of $\Gamma \setminus \{P_n\}$, hence

$$(2.8) \quad \Gamma(\varphi) = \{\mathbf{x}_t K \mid \mathbf{x}_t = \sum_j \mathbf{p}_j t^j,\ t \in K\} \cup \{P_n\}.$$

The map $X(\in \Gamma) \to \mathcal{H}_X(\in u\Pi/(P_0 \vee \ldots \vee P_{n-2}))$ is a bijection. Its inverse map may be regarded as being a *Veronese map* $\mathcal{P}(K^2) \to \mathcal{P}(\mathbf{V})$, where we have to identify the projective line $\mathcal{P}(K^2)$ with a pencil of hyperplanes. We remark that (2.7) yields another way of generating a non-degenerate normal curve: it involves n projectively related pencils of hyperplanes. See [5,275] or [6,323] and the introduction of [28].

2.2.2. Let $\mathcal{M}=\mathcal{P}(\mathbf{M})$ be a k-chord of φ $(k \geq 1)$ with $P_0, P_n \notin \mathcal{M}$ and write $p_{\mathbf{M}}: \mathbf{p}_n K \oplus \mathbf{M} \to \mathbf{M}$ for the projection with kernel $\mathbf{p}_n K$. It follows immediately that $g_{\mathbf{M}} := (g|\mathbf{M})p_{\mathbf{M}} : \mathbf{M} \to \mathbf{M}$ (c.f.(2.1)) induces the trace map $\varphi_{\mathcal{M}}$.

Clearly, $\mathbf{x}_t \in \Gamma \cap \mathcal{M}$ if and only if $\mathbf{x}_t$ is an eigenvector of $g_{\mathbf{M}}$. Since

[3] Upper indices are always in brackets in order to distinguish them from exponents.

(2.9) $\mathbf{x}_t^{g_\mathbf{M}} = (\sum_j \mathbf{p}_j t^j)^{gp_\mathbf{M}} = (\sum_{j=0}^{n-1} \mathbf{p}_j t^{j-1} + \mathbf{p}_n)^{p_\mathbf{M}} = \mathbf{x}_t t,$

the vector $\mathbf{x}_t$ is belonging to the eigenvalue[4] $t \neq 0$ of $g_\mathbf{M}$. Thus the problems to determine the eigenvectors of $g_\mathbf{M}$, the eigenvalues of $g_\mathbf{M}$, the intersection $\Gamma \cap M$, respectively, are equivalent. In the special case of a hyperplane $M = P(\mathbf{H}) = P(\ker \sum_j h_j \mathbf{p}_j^*)$, say, $\mathbf{x}_t K \in \Gamma \cap M$ if and only if

$$<\sum_j h_j \mathbf{p}_j^*, \mathbf{x}_t> = h_n t^n + \ldots + h_0 = 0,$$

where $h_0 h_n \neq 0$. Cf [17], [29].

A theorem by G. Gordon and T.S. Motzkin [14,220] tells us that at most n conjugacy classes of K contain zeros of the equation $h_n t^n + \ldots + h_0 = 0$. On the other hand, we might as well infer from results by P.M. Cohn [13,207] that the spectrum of the linear automorphism $g_\mathbf{H}$ consists of n conjugacy classes of K at most. It is our task to give a translation into the language of geometry.

Two points $U, V \in \Gamma$ are called *conjugate* if, either $U = V$, or $U \neq V$ and the line UV meets Γ in at least three different points. A point of Γ is named *regular* if no other point of Γ is conjugate to it [7], [17].

The definition of regularity in [28]is different, but equivalent. In [33,346] the term "points of first kind" has been introduced for regular points of a non-degenerate conic.

By Proposition 1.1. points P_0, P_n are regular. Suppose that $U = \mathbf{x}_u K$ and $V = \mathbf{x}_v K$ are distinct. The line UV carries any point $W = (\mathbf{x}_u + \mathbf{x}_v w)K \in \Gamma$ $(w \neq 0)$ if and only if W $(\neq U, V)$ is invariant under the trace map φ_{UV}. This is equivalent to $\mathbf{x}_u + \mathbf{x}_v w$, $\mathbf{x}_u u + \mathbf{x}_v v w$ being linearly dependent by (2.9) or, in other

---

[4]Using matrices instead of linear maps would force to speak of *right eigenvalues*; cf. [13,205].

words

(2.10) $u = w^{-1}vw.$

Hence U and V are conjugate points of Γ if and only if u and v are conjugate elements of the ground field K. If U and V are conjugate and distinct, then, by (2.10),

(2.11) $\{(\mathbf{x}_u+\mathbf{x}_v vws)K\} \cup \{\mathbf{x}_v K\}$,

where s is in the centralizer of u, equals the set of invariant points of φ_{UV}. This set (2.11) is always infinite [14,221], [22,409].

2.2.3. Given a point $X \in \Gamma$ $(=\Gamma(\varphi))$ all points conjugate to X form the *conjugacy class of* X *in* Γ. By definition, this conjugacy class does not depend on the normal isomorphism φ which has been used as generating map of Γ.

PROPOSITION 2.2. *Let* M *be a* k*-dimensional subspace of* Π $(1 \leq k \leq n-1)$ *and let* $\{Q_0, \ldots, Q_k\}$ *be a basis of* $\Pi(M)$ *which is contained in the non-degnerate normal curve* $\Gamma = \Gamma(\varphi)$ *as given in 2.1.1. The points* Q_i *are pairwise conjugate if and only if there is an additional point* $Q_{k+1} \in \Gamma$ *such that* $\{Q_0, \ldots, Q_k, Q_{k+1}\}$ *is a frame of* $\Pi(M)$.

Proof. (1) Suppose $P_0 \in M$. Then every point of $\Gamma \cap M$, other than P_0, has to be an element of $M^{\varphi} \neq M$ as well. Hence $P_0 = Q_0$, say, and none of the assertions of the criterion holds in M. Substituting φ by φ^{-1} shows $P_n \in M$ to be impossible in any event.

(2) Assume that neither P_0 nor P_n is in M. The result is

established, by definition, if M is a line. We use induction on k: If $Q_0,\ldots,Q_k$ are pairwise conjugate $(2\leq k\leq n-1)$, then there is a frame $\{Q_0,\ldots,Q_{k-1},Q_k'\}$ of $\Pi(Q_0\vee\ldots\vee Q_{k-1})$ which is contained in Γ. The line joining the conjugate points Q_k and Q_k' carries the point Q_{k+1}, as required.

Conversely, if $\{Q_0,\ldots,Q_{k+1}\}\subset\Gamma$ is a frame of $\Pi(M)$, then M is a chord of φ. Let $Q_k':=Q_kQ_{k+1}\cap(Q_0\vee\ldots\vee Q_{k-1})$; this Q_k' is φ_M-invariant, hence it is in the normal curve Γ and $Q_0,\ldots,Q_{k-1},Q_k,Q_k'$ are pairwise conjugate.□

By Proposition 2.2, any $m\leq n+1$ pairwise inconjugate points of Γ are independent and $m\leq n$ implies that the proper subspace spanned by them meets Γ in no other points. Consequently, any k-dimensional subspace of Π $(1\leq k\leq n-1)$ has non-empty intersection with at most k+1 conjugacy classes of a non-degenerate normal curve in Π. (Cf. the Gordon-Motzkin theorem.)

Regular points are important when transferring classical theorems on conics (e.g. Pascal's theorem [7,75]) to the general case.

2.3. Generating maps

2.3.1. We are still persuing the discussion of the normal isomorphism φ given by an ordered basis $(\mathbf{p}_0,\ldots,\mathbf{p}_n)$ of $\mathbf{V}$ according to 2.1.1, and we shall use all the other notations introduced there.

PROPOSITION 2.3. *Two ordered bases* $(\mathbf{p}_j)$, $(\mathbf{p}_j')$ *of* $\mathbf{V}$ *determine the same normal isomorphism if and only if*

(2.12) $\mathbf{p}_j' = \mathbf{p}az^j$ *(j=0,...,n),*

where $a\in K^\times$ *and* $z\in Z^\times$.

Proof. If $\varphi=\varphi'$, then $\mathbf{p}_j'=\mathbf{p}_jc_j$ with $c_j\in K^\times$, as follows from (2.5). The bundle $u\Pi/P_0$ is fixed elementwise under the projective automorphism $\gamma\gamma'^{-1}$, hence there is an element $z\in Z^{\times}$ such that $\mathbf{p}_j^{gg'^{-1}}=\mathbf{p}_j c_j c_{j-1}^{-1}=\mathbf{p}_j z$ $(j=1,\ldots,n)$. Now $c_0=:a$ yields $\mathbf{p}'_j=\mathbf{p}_j a z^j$ by induction on $j=0,\ldots,n$. Reversing the above arguments completes the proof.□

Observe that $E'=(\sum_j \mathbf{p}'_j)K$ is a regular point of $\Gamma(\varphi)$.

2.3.2. Now, in several steps, all normal isomorphisms yielding the same non-degenerate normal curve will be determined [17].

THEOREM 2.3. *Assume that* $\varphi:u\Pi/P_0\to u\Pi/P_n$ *is a normal isomorphism. There exists a normal isomorphism* $\varphi':u\Pi/P_0\to u\Pi/P_n$ *different from* φ, *but also generating the non-degenerate normal curve* $\Gamma(\varphi)$, *if and only if* $\Gamma(\varphi)\setminus\{P_0,P_n\}$ *is contained in a hyperplane and* $\dim\Pi=n\geq 3$.

Proof. (1) If $\Gamma(\varphi)\setminus\{P_0,P_n\}$ is in a hyperplane and $n\geq 3$, then write N for the subspace spanned by $\Gamma(\varphi)\setminus\{P_0,P_n\}$. By Proposition 2.2, neither P_0 nor P_n is an element of N, and we choose a hyperplane $H\supset N$ that meets P_0P_n in a point other than P_0,P_n. Let σ denote the involutory perspective collineation[5] which has axis H and maps P_0 to P_n. Then $\Gamma(\varphi)^{\sigma}=\Gamma(\varphi)$; under σ the tangent of φ in P_n is not mapped to the tangent of φ in P_0, because these two lines are skew by formula (2.5). Now $\varphi':=\sigma\varphi^{-1}\sigma^{-1}$ generates $\Gamma(\varphi)$ too, however the tangent of φ' in P_0 is not the tangent of φ in P_0, whence $\varphi\neq\varphi'$.

(2) It follows from $n=2$ that every line through P_0 has

[5]This collineation σ is not given correctly in my paper [17].

non-empty intersection with its image under φ, whence $\varphi=\varphi'$.

(3) Suppose $\{Q_0,\ldots,Q_n\}\subset(\Gamma(\varphi)\setminus\{P_0,P_n\})$ to be a basis of Π. Without loss of generality let $Q_0=E$. By Proposition 2.2, the points P_0,P_n are not lying in any face of the basis $\{Q_i\}$. Write $M:=Q_0\vee\ldots\vee Q_{n-1}=:P(\mathbf{M})$, $Q_{n0}:=P_0Q_n\cap M$ and $Q_{nn}:=P_nQ_n\cap$ $\cap M$.

Any normal isomorphism $\varphi':u\Pi/P_0\to u\Pi/P_n$ that generates $\Gamma(\varphi)$ has a trace map φ'_M in M that takes the ordered frame $(Q_0,\ldots,Q_{n-1},Q_{n0})$ to the ordered frame $(Q_0,\ldots,Q_{n-1},Q_{nn})$. In terms of the notations introduced in 2.2.2, the trace map φ'_M is induced by $g'_{\mathbf{M}}$ and formula (2.9), together with the regularity of $Q_0=E=\mathbf{e}K$, implies that $\mathbf{e}$ belongs to a central eigenvalue of $g'_{\mathbf{M}}$. Hence comparing φ and φ' yields that $g_{\mathbf{M}}g'^{-1}_{\mathbf{M}}$ has $\mathbf{e}$ as an eigenvector belonging to a central eigenvalue. Furthermore, $\{Q_0,\ldots,Q_{n-1},Q_{n0}\}$ is elementwise invariant under $\varphi_M\varphi'^{-1}_M$. Thus $g_{\mathbf{M}}g'^{-1}_{\mathbf{M}}$ is a central dilatation on $\mathbf{M}$ and $\varphi_M=\varphi'_M$. This in turn is equivalent to $\varphi=\varphi'$, as required.□

There is a close connection between the decomposition of a projective collineation κ in a product of perspective collineations and the existence of κ-invariant points $\mathbf{f}K$ such that $\mathbf{f}$ is belonging to a central eigenvalue of any inducing linear automorphism for κ. See [32] with remarks given in [17].

The next theorem links geometry with algebra.

THEOREM 2.4. *If* $\varphi:u\Pi/P_0\to u\Pi/P_n$ *is a normal isomorphism, then* $\Gamma(\varphi)\setminus\{P_0,P_n\}$ *is contained in a hyperplane if and only if the ground field* K *has at most* $\dim\Pi+1=n+1$ *elements.*

Proof. (1) If the centre Z of K has cardinality $\geq n+2$, then $\Gamma(\varphi)\setminus\{P_0, P_n\}$ includes at least n+1 regular points which span all of $\mathcal{P}$ by Proposition 2.2.

(2) Let K be a skewfield with finite centre. The "only-if" result will be established if we are able to show the existence of an element $u \in K$ which is transcendental over Z [22].

Suppose that all elements of K are algebraic over Z. Take $a \in K \setminus Z$, whence Z(a), the commutative subfield of K spanned by Z and a, is a Galois-field. There is a non-trivial automorphism of Z(a) that fixes Z elementwise and takes a to $a^q \neq a$, say, where q is a power of a prime number. By the Skolem-Noether theorem (cf. e.g. [13,46] or [26,45] for an elementary proof), this automorphism can be extended to an inner automorphism of K. Hence there is some $u \in K$ such that $u^{-1}yu = y^q$ for all $y \in Z(a)$. According to our assumption this u is algebraic over Z. Therefore the minimal polynomials of u, a, respectively, and $uy^q = yu$ imply that the subfield generated by Z,a,u has finite degree over Z, so that it is a finite field. But, by Wedderburn's theorem, any finite field is commutative and this contradicts $ua \neq au$.

It follows from the existence of a hyperplane which contains $\Gamma(\varphi)\setminus\{P_0, P_n\}$ that there is a central hyperplane $\mathcal{H}$ (with respect to $\{P_0, \ldots, P_n, E\}$) having the same property, because the subspace spanned by $\Gamma(\varphi)\setminus\{P_0, P_n\}$ is central by 1.2.2. Let $\mathcal{H} = \mathcal{P}(\ker \sum_j z_j \mathbf{p}_j^*)$, $z_j \in Z$, $z_0 z_n \neq 0$, then

(2.13) $z_n t^n + \ldots + z_0 = 0$ for all $t \in K$

which is not possible, because of $u \in K$ being transcendental over Z.

(3) The proof is completed by the trivial remark that $|\Gamma(\varphi) \setminus \{P_0, P_n\}| \leq n$ follows from $|K| \leq n+1$. □

There is no unique way from $\Gamma(\varphi)$ back to φ if and only if K satisfies a polynomial identity (2.13) (cf.[13,162]) and $n \geq 3$ which in turn is equivalent to $|K| \leq n+1$ and $n \geq 3$. We add, for the sake of completeness, that only if this "case of small ground field" is excluded "Lemma 6" and "Folgerung 2" in [28] are correct results.

As will be shown in 2.4.2, the group of automorphic collineations of a non-degenerate normal curve operates 3-fold transitively on the set of its regular points. Since a non-regular point cannot be fundamental by Proposition 1, we have the following.

THEOREM 2.5. *Two different points of a non-degenerate normal curve form a fundamental pair if and only if they are regular.*

2.4. Automorphic collineations

2.4.1. Every automorphic collineation of $\Gamma(\varphi)$ (cf.2.1.1) takes (non-) regular points of $\Gamma = \Gamma(\varphi)$ to (non-) regular points of Γ. Those linear automorphisms of $\mathbf{V}$ such that

$$\text{(2.14)} \quad \mathbf{p}_j \mapsto \mathbf{p}_j z^j,$$

$$\text{(2.15)} \quad \mathbf{p}_j \mapsto \mathbf{p}_{n-j}, \qquad \text{with } j=0,\dots,n,\ z \in Z^\times$$

$$\text{(2.16)} \quad \mathbf{p}_j \mapsto \sum_{k=0}^{n-j} \mathbf{p}_{j+k} \binom{j+k}{k} z^k$$

induce automorphic collineations of Γ: this follows immediately from formula (2.12), $\Gamma(\varphi) = \Gamma(\varphi^{-1})$ and, by the binomial theorem,

$$\mathbf{x}_t \mapsto \sum_{j=0}^{n} \sum_{k=0}^{n-j} \mathbf{p}_{j+k} \binom{j+k}{k} z^k t^j = \mathbf{x}_{t+z},$$

respectively.

It is easily seen that these automorphic collineations generate a group which, regarded as transformation group on the set of regular points of Γ, is isomorphic to the group of projectivities of the projective line $P(Z^2)$ over the centre Z of K. Thus this group of automorphic collineations of Γ is sharply 3-fold transitive on the set of regular points of Γ. In view of this, we have to discuss only the stabilizer of any three regular points of Γ within the group G of all automorphic collineations of Γ.

If $\sigma \in G$ fixes P_0, P_n, E, then $\sigma^{-1}\varphi\sigma$ generates Γ. There are two cases:

(1) Let $n=2$ or $|K| \geq n+2$. Then, by Theorem 2.3 and Theorem 2.4, $\sigma^{-1}\varphi\sigma = \varphi$, and therefore the frame $\{P_0, \ldots, P_n, E\}$ is fixed elementwise under σ. Conversely, every collineation which fixes this frame elementwise is in the group G.

(2) Suppose $n \geq 3$ and $|K| \leq n+1$. The non-degenerate normal curve Γ has $|K|+1$ distinct points and either is a frame of Π or is a basis of the subspace spanned by Γ ($|\Gamma| \leq n+1$). Hence every permutation of Γ which fixes P_0, P_n, E can be extended to at least one automorphic projective collineation of Γ. See [18], [20].

To sum up, we have shown:

THEOREM 2.5. *The group G of automorphic collineations of a* *non-degenerate normal curve* Γ *has a subgroup of projective collineations which is sharply three-fold transitive on the set of regular points of* Γ. *If* $\dim\Pi=n=2$ *or* $|\Gamma|\geq n+3$, *then the stabilizer of any three different regular points of* Γ *coincides with a group of collineations fixing a frame of* Π *pointwise. If* $n\geq 3$ *and* $|\Gamma|\leq n+2$, *then every permutation of* Γ *is the restriction of at least one projective collineation of* Π.

A different way to determine the group G of all automorphic collineations of Γ, working only for $[K:Z]=\infty$, can be found in [28] and involves Amitsur's theorem on generalized polynomial identities. However, as pointed out in [17], the original proof [28,440] is correct only under certain additional assumptions.

2.4.2. We finish this chapter with remarks on osculating subspaces. It is possible to associate with every point $X\in\Gamma(\varphi)$ a flag $(X=S_X^{(0)}(\varphi),S_X^{(1)}(\varphi),\ldots,S_X^{(n-1)}(\varphi))$ the elements of which are called *osculating subspaces of* φ *in* X. See [7] (n=2) and [15] ($n\geq 3$) for details.

The definition of tangent of a conic used in [3]is different from the one in [7]. The definition of osculating subspaces in [28] does not make sense for $|\Gamma|\leq n+2$ and $n\geq 3$, because they are not determined uniquely, and fails to work in non-regular points if the characteristic of the ground field is $\neq 0$ and $<n$.

We see that a normal isomorphism φ does generate not only the non-degenerate normal curve $\Gamma(\varphi)$ but also the set of flags

$$\Gamma^{(n-1)}(\varphi) := \{(X,S_X^{(1)}(\varphi),\ldots,S_X^{(n-1)}(\varphi))\,|\,X\in\Gamma(\varphi)\}.$$

The linear automorphisms of $\mathbf{V}$ given by formulae (2.14),

(2.15), (2.16) induce collineations of the group $G^{(n-1)}$, i.e. the group of automorphic collineations of $\Gamma^{(n-1)}$. Clearly, demanding $\Gamma(\varphi)=\Gamma(\varphi')$ is a coarser relation than demanding $\Gamma^{(n-1)}(\varphi)=\Gamma^{(n-1)}(\varphi')$, where φ,φ' are normal isomorphisms. Hence $G^{(n-1)}$ is a subgroup of G. These two groups coincide if and only if $\Gamma(\varphi)=\Gamma(\varphi')$ does always imply $\Gamma^{(n-1)}(\varphi)=\Gamma^{(n-1)}(\varphi')$. This in turn is equivalent to $|\Gamma|\geqq n+3$ or $n=2$. In the latter cases the term *osculating subspaces of a non-degenerate normal curve* does make sense [7], [17].

Most properties of osculating subspaces depend on the characteristic of the ground field K irrespective of whether K is commutative or not. We mention, without proof, one result: If the characteristic of K is a prime number which divides the dimension n of Π, then all osculating hyperplanes of a normal isomorphism belong to a pencil of hyperplanes. See [18], [20], [35], [36].

3. DEGENERATE CONICS

3.1. Degenerate projectivities

3.1.1. Let $\Pi=\Pi(\mathbf{V})$ be a projective plane. The existence of a degenerate projectivity implies that Π is non-pappian or, equivalently, that K is a skewfield. We assume (in this chapter only) that K is a skewfield.

An ordered basis $(\mathbf{p},\mathbf{q};\mathbf{a})$ of $\mathbf{V}$ and a non-central element $a\in K\setminus Z$ give rise to linear automorphisms $g_0, g_1\in GL(\mathbf{V})$ such that

$$(3.1)\qquad \mathbf{p}^{g_0}=\mathbf{p},\quad \mathbf{q}^{g_0}=\mathbf{q},\quad \mathbf{a}^{g_0}=a\mathbf{a},$$

$$(3.2)\qquad \mathbf{p}^{g_1}=\mathbf{q},\quad \mathbf{q}^{g_1}=\mathbf{p},\quad \mathbf{a}^{g_1}=\mathbf{a}.$$

The collineation induced by g_0 is a homology γ_0, say, with centre $A:=\mathbf{a}K$, axis PQ ($P:=\mathbf{p}K$, $Q:=\mathbf{q}K$) and characteristic cross-ratio $CR(X^{\gamma_0},X,XA\cap PQ,A)=\hat{a}$, where $\hat{a}\subset K$ denotes the conjugacy class of a and X is any point of $\mathcal{P}\setminus(\{A\}\cup PQ)$. The map g_1 induces an involutory perspective collination denoted by γ_1 with axis AU ($U:=\mathbf{u}K$, $\mathbf{u}=\mathbf{p}+\mathbf{q}a$) and $P^{\gamma_1}=Q$. Put $g=:g_0g_1$ and $\gamma:=\gamma_0\gamma_1$. This γ is a projective collineation which interchanges P with Q and fixes A. Hence the restricted map

(3.3) $\quad \zeta := \gamma|(u\Pi/P) : u\Pi/P \to u\Pi/Q$

is a projectivity which has PQ as invariant line. Setting $\mathbf{M}:=\mathbf{a}K\oplus\mathbf{u}K$ the trace map ζ_{AU} in $AU=\mathcal{P}(\mathbf{M})$ is induced by the linear automorphism $g_{\mathbf{M}}:=(g|\mathbf{M})p_{\mathbf{M}}$, where $p_{\mathbf{M}}:\mathbf{V}\to\mathbf{M}$ is the projection with kernel $\mathbf{q}K$. Consequently, $\mathbf{a}^{g_{\mathbf{M}}}=\mathbf{a}a$, $\mathbf{u}^{g_{\mathbf{M}}}=\mathbf{u}a$ and

(3.4) $\quad \{(\mathbf{a}+\mathbf{u}s)K\}\cup\{\mathbf{u}K\}$,

where s is in the centralizer of a, is the set of invariant points of ζ_{AU}. The trace map ζ_{AU} is non-identical, because a is a non-central element of K. Thus, firstly, ζ_{AU} has infinitely many fixed points (cf. 2.2.2), secondly, ζ is a degenerate projectivity and, thirdly, $\Gamma(\zeta)$ is a degenerate conic. We shall say that ζ (as well as $\Gamma(\zeta)$) is determined by the ordered basis $(\mathbf{p},\mathbf{q};\mathbf{a})$ of $\mathbf{V}$ and the element $a\in K\setminus Z$.

Letting $a\in Z^{\times}$ the above construction yields a perspectivity ζ and $\Gamma(\zeta)=PQ\cup AU$ is no conic in the sense of our definition. However, sometimes it would be more conveniant to use the term "degenerate conic" in this case as well (cf. 4.1.2); but we shall stick to our previous definitions as given in 1.2.4.

<u>3.1.2.</u> The following proposition illustrates that those examples of degenerate projectivities as introduced in 3.1.1

are, in fact, all degenerate projectivities.

PROPOSITION 3.1. *Any degenerate projectivity* $\zeta: u\Pi/P \to u\Pi/Q$ *is determined by an ordered basis of* $\mathbf{V}$ *and an element of* $K\setminus Z$.

Proof. Let $l, \tilde{l} \neq PQ$ be two different lines passing through P. Then $A := l \cap l^{\zeta}$, $\tilde{\mathbf{A}} := \tilde{l} \cap \tilde{l}^{\zeta}$ are different points of the degenerate conic $\Gamma(\zeta)$. By Proposition 1.1, the line $A\tilde{A}$ meets PQ in a point U other than P,Q. The trace map ζ_{AU} is fixing $A, \tilde{A}, U$ but is not an identity-map, because ζ is no perspectivity. Suppose $A=\mathbf{a}K$, $U=\mathbf{u}K$, $\tilde{A}=(\mathbf{a}+\mathbf{u})K$ and $\mathbf{M}=\mathbf{a}K\oplus\mathbf{u}K$. If $g_{\mathbf{M}}\in$ $\in GL(\mathbf{M})$ induces the trace map ζ_{AU}, then $\mathbf{a}$ and $\mathbf{u}$ are eigenvectors of $g_{\mathbf{M}}$ belonging to the same eigenvalue $a\in K\setminus Z$, say. Choose $\mathbf{p},\mathbf{q}\in\mathbf{V}$ such that $P=\mathbf{p}K$, $Q=\mathbf{q}K$, $U=(\mathbf{p}+\mathbf{q}a)K$. Then $(\mathbf{p},\mathbf{q};\mathbf{a})$ and a determine the degenerate projectivity ζ.□

PROPOSITION 3.2. *Two ordered bases* $(\mathbf{p},\mathbf{q};\mathbf{a})$, $(\mathbf{p}',\mathbf{q}';\mathbf{a}')$ *of* $\mathbf{V}$ *and elements* $a, a'\in K\setminus Z$, *respectively, such that*

$$\begin{aligned} \mathbf{p}' &= \mathbf{p}c_{00} \\ \mathbf{q}' &= \mathbf{q}c_{11} \qquad\qquad (c_{jk}\in K) \\ \mathbf{a}' &= \mathbf{p}c_{02} + \mathbf{q}c_{12} + \mathbf{a}c_{22} \end{aligned}$$

determine the same degenerate projectivity if and only if there exists $z\in Z^{\times}$ *with*

$$(3.4)\qquad c_{00} = c_{11}z,\; c_{12} = c_{02}c_{22}^{-1}ac_{22},\; a' = c_{22}^{-1}ac_{22}z.$$

Proof. In terms of the notation of 3.1.1, we have $\zeta=\zeta'$ if and only if the trace maps ζ_{AU} and ζ'_{AU} coincide or, equivalently,

$$\begin{aligned} \mathbf{a}^{g'_{\mathbf{M}}} = \mathbf{p}(c_{02}a'c_{22}^{-1} - c_{00}c_{11}^{-1}c_{12}c_{22}^{-1}) + \mathbf{q}(*) + \\ + \mathbf{a}c_{22}a'c_{22}^{-1} = \end{aligned}$$

$$= \mathbf{a}^{g_{\mathbf{M}_z}} = \mathbf{a}az$$

and

$$\mathbf{u}^{g_{\mathbf{M}}'} = \mathbf{p}c_{00}c_{11}^{-1}a + \mathbf{q}(**) = \mathbf{u}^{g_{\mathbf{M}_z}} = \mathbf{u}az$$

with $z \in Z^{\times}$.□

Our way of dealing with degenerate projectivities is different from the one used in [27]. We have been aiming at finding natural extensions of a given degenerate projectivity to a collineation of Π. As is shown by Proposition 3.2, our method yields no unique extending collineation, but all such collineations are still closely related.

Degenerate conics are also called *C-configurations* and can be obtained as certain planar sections of a regulus in a three dimensional non-pappian projective space [33,325]. Examples of degenerate conics in translation planes are given in [30].

3.1.3. Using the notation introduced in 3.1.1, we look at the degenerate conic $\Gamma(\zeta)=:\Gamma$. The improper part of Γ with respect to ζ is the line $\Gamma^{\times}(\zeta)=PQ$. This is the *only* line contained in Γ, because ζ is no perspectivity. Furthermore, $\Gamma^{\times}(\zeta)=G(\zeta)$ is the fundamental line of ζ. So it makes sense to call $\Gamma^{\times}(\zeta)$, $\Gamma^{o}(\zeta)$, $G(\zeta)$ the *improper part*, *proper part*, *fundamental line*, respectively, *of the degenerate conic* Γ.

We deduce the parametric representation

$$(3.5) \qquad \Gamma^{o}(\zeta) = \{\mathbf{y}_t K \mid \mathbf{y}_t = \mathbf{p}t + \mathbf{q}ta + \mathbf{a},\ t \in K\},$$

since $\Gamma^{o}(\zeta)=:\Gamma^{o}$ equals the set of all points $\mathbf{a}'K$ with $\mathbf{a}'$ satisfying the conditions in Proposition 3.2.

Consider the projection $\pi: P \to PQ$ through any point $\mathbf{y}_u K$, say. We obtain

$$(3.6) \qquad n := (\Gamma^{o} \setminus \{\mathbf{y}_u K\})^{\pi} = \{(\mathbf{p} + \mathbf{q}(t-u)a(t-u)^{-1})K \mid t \in K \setminus \{u\}\} =$$

$$= \{X \in PQ \mid CR(X,E,P,Q) = \hat{a}\},$$

where $E=(\mathbf{p}+\mathbf{q})K$. This set n is infinite [19]. By the proof of Proposition 3.1, every point of n is the image of infinitely many points of Γ^0 and , by the last equality in (3.6), n does not depend on the choice of $u\in K$. Therefore every line joining two different points of Γ^0 meets PQ in a point belonging to n and, conversely, any line which joins a point of n and a point of Γ^0 can be spanned by two distinct point of Γ^0 as well. Clearly, a point of n never is a fundamental point of Γ.

Next we take any line $l=P(\ker(h_0\mathbf{p}^*+h_1\mathbf{q}^*+h_2\mathbf{a}^*))$ different from PQ, i.e. $(h_0,h_1)\neq(0,0)$, where $\{\mathbf{p}^*,\mathbf{q}^*,\mathbf{a}^*\}$ is the dual basis of $\{\mathbf{p},\mathbf{q},\mathbf{a}\}$. The intersection $l\cap\Gamma^0$ is the set of all points $\mathbf{y}_tK\in\Gamma^0$ with $t\in K$ and

(3.7) $\quad h_0t + h_1ta + h_2 = 0.$

For $h_0=0$ or $h_1=0$ a unique point $\mathbf{y}_tK$ exists in accordance with Proposition 1.1. If $h_0h_1\neq 0$, then we may assume $h_1=1$, hence

(3.8) $\quad h_0t + ta + h_2 = 0.$

See [13,222] for results on the solutions of equation (3.8). Cf. also [33,331].

3.2. Generating maps

3.2.1. The crucial result on degenerate conics is

THEOREM 3.1. *Any two degenerate projectivities ζ,ζ' yield the same degenerate* ***conic*** ***if*** *and only if there are ordered bases* $(\mathbf{p},\mathbf{q};\mathbf{a})$, $(\mathbf{p}',\mathbf{q}';\mathbf{a})$ *of* $\mathbf{V}$ *and elements* $a,a'\in K\setminus Z$, *re-*

spectively, which determine ζ, ζ' *such that, either*

(3.10) $[a:Z] \neq 2, \quad a' = (z_0+z_1a)(z_2+z_3a)^{-1}$

$$\mathbf{p}' = \mathbf{p}z_1w + \mathbf{q}z_0w,$$

$$\mathbf{q}' = \mathbf{p}z_3w + \mathbf{q}z_2w,$$

with $z_i \in Z,\ z_1z_2-z_0z_3 \neq 0,\ w \in K^{\times}$, *or*

(3.11) $[a:Z] = 2, \quad a' = z_0 + z_1a,$

$$\mathbf{p}' = \mathbf{p}(-vz_0+(w-m_1v)z_1) + \mathbf{q}(-wz_0+m_0vz_1),$$

$$\mathbf{q}' = \mathbf{p}v \qquad\qquad + \mathbf{q}w,$$

with $a^2=m_0+m_1a,\ z_i, m_i \in Z,\ v,w \in K,\ (v,w) \neq (0,0),\ v^{-1}w \notin \hat{a}$ *(when* $v \neq 0$*), where* $\hat{a} \subset K$ *denotes the conjugacy class of* a.

Proof. (1) Suppose $\Gamma(\zeta)=\Gamma(\zeta')$. We read off from Proposition 3.1 that ζ, ζ' are determined, respectively, by ordered bases $(\mathbf{p},\mathbf{q};\mathbf{a})$, $(\mathbf{p}',\mathbf{q}';\mathbf{a})$ of $\mathbf{V}$ and elements $a, a' \in K \setminus Z$, say. Since $\mathbf{p}K \vee \mathbf{q}K = \mathbf{p}'K \vee \mathbf{q}'K$, we obtain

(3.12) $\mathbf{p}' = \mathbf{p}c_{00} + \mathbf{q}c_{10}$

$$\mathbf{q}' = \mathbf{p}c_{01} + \mathbf{q}c_{11},$$

where (c_{jk}) is an invertible matrix with entries in K. From formulae (3.5) and (3.12) we deduce

$$\Gamma^0(\zeta') = \{\mathbf{y}'_tK \mid \mathbf{y}'_t = \mathbf{p}(c_{00}t+c_{01}ta')+\mathbf{q}(c_{10}t+c_{11}ta')+\mathbf{a},\ t \in K\}.$$

Hence, by $\Gamma(\zeta)=\Gamma(\zeta')$,

$$(c_{00}t+c_{01}ta')a = (c_{10}t+c_{11}ta')$$

for all $t \in K$ or, equivalently,

(3.13) $c_{10}t - c_{00}ta + c_{11}ta' - c_{01}ta'a = 0$ for all $t \in K$.

The left-side coefficients of this identity (3.13) cannot vanish simultaneously, because (c_{jk}) is a regular matrix.

Then, by Lemma 1.1, there exists a non-trivial linear combination $z_0+z_1a-z_2a'-z_3a'a=0$ $(z_i\in Z)$. Thus

$$(3.14)\quad a' = (z_0+z_1a)(z_2+z_3a)^{-1}$$

is an element of the commutative subfield $Z(a)\subset K$ and substitution in (3.13) implies

$$(3.15)\quad (c_{10}z_2+c_{11}z_0)t + (c_{10}z_3-c_{00}z_2+c_{11}z_1-c_{01}z_0)ta + (-c_{00}z_3-c_{01}z_1)ta^2 = 0 \text{ for all } t\in K.$$

If $1,a,a^2$ are linearly independent over Z, i.e. $[a:Z]>2$, then all left-side coefficients in (3.15) have to vanish simultaneously by Lemma 1.1. Therefore

$$(3.16)\quad (c_{jk}) = \begin{pmatrix} z_1 & -z_3 \\ -z_0 & z_2 \end{pmatrix}\cdot\begin{pmatrix} w & 0 \\ 0 & w \end{pmatrix}$$

with $w\in K^\times$ and $z_1z_2-z_0z_3\neq 0$.

If $[a:Z]=2$, then $a^2=m_0+m_1a$ where $m_0,m_1\in Z$, $m_0\neq 0$. We may assume $z_2=1$, $z_3=0$ in (3.14). Substituting in (3.15) we have

$$(3.17)\quad (c_{10}+c_{11}z_0-c_{01}z_1m_0)t + (-c_{00}+c_{11}z_1-c_{01}(z_0+z_1m_1))ta = 0 \text{ for all } t\in K.$$

The same arguments as before yield

$$(3.18)\quad (c_{jk}) = \begin{pmatrix} w-m_1v & v \\ m_0v & w \end{pmatrix}\cdot\begin{pmatrix} z_1 & 0 \\ -z_0 & 1 \end{pmatrix}$$

with $v,w\in K$. Since (c_{jk}) is invertible, its right column rank equals 2. Thus $v=0$ forces $w\neq 0$ and $w=0$ implies $v\neq 0$. Suppose, finally, that $vw\neq 0$. Then $v^{-1}w-m_1\neq m_0w^{-1}v$ and consequently

$$(3.19)\quad (v^{-1}w)^2 \neq m_0 + m_1(v^{-1}w).$$

By [12,302] or [13,54], the inequality (3.19) is equivalent

to $(v^{-1}w)$ not conjugate to a. (Another proof of this can be given by discussing the intersection of a line and a non-degenerate conic; cf. 2.2.2.)

(2) By reversing the above arguments, it is clear that (3.15) or (3.17) implies $\Gamma(\zeta')\subset\Gamma(\zeta)$ and the reader will easily show that $\Gamma(\zeta')=\Gamma(\zeta)$, as required.□

The proof of Theorem 3.1 follows the same pattern as that of "Satz2" in [27]. In contrast to [27], where most of this chapter's material has been taken from, nearly all of our results will turn out to be immediate consequences of Theorem 3.1. Thus it is possible to omit some lengthy calculations including another application of results on generalized polynomial identities [27,248-249].

3.2.3. A subset c of the fundamental line of a degenerate conic Γ is called a *fundamental chain of* Γ if c is a maximal set with the following property: Any two different points $P',Q'\in c$ form a fundamental pair of Γ.

COROLLARY 3.1. *Let* $\Gamma(\zeta)$ *be a degenerate conic given by* (3.3). *If* $[a:Z]\neq 2$, *then*

$$(3.20)\qquad [P,Q,E]_Z$$

where $E=(\mathbf{p}+\mathbf{q})K$ *is the only fundamental chain of* $\Gamma(\zeta)$. *If* $[a:Z]=2$, *then any point* $Q'=\mathbf{q}'K\in PQ\setminus n$, $\mathbf{q}'=\mathbf{p}v+\mathbf{q}w$, *lies in a fundamental chain* $c_{Q'}$ *of* $\Gamma(\zeta)$ *given by*

$$(3.21)\qquad c_{Q'} = [P',Q',E']_Z$$

where $P'=\mathbf{p}'K$, $\mathbf{p}'=\mathbf{p}(w-m_1v)+\mathbf{q}m_0v$, $E'=(\mathbf{p}'+\mathbf{q}')K$, *and any two different fundamental chains of* $\Gamma(\zeta)$ *have empty intersection.*

Proof. Theorem 3.1 immediately establishes the result, since $\{X\in PQ\,|\,\{X,Q'\}$ is a fundamental pair of $\Gamma(\zeta)\}$ is a is a Z-chain (cf. 1.2.2) as well as a fundamental chain of $\Gamma(\zeta)$.□

We remark that for $[a:Z]=2$ the line PQ is covered (disjointly) by the fundamental chains of $\Gamma(\zeta)$ and the set $\mathcal{N}$. The concept of "ordinary fundamental point" ("gewöhnlicher Grundpunkt"), as has been introduced in [27], will not be used in this article, because "fundamental chain" seems more appropriate.

The following result may be regarded as "Pascal's theorem" for degenerate conics [27].

PROPOSITION 3.3. *Let* Γ *be a degenerate conic. Given three pairwise different points* P_1, P_2, P_3 *in the same fundamental chain of* Γ *and three pairwise different points* A_1, A_2, A_3 *in the proper part of* Γ, *then*

$$P_1A_2 \cap P_2A_1,\ P_1A_3 \cap P_3A_1,\ P_2A_3 \cap P_3A_2$$

are three collinear points.

Proof. By calculation.□

3.4. Automorphic collineations and classification

3.4.1. Suppose that, as before, a degenerate conic $\Gamma=\Gamma(\zeta)$ is given by formula (3.3). We shall frequently adopt notions of affine geometry by regarding PQ as line at infinity. The term $\dot{f}$-*automorphism of* $\mathbf{V}$ will be used as a shorthand for any bijective semi-linear map $\mathbf{V}\to\mathbf{V}$ with respect to an automorphism $\dot{f}$ of the skewfield K.

If a collineation $\kappa \in P\Gamma L(\Pi)$ fixes the line PQ, then κ is induced by one and only one $\dot{f}$-automorphism $f \in \Gamma L(\mathbf{V})$ such that

$$\mathbf{a}^f = \mathbf{p}c_{02} + \mathbf{q}c_{12} + \mathbf{a}. \tag{3.22}$$

This collineation κ is projective if and only if $\dot{f}$ is an inner automorphism of K, whereas $\dot{f}=\mathrm{id}_K$ yields the normal subgroup of those projective collineations which preserve all affine ratios in $P\setminus PQ$. Since $\kappa^{-1}\zeta\kappa$ is a generating map of Γ^{κ} which is determined by $(\mathbf{a}^f,\mathbf{p}^f;\mathbf{q}^f)$ and $a^{\dot{f}}$, Proposition 3.1 and Theorem 3.1 give necessary and sufficient conditions for f to induce an automorphic collineation of Γ:

THEOREM 3.2. *Let* $\Gamma(\zeta)$ *be determined by an ordered basis* $(\mathbf{p},\mathbf{q};\mathbf{a})$ *of* $\mathbf{V}$ *and an element* $a\in K\setminus Z$. *Any* $\dot{f}$*-automorphism* $f\in\Gamma L(\mathbf{V})$ *satisfying* (3.22) *induces an automorphic collineation of* $\Gamma(\zeta)$ *if and only if, firstly, the conditions stated in Theorem 3.1. for* $a',\mathbf{p}',\mathbf{q}'$ *hold when substituting* $a^{\dot{f}},\mathbf{p}^f,\mathbf{q}^f$, *respectively, and, secondly, there exists* $u\in K$ *such that*

$$\mathbf{a}^f = \mathbf{p}u + \mathbf{q}ua + \mathbf{a}.$$

Those linear maps $f\in GL(\mathbf{V})$ whose matrices with respect to $(\mathbf{p},\mathbf{q};\mathbf{a})$ equal

$$\begin{pmatrix} 1 & 0 & u \\ 0 & 1 & ua \\ 0 & 0 & 1 \end{pmatrix} \quad \text{with } u\in K$$

induce a normal subgroup of automorphic translations of $\Gamma=\Gamma(\zeta)$ which operates regularly on the proper part of Γ. The stabilizer of $A=\mathbf{a}K$ within the group of all automorphic collineations of Γ is induced by $\dot{f}$-automorphisms with matrices (written as a product)

$$\begin{pmatrix} w & 0 & 0 \\ 0 & w & 0 \\ 0 & 0 & 1 \end{pmatrix} \cdot \begin{pmatrix} z_1 & -z_3 & 0 \\ -z_0 & z_2 & 0 \\ 0 & 0 & 1 \end{pmatrix}$$

for $[a:Z]\neq 2$, $a^{\dot{f}}=(z_0+z_1a)(z_2+z_3a)^{-1}$ and

$$\begin{pmatrix} w-m_1v & v & 0 \\ m_0v & w & 0 \\ 0 & 0 & 1 \end{pmatrix} \cdot \begin{pmatrix} z_1 & 0 & 0 \\ -z_0 & 1 & 0 \\ 0 & 0 & 1 \end{pmatrix}$$

for $[a:Z]=2$, $a^{\dot{f}}= z_0+z_1a$. Here u,v,w are subject to the conditions stated in Theorem 3.1. If $[a:Z]\neq 2$, then the orbit of $Q=\mathbf{q}K$ under this stabilizer is a subset q of the only fundamental chain of Γ. The "size" of this subset q depends on the existence of "suitable" automorphisms of K. If $[a:K]=$ $=2$, then the orbit of Q under this stabilizer equals $PQ\setminus n$; the orbit of Q under the stabilizer of A within the subgroup of automorphic collineations of Γ which preserve all affine ratios ($\dot{f}=id_K$) is also $PQ\setminus n$ for $[a:Z]=2$.

The matrices

$$\begin{pmatrix} w & 0 & 0 \\ 0 & w & 0 \\ 0 & 0 & 1 \end{pmatrix} \quad \text{with } w\in Z^{\times}$$

yield linear maps inducing a subgroup of automorphic homologies of Γ whose common centre is A.

<u>3.4.2.</u> We turn to the problem of classification [27]. If we are given two degenerate conics, then (possibly after applying a projective collineation on one of them) we may assume that these conics are determined by the same ordered basis of **V** and two elements of $K\setminus Z$. From Theorem 3.2 we deduce immediately

THEOREM 3.3. *Given two degenerate conics* $\Gamma(\zeta), \Gamma(\zeta')$ *which are determined by the same ordered basis of* **V** *and elements* $a, a' \in K \setminus Z$, *respectively, there is a (projective) collineation mapping* $\Gamma(\zeta')$ *onto* $\Gamma(\zeta)$ *if and only if there is an (inner) automorphism* $\dot{f}$ *of* K *such that, either*

(3.23) $[a:Z] \neq 2, \quad a'^{\dot{f}} = (z_0+z_1a)(z_2+z_3a)^{-1}$

with $z_1z_2-z_0z_3 \neq 0, \quad z_i \in Z$, *or*

(3.24) $[a:Z] = 2, \quad a^{\dot{f}} \in Z(a)$.

This brings to an end our discussion of degenerate conics.

4. DEGENERATE NORMAL CURVES

4.1. A few results

4.1.1. Suppose that $\Pi(\mathbf{V})$ is an n-dimensional projective space. Let

$$\zeta : u\Pi/P \to u\Pi/Q \quad (P \neq Q)$$

be a degenerate projective isomorphism whose fundamental subspace $G(\zeta)$ is k-dimensional $(1 \leq k \leq n-1)$.

If k=1, then the improper part $\Gamma^{\times}(\zeta)$ of the normal curve $\Gamma(\zeta)$ with respect to ζ is the line PQ, whereas $k \geq 2$ implies that $\Gamma^{\times}(\zeta)$ is a non-degenerate normal curve in the projective space $\Pi(G)$ $(G := G(\zeta))$, because the restricted map $\zeta|(u\Pi(G))/P$ is non-degenerate.

Next let G be no hyperplane. Then $\zeta|u\Pi/G$ is a projective automorphism of $u\Pi/G$. If $X \in \Gamma^{\circ}(\zeta)$, i.e. the proper part of $\Gamma(\zeta)$ with respect to ζ, then $(X \vee G)^{\zeta} = X \vee G$ and $\Gamma(\zeta) \cap (X \vee G)$

is the point set generated by $\zeta|(u\Pi(X\vee G))/P$. This makes clear that $\Gamma^{o}(\zeta)$ is empty if ζ has no invariant (k+1)-dimensional subspaces.

4.1.2. The preceding discussion tells us that at first projective isomorphisms ζ with a fundamental hyperplane have to be studied. Here the generated point set $\Gamma(\zeta)$ is either the union of two distinct lines (n=2; cf. 3.1.1) or a degenerate normal curve ($n\geq 3$).

According to [28,445] there exists always a basis $\{\mathbf{p}_0,\ldots,\mathbf{p}_{n-1},\mathbf{a}\}$ of $\mathbf{V}$ such that

$$(4.1)\qquad \Gamma^{\times}(\zeta) = \{\mathbf{x}_t K \mid \mathbf{x}_t=\sum_{j=0}^{n-1}\mathbf{p}_j t^j,\ t\in K\} \cup \{\mathbf{p}_{n-1}K\}$$

and

$$(4.2)\qquad \Gamma^{o}(\zeta) = \{\mathbf{y}_t K \mid \mathbf{y}_t=\sum_{j=0}^{n-1}\mathbf{p}_j t a^j+\mathbf{a},\ t\in K\}$$

with $a\in K^{\times}$. A hyperplane $H=P(\ker\sum_{j=0}^{n-1}h_j\mathbf{p}_j^*+h\mathbf{a}^*)$, where $\{\mathbf{p}_0^*,\ldots,\mathbf{p}_{n-1}^*,\mathbf{a}^*\}$ is the dual basis of $\{\mathbf{p}_0,\ldots,\mathbf{p}_{n-1},\mathbf{a}\}$, contains $\Gamma^{o}(\zeta)$ if and only if

$$(4.3)\qquad h_0 t + h_1 ta + \ldots + h_{n-1}ta^{n-1} + h = 0 \text{ for all } t\in K$$

or, equivalently, h=0 and $[a:Z]\leq n-1$ by Lemma 1.1. On the other hand, it is easily seen that the dimension of the subspace spanned by $\Gamma^{o}(\zeta)$ equals $\min\{[a:Z],n\}$. (Use "Satz 1" in [17] and "9.3" in [28].) Hence $a\in Z^{\times}$ if and only if $\Gamma^{o}(\zeta)$ is an affine line. Some examples of automorphic collineations of $\Gamma(\zeta)$ are included in [28].

4.1.3. Returning to the general case and assuming $k\leq n-2$ it follows that every ζ-invariant (k+1)-dimensional subspace has non-empty intersection with $\Gamma^{o}(\zeta)$. Results concerning the case k=n-2 can be found in [28].

4.2. Final remarks

4.2.1. Let $\Pi(\mathbf{V})$ be a 4-dimensional projective space and let K be the Galois-field of order 2. A non-degenerate normal curve in Π is just a triangle by (2.8). On the other hand, it is easy to see that there exists a degenerate projective isomorphism ζ, say, the generated point set of which is a triangle as well: We have to ensure only that the fundamental subspace $G(\zeta)$ is a plane and that none of the seven hyperplanes passing through $G(\zeta)$ is ζ-invariant. This forces $\Gamma^{\times}(\zeta)$ to be a triangle in $G(\zeta)$ and $\Gamma^{o}(\zeta)=\emptyset$, as required.

This example shows that a normal curve may be degenerate as well as non-degenerate, but all normal curves with this property are not known to the author. Cf. however [28,436].

4.2.2. In general, the problems of classification, automorphic collineations and generating maps seem to be unsolved for degenerate normal curves. It should also be interesting to discuss, for example, the group of collineations fixing the point set $\Gamma^{o}(\zeta)$ given by (4.2) irrespective of what happens to $\Gamma^{\times}(\zeta)$ given by (4.1).

REFERENCES

1. Amitsur, S.A.: 'Generalized Polynomial Identites and Pivotal Monomials', *Trans.Amer.Math.Soc.* 114, 210-226 (1965).
2. Amitsur, S.A.: 'Rational Identities and Applications to Algebra and Geometry', *J.Algebra* 3, 304-359 (1966).
3. Artzy, R.: 'The Conic $y=x^2$ in Moufang Planes', *Aequ.Math.* 6, 30-35 (1971).
4. Benz, W.: *Vorlesungen über Geometrie der Algebren*, Grundlehren Bd. 197, Berlin-Heidelberg-New York: Springer,1973.
5. Bertini, E.: *Introduzione alla geometria proiettiva degli iperspazi*, Pisa: E.Spoerri, 1907.
6. Bertini, E.: *Einführung in die projektive Geometrie mehrdimensionaler Räume*, Wien: Seidel&Sohn, 1924.)[6]

[6]Translation of [5] into German.

7. Berz, E.: 'Kegelschnitte in desarguesschen Ebenen', *Math. Z.* 78, 55-85 (1962).
8. Brauner, H.: *Geometrie projektiver Räume I*, Mannheim-Wien-Zürich: BI-Wissenschaftsverlag, 1976.
9. Brauner, H.: *Geometrie projektiver Räume II*, Mannheim-Wien-Zürich: BI-Wissenschaftsverlag, 1976.
10. Burau, W.: *Mehrdimensionale projektive und höhere Geometrie*, Berlin: VEB Dt.Verlag d.Wissenschaften, 1961.
11. Clifford, W.K.: 'On the Classification of Loci', *Phil. Trans.Roy.Soc.II*, 663-681 (1878) (in: *Mathematical Papers*, London: Macmillan & Co., 1882).
12. Cohn, P.M.: *Free Rings and their Relations*, London-New York: Academic Press, 1971.
13. Cohn, P.M.: *Skew Field Constructions*, LMS Lecture Note Ser. 27, Cambridge: Cambridge U.P., 1977.
14. Gordon, B., Motzkin, T.S.:' On the Zeros of Polynomials over Division Rings', *Trans.Amer.Math.Soc.* 116, 218-226 (1965). Correction *ibid.* 122, 547 (1966).
15. Havlicek, H.: 'Normisomorphismen und Normkurven endlichdimensionaler projektiver Desargues-Räume', *Monatsh.Math.* 95, 203-218 (1983).
16. Havlicek, H.: 'Eine affine Beschreibung von Ketten', *Abh. Math.Sem.Univ.Hamburg* 53, 267-276 (1983).
17. Havlicek, H.: 'Die automorphen Kollineationen nicht entarteter Normkurven', *Geom.Dedicata* 16, 85-91 (1984).
18. Havlicek, H.: 'Erzeugnisse projektiver Bündelisomorphismen', *Ber.Math.Stat.Sekt.Forschungszentr.Graz Nr.* 215 (1984).
19. Herstein, I.N.: 'Conjugates in Division Rings', *Proc.Amer. Math.Soc.* 7, 1021-1022 (1956).
20. Herzer, A.: 'Die Schmieghyperebenen an die Veronese-Mannigfaltigkeit bei beliebiger Charakteristik', *J.Geometry* 18, 140-154 (1982).
21. Kötter, E.: 'Die Entwicklung der synthetischen Geometrie', *J.Ber.DMV* 5, 2.Heft, 1-486 (1901).
22. Krüger, W.: 'Regelscharen und Regelflächen in dreidimensionalen desarguesschen Räumen', *Math.Z.* 93, 404-415 (1966).
23. Krüger, W.: 'Kegelschnitte in Moufangebenen', *Math.Z.* 120, 41-60 (1971).
24. Müller, E., Kruppa, E.: *Vorlesungen über Darstellende Geometrie, I.Bd.: Die linearen Abbildungen*, Leipzig-Wien: F.Deuticke, 1923.
25. Ostrom, T. G.: 'Conicoids: Conic-like figures in Non-pappian Planes', in: Plaumann, P., Strambach, K. (Eds.): *Geometry-von Staudt's Point of View*, Dordrecht: D.Reidel, 1981.
26. Pickert, G.: 'Projectivities in Projective Planes', in: Plaumann, P., Strambach K. (Eds.): *Geometry-von Staudt's Point of View*, Dordrecht: D.Reidel, 1981.

27. Riesinger, R.: 'Entartete Steinerkegelschnitte in nichtpapposschen Desarguesebenen', *Monatsh.Math.* **89**, 243-251 (1980).
28. Riesinger, R.: 'Normkurven in endlichdimensionalen Desarguesräumen', *Geom.Dedicata* **10**, 427-449 (1981).
29. Riesinger, R.: 'Geometrische Überlegungen zum rechten Eigenwert-Problem für Matrizen über Schiefkörpern', *Geom.Dedicata* **12**, 401-405 (1982).
30. Rosati, L.A.: 'Su alcune varietà dello spazio proiettivo sopra un corpo non commutativo', *Ann.Mat.pura appl.,IV Ser.* **59**, 213-227 (1962).
31. Rosati, L.A.: 'Su alcuni problemi di geometria non lineare sopra un corpo sghembo, *Atti Acad.naz.Lincei, VIII Ser., Rend., Cl.sci.fis.mat.natur* **36**, 615-622 (1964).
32. Schaal, H.: 'Zur perspektiven Zerlegung und Fixpunktkonstruktion der Affinitäten von $A^n(K)$', *Arch.Math.* **38**, 116-123 (1982).
33. Segre, B.: *Lectures on Modern Geometry*, Roma: Ed. Cremonese, 1962.
34. Segre, C.: 'Mehrdimensionale Räume', in: *Encyklopädie d. Math.Wiss. III, 2,2A,* Leipzig: Teubner,1921-1928.
35. Timmermann, H.: 'Descrizioni geometriche sintetiche di geometrie proiettive con caratteristica p>0', *Ann.Mat. pura appl., IV Ser.* **114**, 121-139 (1977).
36. Timmermann, H.: *Zur Geometrie der Veronesemannigfaltigkeit bei endlicher Charakteristik*, Habilitationsschrift, Hamburg: 1978.
37. Veronese, G.: 'Behandlung der projectivischen Verhältnisse der Räume von verschiedenen Dimensionen durch das Princip der Projicirens und des Schneidens', *Math.Ann.* **19**, 161-234 (1882).

Part II

Hjelmslev Geometries

A TOPOLOGICAL CHARACTERIZATION OF HJELMSLEV'S CLASSICAL GEOMETRIES

J. W. (Michael) Lorimer*
University of Toronto

"When there is really nothing, left to say
I sit and write, to pass the day"
(JWL)
"For evil to succeed in this world
it requires only that good men do nothing".
(Majorie Kinnan Rawlins; The Sojourner)

This article establishes a theoretical beginning for the study of topological Hjelmslev planes. Like the Drake-Jungnickel paper in this book, we are also interested in epimorphisms between incidence structures. But, rather than finite structures, we consider topological incidence structures; that is incidence structures whose point and line sets are topological spaces and where the joining of points and the intersecting of lines (where they exist) are continuous functions. In particular we look for factorizations called "solutions" of open continuous maps $\emptyset : P \rightarrow P'$ where P and P' are topological H-planes. The oldest and most elegant examples are constructed via homeogeneous coordinates over the topo-

* The author gratefully acknowledges the support of the national science and research council of Canada and the Alexander von Humboldt-Stiftung.

R. Kaya et al. (eds.), Rings and Geometry, 81–151.

logical rings $\mathbb{R}[x]/(x^n)$. The maximal chain of ideals $(0) \subsetneq (x^{n-1})/(x^n) \subsetneq ---- \subsetneq (x)/(x^n)$ generates a solution for this plane. It is our intention in these notes to determine all locally compact connected Pappian Hjelmslev Planes and to show that they all have solutions of the classical type. In fact, we prove that they are just the topological geometries over the rings $K[x]/(x^n)$, where K is the reals or complexes.

In sections 1 and 2 we consider the interaction between Hjelmslev rings (H-rings) and H-planes, and give construction methods and examples for H-rings. As all our classical examples have nilpotent radicals, we consider in section 3 the geometric significance of nilpotent radicals. In section four we introduce topological Hjelmslev planes and establish that the canonical projection is open-continuous and hence that the canonical image is an ordinary topological (affine or projective) plane. The technique here is to coordinatize the affine H-plane with a biternary ring and use the fact that they induce topological loops on the lines. We also observe that topological biternary rings generate topological planes. Next, we consider the various types of connectedness possible in our planes. Finally, we study topological desarguesian H-planes and consider the notion of a "topological solution" for such planes. In section 5 we consider locally compact H-planes and prove the important result that all such planes are separable σ-compact metric spaces. We then consider compactness and prove that, in direct contrast to ordinary planes, a proper topological desarguesian PH-plane is never compact. In section six we return to commutative H-rings and discuss structure theorems for H-rings due essentially to M^c^Lean. Finally, in section seven, we prove that the only locally compact connected Pappian H-planes are the ones over the rings $K[x]/(x^n)$, where K is the reals or complexes.

"I don't care if you can prove it,
What can I use it for?"
W.C. Lorimer

§ 1. HJELMSLEV PLANES AND HJELMSLEV RINGS

For an incidence structure $H = \langle \mathbb{P}, \mathbb{L}, I \rangle$ points are denoted by $P, Q, R, \ldots$ and lines by $a, b, c, \ldots$. $[P,Q]$ denotes the number of lines incident with P and Q , and $[\ell, m]$ the number of points incidence with ℓ and m . All rings R have a unit element 1 . For any ring R , J is the *Jacobson radical.*

(1.1) For a ring R , the following are equivalent.

(i) R possesses a unique maximal right (left) ideal.

(ii) The set of non-units forms a proper ideal.

(ii) R/J is a skew field (non-commutative field).

([Lambek])

Any ring R satisfying one of the properties above is a *local ring*. In this case the ideal mentioned in (i) and (ii) is the radical J.

Let L be a local ring. $L^* = L\backslash J$ is the group of units. On the set $L \times L \times L \setminus J \times J \times J$ we define two equivalence relations R_ℓ , R_r with equivalence classes

$$< a\,b\,c > = \{\lambda(a\,b\,c) \; : \; \lambda \in L^*\}$$

$$[u\,v\,w] \;\; = \{(u\,v\,w)\lambda \; : \; \lambda \in L^*\}.$$

The incidence structure $P(L)$ has points $< a\,b\,c >$, lines $[u\,v\,w]$, and incidence defined by

$$< a\,b\,c > I \; [u\,v\,w] \iff au + bv + cw = 0 \; .$$

Two points $< a\,b\,c >$, $< x\,y\,z >$ are *neighbours* $(< a\,b\,c > \approx < x\,y\,z >)$ provided there exists $\lambda \in L^*$ so that $(a\,b\,c) - \lambda(x\,y\,z) \in J \times J \times J$. Dually, we define *neighbouring lines*. $\approx$ is an equivalence relation on both the point and line sets, and $\not\approx$ is its negation.

$P(L)$ satisfies the following properties:

(K1) If $< a\,b\,c > \not\approx < x\,y\,z >$, then $[< a\,b\,c >, < x\,y\,z >] = 1$.

(K2) If $[u\,v\,w] \not\approx [x\,y\,z]$, then $[u\,v\,w] \, , [x\,y\,z] = 1$.

(K3) $P(L/J)$ is the projective plane over the skew field L/J and the quotient map $v : L \to L/J$ induces an epimorphism $\pi : P(L) \to P(L/J)$ satisfying the conditions

$$\pi(< a\,b\,c >) = \pi(< x\,y\,z >) \iff < a\,b\,c > \approx < x\,y\,z >$$

$$\pi([\; a\,b\,c \;]) = \pi([\; x\,y\,z]) \iff [a\,b\,c] \approx [\,x\,y\,z\,] \; .$$

$P(L)$ is the *Projective Klingenberg plane over* L . (See [Klingenberg] or [Bacon, 2]).

The following charts list geometric properties of $P(L)$ on the left, and the equivalent algebraic properties of L on the right.

$P(L)$		L
$[P,Q] \geq 1$ for all points P, Q.	1.	The lattice of right ideals is a chain.
$[\ell,m] \geq 1$ for all lines ℓ, m.	2.	The lattice of left ideals is a chain.
$P \approx Q \Leftrightarrow [P,Q] > 1$ or $[P,Q] = 0$	3.	Every non-unit is a right zero divisor.
$\ell \approx m \Leftrightarrow [\ell,m] > 1$ or $[\ell,m] = 0$	4.	Every non-unit is a left zero divisor.

(See [Bacon, 2] or [Veldkamp])

A ring H satisfying the four algebraic properties above is a *Hjelmslev ring* ([Törner]).

An incidence structure $H = \langle \mathbb{P}, \mathbb{L}, I \rangle$ is a *projective Hjelmslev plane* (PH-plane for short) provided the following axioms hold:

(PH1) For any two points P, Q : $[P,Q] \geq 1$.

(PH2) For any two lines ℓ, m : $[\ell,m] \geq 1$.

(PH3) There exists a projective plane P and an epimorphism $\pi = (\pi_{\mathbb{P}}, \pi_{\mathbb{L}})$ from H to P with the properties:

(a) $\pi_{\mathbb{P}}(P) = \pi_{\mathbb{P}}(Q) \iff [P,Q] > 1$

(b) $\pi_{\mathbb{L}}(\ell) = \pi_{\mathbb{L}}(m) \iff [\ell,m] > 1$

We say two points P,Q are *neighbours* $(P \underset{\mathbb{P}}{\sim} Q) \iff [P,Q] > 1$ and dually for lines.

In addition, P *is a neighbour to* ℓ $(P \sim \ell)$ $\iff$ there is a point $X I \ell$ so that $P \sim X$.

Generally we write $\sim$ for both $\underset{\mathbb{P}}{\sim}$ and $\underset{\mathbb{L}}{\sim}$ and π for both $\pi_{\mathbb{P}}$ and $\pi_{\mathbb{L}}$.

The neighbour relations are equivalence relations with equivalence classes $\bar{P}$ and $\bar{\ell}$ respectively. The incidence structure $\bar{H} = \langle \mathbb{P}/\underset{\mathbb{P}}{\sim}, \mathbb{L}/\underset{\mathbb{L}}{\sim}, \bar{I} \rangle$, where $\bar{P}\bar{I}\bar{\ell} \iff P \sim \ell$, is also a projective plane isomorphic to P. We identify P with $\bar{H}$ and π with the quotient maps of $\sim$. Then, $\bar{H}$ is the *canonical image* of H and π is the *canonical projection.*

Moreover, we consider lines as point sets in H and $\bar{H}$. Lines of $\bar{H}$ are denoted, then, by $\ell/\sim = \{\bar{P} \mid P \sim \ell\}$.

If $P \not\sim Q$, then $P \vee Q$ is the unique line through P and Q, and dually $\ell \wedge m$ is the unique intersection point of ℓ and m.

An *incidence structure with parallelism* is an incidence structure with an equivalence relation, $\|$, on lines, called *parallelism.* Such a structure satisfies the *parallel postulate* if for each point P and each line ℓ there exists a unique line $L(P,\ell)$ incident with P and parallel to ℓ.

An incidence structure with parallelism $H = \langle \mathbb{P}, \mathbb{L}, \|, \in \rangle$ is an *affine-Hjelmslev* plane (AH-plane for short) <=> the following axioms hold:

(AH1) = (PH1).

Again, points P,Q are *neighbours* <=> $[P,Q] > 1$.

However, two lines ℓ,m are *neighbours* ($\ell \sim m$) <=> every point on one line is a neighbour to some point on the other line.

(AH2) If $P \in \ell, m$, then, $\ell \not\sim m \Leftrightarrow [\ell,m] = 1$.

(AH3) The parallel postulate holds.

(AH4) There is an affine plane A and an epimorphism $\pi : H \to A$ with the properties:

(a) $\pi(P) = \pi(Q) \Leftrightarrow P \sim Q$

(b) $\pi(\ell) = \pi(m) \Leftrightarrow \ell \sim m$

(c) If ℓ and m have no points in common, then $\pi(\ell)$ is parallel to $\pi(m)$ in A .

As in the projective case, we identify A with $\bar{H}$ and $\pi : H \to \bar{H}$ is the canonical projection. Thus, lines of H and $\bar{H}$ are point sets.

The equivalence classes of $\|$ (the parallel pencils) are denoted by Δ (or Δ_ℓ with $\ell \in \Delta_\ell$). Then, Δ_ℓ *is a neighbour to* Δ_m <=> $[\ell,m] \neq 1$. This defines an equivalence relation on pencils. (Lüneburg).

If ℓ is any line of a PH-plane H , let $\mathbb{P}^\ell = \{P \in \mathbb{P} \mid P \not\sim \ell\}$, $\mathbb{L}^\ell = \{x \in \mathbb{L} \mid x \not\sim \ell\}$. Then, $H^\ell = \langle \mathbb{P}^\ell, \mathbb{L}^\ell, \|^\ell, \in \rangle$ is an *associated AH-plane* of H, where $x||^{\ell} y \iff x \wedge \ell = y \wedge \ell$. There are examples of AH-planes which are not the associated AH-planes of any PH-plane. Hence, the two theories, unlike the ordinary case, must be treated separately for many topics. Now, let L be a ring which satisfies properties (i), (iii) and (iv) of a H-ring. Then, L is an *AH-ring*. There are AH-rings which are not H-rings ([Lorimer, Lane]). Following Hjelmslev's original construction ([Hjelmslev 1, page 12]) we define ([Lorimer, Lane]) $A(L)$, *the AH-plane over* L, as follows: Points are elements of $L \times L$ and lines are linear equations $xa + yb + c = 0$ where $(a,b) \notin J \times J$. Lines can thus be divided into two types: $[m,n]' = \{(x,y) | x = ym + n\}$ with $m \in J$, $n \in H$, and $[m,n] = \{(x,y) \mid y = xm + n\}$ with $m,n \in H$. The first coordinate in each type is the slope. Then, two lines are parallel $\iff$ they are of the same type and have the same slope. Moreover, if the plane is proper (i.e. $J \neq (0)$) then there are lines which are disjoint but not parallel. For example, if $m \in J$, $m \neq 0$, then, $[m,1] \wedge [0,0] = \emptyset$ or else $1 \in J$.

(1.2) Definition. A PH-plane or AH-plane is *desarguesian* (Pappian) if and only if it is isomorphic to a PH-plane or AH-plane over a H-ring or AH-ring respectively (Commutative H-ring).

We can give equivalent definitions using configuration theorems, biternary ring theorems or (P,ℓ)-transitivity (see

[Bacon, 1], [Lorimer, 2] and [Seier, 1]).

Seier has shown that desargues' theorem implies the minor desargues' theorem ([Seier, 2]), but as yet no analogue of Hessenberg's theorem is known.

As mentioned above there are AH-rings which are not H-rings. These rings generate AH-planes which can not be represented as a derived AH-plane of a PH-plane over a H-ring. However, if the rings are commutative this situation can not arise. The *classical examples of H-rings* due to Hjelmslev ([Hjelmslev, 2, page 48]) are the rings $\mathbb{R}[x]/(x^n)$ where $\mathbb{R}$ is the real numbers.

Notice that the radical, $J = (x)/(x^n)$ of $\mathbb{R}[x]/(x^n)$ is nilpotent with nilpotency index n. It thus behooves us to consider commutative H-rings with nilpotent radicals. We then consider the geometric significance of nilpotent radicals discovered by Artmann in the late sixties.

§2. CONSTRUCTION OF COMMUTATIVE H-RINGS

All rings in this section are *commutative*.

A ring R is an *E-ring* ([Monk]) <=> R possesses an ideal I so that all ideals of R are of the form

$I^{(n)}$ $\left\{ \text{where } I^{(n)}\text{, the ring product of ideals, is not to be confused with } I^2 = I \times I \right\}$.

Clearly, I is the radical and $I^{(n)} = 0$ for some n. A ring R is a *valuation ring* <=> R is an integral domain whose lattice of ideals forms a chain. (This is equivalent to the classical notion of a valuation ring obtained from valuations on a field K .) (See [Ender] and [Nagata].) E-rings and valuation rings are in abundance, as the next result shows. First we recall some definitions.

A ring R is *Noetherian* <=> it satisfies the ascending chain condition on ideals or equivalently all ideals are finitely generated. In our context, a Noetherian valuation ring is equivalent to the classical notion of a discrete valuation ring. A ring R is *Artinian* <=> it satisfies the descending chain condition on ideals or equivalently R is Noetherian and all prime ideals are maximal.

(2.1) Let R be a local ring with $\bigcap_{n\geq 1} J^{(n)} = (0)$ [in particular a Noetherian ring] and $J = Ra \neq (0)$.
Then, all non-zero ideals of R have the form Ra^i and either R is an E-ring or a principal valuation ring whose radical is not nilpotent.

Proof. This is just a restatement of [Bourbaki, 2, page 379]. □

Let I_1 and I_2 be two ideals of a ring R . I_2 is an *upper neighbour* of I_1 ($I_1 \lessdot I_2$) if and only if there is no ideal I with $I_1 \subsetneq I \subsetneq I_2$.

E-rings and valuation rings generate Hjelmslev rings.

(2.2) Let R be a ring.

(a) Every E-ring R with $J \neq (0)$ is a proper H-ring

with nilpotent radical.

(b) If R is a chain ring and I an ideal, then R/I is a H-ring provided I is a principal ideal or possesses an upper neighbour.

Proof. (a) All ideals are of the form $I^{(i)}$ and $I^{(n)} = (0)$. Hence, the ideals form a chain and so R is local, and $J = I$. First, we show that J is principal. Now $J^{(2)} \subsetneq J$ or else $J^{(n)} = J = (0)$. If $b \in J \setminus J^{(2)}$, then $b \neq 0$ and $Rb = J^{(i)}$. Hence, $J^{(2)} \subsetneq J^{(i)} \subseteq J$ and so $J^{(i)} = J$. Thus, $i = 1$ and $Rb = J$ with $b^n = 0$. We need only show that every non-unit is a zero divisor to complete the proof. Let $a \neq 0$ be a non-unit. By (1.1) $a \in J$ and so $a = rb^n$ and $a \cdot b^{n-1} = rb^n = 0$.
(b) This is a special case of [Törner, 2, page 70]. The two cases are related by the fact that $I_1 \subset I_2 \iff I_2 = Ra$ and $I_1 = Ja$ ([Törner, 2]). □

We next state many different characterizations of E-rings.

(2.3) Let R be a ring with $J \neq (0)$. The following are equivalent.

(1) R is an E-ring.

(2) R is a Hjelmslev ring with nilpotent radical.

(3) R is a Noetherian Hjelmslev ring.

(4) R is an Artinian Hjelmslev ring.

(5) R is a Hjelmslev ring, all zero divisors are nilpotent and $J = Ra$.

(6) R is a Hjelmslev ring with exactly one prime ideal and $J = Ra$.

(7) R is an Artinian, Noetherian local principal ideal ring.

(8) R is an Artinian local ring and $J = Ra$.

We observe that the first six conditions are equivalent even in the non-commutative case.

Proof. The equivalence of the first four conditions is a result of 2.2. (a), [Artmann, 2, Satz, 2.6] and [Törner, 2, 5.27] and [(5) <=> (6)] follows from 5.23 of [Törner, 1]. Also, [(5) <=> (1)] is 4.23 Satz of [Törner, 1]. To complete the proof we show (6) => (7) => (8) => (1).

(6) => (7). Since a H-ring is local, the equivalence of the 1st six conditions implies that R is an Artinian, Noetherian H-ring. Hence, every ideal is finitely generated. But a finitely generated ideal of a H-ring is principal [(Törner,2, 5.2]).

(7) => (8) is clear from previous remarks.

(8) => (1). Since R is Artinian, $J^{(n)} = (0)$. By (2.1) , R is an E-ring.

(2.4) EXAMPLES OF COMMUTATIVE H-RINGS WITH NILPOTENT RADICALS

Let L be a local ring. Then, the characteristic of L (char(L)) is either zero or a power of a prime. For if char(L) were rs where r and s are relatively prime then $r\cdot 1$, $s\cdot 1$ are zero divisors and hence in J . Since there are integers m and n so that $mr + ns = 1$, 1 belongs to J, a contradiction.

There are thus four possibilities:

(I) $\mathrm{char}(L) = 0 = \mathrm{char}(L/J)$.

(II) $\mathrm{char}(L) = p = \mathrm{char}(L/J)$ for a prime p .

(III) $\mathrm{char}(L) = 0$, $\mathrm{char}(L/J) = p$ for a prime p .

In this case $J = pL$ is not nilpotent and L is a principal valuation ring.

(IV) $\mathrm{char}(L) = p^r > p = \mathrm{char}(L/J)$ for a prime p ([Matsamura] or [Cohn]).

Case (III) contains no proper Hjelmslev rings , for then J would consist of zero divisors.

We now exhibit examples of our three possibilities for a H-ring H .

(a) Examples where $\mathrm{char}(H) = \mathrm{char}(H/J)$.

(The equicharacteristic case)

Let $\mathbb{R}$ be the reals. Then $H = \mathbb{R}[x]/(x^n)$ are the classical examples of Hjelmslev ([Hjelmslev, 2, page 48]). Later, Klingenberg observed that for any field K , $H = K[x]/(x^n)$ is also a commutative H-ring with nilpotent radical so that $\mathrm{char}(H) = \mathrm{char}(H/J)$.

Proof. $K[[x]]$ is a Noetherian valuation ring ([Bourbaki, 2, page 380]). Hence, by (2.2) (b) and the fact that a quotient ring of a Noetherian ring is Noetherian we conclude that $K[[x]]/(x^n) \cong K[x]/(x^n)$ is a Noetherian H-ring. Moreover, the radical is $J = (x)/(x^n)$ and so $J^{(n)} = (0)$. Finally $H/J \cong K[[x]]/(x^n)/(x)/(x^n) \cong K[[x]]/(x) \cong K$ and so $\mathrm{char}(H) = \mathrm{char}\,(H/J)$. □

Next we consider some examples of H-rings H where $\mathrm{char}(H/J)$ need to be equal to $\mathrm{char}(H)$. But first we recall the following. If L is a local ring, then the family $\{J^{(n)} | n \geq 1\}$ forms a neighbourhood filter for the J-adic topology on L. This topology can also be described by the metric

$$d(\ell,\ell) = 0 \quad \text{and} \quad d(\ell,m) = 2^{-n} \iff \ell - m \in J^{(n)}$$

but $\ell - m \notin J^{(n+1)}$. L is *complete* if L is complete in the J-adic topology. Clearly, if $J^{(n)} = (0)$ for some n, then every cauchy sequence must eventually be constant and so converges or L is complete.

(b) Let V be a complete, Noetherian valuation ring of characteristic zero whose radical is Vp for a prime p, and $\mathrm{char}(V/Vp) = p \neq 0$ with $p^2 \notin Vp$ and $f(x) = x^m + a_{m-1}x^{m-1} + \cdots + a_1$ is an Eisenstein polynomial of degree m ($a_i \in Vp$, $1 \leq i < m$ and $a_1 \notin Vp^2$). (This is a local ring from case III). Then, $H = V[x]/(f(x), x^n)$ is a H-ring with $J^{(n)} = (0)$, $Hp = J^{(m)}$, $1 \leq m \leq n$, and $\mathrm{char}(H) = p^r > p = \mathrm{char}(H/J)$ where $n = m(r-1) + s$, $1 \leq s \leq m$.

Proof. First we see that $H \cong V[[x]]/(f(x), x^n)$ $\cong V[[x]]/(f(x))/(f(x), x^n)/(f(x))$. By [Nagata, page 50] and

[Krull, page 13] $V[[x]]$ is a Noetherian local ring and a unique factorization domain. Also, from [Nagata, page 111] $f(x)$ is an irreducible (Eisenstein) polynomial over V. Thus, $(f(x))$ is a prime ideal of $V[[x]]$ ([Zariski and Samuel, page 27]). Hence, $R = V[[x]]/(f(x))$ is a Noetherian local integral domain. From [Hungerford, page 246] the radical (non-units) of R is $(x)/(f(x))$. By (3.1), R is a valuation ring whose non-zero ideals are $(x^s)/(f(x))$. Hence, $R/(f(x), x^n)/(f(x))$ is a Hjelmslev ring with radical $J = (x)/(f(x))/(f(x), x^n)/(f(x))$. Thus, $J^{(n)} = (0)$ and $J^{(m)} = Hp$.

Since J is nilpotent, $\mathrm{char}(H) \neq 0$, and so must be p^r. Then, $p = a^m \cdot u$ for a unit u ([Hungerford, Proposition 4]) and so $a^{mr} = 0 \neq a^{m(r-1)}$. Hence, $n = m(r-1) + s$ where $1 \leq s \leq m$. □

All finite chain rings are H-rings with nilpotent radicals ([Clarke and Drake, page 149]), but there do exist infinite chain rings whose radicals are not nilpotent ([Törner, 2, page 73] and [Weller, page 45]). As we shall see later such rings are not important from a topological standpoint, as all locally compact H-rings have nilpotent radicals.

Later we shall see that the examples illustrated above constitute all the commutative H-rings with nilpotent radicals. First, we consider the geometric significance of nilpotent radicals discovered by Artmann.

§3. THE GEOMETRIC SIGNIFICANCE OF NILPOTENT RADICALS

(3.1) Definition ([Artmann, 1, 2]). An epimorphism $\psi : H_1 \to H_2$ of PH-planes is a *H-epimorphism* (also called an eumorphism or refined neighbourhood property) if and only if

$$P \sim Q \iff \psi(P) \sim \psi(Q) \quad \text{for all points } P, Q .$$

and

$$\ell \sim m \iff \psi(\ell) \sim \psi(m) \quad \text{for all lines } \ell, m .$$

(3.2) Definition ([Artmann, 1]). An H-epimorphism $\psi : H_1 \to H_2$ of PH-planes is *minimal* if and only if PIg, h; $g \sim h$ always implies

$$\psi^{-1}\psi(P) \cap g = \psi^{-1}\psi(P) \cap h \quad \text{and dually.}$$

Example. Let H be an H-ring. If I_1 and I_2 are ideals of H and $I_1 \subset I_2$ then by ([Törner, 2]) $I_2 = H_a$ and $I_1 = J_a$. By 2.2 (b) H/I_i $(i = 1, 2)$ are H-rings. The obvious ring epimorphism $\nu : H/I_1 \to H/I_2$ induces a minimal H-epimorphism $\psi_\nu : H(H/I_1) \to H(H/I_2)$.

(3.3) Definition ([Törner, 3]). A PH-plane is of *height* n if and only if there is a chain of PH-planes

$$H = H_n \xrightarrow{\psi_{n-1}} H_{n-1} \xrightarrow{\psi_{n-2}} H_{n-2} \cdots \xrightarrow{\psi_2} H_2 \xrightarrow{\psi_1} H_1 = H/\sim$$

where all the ψ_1 are minimal H-epimorphisms.

An ordinary plane is height one and a PH-plane of height two is also called *uniform*.

Now, in a plane of height n, the only H-epimorphic images are the H_i ([Artmann, 1, 2] or [Törner, 2]); that is the sequence of H-epimorphism above is maximal and is called a <u>solution</u> of H. Let $\phi_i = \psi_i \circ \cdots \circ \psi_{n-1} : H \to H_i$ $(1 \le i \le n-1)$. Then, ϕ_1 is essentially the canonical projection.

A PH-plane of height n satisfies the *axiom of reciprocal segments* (RS) if and only if $P \in g, h$ always implies:

$$\phi_i(g) = \phi_i(h) \quad \text{and} \quad \phi_{i+1}(g) \neq \phi_{i+1}(h)$$
$$\Leftrightarrow g \cap h = g \cap \phi^{-1}_{n-i}\phi_{n-i}(P) \quad \text{for each} \quad i \in \{1,2,\ldots,n-1\}.$$

(3.4) Definition ([Törner, 3]). A PH-plane of height n is of *level* n if and only if each H-epimorphic image satisfies (RS).

We then have the following important result.

(3.5) ([Artmann, 1, 2]). Let H be a H-ring, with radical J. The following are equivalent.

(1) $R(H)$ is of level n.

(2) $R(H)$ is of height n.

(3) J is nilpotent with nilpotency index n. □

In our classical example, $H = \mathbb{R}[x]/(x^n)$ has $J = (x)/(x^n)$, all other ideals are of the form $J^{(i)} = (x^i)/(x^n)$ $(1 \le i \le n)$ and $(0) \subset J^{(n-1)} \subset J^{(n-2)} \subset \cdots \subset J$. Hence, the homomorphisms $H \xrightarrow{\nu_{n-1}} H/J^{(n-1)} \cdots \xrightarrow{\nu_2} H/J^{(2)} \xrightarrow{\nu_1} H/J$ induce a chain of minimal H-epimorphisms

$$H(H) \to H(H/J^{(n-1)}) \cdots \to H(H/J^{(2)}) \to H(H/J) .$$

Now, we can endow the rings $\mathbb{R}[x]/(x^n)$ with two natural topologies: the order topology of the lexicographic ordering used by Hjelmslev or the product topology of $\mathbb{R}^n$. If $n = 1$, we obtain, of course, the reals, and here the two approaches coincide. But, if $n \geq 2$, then the order topology is not connected as the radical is open-closed, while the product topology is connected and the radical has a void interior. Now, if the point and line sets of the PH-plane over our rings are endowed with the obvious quotient topologies (See § 1) then the joining of points and the intersection of lines (where they exist) are continuous functions. The order topology leads to the theory of ordered H-planes (See [Baker, Lane, Lorimer]) and the product topology to the theory of topological H-planes ([Lorimer, 2, 4]). The ordered planes are indeed topological planes, but their topologies are disconnected and non-compact, and the neighbour classes $\bar{P}$ are, in fact, open neighbourhoods. On the other hand, the (metric) product topology produces a locally compact connected PH-plane, which is not compact either if $n > 1$, but whose neighbour classes have void interior. This last condition essentially means that on the canonical projection of our plane (which is the real projective plane) the quotient topology of the neighbour relation is the natural topology of the real projective plane; whereas for ordered H-planes the quotient topology is the discrete topology. These comments lead us naturally to the theory of topological Hjelmslev planes.

§4. TOPOLOGICAL HJELMSLEV PLANES

A *topological affine Hjelmslev* plane $H = \langle \mathbb{P}, \mathbb{L}, ||, \in \rangle$ is an AH-plane such that $\mathbb{P}$ and $\mathbb{L}$ are topological spaces, $\underset{\mathbb{P}}{\sim}$ and $\underset{\mathbb{L}}{\sim}$ are closed relations and the following maps are continuous:

$$\vee : \mathbb{P} \times \mathbb{P} \setminus \underset{\mathbb{P}}{\sim} \longrightarrow \mathbb{L}$$

$$\wedge : \mathbb{L} \times \mathbb{L} \setminus \{(\ell,m) : \Delta_\ell \sim \Delta_m\} \longrightarrow \mathbb{P}$$

$$L : \mathbb{P} \times \mathbb{L} \text{——} \mathbb{L} .$$

That is, joining and intersection are continuous, where they define a function, and so is parallelism.

For ordinary affine planes, $\underset{\mathbb{P}}{\sim} = \Delta_{\mathbb{P}}$ or the plane is hausdorff. Moreover, if the point set is locally compact connected, then the continuity of parallelism is a consequence of the continuity of join and intersection ([Salzmann, 3, page 52]). This is because ordinary parallelism is related to intersection. In our situation, as mentioned earlier, this is not the case. For proper AH-planes, there always exist lines which are disjoint but not parallel.

Throughout the rest of this lecture, ℓ will always denote a line of a topological AH-plane $H = \langle \mathbb{P}, \mathbb{L}, ||, \in \rangle$ and P a point.

(4.1) The domain of $\wedge$ is an open set.

Proof. We show that $\mathcal{D} = \{(a,b) \mid [a,b] \neq 1\}$ is closed. Let (a_α, b_α) be a net in $\mathcal{D}$ converging to (a,b). Suppose $[a,b] = 1$ and $a \wedge b = P$. Now $[a_\alpha, b_\alpha] \neq 1$ implies that $[L(P,a_\alpha), L(P,b_\alpha)] \neq 1$ and so $L(P,a_\alpha) \sim L(P,b_\alpha)$.
But $L(P,a_\alpha) \to L(P,a) = a$ and $L(P,b_\alpha) \to L(P,b) = b$. Since $\underset{\mathbb{L}}{\sim}$ is closed and $(L(P,a_\alpha), L(P,b_\alpha)) \in \underset{\mathbb{L}}{\sim}$, we conclude that $(a,b) \in \underset{\mathbb{L}}{\sim}$ or $(a,b) \in \mathcal{D}$. □

Let ℓ and m be two lines and g a line so that $\Delta_g \not\parallel \Delta_m$. A *parallel projection* $\ell \xrightarrow[g]{} m$ $(X \rightsquigarrow L(X,g) \wedge m)$ is bijective $\Leftrightarrow \Delta_g \not\parallel \Delta_\ell$, and is called *non-degenerate* or *degenerate* depending on whether the projection is bijective or not. The non-degenerate projectivities of ℓ (a chain of non-degenerate projections from ℓ to ℓ) form a group of homeomorphism that acts doubly transitive on pairs of non-neighbouring points.
We now develop sufficient tools to prove some elementary facts and establish that the canonical image $H/\sim$ is itself a topological affine plane.

(4.2) Any two lines are homeomorphic. □
Denoting by $\mathbb{L}_p$, the pencil of lines through P, we have

(4.3) If $\ell \in \mathbb{L}_p$, then $\mathbb{L}_p \backslash \bar{\ell}$ is homeomorphic to a line.
Proof. Choose a line a so that $P \not\parallel a$. The map $a \longrightarrow \mathbb{L}_p \backslash \bar{\ell}$ $(X \rightsquigarrow X \vee P)$ is a homeomorphism. □
Now introduce coordinates into H in the usual way: $\{O,X,Y\}$

are 3 points so that

$\{\bar{O}, \bar{X}, \bar{Y}\}$ is a triangle in $H/\sim$.

Let $E = L(X, O \vee Y) \wedge L(Y, O \vee X)$, and $K = O \vee E$. The elements of K are a, b, c... with $O = o$ and $E = 1$.

A point P is assigned coordinates x,y(P ⟶(x,y)) as shown below. We write P = (x,y).

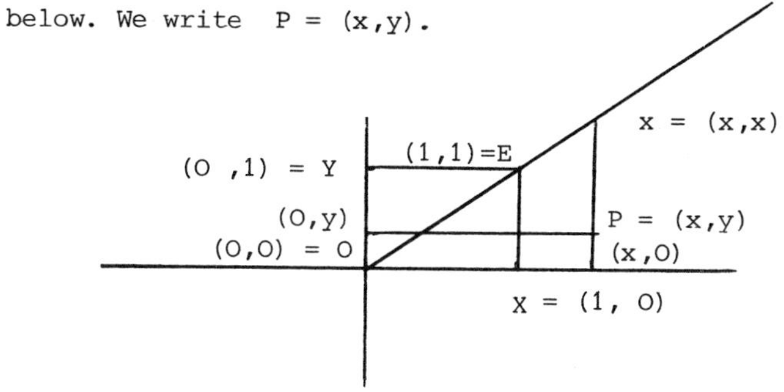

Now, $\mathbb{L} = \mathbb{L}_1 \cup \mathbb{L}_2$ where $\mathbb{L}_1 = \{\ell \in \mathbb{L} \mid [\ell, O \vee Y] \neq 1\}$ are *lines of the first kind* and $\mathbb{L}_2 = \{\ell \in \mathbb{L} \mid [\ell, O \vee Y] = 1\}$ are *lines of the second kind*. A line α in $\mathbb{L}_2$ is assigned coordinates m,n (α ⟶[m,n]) as shown below. We write $\alpha = [m,n]$.

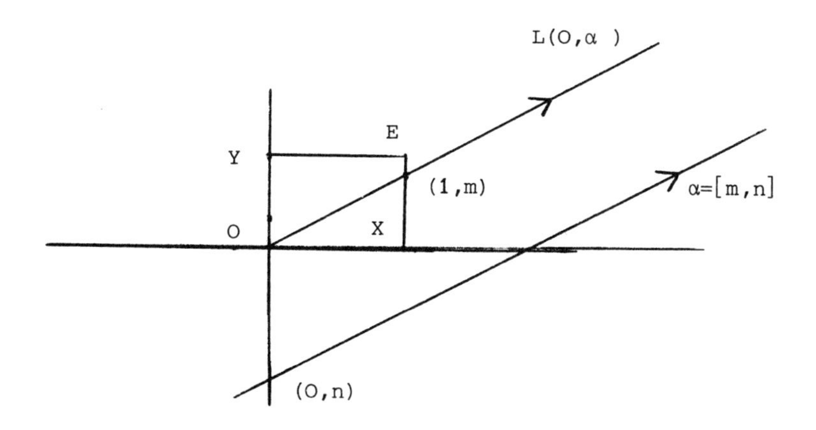

This defines a ternary operation T on K so that $(x,y) \in [m,n] \Longleftrightarrow y = T(x,m,n)$. By interchanging the roles of X and Y in the above discussion we obtain a second ternary operation T' and coordinates $[m,n]'$ for a line in $\mathbb{L}_1$ where $m \sim 0$ ([Lorimer, 2]).

As usual, the addition and multiplication associated with the ternary operator T are given by

$$x + y = T(x,1,y)$$

$$x \cdot y = T(x,y,0).$$

Similarly, $+'$ and $\cdot'$ are the addition and multiplication of T'. Note that a line in $\mathbb{L}_2$ can have two sets of coordinates; that is, if $m \not\sim 0$, then $[m,n] = [u,v]'$ where $u \cdot m = 1 = m \cdot' u$. $K = (K, T, T', o, 1)$ is the *biternary ring* of H with respect to $\{0,X,Y\}$. The many properties of biternary rings may be found in ([Lorimer, 2] or [Bacon, 1]). In particular also $J = \{k \in K : k \sim o\}$ is an ideal and $(x,y) \sim (o,o)$ if and only if $x, y \in J$.

The maps assigning coordinates to points and lines are clearly homeomorphisms. Hence

(4.4) The point space $\mathbb{P}$ and the space of lines of the second kind $\mathbb{L}_2$ are homeomorphic to $\ell \times \ell$ for any line ℓ. Moreover, $\mathbb{L}_2$ is an open set in $\mathbb{L}$. □

T is composed of joins, intersections and parallel maps and so,

(4.5) $T : K^3 \to K$ is continuous and $(K,+)$ is a topological loop. □

This immediately leads to

(4.6) K and hence the point set and line set are regular spaces.

Proof. By ([Hofmann]) every topological loop is regular. □

Now, let π_K be the canonical projection restricted to K.

(4.7) If $K/\sim$ is endowed with the quotient topology of $\pi_K : K \to K/\sim$, then π_K is an open-continuous map and $K/\sim$ is also a topological loop.

Proof. By definition π_K is continuous. Let $J = \{a \mid a \sim o\}$. Then, $\pi_K(a) = a + J$. If $S \subseteq K$ define $M(S) = \{x \in K \mid x + J = a + J$ for some $a \in S\}$. Then, $M(S) = S + J$. Now let V be open in K. Then, from the remarks above, $\pi_K \pi_K^{-1}(V) = M(S) = V + J = \bigcup_{n \in N} (V + n)$. By ([Hofmann]), the maps $x \to x + n$ are homeomorphisms and so $M(S)$ is an open set. □

Let $K = \langle K, T, T', o, 1 \rangle$ be a *biternary ring* of H. An abstract definition of biternary rings can be found in ([Bacon, 1]), and there it is shown that each such ring generates an AH-plane. We now consider a topological analogue to that result. First, define the map $\ddot{x}$ as follows:

$$\ddot{x}(m_1, n_1, m_2, n_2) = x \iff T(x, m_1, n_1) = y,\ T'(y, m_2, n_2) = x;$$
$$\text{for some } y \in K \text{ where } m_2 \sim o .$$

The *inverses of* T are the maps $\dot{n}$, $\dot{m}$ *and* $\dot{x}$ where

$\dot{n}(x,m,y) = n \Leftrightarrow T(x,m,n) = y$

$\dot{m}(a_1,b_1,a_2,b_2) = m \Leftrightarrow T(a_i,m,n) = b_i$;

$i = 1,2$ for some $n \in K$ where $a_1 \not\sim a_2$.

$\dot{x}(m_1,n_1,m_2,n_2) = x \Leftrightarrow T(x,m_i,n_i) = y$;

$i = 1,2$ for some $y \in K$ where $m_1 \not\sim m_2$.

Similarly, T' has inverses $\dot{n}'$, $\dot{s}'$ and $\dot{x}'$.

(4.8) Definition. $K = < K, T, T', o, 1 >$ is a *topological biternary ring* if K is a topological space, $\ddot{x}$, T, T' and their inverses are continuous, and J is closed.

We then have the following result, from ([Lorimer, Baker and Lane]).

(4.9) Every biternary ring of a topological AH-plane is topological, and conversely every topological biternary ring generates a topological AH-ring.

Proof. T, T' and their inverses can all be described as compositions of joins, intersections and parallel maps ([Lorimer, 3, 5.2 Lemma]).

We will not present the details for the converse here. However, we can easily describe the topologies on the AH-plane over the topological biternary ring $K = (K, T, T', o, 1)$: The point set is $K \times K$ and has naturally the product topology; the lines in $\mathbb{L}_2$ also have the product topology; however, for a line $[\hat{u},\hat{v}]'$ $(\hat{u} \sim o)$ in $\mathbb{L}_1$, if U is a neighbourhood of $\hat{u}$ and V of $\hat{v}$, then a neighbourhood is a set

$W(U,V) = \{[u,v]'(u \sim o) \mid (u,v) \in U \times V\} \cup \{[m,n](m \not\sim o) \mid [m,n] = [u,v]'$ and $(u,v) \in U \times V\}$ where, $U \times V \in \Omega((\hat{u},\hat{v}))$. □

Now, as is well known, every *ternary field* $F = \langle F, T, o, 1\rangle$ generates an affine plane, and hence we can construct a symmetric ternary operator T' so that $F = \langle F, T, T', o, 1\rangle$ is a biternary ring. However, $[u,v]'$ $(u \sim o)$ means $u = o$ and so $(x,y) \in [o,v]' \Leftrightarrow x = T'(y,o,v) = v$. Thus, T' is essentially redundant. However, if F is a topological ternary field, then it is not known if F generates a topological affine plane. For we do not have a means to generate a continuous second ternary operator T', and hence can not describe neighbourhoods of lines $[o,r]'$ parallel to the y-axis. However, ([Salzmann, 3]) has shown,

(4.10) If $F = (F, T, o, 1)$ is a locally compact connected ternary field, then the affine plane over F is a topological plane. Hence, the symmetric operator T' and its inverses are continuous and $\langle F, T, T', o, 1\rangle$ is a topological biternary ring. □

Now, we next describe the relationships between the topologies on $\mathbb{P}$ and $\mathbb{L}$.

In any topological space X, $\Omega(x)$ denotes the neighbourhood filter of the point x. If Y is a subspace of X containing x, then $\Omega_Y(x)$ is the relative neighbourhood filter of x in Y.

Now, let $\ell = U \vee V$ and $P \not\mathrel{I} \ell$. If $\mathcal{W}_1 \in \Omega(U)$ and $\mathcal{W}_2 \in \Omega(V)$, then the collection of sets $\mathcal{W}(\mathcal{W}_1, \mathcal{W}_2) = \{R \vee S \mid R \in \mathcal{W}_1, S \in \mathcal{W}_2$ and $R \not\mathrel{I} S\}$ generates a filter $\Omega(\ell : U, V)$. If $\mathcal{V}_1 \in \Omega_{PVU}(U)$ and $\mathcal{V}_2 \in \Omega_{PVV}(V)$ and $P \notin \mathcal{V}_1 \cap \mathcal{V}_2$, then the collection of sets $\mathcal{V}(\mathcal{V}_1, \mathcal{V}_2) = \{R \vee S \mid R \in \mathcal{V}_1, S \in \mathcal{V}_2$ and $R \not\mathrel{I} S\}$ generates a filter $\Omega(\ell : U, V, P)$.

(4.11) $\ell = U \vee V$ and $P \not\mathrel{I} \ell$. Then

$\Omega(\ell) = \Omega(\ell : U, V) = \Omega(\ell : U, V. P)$.

Proof. This follows essentially from the continuity of $\vee$ and the fact $\sim$ is closed. Details are in ([Lorimer, 3, § 3]). □

As an immediate consequence of (4.11) we have

(4.12) $\vee$ is an open map. □

Let $P = \ell \wedge m$. If $\mathcal{U}_1 \in \Omega(\ell)$ and $\mathcal{U}_2 \in \Omega(m)$, then the collection of sets $\mathcal{U}(\mathcal{V}_1, \mathcal{V}_2) = \{u \wedge v \mid u \in \mathcal{V}_1, v \in \mathcal{V}_2$ and $[u,v] = 1\}$ generates a filter $\Omega(P; \ell, m)$.

(4.13) If $P = \ell \wedge m$, then $\Omega(P) = \Omega(P; \ell, m)$. □

Consequently,

(4.14) $\wedge$ is an open map. □

We can prove now that,

(4.15) The canonical image, $H/\sim$, endowed with the quotient topologies of the canonical projection $\pi = (\pi_{\mathbb{P}}, \pi_{\mathbb{L}}) : H \rightarrow H/\sim$, is a topological hausdorff affine plane and $\pi(\pi_{\mathbb{P}}$ and $\pi_{\mathbb{L}})$ is an open-continuous map. Hence we have; $H = \langle \mathbb{P}, \mathbb{L}, \|, \in \rangle$ is a topological AH-plane if and only if H satisfies (AH1), (AH2), (AH3), $\vee$, $\wedge$ and L are continuous maps and the following additional axiom holds:

(AH4)* there exists a hausdorff (topological) affine plane and an open-continuous map $\pi : H \to A$ with the properties

(i) $P \sim Q \iff \pi(P) = \pi(Q)$ for all points P, Q.

(ii) $\ell \sim m \iff \pi(\ell) = \pi(m)$ for all lines ℓ, m.

(iii) $\ell \cap m = \phi$ implies $\pi(\ell) \parallel \pi(m)$ for all lines ℓ, m.

Proof. Let $\bar{K} = K/\sim$, and $\phi : K \times K \longrightarrow \mathbb{P}$ and $\bar{\phi} : \bar{K} \times \bar{K} \longrightarrow \mathbb{P}/\sim$ be the coordinate maps of H and $H/\sim$ respectively. Since $\pi_K : K \longrightarrow \bar{K}$ is open-continuous, then ([Massey, page 240]) implies that the product topology on $\bar{K} \times \bar{K}$ is the identification topology of $\pi^2 : K \times K \longrightarrow \bar{K} \times \bar{K}$. Endow $\mathbb{P}/\sim$ with the identification topology $\mathcal{O}$ from $\bar{\phi}$. Then, ϕ and $\bar{\phi}$ are homeomorphisms and $\pi_K : K \longrightarrow \bar{K}$ is an open-continuous surjection. Since, $\pi_{\mathbb{P}} \circ \phi = \bar{\phi} \circ \pi_K^2$, $\pi_{\mathbb{P}}$ is open-continuous with respect to $\mathcal{O}$ and so $\mathcal{O}$ is the quotient topology of $\mathbb{P}/\sim$. Thus, $\pi_{\mathbb{P}}$ is open-continuous. Next we verify that $\pi_{\mathbb{L}}$ is open. Let Ω be the collection of sets $W(\bar{U}, \bar{V}) = \{X \vee Y \mid X \in \bar{V}, \bar{Y} \in \bar{V}$ and $\bar{X} \neq \bar{Y}\}$ where $\bar{U}$ and $\bar{V}$ are open sets in $\mathbb{P}/\sim$. Then, Ω is a filter base and $\mathbb{L}/\sim = \bigcup W(\bar{U}, \bar{V})$. Then, Ω is a base for a topology, Ω^* , on $\mathbb{L}/\sim$. Since $\pi_{\mathbb{P}}$ is open-continuous, (4.11) implies that $\pi_{\mathbb{L}} : \mathbb{L} \longrightarrow \mathbb{L}/\sim$, where $\mathbb{L}/\sim$ has Ω^*, is also open-continuous. Hence, Ω^* is the quotient topology, and so $\pi_{\mathbb{L}}$ is open. Now, we verify that $H/\sim$ is a topological hausdorff affine plane. Let $\bar{\vee}$ be the join map in $H/\sim$. Then, $\bar{\vee} \circ \pi^2_{\mathbb{P}} = \pi_{\mathbb{L}} \circ \vee$. The domain of $\vee$ is open as $\sim$ is closed. Hence, $\bar{\vee}$ is continuous. Using (4.1), we see that intersection and parallelism are also continuous. Finally, since $\sim$ is closed and $\pi : H \longrightarrow H/\sim$ is open, $H/\sim$ is a hausdorff space. The last assertion now follows immediately from the fact that if $f : X \longrightarrow Y$ is an open-continuous map between topological spaces, then Y is hausdorff if and only if $\{(x,y) : x \in X,\ y \in Y \text{ and } f(x) = f(y)\}$ is a closed set. $\square$

From the proof above we observe that $\bar{\phi} : \bar{K} \times \bar{K} \longrightarrow \mathbb{P}/\sim$ is a homeomorphism with respect to quotient topologies. Now, endow $\bar{K}$ with the subspace topology from $\mathbb{P}/\sim$. Since $H/\sim$ is a topological plane, $\bar{\phi} : \bar{K} \times \bar{K} \longrightarrow \mathbb{P}/\sim$ is also a homeomorphism. We conclude that

(4.16) The subspace topology of $\bar{K}$ coincides with its quotient topology. $\square$

For ordinary planes it is customary to assume the plane is neither discrete nor indiscrete. It then follows that the plane is hausdorff. By (4.15), $H/\sim$ is hausdorff and so is not indiscrete. However, we do have

(4.17) $H/\sim$ is discrete iff the interior of $\bar{P}$ is non-empty for each point P iff the interior of $\bar{P} \cap \ell$ is non-empty for each point-line pair with $P \sim \ell$. □

From now on we shall assume that $H/\sim$ *is endowed with the quotient topologies* and that $H/\sim$ *is not discrete.*

We remark, however, that for an ordered projective Hjelmslev plane H , the quotient topology is discrete and distinct from the order topology on $H/\sim$. Thus, each neighbour class is open([Baker, Lane, Lorimer]).

In a topological space X , let $C(x)$ be the *connected component* of x in X and $Q(x) = \bigcap \{M : M$ is clopen in X, $x \in M\}$ the *quasi-component* of x in X .

The quasi-component has the following useful properties ([Engelking, 1]):

(i) $Q(x)$ is closed and $C(x) \subseteq Q(x)$.

(ii) If $Q(x) = X$, then X is connected.

(iii) If $Z = X \times Y$, for another topological space Y then

$Q(x,v) = Q(x) \times Q(y)$, for all $(x,y) \in X \times Y$.

We may then prove,

(4.18) A line of H is either connected or the quasi-component of any point lies in the neighbour class of the point.

Proof. Assume a line ℓ is not connected, $P \in \ell$, and $Q_\ell(P)$ is the relative quasi-component of P in ℓ . We claim that $\complement(Q_\ell(P)) \cap \complement(\bar{P} \cap \ell) \neq \emptyset$. If this is false, then $\complement(Q_\ell(P)) \subseteq \bar{P} \cap \ell$. Since $Q_\ell(P)$ is closed and the interior of $\bar{P} \cap \ell$ is void, we conclude that $\mathcal{Q}_\ell(P) = \ell$ and so ℓ is connected, a contradiction. Now, we show that $\mathcal{Q}_\ell(P) \subseteq \bar{P} \cap \ell$. If not, take $y \in \mathcal{Q}_\ell(P)$ so that $Y \not\sim P$. By the claim above, there is a point $Z \not\sim P$ with $Z \notin \mathcal{Q}_\ell(P)$. Hence, there is a projectivity ψ mapping Y to Z and fixing P. Thus, $Z = \psi(Y) \in \mathcal{Q}_\ell(P)$, a contradiction. □

Because of 4.4, and property (iii) of a quasi-component, we have

(4.19) $\mathbb{P}$ and any line ℓ are simultaneously connected or the quasi-component of any point lies in its neighbour class. □

A *topological projective Hjelmslev* plane $H = < \mathbb{P}, \mathbb{L}, \varepsilon >$ is a PH-plane such that $\mathbb{P}$ and $\mathbb{L}$ are topological spaces, $\underset{\mathbb{P}}{\sim}$ and $\underset{\mathbb{L}}{\sim}$ are closed relations and the following maps are continuous:

$$\wedge : \mathbb{P} \times \mathbb{P} \setminus \underset{\mathbb{P}}{\sim} \longrightarrow \mathbb{L}$$

$$\vee : \mathbb{L} \times \mathbb{L} \setminus \underset{\mathbb{L}}{\sim} \longrightarrow \mathbb{P}\,.$$

Let ℓ be a line and put $\Sigma(\ell) = \{P : P \sim \ell\}$. The induced incidence structure $H^\ell = < \mathbb{P} \setminus \Sigma(\ell), \mathbb{L} \setminus \bar{\ell}, ||^\ell >$ is a topological AH-plane where $x ||^\ell y \Leftrightarrow x \wedge \ell' = y \wedge \ell$. We write $\mathbb{P}^\ell$ and $\mathbb{L}^\ell$ for the point and line sets of H^ℓ, and again agree that lines are point sets.

We have then

(4.20) H^ℓ is an open topological AH-plane.

Proof. First we observe that $\Sigma(\ell)$ is closed. Let $\{P_\alpha\}$ be a net in $\Sigma(\ell)$ converging to P. Pick $X_\alpha \in \ell$, $X_\alpha \sim P_\alpha$. Choose X so that $X \not\sim P$ and $X \not\sim \ell$. Then, $X \not\sim P_\alpha$, and $h_\alpha = X \vee P_\alpha \longrightarrow X \vee P = h$. Also, h_α, $h \not\sim \ell$ and so $T_\alpha = h_\alpha \wedge \ell \longrightarrow h \wedge \ell = T$. Now $T_\alpha \sim P_\alpha$ or else $h_\alpha \sim \ell$. Hence, $(T_\alpha, P_\alpha) \longrightarrow (T,P)$ and so $T \sim P$. Thus, $P \in \Sigma(\ell)$. $\square$

Now, given a trilateral $\{\ell_i/\sim\}_{i=1}^{3}$ in $H/\sim$, it is easy to see that $\mathbb{P} = \bigcup_{i=1}^{3} \mathbb{P}^{\ell_i}$.

Then (4.20) and duality yield the following analogous results for topological PH-planes.

(4.21) Let H be a topological PH-plane.

(i) H is a regular space.

(ii) All lines are homeomorphic.

(iii) The canonical map is open-continuous in the quotient topology and $H/\sim$ is a topological hausdorff projective plane.

(iv) $H/\sim$ is not discrete <=> each $\bar{P}$ has an empty interior.

From now on *we again assume* $H/\sim$ *is not discrete.*

(4.22) $\Sigma(\ell)$ is nowhere dense and so $\mathbb{P}^{\ell}$ is dense in $\mathbb{P}$.

Proof. Suppose $\Sigma(\ell)$ contains a non-void open set U. Then, $\pi(U) \subseteq \ell/\sim$. But this means $H/\sim$ is discrete ([Salzmann, 2]). $\square$

If $P \not\sim \ell$, then the perspectivity $\wedge_P^{\ell} : \ell \longrightarrow \mathbb{L}_P \; (X \longmapsto X \vee P)$ is a homeomorphism, with inverse $\wedge_{\ell}^{P}$. Consequently, the projectivities of ℓ form a group, $PJ(\ell)$, of homeomorphism (the von Staudt group).

(4.23) $PJ(\ell)$ is triply transitive on non-neighbouring points, i.e. if $(X_i)_1^3$, $(Y_i)_1^3$ are given with $X_i \not\sim X_j$, $Y_i \not\sim Y_j$ $(i \neq j)$ then there is $\psi \in PJ(\ell)$ so that $\psi(X_i) = Y_i$ $1 \leq i \leq 3$ ([Baker, Lane, Lorimer, 3.32]).

We would like to obtain an analogue to (4.18). However, the notion of connectedness does not seem appropriate here. The problem is that, since $PJ(\ell)$ is not triply transitive on *all* points, we could not argue that a connected point set necessarily yields connected lines. The appropriate notion seems to be the following concept.

$\mathbb{P}$ is $\sim$-*disconnected* if $\mathbb{P}$ is a union of non-void open sets so that no point of one is a neighbour to a point in the other. Otherwise, $\mathbb{P}$ is called $\sim$-*connected*. For a point P , $\tilde{\mathcal{Q}}(P) = \bigcap \{C : C$ is a clopen saturated set containing $P\}$ is the $\sim$-quasi-component of P . $\mathbb{P}$ is *totally* $\sim$-*disconnected* if $\tilde{\mathcal{Q}}(P) = \bar{P}$ for all points P .

The significance of these notions is manifested by the following results.

(4.24)

(i) $\mathbb{P}$ is $\sim$-connected $\Leftrightarrow \mathbb{P}/\sim$ is connected.

(ii) $\mathbb{P}$ is either $\sim$-connected or totally $\sim$-disconnected.

(iii) The following sets are simultaneously $\sim$-connected or totally $\sim$-disconnected: $\mathbb{P}$, $\mathbb{L}$, ℓ , $\mathbb{L}_p$, $\ell\backslash\bar{P}$ $(P\in\ell)$, $\mathbb{P}^{\ell}$.

We will not present the proofs of these statements here (see [Lorimer, 4]). □

In order to prove (4.24) we used the fact that $H/\sim$ is not discrete. However, we can prove

(4.25) If $H/\sim$ is discrete, then H is totally $\sim$-disconnected. Hence, an ordered PH-plane is totally $\sim$-disconnected.

Proof. Since $\pi : H \longrightarrow H/\sim$ is open, we conclude that $\bar{P}$ is a clopen saturated set. Since, $\bar{P} \subseteq \tilde{\bar{Q}}(P)$ always, we have $\bar{P} = \bar{Q}(P)$. The last statement comes from ([Baker, Lane, Lorimer]). □

A *topological ring* is a ring R with a topology so that $(a,b) \longrightarrow a-b$ and $(a,b) \longrightarrow ab$ are continuous. R is a *Gelfand ring* if R is a topological ring whose group of units is an open set where $x \longmapsto x^{-1}$ is continuous, i.e. the units are an open (multiplicative) topological group. We then have ([Lorimer, 3]),

(4.26) Let H be a H-ring and $P(H)$ and $A(H)$ the desarguesian PH and AH-planes over H . Then, $P(H)$ and $A(H)$ are topological H-planes if and only if H is a Gelfand

H-ring. Moreover, if $\ell = [001]$, then $P(H)^{\ell}$ is topologically isomorphic to $A(H)$.

Proof. The affine case is just a special case of 4.9. In the projective case, the point set $\mathbb{P} = H^3 \setminus J^3 / R_\ell$ and line set $\mathbb{L} = H^3 \setminus J^3 / R_r$ are endowed with the quotient topologies from R_ℓ and R_r , where R_ℓ and R_r are closed relations whose associated quotient maps are open. It is then relatively easy to show that $P(H)$ is a topological PH-plane.
Now, $P(H)^{\ell}$, with $\ell = [001]$, has point set $\{ < xy1 > : x, y \in H \}$ and so the last statement follows easily. □

Now, let us examine the ideal structure of E-rings and their corresponding geometries.

(4.27) Let H be a Gelfand E-ring. Then, all non-zero ideals I are closed and H/I is a hausdorff Gelfand E-ring. Moreover, if the radical of H is void then so are the radicals of each H/I .

Proof. All ideals are of the form Ha^i ($0 \leq i \leq n$) where $J^{(n)} = (0)$ and $J = Ha$. For any ideal I, $I^r(I^\ell)$ is the right (left) annihilator of I . By ([Kaplansky, 2, page 689]) $I^{r\ell}$ is closed, and by ([Törner, 2, 5.8]) $(Ha^i)^{r\ell} = Ha^i$. Hence, all ideals are closed. Since the quotient maps $\nu_i : H \longrightarrow H/J^{(i)}$ are open-continuous and $J/J^{(i)}$ is the radical of $H/J^{(i)}$, our result follows. □

Now, we obtain a topological version of Artmann's result (3.5).

(4.28) Let H be a Gelfand hausdorff H-ring whose radical has void interior. Then, the following are equivalent.

(1) J has nilpotency index n.

(2) There is a sequence of hausdorff desarguesian PH-planes, $P(H) = P_n \xrightarrow{\psi_{n-1}} P_{n-1} \xrightarrow{\psi_{n-2}} \text{-----} \xrightarrow{\psi_1} P_1$, where each ψ_i is an open-continuous minimal H-epimorphism.

Proof. We need only prove (1) $\Longrightarrow$ (2) because of (3.5). From the proof of (4.27) and previous comments the ideals of H form the chain

$$(0) \subset Ha^{n-1} \subset Ha^{n-2} \subset \text{-----} \subset Ha = J$$

and each ideal is closed. Hence, $H/Ha^i = H_i$ is a hausdorff E-ring with open-continuous quotient map $\nu_i : H \longrightarrow H_i$. The obvious epimorphism $\sigma_i : H_{i+1} \longrightarrow H_i$, with $\sigma_i \circ \nu_i = \nu_{i+1}$, is thus also open-continuous. Let $\psi_i : P(H_{i+1}) \longrightarrow P(H_i)$ be the minimal H-epimorphism induced by σ_i (as mentioned in § 3), and $\phi_i : H_i^3 \setminus J_i^3 \longrightarrow \mathbb{P}_i$ the open-continuous quotient map of R_ℓ (see § 1) which is defined on $\mathbb{P}_i$, the point set of the PH-plane $P(H_i)$. Then, $\psi_i \circ \phi_i = \phi_{i+1} \circ \sigma_i^3$ and so ψ_i is open-continuous. □

If K is any topological field, then $K[x]/(x^n)$ is a Gelfand H-ring under the product topology of K^n. Each field of cardinality α has 2^{2^α} different field topologies ([Kittinen]). Thus, examples of topological H-planes are plentiful, but not much can be said in general. However, all locally compact (commutative) fields are known ([Bourbaki,2]): the disconnected ones are the finite extensions of the p-adic fields $\mathbb{Q}_p$ and the Laurent fields, $F_q((x))$, of power series over the Galois field F_q; the only connected ones are $\mathbb{R}$ and $\mathbb{C}$. We will see that the locally compact H-rings of characteristic zero or prime p are exactly the rings $K[x]/(x^n)$ where K is a locally compact field. Before we do that we examine locally compact H-planes in general.

§5. LOCALLY COMPACT H-PLANES

The condition that $\sim$ is closed means, for ordinary planes, that the plane is hausdorff. The importance of hausdorffness in AH-planes is illustrated below.

(5.1) The following conditions are equivalent in a topological AH-plane.

(a) $\mathbb{P}$ is T_1.

(b) $\mathbb{P}$ is hausdorff.

(c) Every line is hausdorff.

(d) $\mathbb{L}$ is hausdorff.

(e) Every line is closed.

Proof. Since $\mathbb{P}$ is regular, (a) is equivalent to (b). By (4.4), (b) <=> (c). If ℓ and m are two distinct lines in $\mathbb{L}$, then choose a line j so that j meets both ℓ and m only once. Then choose a coordinate system so that $j=[o,o]$. Hence

$\ell, m \in \mathbb{L}_2 \cong [1,o] \times [1,o]$. Thus, $\mathbb{L}_2$ is an open hausdorff subspace and so $\mathbb{L}$ is hausdorff. Thus, (c) implies (d) .

(d) => (e) . Let (P_α) be a net in a line ℓ converging to P . Then, $L(P_\alpha, \ell) = \ell \longrightarrow L(P, \ell)$, and since $\mathbb{L}$ is T_2 , $L(P, \ell) = \ell$ or $P \in \ell$. The last implication is obvious. □

In our discussion, *a locally compact space is assumed to be hausdorff.*

As before, K is a biternary ring of H . The most important consequences of local compactness are metrizability and a countable base of open sets.

(5.2) A locally compact AH-plane is a metric space.

It suffices to prove this result for K . We do this in a series of steps by employing the continuity of addition and multiplication in K .

(a) There exists a sequence $\{a_n\}$ in K , $a_n \not\sim o,1$ so that $a_n \longrightarrow o$.

Proof. Since $K/\sim$ is not discrete, K has no isolated points by (4.17). Thus, each neighbourhood of K has infinitely many points. Take $k \not\sim o,1$ in K . Since K is regular and $K/\sim$ is hausdorff we can choose closed neighbourhoods $U, V \in \Omega(k)$, $W \in \Omega(o)$ and $X \in \Omega(1)$ so that $U \times W \cap (\sim) = \emptyset = V \times X \cap (\sim)$. Thus, $\bar{o} \cap U = \bar{1} \cap V = \emptyset$. Let C' be a compact neighbourhood of k . Then, $C = U \cap V \cap C'$ is also a compact neighbourhood of k , and $(\bar{o} \cup \bar{1}) \cap C = \emptyset$. We can thus select a sequence $\{b_n\}$ in C , $b_n \nmid o,1$. By compactness, there is a subsequence c_n converging to a cluster point b . Now, if $-a$ (the right inverse) is the unique solution of $a + x = o$, then $-: K \longrightarrow K$ is continuous, and so $a_n = c_n - b \longrightarrow o$. □

(b) If $U \in \Omega(o)$ has compact closure C and $a_n \longrightarrow o (a_n \nmid o,1)$, then $\{a_n C\}$ is a neighbourhood basis for o in K .

The proof follows the arguments used for ternary fields in ([Salzmann, 1, page 440]). □

(c) K is metrizable.

Proof. As mentioned in [Salzmann, 4, page 319], any topological loop with a countable base of neighbourhoods of o is metrizable. □

If $a \nmid o$, then, the unique solution of $x \cdot a = 1$, a^{-1} , is the *left inverse* of a . We say *inversion is continuous near zero in* K , if $a_n \longrightarrow o$ and $a_n^{-1} \cdot b_n = c$ implies $b_n \longrightarrow o$ ([Salzmann]).

(5.3) If inversion is continuous near zero, then K is separable and σ - compact.

Proof. Let $\{a_n C\}$ be a neighbourhood basis as in (b) . Then, each a_n has a left inverse a_n^{-1} and we claim $\{a_n^{-1} C\}$ covers

K . If $k \in K$, then let b_n be the unique solution of $a_n^{-1} \cdot x = k$. Hence, $b_n \longrightarrow o$ and so $k = a_n^{-1} \cdot b_n \in a_n^{-1}C$ for some n . Thus, K is σ-compact. By (c) and ([Dugundji, pages 241, 187]), K is separable. □

Comment. It is not known if every locally compact biternary ring has continuous inversion near zero. However, we can prove ([Lorimer, 5]) that every locally compact AH-plane is separable and σ-compact. We will not do this here, as result (5.3) above will serve our purposes.

(5.4) K is not compact, and neither is any topological AH-plane.

Proof. By ([Salzmann, 2]), $K/\sim$ is not compact. □

Using (4.14) and duality we deduce from (5.1) that,

(5.5) In a topological PH-plane H the following are equivalent.

(i) $\mathbb{P}$ is hausdorff.

(ii) All lines are closed in $\mathbb{P}$.

(iii) All pencils, $\mathbb{L}_p$, are closed in $\mathbb{L}$.

(iv) $\mathbb{L}$ is hausdorff.

(5.6) In a topological PH-plane H , $\mathbb{P}$ is locally compact if and only if H possesses a locally compact line.

Proof. The necessity is immediate from (5.5). Now assume ℓ

is a locally compact line. To show $\mathbb{P}$ is locally compact, it suffices to show that any two points lie in an open locally compact subspace. But this is immediate from (4.4) and (4.20). □

(5.7) A locally compact PH-plane $\mathcal{H}$ is a separable metric space.

Proof. An affine H-plane $\mathcal{H}^\ell$ is open and hence locally compact, and metrizable by (5.2). We next show that the biternary rings of $\mathcal{H}^\ell$ have inversion near zero. The result then follows from (6.3) and ([Dugundji, IX, 9.2]). Let $\{O, E, U, V\}$ be chosen so that $\{\bar{O}, \bar{E}, \bar{U}, \bar{V}\}$ is a quadrangle in $\mathcal{H}/\sim$. Put $W = (O \vee E) \wedge (U \vee V)$, $X = (V \vee E) \wedge (O \vee U)$ and $Y = (U \vee E) \wedge (O \vee V)$. Then, $\{O, X, Y\}$ determines a biternary ring $K = (O \vee E) \setminus \bar{W}$ with ternary operator

$$T(x,m,n) = (\{(X \vee n) \wedge (O \vee Y) \vee [\{(X \vee m) \wedge (Y \vee E) \vee O\} \wedge (X \vee Y)\} \wedge (Y \vee x) \wedge (O \vee E) .$$

Now, suppose $a_n \longrightarrow o$, $(a_n \not\sim o)$ $a_n^{-1} \cdot b_n = c$. For each $z \in K$, define $\phi(z) = (O \vee E) \wedge Y \; [(X \vee E) \wedge O(Y \vee E \wedge X \vee z)]$ and $\phi_c(z) = (O \vee E) \wedge \{(Y \vee E) \wedge [(Y \vee z \wedge X \vee c) \vee O]\} \vee X$. Then, $\phi(o) = W$, $\phi_c(W) = O$, and $\phi_c(z)$ is the unique solution of $x \cdot z = c$. Now, put $\psi = \phi_c \circ \phi$. Thus, ψ is continuous, $\psi(o) = O$, and $\psi(a_n) = b_n$ for each n. Hence, $b_n \longrightarrow o$. □

For ordinary topological projective planes, ([Salzmann, 1]) has shown that a locally compact connected or a locally compact desarguesian projective plane is compact. We shall see soon that these results are false for H-planes. At first glance this might suggest we are working in too general a setting. But as we shall see non-compactness in no way precludes us from obtaining significant results.

Now, from (5.5) we have that a topological projective H-plane is locally compact if and only if it possesses a locally compact line.

We can then replace local compactness in (5.6) with compactness for the projective case.

(5.8) A topological hausdorff PH-plane is compact if and only if it possesses a compact line.

Proof. The necessity is obvious from (5.5). Now assume the plane has one (and hence all) line compact. By (5.6) and (5.7) the plane is a locally compact separable metric space. Hence, to show that $\mathbb{P}$ is compact we show that every sequence of points has a convergent subsequence. First we observe that each pencil $\mathbb{L}_P$ is compact ie if $P \nmid \ell$, then $\ell \longrightarrow \mathbb{L}_P$ $(X \rightarrow X \vee P)$ is a homeomorphism; and for any sequence of points $\{A_n\}$ in $\mathbb{P}$, $\bigcup_{i=1}^{\infty} \overline{A}_n \neq \mathbb{P}$, by the Baire category theorem for locally compact spaces ([Engelking, 1, pages 142-145]) and (4.21) (iv).

Now, take a sequence $\{A_n\}$ in $\mathbb{P}$. Then, there is a point P so that $P \not\sim A_n$ for each n. Hence, $p_n = P \vee A_n$ has a convergent subsequence $p_m \longrightarrow p$ in $\mathbb{L}_P$. Because of (4.22) we may also choose a point $Q \not\sim p$ and $Q \not\sim A_n$ for each n. Then $q_m = Q \vee A_m$ has a convergent subsequence $q_i \longrightarrow q$ in $\mathbb{L}_Q$ with $p \not\sim q$. Then, $(p_i, q_i) \longrightarrow (p, q)$ and since $\sim$ is closed we can assume without loss of generality that $p_i \not\sim q_i$ for each i. Then, $p_i \wedge q_i = A_i \longrightarrow p \wedge q$. □

We next show that proper compact desarguesian PH-planes do not exist.

(5.9) Let $P = P(H)$ be a topological desarguesian PH-plane. Then, P is compact if and only if P is an ordinary locally compact desarguesian projective plane. Hence, the lines and point set of a proper locally compact desarguesian PH-plane are never compact.

Proof. The necessity is Salzmann's result mentioned above. Now, suppose P is compact. Then, for any point P, $\bar{P}$ is compact. But $\overline{\langle oo1 \rangle} = \{ \langle xy1 \rangle : (x, y) \in J \times J \}$ is homeomorphic to $J \times J$ and so J is compact. Since $P/\sim$ is not discrete, H has no proper open ideals, and so by ([Goldman and Sah, proposition (1.1)]) H has no non-zero compact right modules. Hence, $J = (o)$ and we are done. The last statement follows immediately from (5.8).

We are now prepared to classify all locally compact Pappian H-planes by determining the locally compact H-rings. We shall see that all locally compact H-rings have nilpotent radicals. So, we next consider H-rings with nilpotent radicals in more detail.

§6. CHARACTERIZATIONS OF COMMUTATIVE H-RINGS WITH NILPOTENT RADICALS

The results here are from ([Lorimer, 8]), but are essentially due to McLean. Let H be a H-ring with nilpotent radical. By (2.3), H is an Artinian, Noetherian principal ideal ring with radical J whose index of nilpotency is n. Then, we may invoke the structure theorems of ([McLean, 3.1, 3.2, 3.3]) to obtain our results.

Now, clearly H is a complete local ring. Hence, from ([McLean, page 255]) the following two cases (not mutually exclusive) cover the possibilities (I), (II) and (IV).

(I)^{o}. $\mathrm{char}(H) = \mathrm{char}(H/J)$.

(II)^{o}. $\mathrm{char}(H/J) = p$ or equivalently, $Hp = J^{(m)}$ for $1 \leq m \leq n$ and p a prime.

In case (II)^{o}, $\mathrm{char}(H) = p^{r}$. Moreover, case (II) with $m < n$ corresponds exactly to the situation with $\mathrm{char}\, H \neq \mathrm{char}(H/J)$, while $m = n$ corresponds to the overlap of (I)^{o} and (II)^{o}.

Case (I)^{o}, as mentioned by McLean, is characterized as a direct consequence of ([Cohn, Theorem 9]).

(6.1) A ring H is a H-ring with nilpotent radical J, $J^{(n)} = (o)$ and $\mathrm{char}(H) = \mathrm{char}(H/J)$ if and only if $H \cong K[x]/(x^n)$ for some field K.

For case $(II)^o$, we use the next result and ideas from ([Hungerford] and [McLean]) to give a more direct proof of the characterization theorem of McLean, which emphasizes the role of Hjelmslev rings.

From ([Matsumura, pages 210-211]) a subring C of a local ring H is a *coefficient ring* of H if C is a noetherian complete local ring with radical $J \cap C$ and $C/C \cap J \cong H/J$ by the canonical map (i.e. $H = C + J$).

From ([Cohen, theorem 11], [Matsumura, pages 210-211], [Ender, page 50], (3.1), and (3.2)) we deduce

(6.2) Let H be a complete local ring with $J^{(n)} = (o)$ and $\mathrm{char}(H) = p^r \geq p = \mathrm{char}(H/J)$ for a prime p. Then, there is a complete Noetherian valuation ring V of characteristic zero whose radical is V_p with $p^2 \notin V_p$ and a homomorphism $\sigma : V \longrightarrow H$ onto a coefficient ring C of H so that the kernel of σ is V_{p^r}. Moreover, C is itself a H-ring with nilpotent radical $C \cap J$.

(6.3) (see [McLean, 3.2]). For a ring H, the following statements are equivalent. p is a prime.

(i) H is a Hjelmslev ring with $J^{(n)} = (o)$, $\mathrm{char}(H/J)=p$, $J^{(m)} = H_p$ and $\mathrm{char}(H) = p^r$ where $n = m(r-1) + s$,

$1 \le s \le m \le n$.

(ii) $H \cong V[x]/(f(x), x^n)$ where V is a complete noetherian valuation ring of characteristic zero whose radical is V_p, $\mathrm{char}(V/V_p) = p$ with $p^2 \notin V_p$ and $f(x) = x^m + a_{m-1}x^{m-1} + \cdots + a_1$ is an Eisenstein polynomial of degree m (i.e., $a_i \in V_p$ $(1 \le i \le m-1)$ and $a_1 \notin V_{p^2}$).

Proof. (ii) $\Longrightarrow$ (i) is 3.4 (b) . Conversely, let H be a H-ring with $J^{(n)} = (o)$, $\mathrm{char}(H/J) = p$ and $J^{(m)} = pH$. By (2.3)(b) , $J = Ha$. If $m = n$, then $\mathrm{char}(H) = \mathrm{char}(H/J)$. By (6.1), $H \cong K[x]/(x^n)$ for a field K . Using (6.2) and following McLean we take $f(x) = x^n - p$ and observe that $K[x]/(x^n) \cong V[x]/(f(x), x^n)$.

Now suppose $m < n$. Let V and $\sigma : V \longrightarrow H$ be chosen as in (6.2). Then $H = C + J$ and by the proof of (2.1) and (2.3)(b), $J = Ha$ where we can choose $a \in C \cap J \backslash C \cap J^{(2)}$ and so $C \cap J = Ca$. Then, $H = C + Ha = C + Ca + J = \cdots = C + Ca + \cdots + Ca^{n-1}$. Then $\phi : V[[x]] \longrightarrow H$ defined by $\phi\left(\sum_{o}^{\infty} r_i x^i\right) = \sum_{o}^{n-1} r_i a^i$ is an epimorphism. By ([Bourbaki, page 392]) every non-zero element of H has the form $a^k u$ where u is a unit. Since $p = \mathrm{char}(H/J) = \mathrm{char}(C/C \cap J) = p$, $p \in C \cap J$ and p is a non-unit of C . Now, since all ideals of C are Ca^i and $Ha^i \cap C$ (as $a \in C$) is an ideal we have that $Ca^i = Ha^i \cap C = (J \cap C)^{(i)}$. Then every non-zero element of C has the form $a^k u$ where u is a unit of C .

Since, $J^{(m)} = pH$, then $Ca^m = pC$ and $p = a^m c$ where c is a unit of C. Then, $\sigma(v) = c$ for some $v \in V$. Moreover, v is a unit of V; or else $\sigma(w) = c^{-1}$ and $\sigma(vw) = 1 = \sigma(1)$ implies $vw - 1$ lies in the radical of V and so vw and thus v is a unit. Let $f(x) = x^m - \frac{p}{v}$.
Since $p^2 \notin V_p$, $f(x)$ is an Eisenstein polynomial over V.
Hence, from the arguments in 2.4 (b), $V[[x]]/(f(x))$ is a valuation ring whose non-zero ideals are $(x^s)/(f(x))$. Moreover, $f(x)$ lies in the kernel of

$$\phi = \mathrm{Ker}(\phi) \text{ as } \phi\left(x^m - \frac{p}{\nu}\right) = a^m - \frac{p}{c} = a^m - \frac{a^m c}{c} = o \text{ . Hence, } \phi$$

induces an epimorphism $\bar{\phi} : V[[x]]/(f(x)) \longrightarrow H$ with $\mathrm{Ker}(\bar{\phi}) \neq o$ as H is not a valutation ring. Then, $\mathrm{Ker}(\bar{\phi})$ $= (x^s)/(f(x))$ for some s, $\bar{\phi}\left(x^s + (f(x))\right) = \phi(x^s)$ $= a^s \in (o) = J^{(n)}$ and so $n \leq s$. Hence, $(x^s) \subseteq (x^n)$. But $(x_n, f(x))/(f(x)) \subseteq (x^s)/(f(x))$ as $\bar{\phi}(\alpha x^n + \beta f(x) + (f(x))$ $= \bar{\phi}\left(\alpha x^n + (f(x))\right) = \phi(\alpha)a^n = o$. Thus, $(x^n)/(f(x))$ $\subseteq (x^n, f(x))/(f(x)) \subseteq (x^s)/(f(x)) \subseteq (x^n)/(f(x))$ and so $\mathrm{Ker}(\bar{\phi}) = (x^n)/(f(x))$.
If $\psi : V[[x]] \longrightarrow V[[x]]/(f(x))$ is the natural homomorphism, then $\phi = \bar{\phi} \circ \psi$ and $\mathrm{Ker}(\phi) = \psi^{-1}(\mathrm{Ker}(\bar{\phi})) = (x^n, f(x))$.
Hence, $H \cong V[[x]]/(x^n, f(x)) \cong V[x]/(x^n, f(x))$. □

Combining (6.1) and (6.3) we obtain

(6.4) For a ring H with radical $J \neq (o)$, the following statements are equivalent.

(1) H is a Hjelmslev ring with $J^{(n)} = (o)$.

(2) $H \cong K[x]/(x^n)$ for a field K or $H \cong V[x]/(f(x), x^n)$ where V is a complete noetherian valuation ring of characteristic zero whose radical is V_p for a prime p, $p^2 \notin V_p$ and $f(x)$ is an Eisenstein polynomial over $V[x]$.

(3) H is a proper homomorphic image of a discrete valuation ring.

Proof. We need only show that (2) <=> (3) . By ([Endler, page 50]) a discrete valuation ring is a principal ideal ring and so Noetherian. Hence, (2.2)(b) and (2.3) prove that (3) => (1) = (2) .

(2) => (3). $K[x]/(x^n) \cong K[[x]]/(x^n)$ and as was mentioned in the proof of (2.4)(a) $K[[x]]$ is a discrete valuation ring. Finally,

$$V[x]/(f(x), x^n) \cong V[[x]]/(f(x), x^n)$$
$$= V[[x]]/(f(x))/(f(x), x^n)/(f(x))$$

and from the proof of (2.4)(b) , $V[[x]]/(f(x))$ is a discrete valuation ring. □

We can now begin our classification of locally compact Pappian H-planes.

§7. LOCALLY COMPACT CONNECTED PAPPIAN HJELMSLEV PLANES

Our object here is to determine all locally compact connected

Pappian PH-planes, and in particular, to give a topological characterization of Hjelmslev's classical geometries over $\mathbb{R}[x]/(x^n)$. More results and details are in ([Lorimer, 8 , 9]). Now to classify locally compact fields, one always assumes that the field is not discrete or equivalently that the field has no proper open ideals. Hence, for topological H-rings we assume that the ring has no proper open ideals or equivalently that the radical has a void interior.

(7.1) Let H be a topological H-ring whose radical has a void interior. Then $P(H)$ is a locally compact PH-plane if and only if H is a locally compact H-ring.

Proof. If $P(H)$ is locally compact, then so is $P(H)^{\ell}$ where $\ell = [oo1]$. Hence, H is locally compact.
Conversely, if H is locally compact, then by ([Kaplansky, 1]), J is closed and so the group of units is an open set, and hence locally compact. By ([Ellis]), the units are a topological group and so H is a Gelfand ring. The result follows from (4.26). □

(7.2) Let H be a locally compact H-ring whose radical J has a void interior. Then, J is nilpotent and H is either connected or totally disconnected.

Proof. From the proof of ([Warner, 1, Theorem 6]) J is nilpotent. By ([Warner, 2, Lemma 7]) it follows that H is either connected or totally disconnected. □

Comment. If the ring H is not commutative, then we can prove, using different techniques, that J is nilpotent if and only if H is connected or totally disconnected. The proof is decidedly more difficult ([Lorimer, 5]). We mention now a few corrections to the proof that a totally disconnected H-ring has a nilpotent radical; namely, we correct <u>"5.21"</u> and <u>"5.23"</u> in [5] as follows:

<u>"5.21" Lemma</u>. Let H be a locally compact totally disconnected H-ring with int $J = \emptyset$. Then, H possesses a compact open subring (containing 1).

Proof. [Goldman and Sah, theorem 1.3].

<u>"5.23" Lemma.</u> Let H be a totally disconnected locally compact H-ring with int $J = \emptyset$. Then, each order (a compact open sub-ring) has an open radical.

Proof. From 5.24 of [5] (which does not depend on 5.23) each order is a local ring. Hence, its radical M is a maximal ideal. By [Kaplansky 1, theorem 1] M is a closed maximal ideal, and hence open by [Numakara, proposition 1].

As mentioned earlier every finite Pappian H-plane is of level n . We have an analogous result for locally compact Pappian H-planes.

(7.3) If H is a locally compact Pappian H-plane, then there is a "topological" solution of H ; namely there is a sequence of locally compact Pappian H-planes,

$$H = H_n \xrightarrow{\psi_{n-1}} H_{n-1} \xrightarrow{\psi_{n-2}} \cdots\cdots H_3 \xrightarrow{\psi_2} H_2 \xrightarrow{\psi_1} H_1 = H/\sim$$

where each ψ_i is an open-continuous minimal H-epimorphism.

Proof. If $\mathcal{H} = \mathcal{H}(H)$ is the pappian AH or PH-plane over the H-ring H, then by (7.1) H is a locally compact Gelfand H-ring. Moreover, since $\mathcal{H}/\sim$ is not discrete, the radical J has a void interior. By (7.2), $J^{(n)} = (o)$ for some n. The result now follows from (4.28) since all ideals of H are closed (4.27) and so their quotient rings are locally compact. □

(7.4) Each locally compact Pappian PH-plane $\mathcal{H}$ is either o-dimensional or connected.

Proof. First we remember ([Engelking, 2]) that for locally compact spaces total disconnectedness is equivalent to o-dimensionality.

Now, by (7.2), $\mathcal{H} = \mathcal{H}(\mathsf{H})$ where H is a locally compact H-ring which is either connected or o-dimensional.

If H is connected, then so is $P(\mathsf{H})^{\ell}$ where $\ell = [oo1]$. Since $P(\mathsf{H})^{\ell}$ is dense in $P(\mathsf{H})$ by (4.21), $P(\mathsf{H})$ is also connected. Finally, assume H is o-dimensional. Hence $P(\mathsf{H})^{[oo1]}$ is o-dimensional ([Engelking, 2, 1.36]). Since J is closed, H/J is o-dimensional ([Hewitt and Ross, 7.11]) and hence so is $P(\mathsf{H}/J)$ ([Salzmann, 2]). Thus $P(\mathsf{H})$ is not connected, and so by (4.18) $C(\langle oo1\rangle) \subseteq \overline{\langle oo1\rangle}$. Now, $\overline{\langle oo1\rangle} = \{\langle xy1\rangle : x, y \in J\}$ lies in $P(H)^{[oo1]}$ and hence is also o-dimensional. By ([Lorimer, 4, 7.4]), $C(\langle oo1\rangle) = C_{\langle oo1\rangle}(\langle oo1\rangle) = \{\langle oo1\rangle\}$. By a change of coordinates, $C(P) = \{P\}$ for all points P, and so $P(H)$ is o-dimensional. □

We next observe that ∿-connectedness is the same as connectedness in our situation.

(7.5) Let $P(H)$ be a proper locally compact Pappian PH-plane. The following statements are equivalents.

(1) $P(H)$ is ∿-connected.

(2) $P(H)$ is connected.

(3) H is connected.

(4) $P(H/J) \cong P(H)/\sim$ is connected.

(5) J is connected.

(6) One neighbour class $\bar{P}$ (and hence all) are connected.

Proof. If $P(H)$ is ∿-connected, then $P(H)/\sim$ is connected by (4.23)(i). If $P(H)$ is not connected, then by (7.4) and its proof we conclude that $P(H)/\sim$ is o-dimensional. Hence, (1) <=> (2). From this and the proof of (7.4), we see that the first four conditions are equivalent; and (5) is equivalent to (6) since $\overline{\langle oo1\rangle} = \{\langle xy1\rangle : x, y \in J\}$. To complete the proof we prove (3) <=> (5). By (7.2) $J^n = (o)$ and so $J = Ha$ by 2.3. Hence, (3) => (5). Finally, suppose J is connected.

If H is not connected, it is o-dimensional and so is J. Hence, $J = \{o\}$ ([Hurewicz and Wallman, page 15]), a contradiction. □

We now begin our task of classifying the locally compact Pappian PH-planes or equivalently the locally compact PH-rings with no proper open ideals.

Let R be a locally compact local ring with radical J. If $J = (o)$, then R is a field and the classification is known if R is not discrete: R is either connected, in which case it is the reals or complexes; or else it is totally disconnected and is a finite extension of a p-adic number field if it is of characteristic zero; and the Laurent field $GF(p^n)((x))$ if it is of non-zero prime characteristic p.

Now what can we say if $J \neq (o)$? The next result gives us a first step.

(7.6) Let R be a non-discrete locally compact local ring with radical J so that $\bigcap_{n\geq 1} J^n = (o)$ and $J = Ra \neq (o)$. Then, all non-zero ideals of R are of the form Ra^i and one of the following occurs:

(i) a (and hence J) is nilpotent, J is closed with a void interior, R/J is a non-discrete locally compact field, and R is a non-compact separable metrizable E-ring which is either connected or totally disconnected. Moreover, all ideals are closed, and R contains an invertible element b that is topologically nilpotent (i.e. $\lim b^n = o$).

(ii) J is nilpotent, J is open-closed, R/J is discrete and R is an E-ring with no invertible topologically nilpotent element and no countable base of open sets. Moreover, R is compact <=> R is finite; and so the structure of R is known ([Clarke-Drake]).

(iii) a is not nilpotent, J is open-closed with R/J discrete, and R is a non-discrete totally disconnected Noetherian valuation ring whose topology is stronger than the (natural) valuation topology. Moreover, the following conditions are equivalent:

(a) R is compact.

(b) R is a separable metric space.

(c) R is a (compact) open valuation ring of a non-discrete locally compact totally disconnected field

(d) R/J is finite.

Proof. By (2.1), all non-zero ideals have the form Ra^i, and a is either nilpotent or $a^n \neq o$ for all n and R is a valuation ring. In fact, since $J \neq (o)$, R is a Noetherian or discrete valuation ring by [Ender, page 50]. From [Kaplanski 1] J is closed in either case. Now, clearly J has a void interior <=> J is open <=> R/J is discrete.

CASE 1. The interior of J is void.

First, we observe that J must be nilpotent. Otherwise, R is an integral domain and so by [Warner, 1, corollary, page 152] must be a field or $J = (o)$. Hence, R is a proper E-ring. The remaining claims of (i) now follow from (2.3), [Lorimer 4, 5.7, 5.8] and (7.2).

CASE 2. The interior of J is non-void.

Then J is open-closed and R/J is discrete. If a is nilpotent, then R is an E-ring but with no invertible nil-potent element. Then, R is Artinian by (2.3) and so [Warner 3, page 56] implies that R is compact $\iff R$ is finite. If R has a countable base of open sets then by [Warner, lemma 3] , Ra^i is open for each i . Thus, $(o) = \bigcap_{i=1}^{n} Ra^i$ is open and so R is discrete, a contradiction. The proof of (ii) is complete.

Finally, assume a is not nilpotent; and so the (natural) topology of R is not discrete. Since $J \neq (o)$, it follows by [Warner 4, theorem 2] that R is totally disconnected. Again by [Ender, page 50] R is a principal ideal domain, and so the remaining claims in (iii) follow from [Warner 4, theorems 7 and 18]. □

We can now easily give a topological characterization of finite H-rings.

Corollary. Let R be a hausdorff ring with radical $J \neq (o)$. The following statements are equivalent.

(1) R is a compact E-ring.

(2) R is a compact H-ring with $J = Ra$.

(3) R is a finite H-ring.

(4) R is a finite chain ring.

Proof. (1) $\Longrightarrow$(2) is immediate from (2.3) and (3) $\Longrightarrow$ (4)$\Rightarrow$ (1) is in [Clarke-Drake]. We need only verify (2) $\Longrightarrow$ (3) . Since J is finitely generated, [Numakura, theorem 12] implies that R is Noetherian, and hence artinian by (2.3). Thus, $J^n = (o)$ [Adamson, 23.6]. Since $J \neq (o)$, J has non-zero zerodivisors, and so R is finite from the theorem. □

From the theorem above we notice that if J has a non-void interior but is still non-discrete, then either R has no countable base of open sets or is a valuation ring. Such rings are clearly inappropriate in the theory of topological H-planes. In any case, the canonical image would be discrete. Hence, as mentioned in [Goldman and Sah], the appropriate hypothesis for topological rings to replace the notion of non-discreteness in topological fields is the assumption that the ring has no open ideals. For local rings this means the following.

(7.7) Let R be a hausdorff local ring. The following statements are equivalent.

(a) R has no proper open ideals.

(b) The radical is not open.

(c) The interior of the radical (int J) is void.

(d) The quotient field $K = R/J$ is non-discrete.

If R is locally compact, then the conditions above are all equivalent to

(e) R contains an invertible element b so that $\lim_n b^n = o$.

Proof. Immediate from (1.1) and [Warner 1, theorem 6]. □

We can now present several characterizations of locally compact H-rings.

(7.8) Let R be a non-discrete locally compact ring with $J \neq (o)$. The following statements are equivalent.

(1) R is a E-plane and $\text{int}\, J = \emptyset$.

(2) R is an H-ring and $\text{int}\, J = \emptyset$.

(3) R is a chain ring, J is finitely generated and $\text{int}\, J = \emptyset$.

(4) R is a local ring, $J = Ra$ and $\text{int}\, J = \emptyset$.

Proof. [(1) ⟹ (2)] is obvious and [(2) ⟹ (1)] follows from (7.2) and (2.3). Thus, (1) ⟹ (3) ⟹ (4) as a finitely generated ideal in a chain ring is clearly principal. Finally, we show that [(4) ⟹ (1)]. By (7.2), $J^n = (o)$. Hence, from (7.6), R is an E-ring. □

The characteristic of a local R (char R) with radical J is either zero or a prime power [Cohn]. There are four possibilities as mentioned on page 13:

I) $\text{char}(R) = o = \text{char}(R/J)$.

II) char($\mathcal{R}$) = p = char($\mathcal{R}/\mathcal{J}$) for a prime p .

III) char($\mathcal{R}$) = o and char($\mathcal{R}/\mathcal{J}$) = p for a prime p .

IV) char($\mathcal{R}$) = $p^n > p$ = char($\mathcal{R}/\mathcal{J}$) for a prime p .

In case (III) , $\mathcal{R}$ is a principal valuation ring and $\mathcal{J}$ is not nilpotent [Matsumura, pages 210-211]. Moreover, if $\mathcal{J}$ is nilpotent, then char($\mathcal{R}$) = o implies that char($\mathcal{R}/\mathcal{J}$) = o . Combining these facts with the previous theorem and the remarks on the structure of non-discrete locally compact fields we arrive at the following result.

(7.9) Let $\mathcal{R}$ be a locally compact H-ring with radical $\mathcal{J}$, $\mathcal{K} = \mathcal{R}/\mathcal{J}$ and int $\mathcal{J} = \emptyset$. Then, exactly one of the following occur:

(A) $\mathcal{R}$ is connected, char($\mathcal{R}$) = o = char($\mathcal{K}$) and $\mathcal{K}$ is the reals or complexes.

(B) $\mathcal{R}$ is totally disconnected, char($\mathcal{R}$) = o = char($\mathcal{K}$) and $\mathcal{K}$ is a finite algebraic extension of the p-adic numbers for some prime p .

(C) $\mathcal{R}$ is totally disconnected, char($\mathcal{H}$) = p = char($\mathcal{K}$) and $\mathcal{K}$ is a Laurent (power series) field over a finite field.

(D) $\mathcal{R}$ is totally disconnected, char($\mathcal{H}$) = $p^k > p$ = char($\mathcal{K}$) and $\mathcal{K}$ is as in (C) . □

We can thus try to classify the topological H-rings with respect to the four possibilities above. The first 3 cases just comprise the (locally compact) equicharacteristic case [char($\mathcal{R}$) = char(K)] or equivalently the locally compact H-rings of characteristic zero or prime p. We consider this case now.

$\mathcal{R}$ is a *Cohn Algebra* if $\mathcal{R}$ is a local algebra and its radical has codimension one.

(7.10) Let $\mathcal{R}$ be a locally compact ring with radical $J \neq (o)$. The following statements are equivalent.

(a) $\mathcal{R}$ is a H-ring of characteristic o or p and int $J = \emptyset$.

(b) $\mathcal{R}$ is a local ring of characteristic o or p, $J = \mathcal{R}a$ and int $J = \emptyset$.

(c) $\mathcal{R}$ is a finite dimensional Cohn algebra over a non-discrete locally compact field and $J = \mathcal{R}a$.

(d) There is a non-discrete locally compact field K, and $\mathcal{R}$ is topologically and algebraically isomorphic to $K[x]/(x^n)$ endowed with the product topology on K^n.

Proof. [(a) $\longrightarrow$ (b)] follows from (7.8). [(b) $\longrightarrow$ (c)]: By (7.2) $\mathcal{R}$ is connected or totally disconnected. Since $\mathcal{R}$ is local o and 1 are the only idempotents. Hence, the proof of [Warner 1, theorem 7] yields (c). [(c) $\longrightarrow$ (d)]: Let $\mathcal{R}$ be a finite dimensional Cohn algebra over a non-discrete locally compact field K. The topology of K is defined by a proper absolute value [Kaplansky 3, theorem 8] and so K contains a non-zero element $\lambda (|\lambda| < 1)$ so that $\lim_n \lambda^n = o$. By (7.7) (e), int $J = \emptyset$. Now J consists of the non-units and so $K \cap J = (o)$. Since J has codimension one we have $R = K \oplus J$. Since $J = Ra$, (7.8) and (2.3) imply that $J^n = (o)$, and R is an E-ring. By [Lorimer 4 , 5.11] $Ja^{i-1} = Ra^i$ for $1 \leq i \leq n$. Hence, $R = K + J$ $= K + Ra = K + (K + J)a = K + Ka + Ja = K + Ka + Ra^2$ $= --- = K + Ka +--+ Ka^{n-1}$ and $\{1, a,---,a^{n-1}\}$ is a basis for R . $K[x]/(x^n)$ also satisfies (d) [Lorimer 4, section 7] and so by [Warner 5, pages 383 and 392] the map

$$\sum_{o}^{n-1} k_i a^i \longrightarrow \sum_{o}^{n-1} k_i x^i$$ is a topological algebra isomorphism

from R to $K[x]/(x^n)$. Finally, from [Lorimer 4, section 7] (d) implies (a) . □

From the above argument we deduce immediately from (7.9),

(7.11) R is a connected locally compact H-ring with int $J = \emptyset$ if and only if R is topologically isomorphic to $K[x]/(x^n)$ where K is the reals or complexes.

We can now characterize Hjelmslev's classical geometries over the rings $\mathbb{R}[x]/(x^n)$.

(7.12) Let H be a topological PH-plane. The following statements are equivalent.

(a) H is topologically isomorphic to one of Hjelmslev's classical geometries over the rings $\mathbb{R}[x]/(x^n)$.

(b) H is a locally compact connected Pappian H-plane so that for each point P, $\dim \bar{P} = \dim \mathbb{P} - 2$ (dim means topological dimension)

(c) $\mathcal{H}$ is a locally compact connected Pappian PH-plane and $\mathcal{H}/\sim$ is topologically isomorphic to the real projective plane.

Proof. [(a)$\Longrightarrow$(b)]. Let H, $= \mathbb{R}[x]/(x^n)$, which is homeomorphic to $\mathbb{R}^n$. The derived AH-plane of $\mathcal{H} = \mathcal{P}(H)$ is open and so $\dim \mathbb{P} = \dim (H \times H) = 2 \cdot n$. Now J is closed and so from the theory of Lie groups $\dim(H/J) = \dim H - \dim J$. Now by (7.10) J has comdimension one and so from the comments in the proof of (7.10) we have that J also has topological dimension $(n - 1)$. Now any translation is a homeomorphism [Lorimer 2] and the translations are transitive on points and preserve the neighbour relation [Lorimer - Lane]. Hence, for any point P, $\dim \bar{P} = \dim \overline{\langle oo1 \rangle}$. Also from (7.9), H/J is homeomorphic to $\mathbb{R}$. Thus, for any point P, $\dim \bar{P} = \dim \overline{\langle oo1 \rangle} = \dim\{(x,y,1) : (x,y) \in J \times J\} = 2 \dim J = 2(\dim H - 1) = \dim \mathbb{P} - 2$. The remaining implications now follow easily using similar arguments combined with (7.9), (7.10) and (7.11). □

We finish these notes by considering finally case (D) of (7.9). To do this we review briefly the notion of a topological quotient ring of a one-dimensional (this is *not* topological dimension.) Macaulay ring from [Warner 1]. For the solution to our problem is essentially there.

Let $\mathcal{B}$ be a topological ring and $\mathcal{R}$ the quotient ring of $\mathcal{B}$ relative to the multiplicative subsemigroup S consisting of all non-zerodivisors b such that the translation $L_b : x \longrightarrow bx$ is open. The *$\mathcal{B}$-topology on* $\mathcal{R} = \mathcal{B}_s$ is the topology obtained by declaring the filter of neighbourhoods of zero in $\mathcal{B}$ to be a fundamental system of neighbourhoods of zero in $\mathcal{B}_s$. Hence, $\mathcal{B}$ is open in the $\mathcal{B}$-topology. Then, $\mathcal{B}_s$, endowed with the $\mathcal{B}$-topology is a topological ring called the *topological quotient ring of* $\mathcal{B}$.

A one-dimensional local Noetherian ring is a *Macaulay ring* if and only if its radical contains a non-zerodivisor or equivalently, the radical is not an associated prime of (o). Now topologize a local Noetherian ring with its natural topology (the powers of its radical M form a fundamental system of neighbourhoods for o). Then, $\{M^{(n)}\mathcal{B}\}_{n\geq 1}$ is a fundamental system of neighbourhoods of zero in $\mathcal{B}_s$. (Recall that the topology of a compact Noetherian local ring $\mathcal{B}$ must be the M-adic topology [Warner 3].)

(7.13) Let $\mathcal{B}$ be a local Noetherian ring.

(a) $\mathcal{B} \subsetneq \mathcal{B}_s$ if and only if $\mathcal{B}$ is a one-dimensional Macaulay ring.

(b) If $\mathcal{B}$ is a one-dimensional Macaulay ring, then $\mathcal{B}_s$ is the total quotient ring of $\mathcal{B}$ i.e. for all non-zerodivisors b of $\mathcal{B}$, $L_b : x \longrightarrow bx$ is an open map.

(c) $\mathcal{B}_s$ is locally compact if and only if $\mathcal{B}$ is compact.

Proof. [Warner 1, theorems 1 and 2] and [Lucke and Warner, Lemma 8].

(7.14) Let $\mathcal{B}$ be a one-dimensional compact Macaulay ring with exactly two proper prime ideals, P and its radical M. Put $A = \mathcal{B}_s$. Then,

(a) A is a locally compact totally disconnected Artinian primary ring with no proper open ideals. Moreover, the unique prime ideal is the radical $J = AP$; and $J \cap \mathcal{B} = P$ is the set of nilpotent elements and the set of zerodivisors. i.e. $A = \mathcal{B}_s = \mathcal{B}_P$ [Zariski, Page 228].

(b) A is a H-ring if and only if J is principal.

(c) If P is principal, then A is a H-ring.

(d) $\mathcal{B}$ is not a chain ring, M is neither nilpotent nor principal and P is not open.

Proof. (a) follows from [Warner 1, theorem 2], [Warner and Luke] and [Zariski, page 214], since M is not an associated prime of (o).

(b) is immediate from (a) and (7.8); and (c) follows from (b) since if $P = \mathcal{B}a$, then $J = AP = Aa$.

(d) First, we observe that $\mathcal{B}$ is neither a H-ring nor a valuation ring, since firstly M contains a non-zerodivisor and secondly, from (a), P (the set of zerodivisors) is distinct from (o). Then, if $\mathcal{B}$ is a chain ring (which being noetherian, implies M is principal) or M is principal, we obtain a contradiction via (7.6) . Clearly, M is not nilpotent. Finally, $AP = J \neq A$ and so P is not open in B by [Goldman and Sah, Lemma 4.3]. □

Every locally compact totally disconnected ring R with no proper open ideals possesses a compact open subring S containing 1.[(Goldman and Sah, page 399]). Any such subring is called an *order of* R .

(7.15) Let R be a locally compact totally disconnected local ring with int $J = \emptyset$. Then every order S of R is a compact open Macaulay ring with exactly two proper prime ideals and R is the topological quotient ring of S .

Proof. Since int $J = \emptyset$, R contains an invertible element b so that $\lim b^n = o$ by (7.7). The result is now contained in the proof of [Warner 1, theorem 4], since o and 1 are the only idempotents. □

Summerizing our results we have,

(7.16) R is a locally compact totally disconnected H-ring if and only if R is the topological quotient ring of a compact local Noetherian ring with exactly two proper prime ideals, P and its radical M ,which contains a non-zerodivisor and the radical of R is principal.

Remark. It would be nice, in view of 7.14 (c), if one could replace the assumption that the radical of R is principal in (7.16) with the condition that P is principal.

Combining (7.16) with (7.10) we have,

(7.17) *The Characterization Theorem for locally compact Pappian PH-planes.*

P is a locally compact Pappian PH-plane if and only if P is topologically isomorphic to $P(H)$ where H is $K[x]/(x^n)$ and K is an non-discrete locally compact field or the radical of H is principal and H is the topological quotient ring of a compact local Noetherian ring with exactly two proper prime ideals whose radical contains a non-zerodivisor. □

From the preceding results we see that the possible dimensions for locally compact Pappian planes are o or $(2n)\ 2^m$ where $m = o,1$ and n is any positive integer. If the dimension is positive then, in the affine case, each line of the plane is, from the proof of (7.12), just $\mathbb{R}^n$ for some n .

In general we have, *correcting the statement of 4.2 theorem in* [Lorimer, 4]*, the following result.

(7.18) If the lines of a topological affine H-plane are topological manifolds, then they are homeomorphic to $\mathbb{R}^n$ for some positive integer n . □

We have not considered here the locally compact desarguesian

(*) The author would like to thank Rainer Löwen for bringing this fact to his attention.

H-planes. Using different techniques one can show [Lorimer, 6],

(7.19) A locally compact connected desarguesian H-plane of topological dimension four is uniform and Pappian. Indeed, it is the classical Hjelmslev plane over the dual numbers $\mathbb{R}[x]/(x^2)$.

Since the writing of this paper I have been able to classify the locally compact (non-commutative) connected H-rings, and to show as well that a locally compact connected AH-ring must be a H-ring. (See Lorimer 8, 9). However, we have no time to present these results here. Finally, we mention that, using (4.9), we can construct examples of locally compact non-desarguesian or non-translation uniform AH-planes. (See [Lorimer-Baker-Lane]). In general, all topological dimensions are possible for H-planes. However, inspired by (7.19), one can try to obtain a classification of locally compact connected uniform translation planes. For, in this situation, the only possible dimensions are 4, 8, 16 and 32.

....."The only excuse for creating something useless is so that one may admire it intently....."

[Oscar Wilde, The picture of Dorian Gray]
"Beauty is in the eye of the beholder."
[Surely (independently) everyone]

REFERENCES

I.J. Adamson "Rings, modules and Algebras" Oliver and Boyd Edinburgh 1971

D.L. Armacost and W.L. Armacost "Uniqueness in structure theorems for L.C.A. groups" Can. J. Math. 30, 593-599 (1978)

B. Artmann 1. "Desarguessche Hjelmslev-Ebenen n-ter Stufe" Mitt. Math. Sem. Giessen 97, 1-19 (1971)

2. "Geometric aspects of primary lattices" Pacific J. Math. 43, 15-25 (1972)

P. Bacon 1. "Coordinatized Hjelmslev Planes" Diss. University of Florida, Gainsville (1974)

2. "An introduction to Klingenberg planes" Vol. I(1976), Vol. II, III (1979), published by P. Y. Bacon, 3101 NW 2nd Avenue, Gainsville, Fla. 32807

C. Baker, N.D. Lane and J.W. Lorimer 1. "Order and topology in projective Hjelmslev planes" Journal of Geometry, Vol. 19, 8-42 (1982)

2. "Preordered affine Hjelmslev planes" Journal of Geometry, Vol. 23, 14-44 (1984)

N. Bourbaki 1. "General topology", Vol. I, Adddison-Wesley Pub. Co., Reading, Mass. 1966

2. "Commutative algebra" 1972

W.E. Clarke and D.A. Drake "Finite chain rings" Abh. Math. Sem. Univ. Hamburg 39, 147-153 (1973).

I. Cohn "On the structure and ideal theory of complete local rings" Trans. Amer. Math. Soc. 59, 54-106 (1946)

D. Drake "Existence of parallelism and projective extensions for strongly n-uniform near affine Hjelmslev planes" Geom. Ded. 3, 295-324 (1974)

J. Dugundji "Topology" Allyn and Bacon Inc. 1966

R. Ellis "A note on the continuity of the inverse"
Proc. Amer. Math. Soc. 8, 372-373 (1957)

O. Ender "Valuation theory" Springer-Verlag, New York 1972

R. Engelking 1. "Outline of general topology" John Wiley and sons Inc., N.Y. 1968

2. "Dimension theory" North Holland Pub. Co., N.Y. 1978

O. Goldman and C. H. Sah "Locally compact rings of special type"
J. of Algebra II, 363-454 (1969)

A. Grothendieck "Topological vector spaces" Gordon and Breach Sci. Pub., N. Y. 1973

J. Hjelmslev 1. "Die natürliche Geometrie"
Abh. Math. Sem. Univ. Hamburg 2,1-36 (1923)

2. "Einleitung in die allgemeine Kongruenzlehre, III KGI"
Dansk. Vid. Selsk. Math. Fys Medd. 19, no. 12, 50 p. (1942)

K.H. Hofmann "Topologische Loops"
Math. Z. 70, 13-37 (1958)

Thomas W. Hungerford "On the structure of principal ideal rings"

W. Hurewitz and H. Wallman "Dimension theory" Princeton Univ. Press 1941

N. Jacobson and O. Taussky "Locally compact rings"
Proc. Nat. Acad. Sci. U.S.A. 21, 106-108 (1935)

N. Jacobson "Theory of Rings"
Math. Surveys 11, Amer. Math. Soc. (1943)

B. Jonsson and G.S. Monk "Representations of primary arguesian lattices"
Pac. J. Math. 30, 95-139 (1969)

I. Kaplansky 1. "Locally Compact rings II"
Amer. Journal of Math. 73, 20-24 (1951)

2. "Dual rings"
Annuls of Math. 49, 689-701 (1948)

3. "Topological methods in valuation theory" Duke Math. J. 14 (1947), 527-541.

J.O. Kiltinen "On the number of field topologies on an infinite field" Proc. AMS 40, 30-36 (1973)

K. Klingenberg "Projective und affine Ebenen mit Nachbarelementen" Math. Z. 60, 384-406 (1954)

W. Krull "Über die Zerlegung der Haupt ideals in allgemeinen Ringen" Math. Ann. 105, 1-14 (1931)

J. Lambek "Lectures on Rings and Modules" Blaisdell Pub. Co. 1966.

J.W. Lorimer and N.D. Lane "Desarguesian affine Hjelmslev planes" Journal für die reine und angewandte Mat. Band 278/279, 336-352 (1975)

J.W. Lorimer 1. "Coordinate theorems for affine Hjelmslev planes" Ann. Math. Pura Appl. 105, 171-190 (1975)

2. "Topological Hjelmslev Planes" Geom. Dedicata 7, 185-207 (1978)

3. "Connectedness in topological Hjelmslev planes" Annali di Mat. pura ed. appl. 118, 199-216 (1978)

4. "Locally Compact Hjelmslev Planes and Rings" Can. J. Math., Vol. XXXIII, No. 4, 988-1021 (1981)

5. "Dual numbers and topological Hjelmslev planes" Can. Math. Bull., Vol. 26 (3), 297-302 (1983)

6. "Compactness in topological Hjelmslev planes" Canad. Math. Bull., Vol. 27 (4), 1984.

7. "Structure theorems for commutative Hjelmslev rings with nilpotent radicals" C.R. Math. Rep. Acad. Sci. Canada, Vol. VI. No. 3, 123-127, June 1984.

8. "Locally compact desarguesian Klingenberg and Hjelmslev planes" (in preparation)

9. "Projective extensions of locally compact desarguesian AH-planes" (in preparation)

J.W. Lorimer, C. Baker and N.D. Lane "Constructions for non-desarguesian topological Hjelmslev planes" (in preparation)

J.B. Lucke and S. Warner "Structure theorems for certain topological rings" Trans. A.M.S., Vol. 186, 65-90 (1973)

H. Lüneburg "Affine Hjelmslev-Ebenen mit transitiver Translationsgruppe" Math. Z. 79, 260-288 (1962)

W.S. Massey "Algebraic topology: An introduction" Graduate Texts in Mathematics 56 Springer-Verlag N.Y. 1967

H. Matsumura "Commutative algebra" W.A. Benjamin Inc. N.Y. 1970

K.R. McLean "Commutative Artinean Principal ideal rings" Proc. London Math. Soc. (3), 26, 249-272 (1973)

D. Montgomery and I. Zippin "Topological transformation groups" R.E. Krieger Pub. Co., Huntington, N.Y. 1974

M. Nagata "Local rings" Interscience, New York and London 1962

K. Numakura "Notes on compact rings with open radicals" Czechoslovak Mathematical Journal, 33 (108) (1983), 101-106.

H. Salzmann 1. "Topologische projektive Ebenen" Math. Z. 67 (436-466) (1957)

2. "Über den Zusammenhang in topologischen projektiven Ebenen" Math. Z. 61, 480-494 (1955)

3. "Topological planes" Advances in Mathematics 2, Fascicle 1 (1967)

4. "Projectivities and the topology of lines" Geometry-von Staudt's Point of View, 313-337 (1981). Edited by P. Plaumann and K. Strambach. D. Reidel Publishing Company

W. Seier "Eine Bemerkung zum Großen Satz von Desargues in affinen Hjelmslev-Ebenen" J. of Geom., Vol. 20, 181-191 (1983)

G. Törner 1. "Hjelmslev-Ringe und Geometrie der Nachbarschaftsbereiche in den zugehörigen Hjelmslev-Ebenen" Giessen Diplomarbeit (1972)

2. "Eine Klassifizierung von Hjelmslev-Ringen und Hjelmslev-Ebenen" Mitt. Math. Sem. Giessen 107 (1974)

3. "Über den Stufenaufbau von Hjelmslev-Ebenen" Mitt. Math. Sem. Giessen 126 (1977)

E.R. van Kampen "Locally compact abelian groups" Proc. Nat. Acad. Sci. U.S. A. 20, 434-436 (1934)

F.D. Veldkamp "Projective Ring Planes: Some special cases" Rendiconti del Seminario Matematico di Brescia, Vol. 7, Atti del Convegno "Geometria combinatoria e di incidenza fondamenti e applicazioni", pub. della Univ. Cattolica Milano, 609-615 (1984)

S. Warner 1. "Locally compact rings having a topologically nilpotent unit" Trans. Amer. Math. Soc., Vol. 139, 145-154, May 1969

2. "Locally compact equicharacteristic semilocal rings" Duke Math. J. 35, 179-189 (1968)

3. "Compact rings" Math. Annalen 145, 52-63 (1962)

4. "Locally compact principal ideal domains" Math. Ann. 188, 317-334 (1970)

5. "Compact and finite-dimensional locally compact vector spaces" Illinois Journal of Mathematics, Vol. 13, no. 2 (1969)

A. Weil "L'integration dans les groupes topologiques et ses applications" Paris, Herman 1940

H. Weller "Zusammenhang zwischen verbandstheoretischen Eigenschaften des Verbandes $L(R^3)$ und arithmetischen Eigenschaften des zugehörigen Ringes"
Mitt. Math. Sem. Giessen 118 (1975)

O. Zariski and P. Samuel "Commutative Algebra" Vol. 1, Springer-Verlag 1975

FINITE HJELMSLEV PLANES AND KLINGENBERG EPIMORPHISMS

David A. Drake and Dieter Jungnickel

The first author acknowledges the financial support of the National Science Foundation.

"When in the Chronicle of wasted time ..."

(Shakespeare, sonnet 106)

In this paper we are concerned with factorizations of certain epimorphisms between finite incidence structures. In particular, we are interested in factorizations called "solutions" of maps $\phi : \Pi \to \Pi'$ where (ϕ, Π, Π') is a "Klingenberg structure". Such a K-structure is called a "PK-plane" when Π' is a projective plane. The most beautiful examples of PK-planes are the "desarguesian" ones; they are obtained by using homogeneous coordinates over local rings. The solutions for these algebraic examples correspond to maximal chains of ideals in the underlying local rings.

In Section 1 we show how to attach an integer to each factor in the solution of a "neighbor cohesive" K-structure. These integers are called "step parameters". The central combinatorial problem for K-structures is the determination of all possible sequences of step parameters. The preceding ideas are illustrated at the end of Section 1

R. Kaya et al. (eds.), Rings and Geometry, 153–231.

by considering a particular class of local rings; desarguesian PK-planes in general are treated in Section 4. Sections 2 and 3 are devoted to non-existence results for K-structures. In Section 5, Lenz's method of "auxiliary matrices" is explained and applied recursively to obtain many step parameter sequences. In Section 6, a symplectic geometry over GF(2) is used to construct a PK-plane with unusial parameters.

"Regular" K-structures are the subject of Sections 7 through 11: one calls a K-structure "regular" if it admits a certain type of point- and line-transitive abelian automorphism group. A desarguesian PK-plane over a commutative local ring is always regular. The nice transitive automorphism groups of regular K-structures simplify the description of their solutions. One also obtains powerful difference methods which lead to recursive constructions and hence to additional step parameter sequences.

There are very few classes of PK-planes whose step parameter sequences have been characterized. All of these are described in this paper in Theorems 4.5, 5.8, 10.10 and in Corollary 10.15.

Finally, in Section 12, we describe 13 open problems which we think important.

1. K-STRUCTURES AND H-STRUCTURES

In this paper we shall be concerned with triples (ϕ,Π,Π') where Π and Π' are incidence structures and $\phi:\Pi\to\Pi'$ is an epimorphism. One calls Π' the gross structure of $\Pi := (\phi,\Pi,\Pi')$. Points p and q are said to be neighbor, and one writes $p \sim q$, if $p^\phi = q^\phi$. The same notation and language are used in the dual setting (of lines G and H). The epimorphism ϕ is called a Klingenberg epimorphism (briefly a K-epimorphism) provided that the following condition and its dual both hold:

For each two non-neighbor points p, q and each line G' that joins p^ϕ to q^ϕ, there is precisely one pre-image G of G' that joins p to q. (1.1)

An incidence structure Π is said to be point connected provided that, to each point pair (p,q), there exists a sequence of flags (p,G_1), (p_1,G_1), (p_1,G_2), ..., (q,G_n). The notion of line connected is defined dually, and Π is said to be connected if it is both point and line connected. A triple (ϕ,Π,Π') is called a K-structure provided that

$\phi:\Pi\to\Pi'$ is a K-epimorphism; (1.2)

Π' is a connected incidence structure with at least three points on every line and at least three lines through each point. (1.3)

A K-structure is said to be neighbor cohesive if the following condition and its dual both hold with $i = 1$; it is called a Hjelmslev structure (briefly an H-structure) if the two conditions hold with $i = 2$.

$p \sim q$ implies $[p,q] \geq i$. (1.4)

Let (ϕ,Π,Π') be a K-structure, and suppose that $\phi = \psi\theta$ where

$\psi: \Pi \to \Pi''$ is an epimorphism and $\theta: \Pi'' \to \Pi'$ is a K-epimorphism. Then ψ is said to be a <u>K-eumorpism</u> of Π ; if Π and Π'' are both H-structures, a K-eumorphism from Π to Π'' is called an <u>H-eumorphism</u>. Eumorphisms are said to be <u>proper</u> if they are not isomorphisms. Let (ϕ, Π, Π') be a K-structure, and let

$$\Pi = \Pi_n \xrightarrow{\psi_{n-1}} \Pi_{n-1} \xrightarrow{\psi_{n-2}} \cdots \xrightarrow{\psi_1} \Pi_1 = \Pi' \tag{1.5}$$

be a factorization of ϕ into epimorphisms ψ_i . Write θ_i to denote $\psi_{i-1} \cdots \psi_2 \psi_1$. Then the sequence of maps and structures of (1.5) is called a <u>K-solution</u> of Π provided that the factorization is maximal subject to the following condition:

$$\psi_{i-1} \text{ is a proper K-eumorphism of } (\theta_i, \Pi_i, \Pi') \text{ for every } i \tag{1.6}$$

If Π is an H-structure, then the sequence (1.5) is called an <u>H-solution</u> of Π provided that the factorization is maximal subject to the following condition:

$$\psi_{i-1} \text{ is a proper H-eumorphism of } (\theta_i, \Pi_i, \Pi') \text{ for every } i . \tag{1.7}$$

Corollary 1.2 below was obtained by Törner in the special case of gross structures that are projective or affine planes (see [68, Satz 1.7] and [69, Sätze 0.12, 0.13]). A generalization of Theorem 1.1 was later obtained by Drake and Jungnickel in [19, Prop. 1.20] .

THEOREM 1.1. (Törner, Drake-Jungnickel). Let (ϕ, Π, Π') be a neighbor cohesive K-structure with K-solution (1.5). Then every K-eumorphism of Π is of the form $\psi_{n-1} \cdots \psi_{i+1} \psi_i \mu$ for some i and some isomorphism μ .

COROLLARY 1.2. Let (ϕ,Π,Π') be an H-structure with H-solution (1.5). Then every H-euphorism of Π is of the form $\psi_{n-1}\cdots\psi_{i+1}\psi_i\mu$ for some i and some isomorphism μ.

Theorem 1.1 and Corollary 1.2 imply that every K-structure has essentially only one K-solution and that every H-structure has essentially one H-solution. If (1.5) is the solution of Π, one says that Π has _K-type_ or _H-type_ n. It is an interesting unanswered question whether or not the H-type and the K-type of an H-structure must coincide. In this paper we shall often omit the prefixes H- and K- from the words "solution" and "type": when the prefixes are omitted, one should assume that H- is meant if and only if Π has been specified to be an H-structure.

Many versions of the following theorem have appeared. The first is due to Kleinfeld [47]; the most general, to Drake and Jungnickel [19, 1.25 and 1.26].

THEOREM 1.3. Let (ϕ,Π,Π') be a neighbor cohesive K-structure with K-solution (1.5) or an H-structure with H-solution (1.5). Then there is a sequence of integers $(q_2, q_3,\ldots, q_n)$ that satisfies the following conditions:

(i) for $2 \leq i \leq n$, $|(p)\psi_{i-1}^{-1}| = |(G)\psi_{i-1}^{-1}| = q_i^2$ for every point p and every line G of Π_{i-1};

(ii) if (p,G) is a flag in Π_i with $(p)\psi_{i-1} = p''$, then $|(p'')\psi_{i-1}^{-1} \cap G| = q_i$ and dually.

One calls $(q_2, q_3,\ldots, q_n)$ the _Klingenberg_ (or _Hjelmslev_) _step parameter sequence_ of Π. One writes $[p]$ to denote the number of lines through p , $[G]$ to denote the number of points on G , and sets k equal to the maximum of all $[p]$ and $[G]$. Then $(k-1 =: q_1, q_2, \ldots, q_n)$ is called the <u>extended Klingenberg</u> (or <u>Hjelmslev</u>) <u>step parameter sequence</u> of Π . Often we shall omit the adjectives "Klingenberg" and "Hjelmslev" with the expectation that the reader will make the appropriate assumption.

One puts t equal to 1 if Π is of type 1, equal to $q_2 \ldots q_n$ otherwise, and says that Π is a (t,q_1) H-structure or a (t,q_1) K-structure. One calls t and q_1 the <u>parameters</u> of Π . Inductive application of Theorem 1.3 yields the following corollary.

COROLLARY 1.4. Let p be a point and G be a line of a neighbor cohesive (t,q_1) K-structure (ϕ,Π,Π') such that $p' := p^{\phi}$ is incident with $G' := G^{\phi}$. Then $|(p')\phi^{-1}| = t^2 = |(G')\phi^{-1}|$. Also $|(p')\phi^{-1} \cap G| = t$, and dually.

We close Section 1 with an example. Let p be a prime number, and let $f(x)$ be a polynomial of degree m that is irreducible over the ring $\mathbb{Z}_p$ of integers modulo p . Let $c = p^n$ for some positive integer n . Interpreting $f(x)$ as a polynomial over $\mathbb{Z}_c$, let R denote the ring $\mathbb{Z}_c[x]/(f(x))$ with $c^m = p^{mn}$ elements. Let R_i denote the quotient ring R/p^iR for $1 \leq i \leq n$, and write R_i^* for the set of units of R_i .

From each R_i one constructs an incidence structure as follows: points are the sets $R_i^*(a,b,c)$ where each of a, b and c is in R_i and at least one of a, b and c is not in pR_i; similarly, lines are sets $(x,y,z)^T R_i^*$. A point and line are incident if and only if

$ax + by + cz = 0$. The natural ring epimorphism from R_i to R_{i-1} induces a eumorphism $\psi_{i-1} : \Pi_i \to \Pi_{i-1}$. It is possible to verify that these eumorphisms give rise to a solution of the H-structure $(\psi_{n-1} \cdots \psi_2 \psi_1, \Pi_n, \Pi_1)$ which, consequently, has extended step parameter sequence $(q, q, \ldots, q)$ and parameter $t = q^{n-1}$.

The H-structures of the preceding paragraph are called <u>projective Hjelmslev planes</u> (briefly <u>PH-planes</u>), because their gross structures Π_1 are projective planes. The construction given is due to Klingenberg [48], [49]; Klingenberg applied the construction to a larger class of rings in order to obtain a characterization of the "desarguesian" PH-planes. These matters will be discussed in more detail in Section 4.

2. NETS AND NON-EXISTENCE RESULTS FOR K-STRUCTURES

We define an <u>r-net</u> Σ of <u>order</u> q (see [10] or [7]) to be an incidence structure with q^2 points, $q > 1$, that has the following properties: every line contains q points; the lines are decomposed into r <u>parallel</u> classes, each of which partitions the point set; any two lines from different classes intersect (in exactly one point). The existence of an r-net of order q is equivalent to the existence of a set of $r - 2$ <u>mols</u> (mutually orthogonal Latin squares) of order q . Consequently, $r \leq q + 1$; and one calls $d := q + 1 - r$ the <u>deficiency</u> of the net. An <u>affine plane</u> of order q is a $(q + 1)$-net of order q . A set S of q points of Σ is called a <u>transversal</u> to Σ if S contains one point of each line of Σ .

We shall soon see (Lemma 2.1 below) that K-structures contain many nets. As a consequence, known non-existence results for nets yield non-existence results for (certain step parameter sequences of) K-structures. The first to recognize the existence of nets (actually, affine planes) in K-structures was Lüneburg [54]. The following development is based on and generalizes the exposition of Törner in chapter 1 of [69]; Törner's exposition, in turn, had generalized to arbitrary step parameters ideas that Drake [15] had first applied only to t and to q_2.

In order to describe the nets of K-structures, we must first introduce some notation. Let (ϕ,Π,Π') be a neighbor cohesive K-structure with solution (1.5). We write $\bar{p}$ for the set of all points x of Π with $(x)\psi_{n-1} = (p)\psi_{n-1}$.

LEMMA 2.1 Let p be a point of a neighbor cohesive K-structure (ϕ,Π,Π') with step parameter $q := q_n$. Let s denote $[(p)\phi]$. Then it is possible to choose q lines from each of s neighbor classes of lines of Π which induce an s-net Σ of order q on $\bar{p}$.

<u>Proof</u>. Let $G_1, G_2, \ldots, G_s$ be mutually non-neighbor lines that are incident with p. On each G_i choose a point p_i that is not neighbor to p. By Theorem 1.3 and the definition of K-structure, each p_i is joined to the points of $\bar{p}$ by q neighbors of G_i which induce a parallel class on $\bar{p}$. These s classes form an s-net of order q. ☐

COROLLARY 2.2. The existence of a neighbor cohesive K-structure with extended step parameter sequence $(k-1 = q_1, q_2, \ldots, q_n)$ implies the existence of k-nets of order q_i for $2 \leq i \leq n$.

<u>Proof</u>. Let the K-structure have solution (1.5). Then, by (1.6), (θ_i,Π_i,Π') is a neighbor cohesive K-structure with extended step

parameter sequence $(q_1,\ldots,q_i)$. The asserted conclusion follows from Lemma 2.1 if Π' has a point p' with $[p'] = k$. Otherwise, Π' has a line G' with $[G'] = k$, and one applies the dual of Lemma 2.1 to obtain the dual of a k-net of order q_i . □

We write $N(q)$ to denote the maximum number of mols of order q .

COROLLARY 2.3. Let Π be a neighbor cohesive K-structure with extended step parameter sequence $(q_1,\ldots, q_n)$. Then $N(q_1) \geq q_1 - 1$ for $i \geq 2$. If $q_1 \neq 2$, then $q_i \neq 6$ for every $i \geq 2$.

Proof. If $q_1 \neq 2$ and $q_i = 6$ for some i , Corollary 2.2 would give the existence of a 4-net of order 6; i.e., the existence of a pair of orthogonal Latin squares of order 6. This conclusion contradicts a result due to Tarry (see [7] or [8], and [66] for recent brief proofs by Betten and Stinson). □

LEMMA 2.4. Let p be a point of a neighbor cohesive K-structure Π . Suppose that there are no multiply intersecting lines in the incidence structure Σ induced by Π on $\bar{p}$. Then Σ is an affine plane of order $q := q_n$.

Proof. In Σ any two points are joined by exactly one line; the number of points is q^2 ; and every line contains q points. It follows that each point external to a line G of Σ lies on a unique line H of Σ which is disjoint from G . □

LEMMA 2.5. (R.H. Bruck, Corollary to Lemma 3.2 in [10]). Let Σ be an r-net of order q . If $q - \sqrt{q} < r$, then any two transversals of Σ

have at most one common point.

PROPOSITION 2.6. (Törner, Satz 1.4 in [69]). Let p be a point of a neighbor cohesive K-structure (ϕ,Π,Π'), and suppose that $q - \sqrt{q} < [(p)\phi] - 1$ where q is one of the step parameters of Π. Then q is the order of a projective plane.

Proof. It suffices to treat the case $q = q_n$. Let Σ be the incidence structure induced by Π on $\bar{p}$. By Lemma 2.4 it is enough to prove that no two distinct lines of Σ have multiple intersections. Suppose that G_0 and G_1 are lines of Π with $|G_0 \cap G_1 \cap \bar{p}| \geq 2$. Let $G_1, G_2, \ldots, G_s$, $s = [(p)\phi]$, be mutually non-neighbor lines of Π that are incident with p. Proceeding as in the proof of Lemma 2.1, one obtains an $(s - 1)$-net of order q with transversals $G_0 \cap \bar{p}$ and $G_1 \cap \bar{p}$. Lemma 2.5 guarantees that $G_0 \cap \bar{p}$ and $G_1 \cap \bar{p}$ are the same line of Σ and, thus, completes the proof of Proposition 2.6. □

The celebrated Bruck-Ryser Theorem (see, for example, [11], [32] or [7]) asserts that there is no projective plane of order q if $q \equiv 1$ or 2 modulo 4 unless q is the sum of the squares of two integers.

COROLLARY 2.7. Let (ϕ,Π,Π') be a neighbor cohesive K-structure with a step parameter $q := q_i$ which is a Bruck-Ryser number. Then $q - \sqrt{q} \geq [(X)^{\phi}] - 1$ for every point X and every line X of Π.

For example, 30 is a Bruck-Ryser number. Thus, if 30 is a step parameter of a neighbor cohesive K-structure, then every point of the gross structure π' lies on at most 25 lines; and every line of Π' is incident with at most 25 points.

3. NETS AND NON-EXISTENCE RESULTS FOR H-STRUCTURES

In this section of the paper we shall obtain stronger restrictions on the step parameter q_2 in the case of H-structures. If (ϕ, Π, Π') is an H-structure with H-solution (1.5), then $(\theta_2 = \psi_1, \Pi_2, \Pi')$ is an H-structure of type 2. If Π has step parameter sequence $(q_2, \ldots, q_n)$, then Π_2 has step parameter sequence (q_2). Hence, it suffices to consider H-structures of type 2; since these structures have only one step parameter, we shall write q for q_2.

PROPOSITION 3.1. Let (ϕ, Π, Π') be an H-structure of type 2, and let X be any point or line of Π'. Then either $[X] = q + 1$ or $[X] \leq q - \sqrt{q} + 1$.

Proof. Let $X = p$ be a point of Π', and suppose that $[p] >$ $q - \sqrt{q} + 1$. We make minor adaptations in the proof of Proposition 2.6. Suppose that G_0 and G_1 are neighbor lines of Π with $|G_0 \cap G_1 \cap \bar{p}| = i \geq 1$. Since Π is an H-structure of type 2, it follows that $i \geq 2$. Arguing as in the proof of Proposition 2.6, one concludes that Π induces on $\bar{p}$ an affine plane of order q. In this affine plane any pre-image of p must lie on $[p] = q + 1$ lines. Duality considerations yield the asserted conclusion in the case that X is a line. □

We shall need the following two lemmas due to Bruck (Lemma 3.1 (iv') and Lemma 3.2 in [10]).

LEMMA 3.2. (Bruck). Let Σ be an r-net of order q. Let x and y be unjoined points of Σ. Then the number of points of Σ which are joined neither to x nor to y is $q - 2 + (q - r)(q - r - 1)$.

LEMMA 3.3. (Bruck). Let Σ be an r-net of order q with distinct transversals G and H. Then $|G \cap H| \leq q - r$.

For the following argument, we shall need to define a function F on the pairs of positive integers (q,j) by the equation

$$4F(q,j) = 4q + j^2 + j + 2 - (16jq + j^4 + 2j^3 + 13j^2 - 20j + 4)^{1/2}.$$

The function f of [15] satisfies $f(q,j) = F(q,j) - 1$, so the results of [15] must be formulated slightly differently in the current paper.

LEMMA 3.4. Let (ϕ,Π,Π') be an H-structure of type 2 with step parameter q. Let x and y be distinct neighbor points of Π; let $G_1, \ldots, G_\ell$ be the lines that join x to y; and let s denote $[(x)\phi]$. If $s > F(q,j)$ for some integer $j > 1$, then there are at most j distinct point sets among the sets $\bar{x} \cap H_i =: H_i'$.

Proof. Assume to the contrary, that $G_1', \ldots, G_{j+1}'$ are distinct point sets. The G_i are all in one neighbor class of lines. From every other neighbor class which contains a line incident with x, we select q lines which induce a parallel class of lines on $\bar{x}$. This selection is possible by Lemma 2.1; the selected lines induce an $(s - 1)$-net Σ of order q on $\bar{x}$, and each of the G_i' is a transversal to this net. Writing r for $s - 1$ and applying Lemma 3.2, one obtains

$$\left|\bigcup_{i=1}^{j+1} G_i'\right| < q + (q - r)(q - r - 1). \qquad (3.1)$$

By Lemma 3.3, $|G_m' \cap G_n'| \le q - r$ if $m \ne n$. Then for $m \ge 2$, G_m' contains at least $q - 2 - (m - 1)(q - r - 2)$ points that are not on any of the lines $G_1', \ldots, G_{m-1}'$. Thus

$$\left|\bigcup_{m=1}^{j+1} G_m'\right| \ge q + \sum_{m=2}^{j+1} [(m-1)r + (m-2)(2-q)] \text{ ; i.e.,}$$

$$2 \cdot \left|\bigcup_{m=1}^{j+1} G_m'\right| \ge 2q + (j+1)jr + j(j-1)(2-q). \qquad (3.2)$$

From (3.1) and (3.2) one obtains

$$0 \le 2r^2 - r(4q+j^2+j-2) + 2q^2 + q(j^2-j-2) - 2j^2 + 2j ,$$

hence

$$0 \le ((q+\frac{j^2+j-2}{4})-r)^2 - jq - (j^4+2j^3+13j^2-20j+4)/16 .$$

Since $q \ge r$ and $j \ge 2$, one may take square roots to see that $r \le F(q,j) - 1$. This contradicts the assumption that $r + 1 = s > F(q,j)$ and completes the proof of Lemma 3.4. □

LEMMA 3.5. Let (ϕ,Π,Π') be an H-structure of type 2 with step parameter q. Let x and y be distinct neighbor points of Π incident with some line G; let s denote $[(x)\phi]$, S denote $[(G)\phi]$. If $s > \max(F(q,j), q + 1 - (q/\ell))$ and $S > F(q,\ell)$, then either $[x,y] \le j$ or $[x,y] = q$.

Proof. Assume that $[x,y] > j$. Lemma 3.4 implies the existence of a pair of lines H and K which join x to y and satisfy $H' := H \cap \bar{x} = K \cap \bar{x} =: K'$. Let $x = x_1, \ldots, x_q$ be the points of H'. By the dual of Lemma 3.4 there are at most ℓ distinct sets x_i' where

x_i' denotes the set of neighbors of G that are incident with x_i. Then there is some set of at least q/ℓ points x_i, each of which lies on the same q neighbors of G, say on $H = G_1, \ldots, G_q$. Thus $|G_m' \cap G_n'| \geq q/\ell$ for all $m, n \leq q$. If $G_m' \neq G_n'$, Lemma 3.3 implies that $q/\ell \leq q - s + 1$. From this contradiction we conclude that $G_m' = G_1' = H'$ for all $m \leq q$. Then x and y are joined by all q lines G_i. □

THEOREM 3.6. Let (ϕ, Π, Π') be an H-structure of type 2 with step parameter q. Let p be a point of Π; s denote $[(p)\phi]$; S denote $\min [(G)\phi]$ as G ranges over all lines through p. Suppose that $q + 1 \neq s > q + 1 - (q/\ell)$ for some integer $\ell \geq 2$ and that $q \neq 4, 8$. Then either $s \leq q + 2 - (2q+3)^{1/2}$, or $S \leq F(q,\ell)$.

Proof. Assume, by way of contradiction, that $s > q + 2 - (2q + 3)^{1/2} = F(q,2)$ and that $S > F(q,\ell)$. By Lemma 3.5 the point p is joined to each of its $q^2 - 1$ neighbor points either by two or by q lines. Let w denote the number of neighbors to which p is joined by two lines. Count in two different ways the number of flags (x,G) with p in G, $p \neq x \sim p$; one obtains

$$2w + q(q^2 - 1 - w) = qs(q - 1) ;$$

hence $w(q - 2) = q(q - 1)(q - s + 1)$. Since $s - 1 < q$, the right hand side of the preceding equation is positive. Then $(q - 2) \mid q(q - s + 1)$, so $(q - 2) \mid 2(q - s + 1)$. It follows that $q - 2$ divides $2(q - 2) - 2(q - s + 1) = 2(s - 3)$. Since $s - 3 < q - 2$, either $s = 3$ or $q - 2 = 2(s - 3)$. If $s = 3$, the assumption that $s > F(q,2)$ implies that $q = 3$ or 4. The case $q = 4$

is an exception, and Proposition 3.1 yields a contradiction if $q = 3 = s$. Therefore $q - 2 = 2(s - 3)$, and q is even. Also $q + 4 = 2s > 2F(q,2)$, so $q < 10$. Having excluded $q = 4$ and $q = 8$, we must conclude that $q = 2$ or 6. The value $q = 6$ is excluded by Corollary 2.3. Then $q = 2 > s - 1$. This contradicts the definition of H-structure and thus completes the proof of Theorem 3.6. □

COROLLARY 3.7. Let (ϕ,Π,Π') be an H-structure of type 2 with step parameter q. Suppose that every point of Π' lies on k lines and that every line of Π' contains k points. Suppose also that $k \neq q + 1$ and that $q \neq 4, 8$. Then $k \leq q + 2 - (2q + 3)^{1/2}$.

Proof. Apply Theorem 3.6 with $s = S = k$ and $\ell = 2$. One concludes that $k \leq q + 1 - (q/2)$ or that $k \leq F(q,2) = q + 2 - (2q + 3)^{1/2}$. The greatest integer in $(q/2) + 1$ is less than or equal to the greatest integer in $F(q,2)$ for all q except $q = 2$ and $q = 4$. Thus the asserted conclusion holds with the possible exception of the case $q = 2$. If $q = 2$, Corollary 2.3 implies $1 = N(2) \geq q_1 - 1 = k - 2$; then $k = 3 = q + 1$, so the case $q = 2$ is vacuously satisfied. □

REMARKS. Theorem 3.6 and Corollary 3.7 both hold for H-structures of arbitrary type $n \geq 2$ if one replaces q by either of q_2 or $t = q_2 \ldots q_n$. The results with q_2 are immediate corollaries. The results with t may be established by making minor modifications in the preceding proofs. The assumption $q \neq 4$ is essential in the versions with t. Otherwise, Corollary 3.7 would imply that $k \leq 2$ when $t = 4$ and $k \neq 5$. The examples at the end of Section 1 yield a counterexample, however; one takes $p = 2$ and $n = 3$ to obtain an H-structure Π with $k = 3$ and $t = 4$. This example is not a counter-example to the version of Corollary 3.7 with q_2 ; for Π is of type 3 with $q_2 = 2$. Corollary 3.7 generalizes the part of Theorem 2.18 in [15] which treats projective H-planes. The new result does not handle affine H-planes, however: the reason is that affine H-planes have non-intersecting neighbor lines and, hence, are not H-structures.

PROPOSITION 3.8. Let (ϕ,Π,Π') be an H-structure of type 2 with step parameter q. If $[p'] = 3$ for some point p' of Π', then $q \neq 3, 5, 7$.

Proof. The proof of Proposition 3.2 in [15] also yields Proposition 3.8 above. □

4. DESARGUESIAN K-PLANES

The classical examples of projective planes are constructed by using homogeneous triples over a (not necessarily commutative) field. Imitating this construction with a local ring, one obtains the so-called desarguesian projective Klingenberg planes. (We shall not consider the affine case in this paper.) Thus let R be a (not necessarily commutative) *local ring*, i.e., the non-units of R form an ideal N. Write $E = R \setminus N$ and define an incidence structure $\Pi = \Pi(R)$ as follows. The points of Π are the homogeneous triples $Ex = E(x_0,x_1,x_2)$, where not all the x_i are in N ; lines, the homogeneous triples $u^TE = (u_0,u_1,u_2)^TE$, where not all the u_i are in N ; incidence is given by $Ex \; I \; u^TE$ if $xu^T = 0$. Note that Π is the desarguesian projective plane over R when R is a field. If R is a proper local ring, denote by ϕ the natural epimorphism of R onto the field $F = R/N$. Then ϕ induces an epimorphism from $\Pi = \Pi(R)$ onto the desarguesian projective plane $\Pi' = \Pi(F)$ which is again denoted by ϕ. One obtains the following result.

THEOREM 4.1. (Klingenberg [50]). The triple (ϕ,Π,Π') is a PK-plane.

Proof. We have to show that ϕ is a K-epimorphism. Thus consider points Ex and Ey with distinct images under ϕ. We may assume that $x_0 \notin N$ and then that $x_0 = 1$. Let $u^T E$ be a line of Π. Then Ex, Ey I $u^T E$ if and only if

$$(y_0x_1-y_1)u_1 + (y_0x_2-y_2)u_2 = 0 \text{ and } u_0 = -x_1u_1-x_2u_2 \qquad (4.1)$$

We first claim that at least one of $y_0x_1-y_1$ and $y_0x_2-y_2$ may be assumed to be in E. In case $y_0 \notin N$ we may assume $y_0 = 1$; then the claim follows from the condition $(Ex)^\phi \neq (Ey)^\phi$, i.e., $x - y \notin N^3$. In case $y_0 \in N$, one of y_1, y_2 is in E, say $y_1 \notin N$; then $y_0x_1 \in N$ and thus $y_0x_1-y_1 \notin N$, since N is an ideal. This verifies the claim, and we may now assume $y_0x_1-y_1 \notin N$. Then (4.1) implies $u_2 \notin N$, since otherwise also $u_1, u_0 \in N$. Hence we may assume $u_2 = 1$; then u_1 and u_0 are uniquely determined from (4.1), and thus (4.1) has a unique solution (up to homogeneous equivalence). □

One calls (ϕ,Π,Π') the desarguesian PK-plane over R; by abuse of notation, we simply write $\Pi = \Pi(R)$ instead of (ϕ,Π,Π').

At this point it may be helpful to indicate why one requires R to be a local ring. Thus consider any finite commutative ring R; let E be its group of units, and define an incidence structure $\Pi(R)$ as before. Now choose any maximal ideal N ; then $F = R/N$ is a field, and we again obtain a natural epimorphism $\phi : \Pi(R) \to \Pi(F)$. Assume that ϕ is a K-epimorphism; we claim that this forces R to be local. It suffices to show that every $x \in R \setminus N$ is a unit. Thus choose such an x and consider the points $E(1,0,0)$ and $E(1,x,0)$. Clearly these points have distinct images under ϕ and thus should be on a unique line of $\Pi(R)$. If x is not a unit, then x is a zero divisor, as will be seen below. Choose $y \neq 0$ with $xy = 0$. Then $(0,0,1)^T E$ and $(0,y,1)^T E$ are two distinct lines joining $E(1,0,0)$ and $E(1,x,0)$, a contradiction. So R has indeed to be a local ring if ϕ is to be a K-epimorphism.

We next mention a simple consequence of Theorem 4.1.

COROLLARY 4.2. Let R be a finite local ring with maximal ideal N . Then $\Pi(R)$ has parameters $q_1 = r = |R/N|$ and $t = |N|$. It is known that r is a power of a prime and that t is a power of r .

Klingenberg has also determined the conditions under which $\Pi(R)$ is an H-plane. A local ring R is called a <u>Hjelmslev ring</u> (briefly an H-ring) if it satisfies the following two conditions:

N consists of two-sided zero divisors; (4.2)

for all $a,b \in N$, one has $a \in bR$ or $b \in aR$, and also $a \in Rb$ or $b \in Ra$. (4.3)

THEOREM 4.3 (Klingenberg [48]). Let R be a local ring; $\Pi = \Pi(R)$ be the desarguesian PK-plane over R . Then R is an H-ring if and only if

Π is a PH-plane.

The proof is similar to that of Theorem 4.1. More generally, R satisfies (4.3) if and only if $\Pi(R)$ is neighbor cohesive. In the finite case, condition (4.2) is trivially satisfied: given $n \in N$, the set Rn does not contain 1 (by definition of a local ring). Thus there are distinct elements r, s of R with $rn = sn$, i.e. $(r-s)n = 0$ and $r-s \neq 0$. This yields the following result.

COROLLARY 4.4 (Bacon [4]). A finite desarguesian neighbor cohesive PK-plane is a PH-plane.

One can also show that one of the requirements in condition (4.3) may be omitted in the finite case, see Clark and Drake [12]; in the infinite case, there are counter-examples due to Baer [6], see also Skornyakov [64].

We next strengthen Corollary 4.2 considerably by producing the solution of a desarguesian H-plane.

THEOREM 4.5 (Drake [14]). Let R be a finite H-ring with maximal ideal N, and write $r = |R/N|$. Then $|R| = r^n$ for some positive integer n, and $\Pi(R)$ is an n-uniform PH-plane.

Proof. Let I be an ideal of R. Since R is local, we have $I \leq N$, und thus R/I is local with maximal ideal N/I. It is easily seen that condition (4.3) is inherited by R/I, and thus R/I is in fact an H-ring. One checks that the natural epimorphism from R onto R/I induces a eumorphism from $\Pi(R)$ onto $\Pi(R/I)$. Following Lemma 5.2 of

[14], we now produce a chain of ideals of R. We first show the existence of an element z in N satisfying $Rz = N$. This is obvious if $N = 0$. Otherwise choose $z_1 \neq 0$ in N and assume, w.l.o.g., that $Rz_1 \neq N$. Then choose $z_2 \in N \setminus Rz_1$; by (4.3), one obtains $z_1 \in Rz_2$ and thus $Rz_1 \lneqq Rz_2$. Continuing this process, we obtain an element z with $Rz = N$, since N is finite. Define n as the degree of nilpotency of z. We now have the chain of ideals

$$0 = Rz^n \lneq Rz^{n-1} \lneq \ldots \lneq Rz^2 \lneq Rz = N \qquad (4.4)$$

of R. All inclusions in (4.4) are proper: assume e.g. $Rz^i = Rz^{i+1}$ with $i \leq n-1$. Then in particular $z^i = rz^{i+1}$ for some $r \in R$, and thus $z^{n-1} = rz^n = 0$, contradicting the definition of n. Now put $R_i = R/Rz^i$. We have already seen that R_i is an H-ring and that the natural epimorphism ψ_{i-1} from R_i onto R_{i-1} induces an H-eumorphism from $\Pi(R_i)$ onto $\Pi(R_{i-1})$. Writing $\Pi_i = \Pi(R_i)$, we obtain from (4.4) a chain

$$\Pi = \Pi_n \xrightarrow{\psi_{n-1}} \Pi_{n-1} \xrightarrow{\psi_{n-2}} \ldots \xrightarrow{\psi_2} \Pi_2 \xrightarrow{\psi_1} \Pi_1 = \Pi' \qquad (4.5)$$

of H-eumorphisms. We shall prove that (4.5) is the H- and K-solution of Π. It will suffice to show that $|R| = r^n$: in this case the solution of Π has length at most n by Corollary 2.3, since $r = |R/N|$ is the order of Π'. Consider the homomorphism α from the additive group of R onto that of N, defined by $r^\alpha = rz$. It is easily seen that each element y of N has a representation of the form $y = rz^i$ where $r \in R \setminus N$ and $1 \leq i \leq n$. Then $y \in \ker\alpha$ if and only if $i \geq n-1$. Thus $\ker\alpha = Rz^{n-1}$ and $|Rz^{n-1}| = |\ker\alpha| = |R|/|N| = r$. Since the restriction of α to Rz^i is a homomorphism onto Rz^{i+1} (with kernel $Rz^{n-1} \leq Rz^i$), one obtains $|R| = r\,|Rz| = r^2|Rz^2| = \ldots = r^n|Rz^n| = r^n$, as desired. Hence $\Pi(R)$ has parameters r and $t = r^{n-1}$ and is of type n with extended step parameter sequence $(r,r,\ldots,r)$. Using the now current definition of n-uniformity (which is possible by a result of Törner [70]), we conclude that $\Pi(R)$ is n-uniform. □

We remark that the ideals which occur in (4.4) are just the powers of N. We have already noted that $Rz = N$. Obviously Rz^2 is contained in N. Conversely, let a,b be in N, say $a = rz$ and $b = sz$. Then $ab = (rzs)z \in Rz^2$, since rzs is in N. Continuing in this way, one sees that $Rz^i = N^i$ for all i.

For the case of general desarguesian PK-planes, one may show that the eumorphisms of $\Pi(R)$ correspond to the ideals of R, cf. Bacon [4]. Thus a desarguesian PK-plane has a unique solution if and only if the underlying local ring is a chain ring. In the finite case, this means that R is in fact an H-ring, see Clark and Drake [12]. Thus Theorem 1.1 is definitely not true if one omits the hypothesis of neighbor cohesiveness. We now mention results of Bacon [4] which show that desarguesian PK-planes have large automorphism groups - as is to be expected. Denote by $GL(3,R)$ the group of all invertible 3 x 3-matrices over R and let $A \in GL(3,R)$. Then A induces a collineation of $\Pi(R)$ as follows:

$$Ex \mapsto E(xA), \quad u^TE \mapsto (A^{-1}u^T)E \tag{4.6}$$

The induced automorphism group of $\Pi(R)$ is denoted by $PGL(3,R)$.

LEMMA 4.6 (Bacon [4]). Let R be a local ring, and denote by $\bar{A}$ the image of the matrix $A \in GL(3,R)$ under the natural epimorphism from R onto $F = R/N$ (applied coordinatewise). The map $A \mapsto \bar{A}$ induces an epimorphism from $PGL(3,R)$ onto $PGL(3,F)$.

The proof of 4.6 in the commutative case uses determinants. The proof of the general case is less obvious but also not difficult. We call a quadruple $p_1, \ldots, p_4$ of points of a K-plane a quadrangle if their images in the gross structure form a quadrangle. Using 4.6 one obtains the following result.

THEOREM 4.7 (Bacon [4]). Let R be a local ring. Then $PGL(3,R)$ is transitive on the set of quadrangles of $\Pi(R)$.

Using ring automorphisms, one can modify (4.6) to obtain collineations induced by semilinear mappings of R ; then an analogue of the fundamental theorem for desarguesian projective planes holds, cf. Bacon [4]. For the case of uniform H-rings ($n = 2$ in 4.5), these results are also in Cronheim [13]. It seems to be an open question whether or not a PK-plane with a quadrangle-transitive collineation group is desarguesian. Some characterizations of desarguesian Klingenberg and Hjelmslev planes are given by Klingenberg [48], [49], [50], Bacon [4], [5], Dugas [26], [27], [28] and Lorimer and Lane [53]. More generally, every PK-plane may be coordinatized by using a "sexternary ring": the procedure is similar to that of coordinatizing ordinary projective planes by ternary rings. The reader is referred to Bacon [4] for this approach.

We now give some examples of H-rings, cf. also Section 1.

EXAMPLES 4.8. a) Let F be a field and put $R = F[x]/(x^n)$. Then R is an H-ring with maximal ideal $N = Rx$. Note that $N^n = 0$ (Klingenberg [48]).

b) $\mathbb{Z}_{p^n}$ is an H-ring with maximal ideal $p\mathbb{Z}_{p^n}$ for all primes p.

c) Let F be a field and σ be an automorphism of F. Define a multiplication on $R = F \times F$ by the rule

$$(x,y)(z,w) = (xz, xw + yz^{\sigma}) \tag{4.7}$$

Then R is an H-ring with maximal ideal $N = \{(0,y) : y \in F\}$. R is commutative if and only if $\sigma = 1$. These rings are due to Kleinfeld [47] and for $\sigma = 1$ already to Study [67]. R is called the ring of <u>σ-dual numbers</u> over F. Here $N^2 = 0$, and $\Pi(R)$ is uniform.

d) Let F be a finite field or, more generally, a perfect field of characteristic p. Define a polynomial $\Phi(X,Y)$ over $\mathbb{Z}$ by the identity

$$p\Phi(X,Y) = (X + Y)^p - X^p - Y^p \tag{4.8}$$

Then $F \times F$ is made into a commutative H-ring R by defining addition and multiplication as follows:

$$(x,y) + (z,w) = (x + z, y + w - \Phi(x,z));$$

$$(x,y)(z,w) = (xz, x^p w + yz^p).$$

Again $N = \{(0,y) : y \in F\}$ and $\Pi(R)$ is uniform. These rings are the <u>Witt rings</u> of <u>length</u> 2, cf., e.g., Jacobson [34].

The following characterization of <u>uniform</u> H-rings (H-rings with $N^2 = 0$ or, equivalently, H-rings for which $\Pi(R)$ is uniform) appears in [13].

THEOREM 4.9 (Cronheim). Let R be a finite uniform H-ring with $R/N \cong GF(q)$, $q = p^n$. Then R is either a ring of σ-dual numbers or a Witt ring of length 2 over $GF(q)$. If R is non-commutative, then R is a ring of σ-dual numbers with $\sigma \neq 1$. If R is commutative, then R is the ring of dual numbers over $GF(q)$ if $p = 0$ in R and the Witt ring otherwise.

Cronheim also determined all automorphisms of the uniform H-rings and the collineation groups of the uniform desarguesian H-planes. We mention one further result:

THEOREM 4.10 (Cronheim [13]). There are exactly $n + 1$ uniform desarguesian H-planes with projective image plane $PG(2,p^n)$.

The structure of H-rings in general is considered by Clark and Drake [12] and by Neumaier [57].

We conclude Section 4 by mentioning a beautiful characterization of the Desarguesian PK-planes due to Bacon [5]. To state the theorem, a few more concepts are needed. Let L be a line of a PK-plane Π. An incidence structure Π_L is obtained from Π by removing all lines neighbor to L together with all their points. Two lines G,H of Π_L are said to be _parallel_ (and one writes $G \parallel H$) if $G \cap L$ equals $H \cap L$. The pair $(\Pi_L, \parallel)$ is called a _derived_ _AK-plane_ of Π. (There is an abstract notion of AK-plane. Every derived AK-plane is an AK-plane, but not conversely. See, e.g., [4] or [55]). One may define Desarguesian AK-planes by using coordinate pairs from local rings (see [50] and [5]).

TRIANGLE THEOREM 4.11 (Bacon [5]) . Let G, H, K be lines of a PK-plane (ϕ, Π, Π') whose images in Π' form a triangle. Then Π is desarguesian if and only if each of the three derived AK-planes is desarguesian.

Bacon has generalized the preceding desarguesian triangle theorem to a "Moufang triangle theorem" in [4]. The statement of the result, however, is more complicated; and the proof of this deep generalization is considerably more involved. Another important result on Moufang PH-planes is due to Dugas [26], [28]. He has proved that every finite Moufang PH-plane is desarguesian except possibly when Π' is of order 2.

5. AUXILIARY MATRICES

In Section 5 we shall learn how one utilizes "auxiliary sets of matrices" to produce K-structures and H-structures whose step parameter sequences do not consist of a single repeated integer. In Theorem 5.7 we obtain H-structures for many non-constant sequences. In Theorem 5.8 the (non-constant) step parameter sequences of a large class of H-structures are characterized.

One of the basic techniques in the construction of K-structures over a given gross structure Π' is the enlargement of an incidence matrix M' for Π' : the enlargement is carried out by replacing the entries of M' by submatrices chosen from a collection $\mathcal{A} = \{A^1, \dots, A^d\}$ of $a \times a$ (0,1)-matrices. One calls $\mathcal{A}$ a <u>Klingenberg auxiliary set</u> (briefly, a <u>K-set</u>) provided that

$$A^\lambda (A^\mu)^T = (A^\lambda)^T A^\mu = J \quad \text{when} \quad \lambda \neq \mu. \tag{5.1}$$

Here, as usual, J denotes the all 1's matrix and B^T denotes the transpose of the matrix B . A K-set is said to be <u>cohesive</u> (<u>doubly cohesive</u>) if it satisfies the following two conditions with $i = 1$ (with $i = 2$):

$$\sum_\lambda A^\lambda (A^\lambda)^T \geq iJ; \quad \sum_\lambda (A^\lambda)^T A^\lambda \geq iJ. \qquad (5.2)$$

Here we have used the symbolism $B := [b_{ij}] \geq C := [c_{ij}]$ to indicate that B and C are both $s \times t$ for some integers s and t and that $b_{ij} \geq c_{ij}$ for all i and j . A doubly cohesive K-set is said to be an <u>H-set</u>.

An incidence matrix of an incidence structure Π is a (0,1)-matrix $M = [m_{ij}]$ whose rows (columns) are in bijective correspondence with the lines (points) of Π . The entry m_{ij} is 1 if and only if the i-th line contains the j-th point. Let (ϕ, Π_2, Π_1) be a K-structure, and let M_i be an incidence matrix for Π_i for each i . One says that the matrix $M_3 = [M^3_{ij}]_{b\times v}$ is an <u>expansion</u> of $M_2 = [m^2_{ij}]_{b\times v}$ by the K-set A if the following conditions hold:

each M^3_{ij} is $a \times a$; (5.3)

M^3_{ij} is the zero matrix if $m^2_{ij} = 0$; (5.4)

M^3_{ij} is in A if $m^2_{ij} = 1$ (5.5)

$M^3_{ij} \neq M^3_{i\ell}$ if columns j and ℓ of M_2 represent non-neighbor points of Π_2 that are joined by the line of row i ; (5.6)

the dual of condition (5.6) holds. (5.7)

If Π_3 is the incidence structure with incidence matrix M_3 , there is a natural epimorphism $\psi_2 : \Pi_3 \to \Pi_2$. One says that Π_3 is an <u>expansion</u>

of Π_2 by A .

LEMMA 5.1. Let (ϕ,Π_2,Π_1) be a K-structure. Let Π_3 be an expansion of Π_2 by a K-set. If $\psi_2 : \Pi_3 \to \Pi_2$ is the natural epimorphism, then $(\psi_2\phi,\Pi_3,\Pi_1)$ is a K-structure with K-eumorphism ψ_2 .

If one wishes to construct neighbor cohesive K-structures or H-structures by means of expansion, one must carry out the expansions with some care. An expansion is said to be differentiated (doubly differentiated) if the following condition and its dual both hold with $d = 1$ (with $d = 2$) :

> if columns j and ℓ of M_2 represent neighbor points of Π_2 , then $M^3_{\cdot j} \neq M^3_{i\ell}$ for at least d values of i for which $m^2_{ij} = m^2_{i\ell} = 1$. (5.8)

One calls an expansion exhaustive if all of the matrices of A are used to replace the 1's of each row and each column of M_2 .

LEMMA 5.2. Under the assumptions of Lemma 5.1, $\Pi_3 := (\psi_2\phi,\Pi_3,\Pi_1)$ is neighbor cohesive if (1) Π_2 is neighbor cohesive, (2) A is cohesive and (3) the expansion is both differentiated and exhaustive. Also Π_3 is an H-structure if (1) Π_2 is an H-structure, (2) A is doubly cohesive and (3) the expansion is exhaustive and doubly differentiated.

Of course, one may apply Lemmas 5.1 and 5.2 with $\Pi_2 = \Pi_1$ and ϕ equal to the identity map. One cannot apply these lemmas, however, without a supply of K-sets. The following examples of K-sets are due to H. Lenz [22].

PROPOSITION 5.3. If $N(q) \geq r-2$, there exists a K-set $\mathcal{A}$ of r matrices of order q^2. If $r = q+1$, $\mathcal{A}$ is an H-set.

Proof. If $N(q) \geq r-2$, there exists an r-net Σ of order q with points $p_1, \ldots, p_{q^2}$ and parallel classes $\Pi_1, \ldots, \Pi_r$. Define (0,1)-matrices $A^\lambda = [a^\lambda_{ij}]$ by the rule $a^\lambda_{ij} = 1$ if and only if p_i and p_j lie on a common line of Π_λ. Clearly $\mathcal{A} := \{A^1, \ldots, A^r\}$ is a K-set of symmetric matrices, and $A^\lambda A^\lambda = A^\lambda (A^\lambda)^T = qA^\lambda$ for each λ. If $r = q+1$, Σ is an affine plane; so each pair of points is joined by some line. Then (5.2) holds with $i = q$. □

The following easily proved lemma is useful for the construction of additional examples of K-sets.

LEMMA 5.4. Let $\mathcal{A} = \{A^1, \ldots, A^d\}$ and $\mathcal{B} = \{B^1, \ldots, B^d\}$ be sets of $a \times a$ (0,1)-matrices. For each $\lambda \leq d$, suppose that there are permutation matrices P^λ and Q^λ with $P^\lambda A^\lambda = B^\lambda Q^\lambda$. Let C^λ denote $P^\lambda A^\lambda$, $\mathcal{C}$ denote $\{C^1, \ldots, C^d\}$. Then $\mathcal{C}$ is a K-set if $A^\lambda (A^\mu)^T = (B^\lambda)^T B^\mu = J$ whenever $\lambda \neq \mu$; $\mathcal{C}$ is i-fold cohesive if the following two conditions also hold:

$$\sum_\lambda (A^\lambda)^T A^\lambda \geq iJ \quad \text{and} \quad \sum_\lambda B^\lambda (B^\lambda)^T \geq iJ .$$

Putting $B^\lambda = (A^\lambda)^T$ in Lemma 5.4 gives the following result.

COROLLARY 5.5. Let $\mathcal{A} = \{A^1, \ldots, A^d\}$ be a set of $a \times a$ (0,1)-matrices. For each $\lambda \leq d$, suppose that there are permutation matrices P^λ and Q^λ with $P^\lambda A^\lambda = A^T Q^\lambda$. Then $\mathcal{C} := \{P^1 A^1, \ldots, P^d A^d\}$ is a K-set if

$A^{\lambda}(A^{\mu})^T = J$ whenever $\lambda \neq \mu$; $\mathcal{C}$ is i-fold cohesive if $\sum_{\lambda} (A^{\lambda})^T A^{\lambda} \gtrsim iJ$.

The H-sets constructed in the proof of the following proposition are due to Drake and Lenz [22, Theorem 5.1].

PROPOSITION 5.6. If q is a prime power and ℓ is a positive integer, there exists an H-set $\mathcal{A}$ of $q+1$ matrices of order $q^{2\ell}$.

Proof. Let (ϕ,Π,Π') be one of the PH-planes constructed at the end of Section 1. One defines p, m, n by setting $q = p^m$ and $\ell = n-1$. Then Π has parameter $t = q^{\ell}$, and Π' is the desarguesian projective plane of order q . Then Π has an incidence matrix

$$M = \begin{bmatrix} 0 & B^0 & B^1 \dots B^q & 0 \dots 0 \\ A^0 & & & \\ \vdots & & & \\ A^q & & * & \\ 0 & & & \\ \vdots & & & \\ 0 & & & \end{bmatrix}$$

where the A^{λ} and B^{μ} are all $t^2 \times t^2$ non-zero submatrices. Since desarguesian PH-planes are flag-transitive, the substructures of Π with incidence matrices A^{λ} and B^{μ} are mutually isomorphic. This means that, for each λ , there are permutation matrices P^{λ}, R^{λ} so that $B^{\lambda} = P^{\lambda}A^{\lambda}R^{\lambda}$. Condition (1.1) and its dual give $(B^{\lambda})^T B^{\mu} = J = A^{\lambda}(A^{\mu})^T$ for $\lambda \neq \mu$. Since Π is an H-structure,

$$\sum_{\lambda} (A^{\lambda})^T A^{\lambda} \gtrsim 2J \quad \text{and} \quad \sum_{\lambda} B^{\lambda}(B^{\lambda})^T \gtrsim 2J .$$

By Lemma 5.4 , the set $\{P^0A^0, \dots P^qA^q\}$ is an H-set of matrices of order $t^2 = q^{2\ell}$. □

Repeated use of Lemma 5.2 and Proposition 5.6 will produce H-structures with many different step parameter sequences. Let $\Pi' = \Pi_1$ be a square tactical configuration; i.e., an incidence structure with $k = q_1 + 1$ lines through each point and k points on each line. Suppose also that Π' is connected and that $k \geq 3$. If q_1 is a prime power and ℓ is a positive integer, we obtain an H-set A of $q_1 + 1$ matrices of order $q_1^{2\ell}$. König's Lemma guarantees that any incidence matrix for Π' is the sum of k permutation matrices. Consequently, the H-structure (ϕ = identity map, Π' , Π') has an exhaustive expansion by A . Since ϕ is one-to-one, every expansion of Π' is (vacuously) doubly differentiated. Thus one obtains an H-structure (ϕ, Π_2, Π') with invariant $t = q_1^{\ell}$. It can be proved that $(q_1, \ldots, q_1)$ is the Hjelmslev (as well as the Klingenberg) step parameter sequence of Π_2 .

One may reiterate the procedure of the preceding paragraph. Suppose that several iterations have produced the H-structure (ϕ, Π_i, Π') with step parameter sequence equal to $(q_2, \ldots, q_i)$. Then let

$$2(q_2 + 1)q_2 \ldots q_{i-1} \leq q_{i+1} + 1 \leq (q_2 + 1)q_2 \ldots q_i \qquad (5.9)$$

It can be proved that there is a doubly differentiated, exhaustive expansion of Π_2 by any H-set A of $q_{i+1} + 1$ matrices. Call a finite sequence $(q_2, \ldots q_n)$ well behaved if, for $2 \leq i < n$, q_{i+1} either is equal to q_i or satisfies the inequalities in (5.9). (To interpret the lower bound for q_3 , one makes the usual assumption that the product of zero factors is 1.) Modulo the many details that we have not furnished here, one obtains the following theorem.

THEOREM 5.7. Let Π' be a connected, square tactical configuration with replication number $q_1 + 1 = q_2 + 1 \geq 3$, and let $(q_2, \ldots, q_n)$ be a well behaved sequence of prime powers. Then there exists an H-structure (ϕ, Π, Π') whose Klingenberg and Hjelmslev step parameter sequences coincide and are equal to $(q_2, \ldots, q_n)$.

We comment that a stronger version of Theorem 5.7 holds; namely, one may replace the requirement that the q_i be prime powers by the possibly weaker assumption that they be orders of projective planes (see Theorem 5.16 in [18]). The proof of the stronger version of Theorem 5.7 requires a stronger version of Proposition 5.6: one needs to prove the existence of "n-uniform" PH-planes (ϕ, Π, Π') over arbitrary finite projective planes Π'. As mentioned in the proof of Theorem 4.5, an <u>n-uniform PH-plane</u> is a PH-plane of type n with all step parameters equal to q where q is the order of Π'. The first proof of the existence on n-uniform PH-planes over arbitrary finite projective planes Π' is due to Artmann [2]. For our purposes, however, the n-uniform PH-planes must have incidence matrices M so that the submatrices A^λ and B^λ (which appear in the proof of Proposition 4.6) are suitably related. Specifically, there must be some point neighborhood in Π that is the dual of some line neighborhood in Π. The existence of such n-uniform PH-planes was established by Drake in [17].

The existence of such n-uniform PH-planes is also necessary for the proof of the Drake-Törner Theorem which we state below. In order to state this theorem (which characterizes the step parameter sequences of a certain class of H-structures), we need to give several definitions.

Let (ϕ,Π,Π') be a neighbor cohesive K-structure with solution (1.5). We set ν_i equal to $\psi_{n-1}\cdots\psi_{i+1}\psi_i$ for $i < n$, ν_n equal to the identity map. We say that points x and y are <u>(at least) i-related</u> and write $x\ (\sim i)\ y$ if $(x)\nu_i = (y)\nu_i$ for $i > 0$; we write $x\ (\sim 0)\ y$ for all points x and y of Π. Points x and y are said to be <u>exactly i-related</u>, and one writes $x\ (\simeq i)\ y$ provided that $x\ (\sim i)\ y$ holds, but $x\ (\sim i+1)\ y$ does not. The same language and notation are used for pairs of lines.

If each Π_i induces an affine plane on the point set $((p)\psi_{i-1})\psi_{i-1}^{-1}$ for each point p of Π_i, then Π is said to be <u>minimally uniform</u>. One says that Π is <u>point balanced</u> if the value of $[p,q]$ is determined by the value of i in the relation $p\ (\simeq i)\ q$. One calls Π <u>balanced</u> if both Π and its dual are point balanced.

It turns out that balanced, minimally uniform H-structures have well behaved step parameter sequences that are extreme in the sense that each q_i is as large or as small as possible. Call a finite sequence $(q_2,\ldots, q_n)$ of integers a <u>Lenz sequence</u> if $q_2 \geq 2$ and if each q_{i+1}, $i \geq 2$, assumes one of the values q_i or $(q_2 + 1)q_2\cdots q_i$.

THEOREM 5.8. (Drake-Törner). Let Π' be a connected, square tactical configuration with replication number $q_2 + 1 \geq 3$. Then there exists a balanced, minimally uniform H-structure (ϕ,Π,Π') with step parameter sequence $S = (q_2,\ldots, q_n)$ if and only if S is a Lenz sequence.

The first version of the preceding theorem was obtained by Drake and Törner in [25] and [17] in the special case that Π' is a projective plane. The observation that the theorem could be generalized to the case of square tactical configurations is due to Drake and Jungnickel [19, Theorem 3.9]. Both of these versions only give the value of $t = q_2 \dots q_n$. The proof of Theorem 5.16 in [18] makes it clear, however, that the Drake-Törner-Jungnickel construction gives the desired step parameter sequences as well as the values of t.

6. QUADRATIC FORMS AND A PH-PLANE WITH $q_2 \neq q_1$.

It is a long standing open question whether or not q_1 divides q_2 for every finite PH-plane. In fact, there is, to date, only one example of a finite PH-plane of type 2 with $q_2 \neq q_1$. This example (with $q_1 = 2$ and $q_2 = 8$) was constructed by Drake and Shult [24] in 1976. The goal of Section 6 is to reproduce their construction.

We shall apply Lemma 5.2 to (ϕ, Π_2, Π_1) where $\Pi_2 = \Pi_1 =: \Pi$ is the projective plane of order 2 and ϕ is the identity map. Since ϕ is one-to-one, the neighbor relation is the identity relation; thus, Π_2 is an H-structure, and every expansion of Π_2 is (vacuously) doubly differentiated. Since any incidence matrix of Π_2 is the sum of three permutation matrices, it is easy to obtain an exhaustive expansion of Π_2 by any set A of three matrices. Lemma 5.2 will, therefore, yield an expansion of Π_2 to a PH-plane Π_3 with $q_1 = 2$ and $t = 8$ for any H-set A of three matrices of order 64.

Since our goal is to guarantee that $q_2 = 8$, however, we must exercise enough care in the construction of A to guarantee that Π_3 is of type 2. If one applies the construction from the proof of Proposition 5.6, for example, one obtains an H-set A of three matrices of order 64, but the resultant PH-plane Π_3 is of type 4.

Rather than constructing the desired H-set $\mathcal{C}$ directly, we shall apply Corollary 5.5 to a set of matrices $\mathcal{A} = \{A^1, A^2, A^3\}$. Thus, it suffices to construct matrices A^i and permutation matrices P^i, Q^i such that (1) $P^iA^i = A^TQ^i$ for $i = 1, 2, 3$; (2) $A^i(A^j)^T = J$ for $i \neq j$; and (3) $\sum_i (A^i)^TA^i \geq 2J$. Let Σ be the incidence structure with incidence matrix

$$\begin{bmatrix} A^1 \\ A^2 \\ A^3 \end{bmatrix}$$

Σ_i be the substructure of Σ with incidence matrix A^i. Condition (1) above is equivalent to the existence of a <u>duality</u> for each Σ_i; i.e., an isomorphism of Σ_i onto the dual of Σ_i. In fact, we shall construct Σ_i with <u>polarities</u>; i.e., with dualities that are involutions. Condition (2) means that each line of Σ_i intersects each line of Σ_j in a single point if $i \neq j$. Condition (3) means that each two points of Σ are joined by two or more lines.

As point set for Σ we take the set of all vectors of the vector space V of dimension 6 over $F = GF(2)$. Let $\mathcal{J}_i$ be the set of all lines of Σ_i, $\mathcal{L}_i$ be the set of lines of Σ_i that are incident with the zero vector 0. We shall choose the $\mathcal{L}_i$ so that each consists of <u>half spaces</u>; i.e., of subspaces of V of dimension 3 and shall take each $\mathcal{J}_i$ to be the collection of all cosets of all members of $\mathcal{L}_i$. Condition (3) will hold if each vector lies in at least two members of $\mathcal{L} := \mathcal{L}_1 \cup \mathcal{L}_2 \cup \mathcal{L}_3$. It is also easy to see that condition (2) holds if $L \cap K = \{0\}$ whenever L and K are members of distinct $\mathcal{L}_i$'s. We shall take L_3 to consist of eight copies of a single half space. The members of L_1 (and of L_2) will twice cover each of $8 \cdot 7/2 = 28$ non-zero vectors of V .

To describe the half spaces in L and to prove the desired properties, it is helpful to endow V with a nondegenerate quadratic form Q of maximum Witt index 3. (The reader may consult [33] or Appendix B of [51] for the necessary background on quadratic forms.) Call a vector b of V <u>singular</u> if $Q(b) = 0$; <u>non-singular</u> if $Q(b) = 1$. Let S denote the set of singular vectors; N , the set of non-singular ones. As usual define a bilinear form B by the rule $B(a,b) = Q(a + b) - Q(a) - Q(b)$. Clearly, for $a \neq b$,

$$B(a,b) = 0 \Longleftrightarrow a + b \text{ is in } S \text{ whenever } a \text{ and } b \text{ are both in } S \text{ or both in } N ;$$

$$B(a,b) = 0 \Longleftrightarrow a + b \text{ is in } N \text{ whenever one of } a \text{ and } b \text{ is in } S \text{ and the other is in } N . \quad (6.1)$$

Since Q is of maximum Witt index, one may express V as an orthogonal direct sum $P_1 \perp P_2 \perp P_3$ where each P_i has a basis of singular vectors $\{s_i, t_i\}$ with $B(s_i,t_i) = 1$.

A <u>totally singular subspace</u> of V is a subspace whose nonzero vectors are all singular. The space V has 30 totally singular half spaces that fall into two disjoint collections U_1, U_2 ; each U_i consists of 15 of the 30, and each two members of the same U_i intersect in a subspace of dimension one. We denote the collection that contains $H := \langle s_1,s_2,s_3 \rangle$ by U_1 . Then precisely eight members of U_2 intersect H in the zero vector alone. They are

$$T_0 = \langle t_1, t_2, t_3 \rangle$$

$$T_{t_1} = \langle t_1, t_2+s_3, t_3+s_2 \rangle$$

$$T_{t_2} = \langle t_1+s_3, t_2, t_3+s_1 \rangle$$

$$T_{t_3} = \langle t_1+s_2, t_2+s_1, t_3 \rangle$$

$$T_{t_1+t_2} = \langle t_1+s_3, t_2+s_3, t_3+s_1+s_2 \rangle$$

$$T_{t_1+t_3} = \langle t_1+s_2, t_2+s_1+s_3, t_1+t_3 \rangle$$

$$T_{t_2+t_3} = \langle t_1+s_2+s_3, t_2+t_3, t_3+s_1 \rangle$$

$$T_{t_1+t_2+t_3} = \langle t_1+s_2+s_3, t_2+s_1+s_3, t_3+s_1+s_2 \rangle \qquad (6.2)$$

One may use (6.1) to see that the basis displayed for each T_a in (6.2) consists of mutually orthogonal singular vectors, hence that each T_a is totally singular. If the s_i's are suppressed in the basis elements of any one T_a, the space T_0 is reproduced. Thus dim $(T_a + H) = 6$, so $T_a \cap H = \{0\}$. We define L_1 to be $\{T_0, \ldots, T_{t_1+t_2+t_3}\}$. The reader may verify that

each of the 28 vectors in $S \setminus H$
lies in precisely two members of L_1. (6.3)

As indicated above, we set J_1 equal to the collection of all cosets of the T_a in L_1; Σ_1 equal to the incidence structure (V, J_1). We now wish to obtain an incidence structure $\Sigma_2 = (V, J_2)$ that is isomorphic to Σ_1. To this end, define a non-singular linear transformation τ on V by setting

$$\tau(s_1) = t_1, \quad \tau(t_1) = s_1$$
$$\tau(s_2) = s_2, \quad \tau(t_2) = t_2 + s_2 + s_3$$
$$\tau(s_3) = s_3, \quad \tau(t_3) = t_3 + s_3 \tag{6.4}$$

Setting K_a equal to $\tau(T_a)$, we obtain the following eight half spaces of V:

$$K_0 = \langle s_1, t_2+s_2+s_3, t_3+s_3 \rangle$$
$$K_{t_1} = \langle s_1, t_2+s_2, t_3+s_2+s_3 \rangle$$
$$K_{t_2} = \langle s_1+s_3, t_2+s_2+s_3, t_1+t_3+s_3 \rangle$$
$$K_{t_3} = \langle s_1+s_2, t_1+t_2+s_2+s_3, t_3+s_3 \rangle$$
$$K_{t_1+t_2} = \langle s_1+s_3, t_2+s_2, t_1+t_3+s_2+s_3 \rangle$$
$$K_{t_1+t_3} = \langle s_1+s_2, t_1+t_2+s_2, t_3+s_1+s_3 \rangle$$
$$K_{t_2+t_3} = \langle s_1+s_2+s_3, t_2+t_3+s_2, t_1+t_3+s_3 \rangle$$
$$K_{t_1+t_2+t_3} = \langle s_1+s_2+s_3, t_1+t_2+s_2, t_1+t_3+s_2+s_3 \rangle \tag{6.5}$$

We claim that, for all a in T_0,

$$(K_a \cap H) \setminus \{0\} \subseteq H \setminus \langle s_2, s_3 \rangle \quad \text{and}$$
$$K_a \setminus (K_a \cap H) \subseteq N \tag{6.6}$$

To verify (6.6), begin by observing that the first basis vector in each K_a is a vector in $H \setminus \langle s_2, s_3 \rangle$. Each of the other two basis elements is the sum of a vector in T_0 and a vector in H which have inner product one; by (6.1) the second and third basis elements for each K_a are non-singular. Since the second and third basis elements are not perpendicular, their sum is also non-singular for each K_a. It follows that the only singular vector in each K_a is the first listed basis vector. This completes the proof of (6.6) and also yields the conclusion that

$$T_a \cap K_b = \{0\} \text{ for all } a \text{ and } b \text{ in } T_0 . \qquad (6.7)$$

Since τ induces an isomorphism between Σ_1 and Σ_2, condition (6.3) implies that

each of 28 nonzero vectors

lies in two members of $\mathcal{L}_2$. (6.8)

By (6.3), (6.7) and (6.8), the members of $\mathcal{L}_1 \cup \mathcal{L}_2$ cover 56 nonzero vectors of V. It follows from (6.7) that τ fixes none of these 56 vectors; from (6.4), that τ fixes each vector in the subspace $L := \langle s_2, s_3, s_1+t_1 \rangle$. Then

$$L \cap T_a = L \cap K_a = \{0\} \text{ for all } a . \qquad (6.9)$$

We let $\mathcal{L}_3$ consist of eight copies of L, $\mathcal{J}_3$ consist of eight copies of each coset of L, Σ_3 be the incidence structure $(V, \mathcal{J}_3)$, Σ be the incidence structure $(V, \mathcal{J}_1 \cup \mathcal{J}_2 \cup \mathcal{J}_3)$. Every nonzero vector lies either in two lines of $\mathcal{J}_1$ or in two lines of $\mathcal{J}_2$ or in eight lines of $\mathcal{J}_3$, so conditions (2) and (3) both hold. To prove that Σ yields an H-set, it thus suffices to prove the existence of a polarity for each Σ_i. The existence of a polarity for Σ_3 is clear. Since Σ_1 and Σ_2 are isomorphic, it suffices to obtain a polarity ϕ for Σ_1. We define ϕ below but refer the reader to [24] for the proof that ϕ is a polarity.

$a \leftrightarrow T_a$ for a in T_0;

$h \leftrightarrow T_0 + h$ for h in $H \setminus \{0\}$;

$s \leftrightarrow T_{a+b} + a$ for s in $S \setminus (T_0 \cup H)$ where a and b are defined by: s is in $T_a \cap T_b$;

$n \longleftrightarrow T_w + h$ for n in N where $n = w + h$, $0 \neq w$ is in T_0 , $0 \neq h$ is in H and $w \not\perp h$.

(6.10)

Aside from the missing proof that the map ϕ defined in (6.10) is a polarity, we have now demonstrated how to obtain an H-set of three matrices of order 64 which will yield a PH-plane (ϕ,Π_3,Π_1) with invariant $t = 8$ and Π_1 or order $q_1 = 2$. It only remains to verify that Π_3 is of type 2. Let $(q_2,\ldots, q_n)$ be the step parameter sequence of Π_3 . Let (p,G) be a flag of Π_3 . By Theorem 1.3, each line through p contains $q_n - 1$ points of $\bar{p}$. It is thus clear from the construction of Σ that p is joined to precisely $q_n - 1$ points of $\bar{p}\setminus\{p\} =: p^*$ by the lines of one neighbor class and to $(q_n - 1)8/2$ points of p^* by the lines of each of the other two neighbor classes of lines through p . It follows that $q_n^2 - 1 = |p^*| = 9(q_n - 1)$, hence $q_n = 8 = q_2\ldots q_n$, and therefore that Π_2 is of type $n = 2$.

7. REGULAR K-STRUCTURES

In the next five sections of this survey we shall consider "regular" K-structures. The special case of regular PK-planes and PH-planes was introduced in 1976 in the second authors's Ph.D. thesis. In Section 7 we show how to describe such structures by difference methods. The construction of regular K-structures thus reduces to a combinatorial problem in finite group theory. The first examples are obtained in Lemma 7.6 and Corollary 7.8.

We begin by considering the notion of regularity for the gross structures to be used. Thus let Ω be a square incidence structure; i.e., one with equally many points and blocks. Ω is called regular if it satisfies (1.3) and if it admits an abelian[1)] collineation group Z acting regularly on both the point and line set of Ω. For example, Ω may be a desarguesian projective plane. We next introduce the notion of a generalized difference set in an abelian group Z. This is a subset D of Z of cardinality ≥ 3 such that the collection $\Delta = \{d - d' : d \neq d' ; d,d' \in D\}$ of differences of D generates Z. It is well-known that regular symmetric (v,k,λ)-designs and abelian (v,k,λ)-difference sets are equivalent concepts (see, e.g., Hall [30] or Beth, Jungnickel and Lenz [7]); in the same way, one routinely verifies the following Lemma.

LEMMA 7.1. Let D be a generalized difference set in Z. Then the development dev $D = (z, \{D + x : x \in Z\}, \in)$ is a square regular incidence structure. (Here blocks $D + x$ and $D + y$ are to be considered as distinct whenever $x \neq y$, so "repeated blocks" may occur.) Conversely, every square regular incidence structure may be represented in this way.

We now call a K-structure (ψ,Π,Π') regular if it admits an abelian collineation group $G = Z \oplus N$ such that

$$\Pi' \text{ is a square incidence structure regular with respect to } Z ; \qquad (7.1)$$

1) The following concepts also make sense for non-abelian groups. Examples could be obtained by using non-abelian difference sets. The standard terminology and some of the following results require the condition "abelian", though.

N acts regularly on each neighbor class (of points or lines) of Π . (7.2)

Of course, G then acts regularly on both the point and line set of Π . We next introduce the corresponding combinatorial concept, a <u>generalized t-difference set</u> in $G = Z \oplus N$. This is a subset $D = \{(d_i, d_{ij}) : i = 1, \ldots, k ; j = 1, \ldots, t\}$ of G such that

$D' = \{d_1, \ldots, d_k\}$ is a generalized difference set in Z ; (7.3)

every $y \in N$ has a unique representation of the form $y = d_{hk} - d_{ij}$ (for all pairs h,i with $h \neq i$). (7.4)

In particular, N has order t^2 . D is called <u>cohesive</u> (<u>special</u>) provided that

each non-zero $y \in N$ has at least one (at least two) representation(s) of the form $y = d_{ij} - d_{ik}$. (7.5)

One than has

PROPOSITION 7.2 (Drake and Jungnickel [20]). Let $\phi : \Pi \to \Pi'$ be an incidence structure epimorphism. Then (ϕ, Π, Π') is a regular K-structure with parameter t if and only if it can be described as follows:

(i) the point set of Π is $G = Z \oplus N$ for some abelian group G ;

(ii) the line set of Π is $\{D+(x,y) : (x,y) \in G\}$ for some generalized t-difference set D in G ;

(iii) incidence in Π is given by containment;

(iv) Π' is regular with respect to Z , i.e., $\Pi' = \text{dev } D'$;

(v) ϕ is defined by $(x,y)^\phi = x$, and $(D+(x,y))^\phi = D'+x$.

Moreover, (ϕ,Π,Π') is neighbor cohesive, resp., an H-structure if and only if D is cohesive, resp., special.

Proof. First let D be a generalized t-difference set and let (ϕ,Π,Π') be constructed as indicated. Clearly, properties (7.1) and (7.2) hold. Thus it suffices to show that ϕ is a K-epimorphism. Thus let $(x,y)^{\phi} \neq (x',y')^{\phi}$, i.e., $x \neq x'$; and assume that x,x' I $D' + a$, say $x = d_i + a$, $x' = d_h + a$. Then $(x,y),(x',y')$ I $D + (a,b)$ if and only if there exist indices j,k such that $y = d_{ij} + b$, $y' = d_{hk} + b$; i.e., precisely when there exist indices j,k such that $y' - y = d_{hk} - d_{ij}$ and $b = y - d_{ij}$. By (7.4), we see that (x,y) and (x',y') are on a unique line of the form $D + (a,b)$. The proof of the dual flag-lifting property is similar. Clearly (ϕ,Π,Π') has parameter t, since $|N| = t^2$.

To prove the converse, let (ϕ,Π,Π') be a K-structure regular with respect to $G = Z \oplus N$. Choose a "base point" p in Π and coordinatize the points of Π by identifying q with the unique element of G mapping p to q (so, in particular, p corresponds to 0). Similarly, choose a base block D and identify the image of D under $(x,y) \in G$ with $D + (x,y)$. Then (i),(ii) and (iii) hold, though we must still verify that D is a generalized t-difference set. By (7.1) and (7.2), we see that we may so coordinatize Π' that (iv) and (v) are satisfied. By Lemma 7.1, D' is a generalized difference set. By Corollary 1.4 and by (7.2), N has order t^2, and for each $d_i \in D'$ there are precisely t elements of the form $(d_i,d_{ij}) \in D$. It remains to verify (7.4). Thus let h,i with $h \neq i$ be given. Then $0 \in D - d_h, D - d_i$; since these two lines are unequal, there is precisely one pre-image of 0 (i.e., precisely one point $(0,z)$) incident with both $D + (-d_h, 0)$ and $D + (-d_i, y)$. Then $z = d_{hk} = d_{ij} + y$, i.e., $y = d_{hk} - d_{ij}$, and this is the only such representation of y .

Finally, consider neighbor points (x,y) and (x,y') . They will be joined by a line $D + (a,b)$ if and only if $x = d_i + a$, $y = d_{ij} + b$, $y' = d_{ik} + b$ for some indices i,j,k; i.e., if and only if $y - y' = d_{ij} - d_{ik}$ and $b = y - d_{ij}$. This shows that (ϕ,Π,Π') is neighbor point cohesive if and only if D is cohesive. Similar arguments apply for neighbor lines and for the case that D is special. □

The definition of a generalized t-difference set suggests that one should construct examples of such sets by using generalized difference sets in a group Z together with suitable matrices (d_{ij}) in a group N . We therefore introduce the notion of a (t,r)-<u>K-matrix</u> (due to Jungnickel [38]). This is a matrix $A = (a_{ij})$ $(i = 0,\ldots, r ; j = 1,\ldots, t)$ with entries from an abelian group N (of order t^2) such that the following conditions hold:

$$\text{if } a_{ij} = a_{ik} \text{, then } j = k ; \tag{7.6}$$

for each pair h,i with $h \neq i$, the list of differences $a_{hk} - a_{ij}$ contains each element of N exactly once.

(7.7)

A is called <u>cohesive</u> (or a <u>CK-matrix</u>), resp., an <u>H-matrix</u> if also

each non-zero element of N occurs at least once, resp., at least twice among the $(r + 1)t(t - 1)$ differences of the form $a_{ij} - a_{ik}$ with $j \neq k$. (7.8)

One immediately obtains

LEMMA 7.3. Let D be a generalized t-difference set in $G = Z \oplus N$, and let $r = |D'| - 1$. Then the $r + 1$ sets $\{d_{ij} : j = 1,\ldots, t\}$ $(i = 0,\ldots, r)$ form the rows of a (t,r)-K-matrix over N (the ordering within the rows being immaterial). Conversely, given a generalized difference set $D' = \{d_o,\ldots, d_r\}$ in Z and a (t,r)-K-matrix $A = (a_{ij})$ over N, one obtains a generalized t-difference set D by putting

$$D = \{(d_i, a_{ij}) : i = 0,\ldots, r ; j = 1,\ldots, t\} .$$

Moreover, D is cohesive (special) if and only if A is a CK-matrix (H-matrix).

COROLLARY 7.4. Let Π' be a square regular incidence structure with block size $r + 1$. Then there exists a regular K-structure (ϕ,Π,Π') with parameter t over Π' if and only if there exists a (t,r)-K-matrix; and similarly for neighbor cohesive K-structures and for H-structures.

COROLLARY 7.5. There exists a regular (t,r)-K-structure, neighbor cohesive K-structure or H-structure, respectively, with parameter t for which the gross structure is a symmetric $(v,r + 1,\lambda)$ - design if and only if there exist an abelian $(v,r + 1,\lambda)$ - difference set and a (t,r)-K-matrix, CK-matrix or H-matrix, respectively.

We conclude this section by giving some examples. It is well-known that the projective plane Π' of order 2 may be represented by the difference set $\{0,1,3\}$ in $\mathbb{Z}_7$. The reader may easily verify that

$$\begin{pmatrix} (1,1) & (0,0) \\ (1,0) & (0,0) \\ (0,1) & (0,0) \end{pmatrix}$$

is a (2,2)-H-matrix over $N = EA(4) = \mathbb{Z}_2 \oplus \mathbb{Z}_2$. Hence

$$D = \{(0,0,0),(0,1,1),(1,0,0),(1,1,0),(3,0,0),(3,0,1)\}$$

describes a regular (2,2)-PH-plane. This example is easily generalized: one takes the $q+1$ 1-dimensional subspaces of the 2-dimensional vector space over $GF(q^2)$ as the rows of a matrix over $N = EA(q^2)$ to obtain a (q,q)-H-matrix. Then one uses a Singer difference set for $PG(2,q)$ (see, e.g., [7] or [30]) to obtain a regular (q,q)-H-plane.

LEMMA 7.6. There exists a regular (q,q)-H-plane for each prime power q .

In the next section, we shall strengthen 7.6 considerably by showing that every (finite) desarguesian PK-plane corresponding to a commutative local ring is in fact regular. Using Corollary 7.5, this will then furnish us with a large supply of H-matrices to be used later as the starting material for recursive constructions. Before doing so, we use the examples just constructed to produce further K-matrices. The following lemma is an easy exercise for the reader, cf. [38].

LEMMA 7.7. Assume the existence of a (t,r)-K-matrix over N and of a (t',r)-K-matrix over N' . Then there also exists a (tt',r)-K-matrix over $N \oplus N'$.

Omitting rows from a K-matrix obviously yields a K-matrix (but the properties CK and H are not respected!). Thus 7.6 and 7.7 imply

COROLLARY 7.8. Let $t = q_1 \ldots q_n$ be the prime power factorization of the positive integer t. If $r \leq q_i$ for $i = 1, \ldots, n$, then there exists a (t,r)-K-matrix with entries from $EA(q_1^2) \oplus \ldots \oplus EA(q_n^2)$. Moreover, there also exists a regular (t,r)-K-structure under these conditions.

The last assertion follows from the (trivial) existence of a generalized difference set of cardinality $r + 1$. It is an interesting open question whether the existence of a (t,r)-K-matrix implies $r \leq q_i$ for each factor q_i in the prime power factorization of t. This is true if one restricts attention to K-matrices for which every row is a subgroup (of order t) of N : see Theorem 8.12 of Drake and Jungnickel [20]. Such K-matrices have been called <u>uniform</u> in [20]. They correspond to translation nets and are called partial congruence partitions in this context: see Jungnickel [45], [46] and Sprague [65] for results on these structures. Particular examples are obtained from partial t-spreads in finite projective spaces; these partial t-spreads were introduced by Mesner [56] and have been much studied ever since: see e.g., Hirschfeld [31].

8. GENERALIZATIONS OF SINGER'S THEOREM AND A RECURSIVE CONSTRUCTION

The well-known theorem of Singer [63] states that the symmetric designs formed by the points and hyperplanes of finite projective spaces are regular; in present terminology, they admit cyclic Singer groups (see e.g. [7], [30] or [32]). We shall now generalize this result to finite <u>pappian</u> K-planes, i.g., PK-planes $\Pi(R)$ belonging to a commutative local ring R.

THEOREM 8.1 (Hale and Jungnickel [29]). Any finite pappian PK-plane is regular.

Proof. Let R be a finite commutative local ring, and define $\Pi = \Pi(R)$ as in Section 4. Denote the maximal ideal of R by N, and assume $|N| = q^n$, $|R| = q^{n+1}$; thus $F = R/N \cong GF(q)$. We first define a local ring structure on $S = R^3$ extending R and having $M = N^3$ as its maximal ideal. To this end, choose an irreducible monic polynomial $\bar{f}$ of degree 3 in $F[x]$ and a monic irreducible pre-image f of $\bar{f}$ in $R[x]$ (with respect to the natural epimorphism from R onto F). We claim that $S := R[x]/(f)$ has the desired properties. Clearly S is a commutative extension ring of R, and R is contained in S as the image of the constant polynomials. Also, the additive group of S is isomorphic to R^3, since f has degree 3. It remains to consider the ideal $M = (N) = SN$. One has

$$S/M = (R[x]/(f)) \,/\, ((N[x]+(f))/(f)) \cong R[x]/(N[x]+(f))$$
$$\cong (R[x]/N[x]) \,/\, ((N[x]+(f))/N[x]) \cong F[x]/(\bar{f}) .$$

Since the last quotient is the 3-dimensional field extension $GF(q^3)$ of F, we conclude that M is a maximal ideal of S. But N is a nil ideal of R, since every non-unit of R is a zero divisor (see the remarks preceding Corollary 4.4). Then $M = SN$ is also a nil ideal, since S is commutative. Thus M is contained in every maximal ideal of S (see, e.g., Bourbaki [9, § 6 No. 3]), i.e., M is the unique maximal ideal of S, and S is local.

Now let be any element of $S^* := S \setminus M$. Then the map

$$\gamma_a : S \to S \quad \text{with} \quad \gamma_a(x) = ax$$

is a bijective linear map of R^3 (identifying S with R^3). Thus S^* is a subgroup of $GL(3,R)$. By definition, $\Pi(R)$ is coordinatized by homogeneous triples $R^* x$ and $u^T R^*$, writing $R^* = R \setminus N$. Hence the points of $\Pi(R)$ correspond to the elements of S^*/R^* (and similarly for lines). Thus $G := S^*/R^*$ is a transitive abelian (hence regular) collineation group of $\Pi(R)$; its order is $q^{2n}(q^2 + q + 1)$. Since G is abelian and since q^{2n} and $q^2 + q + 1$ are coprime, G splits as $G = Z \oplus K$, where Z has order $q^2 + q + 1$ and K has order q^{2n}. The natural epimorphism of S onto $S/M \cong GF(q^3)$ induces an epimorphism of $\Pi = \Pi(R)$ (coordinatized by G) onto $\Pi' = \Pi(F)$ (coordinatized by $GF(q^3)^*/F^*$). The image of S^* induces a Singer cycle of Π', i.e., a regular cyclic group of order $q^2 + q + 1$. For reasons of cardinality, this group is the (isomorphic) image of Z. Thus G satisfies (7.1) and acts regularly on Π. Hence K satisfies (7.2), and Π is regular with respect to G. □

For the special case of the H-rings of Example 4.8.a, Theorem 8.1 was proved by a rather lengthy computational argument in [38]. This earlier argument was modified and combined with Cronheim's classification of the uniform H-rings (see Theorem 4.10) to obtain the following result.

THEOREM 8.2 (Hale and Jungnickel [29]). Every finite uniform Desarguesian Hjelmslev plane is regular.

Using the H-rings of Example 4.8.a in Theorem 8.1 and Corollary 7.5, we obtain the following important result which is crucial to the recursive constructions of Section 11.

THEOREM 8.3 (Jungnickel [38]). Let q be a prime power and n be a positive integer. Then there exists a (q^n,q)-H-matrix belonging to a pappian PH-plane.

We conclude this section with a first recursive construction. Let us start with an example. Consider the following (5,5)-H-matrix

$$\begin{pmatrix} (2,2) & (4,4) & (3,3) & (1,1) & (0,0) \\ (2,4) & (4,3) & (3,1) & (1,2) & (0,0) \\ (2,3) & (4,1) & (3,2) & (1,4) & (0,0) \\ (2,1) & (4,2) & (3,4) & (1,3) & (0,0) \\ (2,0) & (4,0) & (3,0) & (1,0) & (0,0) \\ (0,2) & (0,4) & (0,3) & (0,1) & (0,0) \end{pmatrix}$$

which is obtained from the 1-dimensional subspaces of $\mathbb{Z}_5^2$. Using this matrix and the (2,2)-H-matrix of Section 7, we obtain a (10,2)-H-matrix as follows (where we write xyzw for $(x,y,z,w) \in \mathbb{Z}_2 \oplus \mathbb{Z}_2 \oplus \mathbb{Z}_5 \oplus \mathbb{Z}_5$) :

$$\begin{pmatrix} 1122 & 1144 & 1133 & 1111 & 1100 & 0024 & 0043 & 0031 & 0012 & 0000 \\ 1023 & 1041 & 1032 & 1014 & 1000 & 0021 & 0042 & 0034 & 0013 & 0000 \\ 0120 & 0140 & 0130 & 0110 & 0100 & 0002 & 0004 & 0003 & 0001 & 0000 \end{pmatrix}$$

This example may be generalized as follows.

THEOREM 8.4 (Jungnickel [38]). Assume the existence of (t,r)- and (s,q)-H-matrices, where $q = t(r + 1) - 1$. Then there also is an (st,r)-H-matrix.

<u>Proof</u>. Let $A = (a_{ij})$ $(i = 0,\dots, r;\ j = 1,\dots, t)$ be a (t,r)-H-matrix, and let B_{ij} $(i = 0,\dots, r;\ j = 1,\dots, t)$ be the $q + 1 = t(r + 1)$ rows of an (s,q)-H-matrix. Define an $(r+1) \times st$-matrix C by choosing

$$C_i = (a_{i1} \times B_{i1} \quad a_{i2} \times B_{i2} \quad \ldots \quad a_{it} \times B_{it})$$

as the i'th row of C. To show that C is an (st,r)-K-matrix, we have to consider the differences arising from C_i and C_h. But, as A is a K-matrix, the differences $a_{ij} - a_{hk}$ contain each possible first coordinate x exactly once; and, as B is a K-matrix, the differences from B_{ij} and B_{hk} then contain each possible second coordinate y exactly once (for given j and k). It remains to verify that C is actually an H-matrix. First consider differences of the type (0,y). The list of the $(q + 1)s(s - 1)$ differences arising from entries in the same row of B contains each $y \neq 0$ at least twice; thus each (0,y) occurs at least twice from C. Finally, let $x \neq 0$ occur in the form $x = a_{ij} - a_{ik}$; then the rows B_{ij} and B_{ik} of B are distinct, and therefore the differences from $a_{ij} \times B_{ij}$ and $a_{ik} \times B_{ik}$ contain all pairs (x,y). Since x has at least two representations of the given type, we see that each difference (x,y) with $x \neq 0$ occurs at least twice from C. □

The corresponding result for PH-planes in general is contained in the recursive construction discussed in Section 5 and was first proved by Drake and Lenz [22]. Using 8.4 and the examples provided in 8.3, we obtain many H-matrices whenever both r and q are prime powers. We give a few examples:

r	2	2	2	2	3	4	5	
t	2	4	8	10	3	4	5	
q	5	11	23	29	11	19	29	
s	5^n	11^n	23^n	29^n	11^n	19^n	29^n	
st	$2 \cdot 5^n$	$4 \cdot 11^n$	$8 \cdot 23^n$	$10 \cdot 29^n$	$3 \cdot 11^n$	$4 \cdot 19^n$	$5 \cdot 29^n$	....

COROLLARY 8.5. Let $t = q_2 \dots q_n$, where all q_i are prime powers and where

$$q_{i+1} = q_i \quad \text{or} \quad q_{i+1} = (q_2+1)q_2 \dots q_i \tag{8.1}$$

for $i = 2, \dots, n-1$. Then there exists a (t,q_2)-H-matrix.

This follows by repeated application of 8.4, using the H-matrices of 8.3. The sequences $(q_2, \dots, q_n)$ described in 8.5 are called _special Lenz sequences_, since the existence of PH-planes for these sequences was first proved by Lenz. (Compare Theorem 5.8.) The notation of 8.5 suggests that the corresponding PH-planes have step parameter sequences $(q_2, \dots, q_n)$ which is in fact true. We shall investigate such PH-planes later, after considering eumorphisms of regular PH-planes in the next section. It may be useful to point out that the use of K- and H-matrices does not yield any existence results for PK- and PH-planes that could not be obtained as well by using auxiliary matrices as in Section 5. Their advantage is to produce examples having large and well-behaved collineation groups; consequently, regular structures are much easier to describe and to construct. The following result is due to Lenz (see [41]) .

THEOREM 8.6 (Lenz). Assume the existence of a (t,r)-K-matrix or H-matrix. Then there also exist $r+1$ symmetric $(t^2 \times t^2)$-auxiliary matrices satisfying (5.1), respectively, (5.1) and (5.2) with $i = 2$.

Proof. Let $A = (a_{ij})$ be a (t,r)-K-matrix over the group N . For $i = 0, \dots, r$, denote the i'th row of A by A_i ; and let χ_i denote the characteristic function of A_i , i.e.,

$$\chi_i(x) = \begin{cases} 1 & \text{if } x \in A_i \\ 0 & \text{if } x \in N \setminus A_i \end{cases} \tag{8.2}$$

Define $M_i = (m^i_{jk})$ by putting $m^i_{jk} = \chi_i(x_j+x_k)$ where $N = \{x_1,\ldots,x_{t^2}\}$. Consider the inner product of row a of M_h with row b of M_i, i.e.,

$$\sum_{k=1}^{t^2} \chi_h(x_a + x_k)\chi_i(x_b + x_k) . \tag{8.3}$$

The k'th term in this sum is equal to 1 if and only if $x_a+x_k = a_{hj}$ and $x_b+x_k = a_{i\ell}$ for some indices j,ℓ (and equal to 0 otherwise). Since there is a unique representation of the form $x_a-x_b = a_{hj}-a_{i\ell}$, we conclude that the sum in (8.3) is 1, i.e., that $M_0,\ldots,M_r$ satisfy equation (5.1). Clearly the M_i are symmetric. Next consider $\sum_{i=0}^{r} M_iM_i^T$; i.e., consider sums $\sum_{i,k} \chi_i(x_a+x_k)\chi_i(x_b+x_k)$. Now such a term contributes 1 to the sum if and only if $x_a+x_k = a_{ij}$ and $x_b+x_k = a_{ih}$ for some indices h,j . Assume $a \neq b$; then x_a-x_b has at least two representations of the type $x_a-x_b = a_{ij}-a_{ih}$ by the definition of an H-matrix. Thus the M_i also satisfy (5.2) if A is an H-matrix (the case $a = b$ is trivial). □

9. EUMORPHISMS OF REGULAR K-STRUCTURES

We shall now consider eumorphisms of regular K-structures. These correspond (in the neighbor cohesive case) to certain subgroups of the group N of (7.2). In particular, we shall see that all the K-structures occurring in the K-solution of a regular H-structure are in fact regular H-structures.

We begin by constructing new K-matrices from a given K-matrix. Thus let A be a (t,r)-K-matrix over N, say $A = (a_{ij})$ $(i = 0,\dots,r$; $j = 1,\dots,t)$. A subgroup U of N of order u^2 is called a <u>K-subgroup</u> for A if the following condition holds:

$$|\{a_{im} : a_{im} \equiv a_{ik} \bmod U\}| = u \quad \text{for all indices } i,k. \qquad (9.1)$$

We then define a matrix A/U over N/U by taking as the i'th row of A/U the set $\{a_{im} + U : m = 1,\dots,t\}$ (for $i = 0,\dots,r$). Because of (9.1) each row of A/U contains exactly t/u cosets in N/U. We claim that A/U is actually a K-matrix. Thus let $h,i \in \{0,\dots,r\}$ with $h \neq i$, and let $x + U$ be any element of N/U. Since A is a K-matrix, there are indices j,k with $x = a_{hk} - a_{ij}$; then $x + U$ arises as the difference $x + U = (a_{hk} + U) - (a_{ij} + U)$ from rows h and i of A/U. For reasons of cardinality, the difference representation of $x + U$ is unique, proving our claim. Obviously, A/U will be a CK-matrix provided A is cohesive. Next assume that A is even an H-matrix; we shall show that then A/U is likewise an H-matrix. Thus choose $\bar{x} \in N/U$ with $\bar{x} \neq 0$. There are u^2 elements $x \in U$ with $x + U = \bar{x}$, each of which has at least two difference representations of the form $x = a_{ij} - a_{ik}$. Each of these representations induces a difference representation of $\bar{x}$ from A/U. Because of (9.1), at most u^2 of these representations may coincide; i.e., $\bar{x}$ has indeed at least 2 difference representations from A/U. Thus we have proved

LEMMA 9.1. Let A be a (t,r)-K-matrix over N, and let U be a K-subgroup of order u^2 for A. Then A/U is a $(t/u,r)$-K-matrix. If A is in fact a CK- or H-matrix, then A/U is a matrix of the same type.

We now consider the corresponding regular K-structures. Thus let

(ϕ,Π,Π') be the regular (t,r)-K-structure corresponding to the generalized t-difference set D constructed from A and the generalized difference set $D' = \{d_o,\ldots,d_r\}$ as in 7.3 and 7.2. By 7.3 and 9.1, we may also construct a generalized (t/u)-difference set $D'' = \{(d_i,a_{ij} + U) : i = 0,\ldots,r ; j = 1,\ldots,t\}$ from D' and A/U, where U is a K-subgroup of order u^2 for A. By 7.2, D" gives rise to a regular K-structure (Θ,Π'',Π'). By definition, the mapping $\psi : \Pi \to \Pi''$ given by

$$(x,y)^{\psi} = (x,y+U) \quad \text{and} \quad (D+(x,y))^{\psi} = D''+(x,y+U) \qquad (9.2)$$

is a K-eumorphism. Thus we have

COROLLARY 9.2. Let (ϕ,Π,Π') be a regular (t,r)-K-structure with K-matrix A over N. Then any K-subgroup of N for A induces a eumorphism ψ onto a K-structure (Θ,Π'',Π'). If Π is neighbor cohesive or an H-structure, then so is Π''.

In order to prove the converse of 9.2, we now assume that (ϕ,Π,Π') is in fact neighbor cohesive. Thus let ψ be a K-eumorphism from Π to a K-structure (Θ,Π'',Π'). We first show that ψ is compatible with isomorphisms of (ϕ,Π,Π'), i.e., that

$$p^{\psi} = q^{\psi} \Longrightarrow (p^{\alpha})^{\psi} = (q^{\alpha})^{\psi}, \text{ and dually} \qquad (9.3)$$

(where an isomorphism α of a K-structure is an incidence structure isomorphism respecting the neighbor relation). We may assume that Π'' occurs in a K-solution of (ϕ,Π,Π') as described in (1.5), say $\Pi'' = \Pi_i$ and $\psi = \psi_{n-1}\ldots\psi_i$. Clearly $\alpha\psi$ is also a K-eumorphism. Since Π is neighbor cohesive, Theorem 1.1 implies that $\alpha\psi$ is of the form $\psi\beta$ for some isomorphism β. Thus

$p^\psi = q^\psi \Longrightarrow p^{\psi\beta} = q^{\psi\beta} \Longrightarrow p^{\alpha\psi} = q^{\alpha\psi}$, proving (9.3). Let Π be regular w.r.t. $G = Z \oplus N$; then each element of G acts as an automorphism γ of Π. We claim that if $p^\psi = p^{\gamma\psi}$ for some point p (and such a γ), then $q^\psi = q^{\gamma\psi}$ for all points q. Since G is regular on the points of Π, we may choose $\delta \in G$ with $p^\delta = q$. By (9.3), $p^\psi = p^{\gamma\psi}$ implies $p^{\delta\psi} = p^{\gamma\delta\psi} = p^{\delta\gamma\psi}$ (since G is abelian), i.e., $q^\psi = q^{\gamma\psi}$. Coordinatizing Π as in 7.2 we now obtain a subgroup U of N by

$$U = \{\gamma \in N : p^\psi = p^{\gamma\psi} \text{ for all } p\} = \{x \in N : (0,0)^\psi = (0,x)^\psi\}. \quad (9.4)$$

Because of 1.3, the parameter of (Θ,Π'',Π') may be written in the form $t' = t/u$. Then U has order u^2 by Corollary 1.4. It remains to check condition (9.1). Thus consider the point (a_i,a_{ik}) of Π. Because of (9.3), we have $(a_i,a_{ik})^\psi = (a_i,a_{im})^\psi$ if and only if $(0,0)^\psi = (0,a_{im} - a_{ik})^\psi$, i.e., if and only if $a_{im} - a_{ik} \in U$. Again using Corollary 1.4, there are exactly u points (a_i,a_{im}) with $(a_i,a_{ik})^\psi = (a_i,a_{im})^\psi$; thus U satisfies (9.1). Because of (9.4), one now easily sees that (Θ,Π'',Π') may be described by using the K-matrix A/U (together with the generalized difference set D' belonging to Π') and that ψ may be described as in (9.2). We thus have

LEMMA 9.3. Let (ϕ,Π,Π') be a regular neighbor cohesive K-structure. Then any K-eumorphism ψ of Π induces a K-subgroup for the CK-matrix A describing Π and may be represented as in (9.2).

Combining the previous results with those of Section 1, we arrive at the following description of the solution of a regular cohesive K-structure.

THEOREM 9.4 (Jungnickel [36]). Let (ϕ,Π,Π') be a neighbor cohesive regular K-structure, described by the generalized t-difference set

$$D = \{(d_i, a_{ij}) : i = 0,\dots,r \ ; \ j = 1,\dots,t\} \subset Z \oplus N$$

as in 7.2 (so $A = (a_{ij})$ is the corresponding (t,r)-CK-matrix). Then the K-subgroups of N for A form a chain

$$E = U_n \lneqq U_{n-1} \lneqq \dots \lneqq U_2 \lneqq U_1 = N \tag{9.5}$$

and the essentially unique K-solution

$$\Pi = \Pi_n \overset{\psi_{n-1}}{\to} \Pi_{n-1} \overset{\psi_{n-2}}{\to} \dots \overset{\psi_2}{\to} \Pi_2 \overset{\psi_1}{\to} \Pi_1 = \Pi' \tag{9.6}$$

may be described as follows. Π_i is the development of

$$D_i = \{(d_h, a_{hk} + U_i) : h = 0,\dots,r \ ; \ k = 1,\dots,t\},$$

and the K-eumorphism $\psi_i : \Pi_{i+1} \to \Pi_i$ is given by

$$(x,y+U_{i+1}) \mapsto (x,y+U_i) \text{ and } D_{i+1}+(x,y+U_{i+1}) \mapsto D_i+(x,y+U_i). \tag{9.7}$$

All Π_i are neighbor cohesive regular K-structures (with corresponding CK-matrices $A_i = A/U_i$). Let $(r,q_2,\dots,q_n)$ be the extended step parameter sequence of (ϕ,Π,Π'). Then Π_i has parameter $t_i=q_2\dots q_i$, and U_i has order u_i^2, where $u_i = q_{i+1}\dots q_n$.

Using 9.2 once again, one obtains

COROLLARY 9.5. Let (ϕ,Π,Π') be a regular H-structure with K-solution as described in 9.4. Then this solution is already the H-solution of (ϕ,Π,Π'), and each Π_i is a regular H-structure.

We call the chain (9.5) the _chain of K-subgroups_ for A ; moreover, n is called the _type_ of A, and the q_i $(i = 2,\dots,n)$ are the _step parameters_ of A. For example, we shall see in the next section that the H-matrices constructed in Corollary 8.5 are of type n with step parameters $q_2,\dots q_n$. By Theorem 4.5, the H-matrices belonging to pappian H-planes with parameters (q^n,q) (see 8.3) have type $n+1$ and step parameters $(q,\ldots,q)$.

10. BALANCED H-MATRICES

We now want to determine the average number of occurences of an element $x \neq 0$ of N as a difference from a CK-matrix A of type n, where x runs over all elements in $U_i \setminus U_{i+1}$ (for $i = 1,\ldots,n-1$) . We then study matrices where these average values are always the exact values. These preliminary considerations allow us to characterize the extended step parameter sequences of two classes of regular PH-planes in Theorem 10.10 and Corollary 10.15.

LEMMA 10.1. Let A be a (t,r)-CK-matrix of type n with chain of K-subgroups (9.5) and step parameters $q_2,\ldots,q_n$. Then the average number of representations of $x \in U_i \setminus U_{i+1}$ as a difference $x = a_{hj} - a_{hk}$ $(h = 1,\ldots,n-1)$ is given by

$$\lambda_i = (r+1)q_2\ldots q_{i+1} / (1+q_{i+1}) \; ; \tag{10.1}$$

in particular,

$$\lambda_1 < \lambda_2 < \ldots < \lambda_{n-1} \; . \tag{10.2}$$

<u>Proof</u>. We count all difference representations of the type described in two ways. Since there are $u_i^2 - u_{i+1}^2$ choices for x , we obtain $\lambda_i(u_i^2 - u_{i+1}^2)$ such representations. On the other hand, given any of the $t(r+1)$ entries a_{hj} of A , (9.1) implies that there are exactly $u_i - u_{i+1}$ entries a_{hk} with $a_{hj} - a_{hk} \in U_i \setminus U_{i+1}$. Thus $\lambda_i(u_i + u_{i+1}) = t(r+1)$, which yields (10.1), as $u_j = q_{j+1}\ldots q_n$. Then (10.2) is an easy consequence. □

Note that λ_i is also the average joining number of exactly i-related points (see the definitions preceding Theorem 5.8) in a corresponding regular (t,r)-CK-structure. This fact is a consequence of the following simple lemma.

LEMMA 10.2. Let (ϕ,Π,Π') be a regular K-structure with K-matrix A. Then the number of lines joining the distinct neighbor points (x,y) and (x,y') equals the number of occurences of $y - y'$ as a difference of the form $a_{ij} - a_{ik}$.

We now call a (t,r)-CK-matrix A (with step parameters and K-subgroups as above) <u>balanced of type</u> n, if every $x \in U_i \setminus U_{i+1}$ has exactly λ_i difference representations from A. Recall that $q_2 \geq r$ by Corollary 2.3; then (10.1) implies $\lambda_1 \geq r$, and by (10.2) we obtain

COROLLARY 10.3. A balanced CK-matrix is an H-matrix.

Recall that an H-structure of type n is called <u>balanced</u> if the joining number of exactly i-related points is a constant λ_i. Using 10.1, 10.2 and Theorem 9.4, one easily obtains the following result.

THEOREM 10.4. Let (ϕ,Π,Π') be a regular (t,r)-H-structure of type n, with step parameters $q_2,\dots,q_n$ and with H-matrix A. Then (ϕ,Π,Π') is balanced if and only if A is balanced. In this case, the joining numbers λ_i are given by (10.1), and they satisfy (10.2).

This was proved for balanced regular PH-planes in [39] and later generalized to the present version in [20]. Now let A be a balanced H-matrix of type n, as before, and put $A_i := A/U_i$. Consider the chain

$$E = V_i \lneqq V_{i-1} \lneqq \cdots \lneqq V_1 = N/U_i \qquad (10.3)$$

of subgroups of N/U_i , where we define $V_j := U_j/U_i$ for $j = 1,\ldots,i$. It is not difficult to see that the V_j are K-subgroups of N/U_i for A_i and that A_i is balanced of type i with step parameters $q_2,\ldots,q_i$. We thus obtain the following

LEMMA 10.5. Let A be a balanced H-matrix of type n . Then each matrix $A_i = A/U_i$ is a balanced H-matrix of type i . Hence each K-structure occurring in the solution of a regular balanced H-structure is in fact a regular balanced H-structure.

LEMMA 10.6. Let A be a balanced H-matrix of type n . Then $q_2 = \lambda_1 = r$; in particular, r divides t .

Proof. By Lemma 10.5, it suffices to prove the assertion for $n = 2$. Then (with $\lambda = \lambda_1$) we have $(r+1)t = \lambda(t+1)$ from (10.1). As t and $t+1$ are coprime, $t+1$ divides $r+1$. Since $t \gtrsim r$, the assertion follows. □

Note that the uniform PH-planes are precisely the balanced PH-planes of type 2. Regarding regularity, we first prove

LEMMA 10.7. Any (r,r)-K-matrix is a balanced H-matrix of type 2. Such matrices exist if and only if r is a prime power and $N = EA(r^2)$.

Proof. If r is a prime power, then an (r,r)-H-matrix exists by 8.3. Thus let $A = (a_{ij})$ be an (r,r)-K-matrix, and let $x \neq 0$ be an element of N . Then all difference representations of x from A belong to the same row of A (for $x = a_{ij} - a_{ik} = a_{hm} - a_{hn}$ with $h \neq i$ would imply $a_{ij} - a_{hm} = a_{ik} - a_{hn}$, and thus $j = k$ by definition of a K-matrix). But each row of A gives rise to at least $r-1$ differences $x \neq 0$, and there are $r+1$ rows and only r^2-1 elements $x \neq 0$. Thus each row of A gives rise to exactly $r-1$ distinct differences. Let N_i consist of the elements of row i of A, and assume, w.l.o.g., that $a_{i1} = 0$. Then the differences arising from row i are precisely $a_{i2},\ldots,a_{im}$, and thus the difference $a_{ij} - a_{ik}$ of any two elements of N_i is again in N_i. Hence N_i is a subgroup of N. The K-property of A implies that $N_i + N_j = N_i - N_j = N$, whenever $i \neq j$. For reasons of cardinality, N is the union of the N_i. Hence $N_o,\ldots, N_r$ is a congruence partition of N (in the sense of André [1]); thus r is a prime power and N is elementary abelian. Clearly A is balanced (of type 2). ◻

COROLLARY 10.8. Let A be a balanced (t,r)-H-matrix. Then r is a prime power, and r divides t.

This is a direct consequence of 10.6 and 10.7. We next describe the balanced H-matrices for which t is a power of r.

LEMMA 10.9. A balanced (r^k,r)-H-matrix A of type n exists if and only if r is a prime power and $k = n - 1$. Then A belongs to an n-uniform regular PH-plane.

Proof. Let A be a balanced (r^k,r)-H-matrix of type n. By 10.8, r and (thus) $t = r^k$ are prime powers. By Theorem 10.4, A gives rise to a balanced (t,r)-PH-plane Π of type n. Then Π is n-uniform and thus

$k = n - 1$ by Satz 1.19 of Drake and Törner [25]. Conversely, for any prime power r, Theorem 8.3 implies that there is an (r^{n-1}, r)-H-matrix A belonging to a pappian PH-plane. By Theorem 4.5, this PH-plane is n-uniform. It is well-known that any n-uniform PH-plane is balanced: see [25]. In the special case of a desarguesian PH-plane Π, balance also follows from the obvious fact that $\mathrm{Aut}\,\Pi$ acts transitively on pairs of exactly i-related neighbor points. The assertion follows by another application of 10.4. □

THEOREM 10.10 (Jungnickel [39]). A regular n-uniform (t,r)-PH-plane exists if and only if r is a prime power and $t = r^{n-1}$.

We next show that the construction of Theorem 8.4 preserves balance.

LEMMA 10.11. Assume the existence of a balanced (t,r)-H-matrix A of type n and of a balanced (s,q)-H-matrix B of type m, where $q = t(r+1)-1$. Then there also exists a balanced (st,r)-H-matrix C of type $n + m - 1$.

Proof. Construct C as in the proof of Theorem 8.4. Then C is an (st,r)-H-matrix. We have to show that C is balanced of type $n + m - 1$. Let (9.5) be the chain of K-subgroups for A, and let λ_i, q_i, u_i have their usual meaning. Similarly, let V_j $(j = 1,\dots,m)$ be the K-subgroups for B (B being defined over the group M), with "parameters" λ'_j, a'_j, u'_j. Define a chain of subgroups

$$E = W_{m+n-1} \lneqq W_{m+n-2} \lneqq \dots \lneqq W_2 \lneqq W_1 = K \qquad (10.4)$$

of $K := N \oplus M$ by putting $W_i = U_i \oplus M$ for $i = 1,\dots,n$ and

$W_i = V_{i-n+1}$ for $i = n,\ldots,n+m-1$. Moreover, write $p_i = q_i$ for $i = 1,\ldots,n$ and $p_i = q'_{i-n+1}$ for $i = n+1,\ldots,n+m-1$. Then W_i has order w_i^2 with $w_i = p_{i+1}\ldots p_{n+m-1}$; it is easily checked that one also has $|\{c_{jm} : c_{jh} \equiv c_{jk} \bmod W_i\}| = w_i$ for all i,j,k . Writing $\mu_i = \lambda_i$ for $i = 1,\ldots,n-1$ and $\mu_i = \lambda'_{i-n+1}$ for $i = n,\ldots,m+n-2$, one may check that each $x \in W_i \setminus W_{i+1}$ occurs precisely μ_i times as a difference from A (the proof is similar to that of 8.4). It is also not difficult to see that C has type $n+m-1$; in fact, this is a special case of the more general result 11.1 below. Thus C is balanced of type $n+m-1$. □

Recursive application of 10.11 using the H-matrices of 10.9 yields the following

COROLLARY 10.12. Let $(r = q_2,\ldots,q_n)$ be a special Lenz sequence, and put $t = q_2\ldots q_n$ (as in 8.5). Then there exists a balanced (t,r)-H-matrix and hence a balanced regular (t,r)-H-structure of type n with step parameters $q_2,\ldots,q_n$.

We remark that a balanced H-structure constructed over a symmetric design (or, more generally, a symmetric divisible partial design) is a symmetric divisible partial design, which yields a method of constructing many examples of this interesting type of partial design (even with regular automorphism groups in the case of regular H-structures). We refer the reader to Drake and Jungnickel [19],[20] and, for a related construction, to Jungnickel [43]. We also remark that n-uniform PH-planes may be characterized in terms of divisible partial designs: see Limaye and Sane [52] and, for $n = 2$, Jungnickel [42]. It is an interesting open problem whether there exist balanced H-matrices for sequences other than special Lenz sequences. To obtain a complete characterization of the possible parameter sequences, one therefore adds a certain technical condition; a balanced H-matrix satisfying this condition is called "uniformly balanced". Geometrically, such H-matrices correspond to minimally uniform balanced PH-planes (recall the definition preceding Theorem 5.8). Since the details are of a more technical nature, we will not present them here; we refer the reader to [39] and [20]. We just mention the main results. One calls (t,r) a <u>(special) Lenz pair</u> if there exists a (special) Lenz sequence $(q_2,\dots,q_n)$ with $r = q_2$ and $t = q_2 \dots q_n$.

THEOREM 10.13 (Jungnickel [39]). A uniformly balanced (t,r)-H-matrix exists if and only if (t,r) is a special Lenz pair. Any such matrix is constructed from H-matrices belonging to m-uniform regular PH-planes (for various m) as in the proof of Lemma 10.11.

COROLLARY 10.14 (Drake and Jungnickel [20]). Let Π' be a symmetric regular incidence structure with block size $r + 1$. Then there exists a regular, balanced, minimally uniform H-structure (ϕ,Π,Π') with parameter t over Π' if and only if (t,r) is a special Lenz pair.

COROLLARY 10.15 (Jungnickel [39]). The parameter spectrum of regular, balanced, minimally uniform PH-planes is the set of all special Lenz pairs.

11. RECURSIVE CONSTRUCTIONS

In this section we consider further recursive constructions for H-matrices; the results to be presented are taken from Jungnickel [35], [37]. We begin by analyzing the proof of Theorem 8.4. We want to generalize the construction given there by dropping the requirement that all the B_{ij} are distinct rows of B. It is clear that C will still be a K-matrix providing only that B_{ij} and B_{hk} are distinct rows of a K-matrix B, whenever $h \neq i$. If B is an H-matrix and if we use all rows of B, then we will also obtain each difference of the type $(0,y)$ from C at least twice. The only difficulty arises in guaranteeing that each difference of the type (x,y) with $x \neq 0$ occurs at least twice. This will hold provided that, given any $x \neq 0$, there are at least two difference representations $x = a_{ij} - a_{ik}$ for which B_{ij} and B_{ik} are distinct. This leads to the concept of a <u>z-partition</u> for A; i.e., a collection of partitions $\underline{P}_i = \{P_{i1},\dots,P_{iz}\}$ of the rows of A into z subsets, satisfying the following condition:

> each nonzero $x \in N$ occurs at least twice as a difference of the type $x = a_{ij} - a_{ik}$ where a_{ij} and a_{ik} are in distinct members of some $\underline{P}_i$. (11.1)

(One could also consider variable sizes for the $\underline{P}_i$, but there are no applications of this generalization up to now.) Assume now that A is a (t,r)-H-matrix with a z-partition, and let B be an (s,q)-H-matrix satisfying $z(r+1) \leq q+1 \leq t(r+1)$. Then we may construct an (st,r)-H-matrix C by replacing each a_{ij} by $a_{ij} \times B_{ij}$, where the B_{ij} are rows of B chosen subject to the following two requirements:

(i) all rows of B are actually used;

(ii) $B_{ij} \neq B_{hk}$ whenever either $h \neq i$ or $h = i$ and a_{ij} and a_{ik} are in distinct members of $\underline{P}_i$.

The preceding discussion shows that C is indeed an H-matrix. We shall require the following more precise version of this result.

THEOREM 11.1 (Jungnickel [35]). Assume the existence of a (t,r)-H-matrix A of type n with an x-partition and of an (s,q)-H-matrix B of type m with a y-partition. Let $q_2,\dots,q_n$ and $p_2,\dots,p_m$ be the step parameters of A and B, respectively, and assume moreover that

$$x(r+1) \leq q+1 \leq t(r+1) . \qquad (11.2)$$

Then there also exists an (st,r)-H-matrix C of type $n+m-1$ with an xy-partition and with step parameter sequence $(q_2,\dots,q_n,p_2,\dots,p_m)$.

Proof. We have already seen how to construct C. It remains to show that C admits an xy-partition and has step parameters as asserted. Let $\underline{P}_i = \{P_{i1},\dots,P_{ix}\}$ be the partition of the i'th row of A; similarly, let $\underline{Q}_{ij} = \{Q_{ij1},\dots,Q_{ijy}\}$ be the partition of row B_{ij} of B. Then define a partition $\underline{R}_i$ of row i of C by putting

$$\underline{R}_i = \{R_{ijk} : j = 1,\dots,x \ ; \ k = 1,\dots,y\} ,$$

where R_{ijk} is the union of all sets $a_{im} \times Q_{imk}$ with $a_{im} \in P_{ij}$. The reader may check that this defines an xy-partition of C. It remains to consider the chain of K-subgroups for C. Let (9.5) be the chain of K-subgroups for A (with "parameters" q_i, u_i), and let V_j be the K-subgroups for B (as in the proof of Lemma 10.11). We construct a chain (10.4) as in this proof and show, in exactly the same way, that each W_i is a K-subgroup of order w_i^2 for C. It remains to show that the chain (10.4) may not be refined by inserting a further K-subgroup. Thus assume $W_{i+1} \lneqq W \lneqq W_i$ where W is a K-subgroup for C. First let $i \geq n$. This means $V_{i-n+2} \lneqq W \lneqq V_{i-n+1}$; clearly W would be a K-subgroup for B, contradicting our choice of the V_j. Next let $i < n$; then $U_{i+1} \oplus M \lneqq W \lneqq U_i \oplus M$, and thus $W = U \oplus M$ for some subgroup U of N; then U would be a K-subgroup for A, contradicting our choice of the U_i. This proves the assertion. □

To apply Theorem 11.1 we now need examples of H-matrices with partitions. Clearly, each (t,r)-H-matrix has a t-partition; thus Theorem 8.4 is indeed a special case of Theorem 11.1. Further simple examples are obtained from uniform H-matrices. Thus let A be an (r,r)-H-matrix over N. We have seen in the proof of Lemma 10.7 that each row of A is a subgroup N_i of N. Thus it is obvious that one obtains a 2-partition for A by splitting each N_i into a singleton and its complement.

LEMMA 11.2. Each (r,r)-H-matrix has a 2-partition. Such matrices exist if and only if r is a prime power.

COROLLARY 11.3. Let r be a prime power, and assume the existence of an (s,q)-H-matrix with a y-partition where $2(r+1) \leq q+1 \leq r(r+1)$. Then there exists an (sr,r)-H-matrix with a 2y-partition.

By 8.3, we may always choose q to be another prime power and $s = q^n$ for some positive integer n. This yields the following result.

COROLLARY 11.4 (Drake and Lenz [22]). Let q and r be prime powers satisfying $2(r+1) \leq q+1 \leq r(r+1)$, and let n be a positive integer. Then there exists an (rq^n,r)-PH-plane.

Comparing the original proof with the one using H-matrices, one obtains a good idea of how much the difference technique simplifies all the arguments required. Before constructing further examples of H-matrices with partitions, we show how we may use the previous results to produce examples of H-planes with non-increasing step parameter sequence. For instance, there exists a (125,5)-H-matrix A with a 4-partition by Corollary 11.3 (which is constructed from (5,5)- and (25,25)-H-matrices). A has step parameters (5,25). Using A and a $(23^m,23)$-H-matrix of type $m+1$, Theorem 11.1 yields the existence of an H-matrix with step parameter sequence (5,25,23,...,23). This prodecure is easily generalized to obtain

THEOREM 11.5 (Jungnickel [36]). Let r be a prime power ≥ 5 . Then there exists a $(q^m r^3,r)$-H-matrix (and hence a $(q^m r^3,r)$-PH-plane) of type m+3 with step parameter sequence $(r,r^2,q,\ldots,q)$, $q < r^2$.

From 11.5, it is clear that there exists a Hjelmslev plane with non-increasing step parameter sequence for every type $n \geq 4$ (for every prime power $r \geq 5$). Examples of type 2 cannot exist, since $q_2 \geq r$; no example of type 3 is known. There are also examples of type 4 with $r = 2,3,4$ (see Sane [59], [61] and Drake and Sane [23]); the constructed sequences include (2,5,2), (3,11,3) and (4,19,17). These constructions are, however, very involved. If we settle for types ≥ 6, we can give simple examples as before. For instance, let $r = 2$ and $n = 6$. First construct a (10,2)-H-matrix with a 4-partition and step parameters (2,5). Using this matrix together with a matrix of type 4 with step parameters $(11,11^2,q)$ as in 11.5 (e.g., we may take $q = 47$), we obtain an H-matrix of type 6 with step parameter sequence $(2,5,11,11^2,q)$. Similar constructions yield

COROLLARY 11.6. Let r be any prime power and n be any integer ≥ 6. Then there exists a regular (t,r)-PH-plane of type n with non-increasing step parameter sequence. For $r \geq 5$, this also holds for $n = 4$ and $n = 5$.

We now construct partitions of the H-matrices in 8.3.

LEMMA 11.7. Let A be a (q^{n-1},q)-H-matrix of type n (which exists for every prime power q and every $n \geq 2$). Then A has a $(q^{n-2}+1)$-partition.

Proof. Let A_i be the i'th row of A and assume, w.l.o.g., that $a_{i1} = 0$. By the results of Section 9 (since A has step parameters $q,\dots,q$), the smallest non-trivial K-subgroup $U := U_{n-1}$ for A has order q. Let A_i' consist of the q elements $\equiv a_{i1} \bmod U$ (by (9.1)), i.e., $A_i' \subset U$. Let A' be the $(q+1) \times q$-matrix with rows A_i'. Since A' is a sub-matrix of A, differences from distinct rows of A' are pairwise distinct. Thus A' is a (q,q)-K-matrix over U; by 10.7 and 11.2, A' is an H-matrix with a 2-partition $\underline{P}_i'$ $(i=0,\dots,q)$. For each i, partition row i of A into the two members of $\underline{P}_i'$ and into the $q^{n-2}-1$ distinct sets $\{a_{im} : a_{im} \equiv a_{ik} \bmod U\}$ with $a_{ik} \notin U$: see (9.1). Call this partition $\underline{P}_i$.

Now consider any difference representation $x = a_{ij} - a_{ik}$. For $x \in U$, there are at least two such representations with a_{ij} and a_{ik} in distinct members of $\underline{P}_i' \subset \underline{P}_i$. For $x \notin U$, we have $a_{ij} \not\equiv a_{ik} \bmod U$; then a_{ij} and a_{ik} are in distinct members of $\underline{P}_i$ by definition. Thus $\underline{P}_o, \ldots, \underline{P}_q$ is the desired $(q^{n-2}+1)$-partition of A . □

One may now use this result in Theorem 11.1 to obtain analogues of 11.3. Still further new invariant pairs may be obtained if one first uses 11.1 and 11.2 to obtain a recursive construction for (q^x,q)-H-matrices with smaller partitions. For example, let r be a prime power $\neq 2$. Then $2(r+1) \leq r^2 + 1 \leq r^2(r+1)$, and thus we may use (r,r)- and (r^2,r^2)-H-matrices with 2-partitions to obtain an (r^3,r)-H-matrix (of type 3) with a 4-partition; observe that 11.7 only guarantees the existence of an (r^3,r)-H-matrix with an (r^2+1)-partition. In a next step, we use such an (r^3,r)-H-matrix with a 4-partition and a (q,q)-H-matrix with a 2-partition to obtain an (r^3q,r)-H-matrix with an 8-partition. For $r \geq 5$, we can take $q = r^2, r^3, r^4$; for $r = 3$ and 4 we can still take $q = r^3$ and r^4 , since the required inequality (11.2) is

$$4(r+1) \leq q+1 \leq r^3(r+1) .$$

This yields (r^x,r)-H-matrices with 8-partitions with $x = 5,6,7$ for most values of r . Continuing in this way and using more detailed considerations (in particular, using 11.7) for some special values (e.g. $r < 5$, $x = 2$ or 4), one obtains the following result which slightly strengthens a result of [35].

THEOREM 11.8 (Jungnickel [37]) . Let r be a prime power. Then there exists an (r^x,r)-H-matrix of type $g(r,x)$ with an $f(r,x)$-partition, where the functions f and g are defined as follows. For any positive integer x, define $n = n(x)$ by the requirement $2^{n-1} \leq x < 2^n$. Then put

$$f(2,1)=2,\ f(2,6)=33,\ f(2,8)=27 \text{ and } f(2,x)=5\cdot 2^{n-2}$$

$$\text{for } x \neq 1,6,8; \tag{11.3}$$

$$g(2,6)=7 \text{ and } g(2,x)=n+2 \quad \text{for} \quad x \neq 6\ ; \tag{11.4}$$

for $r \neq 2$ put

$$f(r,2)=r+1,\ f(r,4)=r^3+1,\ f(4,5)=10,\ f(4,8)=20,\ f(3,8)=112$$

$$\text{and} \quad f(r,x)=2^n \quad \text{otherwise}; \tag{11.5}$$

$$g(r,4)=5 \text{ and } g(r,x)=n+1 \text{ otherwise.} \tag{11.6}$$

COROLLARY 11.9. Let q and r be prime powers, x and w be positive integers satisfying

$$f(r,x)(r+1) \leq q+1 \leq r^x(r+1)\ . \tag{11.7}$$

Moreover assume the existence of an (s,q)-H-matrix of type m with a z-partition (e.g., take $s=q^w$, $m=g(q,w)$, $z=f(q,w)$). Then there also exists an (sr^x,r)-H-matrix of type $g(r,x)+m-1$ with a $z \cdot f(r,x)$-partition.

Clearly, this recursive procedure may be iterated. The results presented in this section yield the existence of PH-planes for almost all the invariant pairs (t,r) constructed in [22] and [16] by using auxiliary matrices. These results also yield many pairs not previously constructed; e.g., the lower bound in (11.7) is asymptotically $\log_2 x$, whereas that of [16] is asymptotically $r^{x/2}$. However, there are invariant pairs constructed by more elaborate methods which have not yet succumbed to the H-matrix attack: this holds for most of the pairs constructed by Sane [58],[59],[60],[61] and by Drake and Sane [23].

We finally remark that recursive techniques similar to the ones presented here may also be used to construct proper CK-matrices (ones that are not H-matrices) and neighbor cohesive PK-planes. We conclude by mentioning the following result (cf. 4.4).

THEOREM 11.10 (Jungnickel [40]). Let q be a prime power and let n be a positive integer $\neq 1,2,3,4,6$. Then there exists a proper (q^n,q)-CK-matrix.

12. OPEN PROBLEMS

We conclude this paper by listing some open questions on K-structures that we think important. Before doing so, let us mention one further result. Let Π be an incidence structure and assume that (ϕ,Π,Π') is a PK-plane for some choice of ϕ and Π'. Then Π alone determines Π' and ϕ (up to isomorphism). This is, of course, trivial for PH-planes, since then two points x,y are neighbor if and only if $[x,y] \geq 2$, and dually. The beautiful general case is due to Sane and Singhi [62]. We do not think that this result may be generalized to arbitrary finite K-structures though no explicit counter-example seems to be known.

PROBLEM 1. Are the K-solution and the H-solution of an H-structure necessarily identical? (The answer to this question is positive in the case of regular H-structures: see Theorem 9.4.)

PROBLEM 2. Let Π be an H-structure with extended step parameter sequence $(q_1,\ldots,q_n)$. Are the q_i necessarily orders of projective planes? (There is a positive answer in the special case of balanced minimally uniform H-structures: see Theorem 4.8 and Corollary 10.14. It would also be interesting to find further classes of H-structures for which all q_i have to be orders of planes.)

PROBLEM 3. Find further classes of H-structures for which the step parameter sequences can be characterized.

PROBLEM 4. Is it true that q_1 always divides q_2 in the extended step parameter sequence of an H-structure? (One of the smallest conceivable counter-examples would have $q_1 = 5$ and $q_2 = 8$: see Drake and Jungnickel [21].)

PROBLEM 5. The construction of the only known example of an extended step parameter sequence with $q_1 \neq q_2$ is presented in Section 6. Find further examples.

PROBLEM 6. Find further constructions and further non-existence results. (As an illustration, let us mention that the existence of (t,2)-PH-planes has been proved for just 75 values of $t \leq 1000$, cf. Drake and Sane [23]; and, on the other hand, only three cases are known to be impossible, cf. Proposition 3.8.)

PROBLEM 7. Find H-matrices with better partition sizes to strengthen the results of Section 11.

PROBLEM 8. Find further difference methods, i.e., further H-matrices.

Can the step parameter sequences of Sane [58], [59], [60], [61] and of Drake and Sane [23] be obtained in this way? In particular, is there an H-matrix with step parameter sequence (2,5,2)? Up to now, no example of an extended step parameter sequence for an H-structure is known for which the non-existence of a corresponding H-matrix has been proved. Find such an example. (Possible candidates include (2,2,5,2), (5,8) and (2,8); cf. Drake and Jungnickel [21].)

PROBLEM 9. Is a balanced H-structure necessarily minimally uniform? If not, characterize the extended step parameter sequences of balanced H-structures.

PROBLEM 10. Artmann [2] has proved that any projective plane Π' occurs in an infinite sequence

$$\ldots \overset{\psi_{n+1}}{\to} \Pi_{n+1} \overset{\psi_n}{\to} \Pi_n \overset{\psi_{n-1}}{\to} \ldots \overset{\psi_2}{\to} \Pi_2 \overset{\psi_1}{\to} \Pi_1 = \Pi'$$

of PH-planes, where each ψ_i is a eumorphism. Artmann has also proved that the inverse limit of the Π_i exists and is a projective plane. Is any desarguesian PH-plane contained in a sequence of desarguesian PH-planes for which an inverse limit exists? For example, the chain of H-rings

$$\ldots \to \mathbb{Z}_{p^{n+1}} \to \mathbb{Z}_{p^n} \to \ldots \to \mathbb{Z}_p$$

gives rise to such a sequence; here the inverse limit is the projective plane over the field $\mathbb{Q}_p$ of p-adic numbers. Similarly,

$$\ldots \to GF(q)[x]/(x^{n+1}) \to GF(q)[x]/(x^n) \to \ldots \to GF(q)$$

yields an example with inverse limit defined over the field of formal Laurent series over $GF(q)$.

PROBLEM 11. Find a more explicit description of all finite H-rings which allows one to settle the isomorphism problem for such rings. (A solution for the uniform case was given by Cronheim [13]: see Theorem 4.9.)

PROBLEM 12. Is every desarguesian PK-plane regular? (The answer is positive if the underlying local ring is commutative or uniform: see Theorems 8.1 and 8.2.)

PROBLEM 13. Is every regular n-uniform PH-plane necessarily desarguesian? (The converse holds by Theorems 4.5 and 8.1.)

The reader may also wish to consult Chapter 8.7.b of the new edition of Dembowski's "Finite Geometries" which is presently being prepared by Professor J.C.D.S. Yacub for further results on H-planes (in particular, for the case of affine H-planes). Many more papers dealing with H-planes and related areas are mentioned in the bibliographies by Artmann, Dorn, Drake and Törner [3] and by Jungnickel [44].

Oh sure I am the wits of former daies,
To subjects worse have given admiring praise.

(Skakespeare, sonnet 59)

BIBLIOGRAPHY

1. J. André: Über nicht-Desarguessche Ebenen mit transitiver Translationsgruppe. Math. Z. 60 (1954), 156 - 186.

2. B. Artmann: Existenz und projektive Limiten von Hjelmslev-Ebenen n-ter Stufe. In: Atti Conv. Geom. Comb. Perugia (1971), 27 - 41

3. B. Artmann, G. Dorn, D.A. Drake and G. Törner: Hjelmslev'sche Inzidenzgeometrie und verwandte Gebiete - Literaturverzeichnis. J. Geom. 7 (1976), 175 - 191.

4. P.Y. Bacon: An introduction to Klingenberg planes. I (1976), II and III (1979). Published by the author. Gainesville, Fla.

5. P.Y. Bacon: Desarguesian Klingenberg planes. Trans. Amer. Math. Soc. 241 (1978), 343 - 355.

6. R. Baer: A unified theory of projective spaces and finite abelian groups. Trans. Amer. Math. Soc. 52 (1942), 283 - 343.

7. T. Beth, D. Jungnickel und H. Lenz: Design Theory. Bibliographisches Institut, Mannheim/Wien/Zürich (1985).

8. D. Betten: Zum Satz von Euler-Tarry. Math. Nat. Unt. 36 (1983), 449 - 453.

9. N. Bourbaki: Eléménts de mathématique. Fasc. XXVII, Algèbre commutative. Hermann, Paris (1961).

10. R.H. Bruck: Finite nets II. Uniqueness and imbedding. Pacific J. Math. 13 (1963), 421 - 457.

11. R.H. Bruck and H.J. Ryser: The nonexistence of certain finite projective planes. Canadian J. Math. 1 (1949), 88 - 93.

12. W.E. Clark and D.A. Drake: Finite chain rings. Abh. Math. Sem. Hamburg 39 (1973), 147 - 153.

13. A. Cronheim: Dual numbers, Witt vectors and Hjelmslev planes. Geom. Ded. 7 (1978), 287 - 302.

14. D.A. Drake: On n-uniform Hjelmslev planes. J. Comb. Th. 9 (1970), 267 - 288.

15. D.A. Drake: Nonexistence results for finite Hjelmslev planes. Abh. Math. Sem. Hamburg 40 (1974), 100 - 110.

16. D.A. Drake: More new integer pairs for finite Hjelmslev planes. Ill. J. Math. 19 (1975), 618 - 627.

17. D.A. Drake: Construction of Hjelmslev planes. J. Geom. 10 (1977), 183 - 197.

18. D.A. Drake: The use of auxiliary sets of matrices in the construction of Hjelmslev and Klingenberg structures. Lecture Notes in Pure and Appl. Math. 82 (1983), 129 - 153.

19. D.A. Drake and D. Jungnickel: Klingenberg structures and partial designs I. Congruence relations and solutions. J. Stat. Planning Inf. 1 (1977), 265 - 287.

20. D.A. Drake and D. Jungnickel: Klingenberg structures and partial designs II.Regularity and uniformity. Pacific J. Math. 77 (1978), 389 - 415.

21. D.A. Drake and D. Jungnickel: Das Existenzproblem für projektive (8,5)-Hjelmslevebenen. Abh. Math. Sem. Hamburg 50 (1978), 118 - 126.

22. D.A. Drake and H. Lenz: Finite Klingenberg planes. Abh. Math. Sem. Hamburg 44 (1975), 70 - 83.

23. D.A. Drake and S.S. Sane: Auxiliary sets of matrices with new step parameter sequences. Linear Alg. Appl. 46 (1982), 131 - 153.

24. D.A. Drake and E.E. Shult: Construction of Hjelmslev planes from (t,k)-nets. Geom. Ded. 5 (1976), 377 - 392.

25. D.A. Drake and G. Törner: Die Invarianten einer Klasse von projektiven Hjelmslev-Ebenen. J. Geom. 7 (1976), 157 - 174.

26. M. Dugas: Charakterisierungen endlicher desarguesscher uniformer Hjelmslev-Ebenen. Geom. Ded. 3 (1974), 295 - 324.

27. M. Dugas: Eine Kennzeichnung der endlichen desarguesschen Hjelmslev-Ebenen. Arch. Math. 27 (1976), 556 - 560.

28. M. Dugas: Moufang Hjelmslev-Ebenen. Arch. Math. 28 (1977), 318 - 322.

29. M.P. Hale and D. Jungnickel: A generalization of Singer's theorem. Proc. Amer. Math. Soc. 71 (1978), 280 - 284.

30. M. Hall: Combinatorial theory. Blaisdell, Waltham, Mass. (1967).

31. J.W.P. Hirschfeld: Projective geometries over finite fields II. Oxford University Press (to appear).

32. D.R. Hughes and F.C. Piper: Projective planes. Springer, Berlin/ Heidelberg/New York (1973).

33. B. Huppert: Endliche Gruppen I. Springer, Berlin/Heidelberg/New York (1967).

34. N. Jacobson: Lectures in abstract algebra III. Springer, New York/ Hedelberg/Berlin (3rd ed. 1980).

35. D. Jungnickel: Regular Hjelmslev planes II. Trans. Amer. Math. Soc. 241 (1978), 321 - 330.

36. D. Jungnickel: On the congruence relations of regular Klingenberg structures. J. Comb. Inf. System Sc. 3 (1978), 49 - 57.

37. D. Jungnickel: Reguläre Klingenberg-Strukturen und Hjelmslevebenen. Habilitationsschrift, Freie Universität Berlin (1978).

38. D. Jungnickel: Regular Hjelmslev planes. J. Comb. Th. (A) 26 (1979), 20 - 37.

39. D. Jungnickel: On balanced regular Hjelmslev planes. Geom. Ded. 8 (1979), 445 - 462.

40. D. Jungnickel: Construction of regular proper CK-planes. J. Comb. Inf. System Sc. 4 (1979), 14 - 18.

41. D. Jungnickel: Die Methode der Hilfsmatrizen. In: Contributions to geometry, pp. 388 - 394. Birkhäuser, Basel (1979)

42. D. Jungnickel: On an assertion of Dembowski. J. Geom. 12 (1979), 168 - 174.

43. D. Jungnickel: A recursive construction of divisible partial designs from symmetric transversal designs. J. Comb. Inf. System Sc. 5 (1980), 173 - 183.

44. D. Jungnickel: Hjelmslev'sche Inzidenzgeometrie und verwandte Gebiete. Literaturverzeichnis II. J. Geom. 16 (1981), 138 - 147.

45. D. Jungnickel: Existence results for translation nets. In: Finite geometries and designs. LMS Lecture Notes 49 (1981), 172 - 196. Cambridge University Press.

46. D. Jungnickel: Maximal partial spreads and translation nets of small deficiency. J. Alg. 90 (1984), 119 - 132.

47. E. Kleinfeld: Finite Hjelmslev planes. Ill. J. Math. 3 (1959), 403 - 407.

48. W. Klingenberg: Projektive und affine Ebenen mit Nachbarelementen. Math. Z. 60 (1954), 384 - 406.

49. W. Klingenberg: Desarguessche Ebenen mit Nachbarelementen. Abh.

Math. Sem. Hamburg 20 (1955), 97 - 111.

50. W. Klingenberg: Projektive Geometrien mit Homomorphismus. Math. Ann. 132 (1956), 180 - 200.

51. E.S. Lander: Symmetric designs. An algebraic approach. LMS Lecture Notes 74. Cambridge University Press (1983).

52. B.V. Limaye and S.S. Sane: On partial designs and n-uniform Hjelmslev planes. J. Comb. Inf. System Sc. 3 (1978), 223 - 237.

53. J.W. Lorimer and N.D. Lane: Desarguesian affine Hjelmslev planes. J. Reine Ang. Math. 278/279 (1975), 336 - 352.

54. H. Lüneburg: Affine Hjelmslev-Ebenen mit transitiver Translationsgruppe. Math. Z. 79 (1962), 260 - 288.

55. F. Machala: Über projektive Erweiterungen affiner Klingenbergscher Ebenen. Czech. Math. J. 29 (1979), 116 - 129.

56. D.M. Mesner: Sets of disjoint lines in PG(3,q). Canadian J. Math. 19 (1967), 273 - 280.

57. A. Neumaier: Nichtkommutative Hjelmslev-Ringe. In: Festband für H. Lenz, Preprint No. 9, Freie Universität Berlin (1976), pp. 200 - 213.

58. S.S. Sane: New integer pairs for Hjelmslev planes. Geom. Ded. 10 (1981), 35 - 48.

59. S.S. Sane: Some new invariant pairs (t,3) for projective Hjelmslev planes. J. Geom. 15 (1981), 64 - 73.

60. S.S. Sane: On the theorems of Drake and Lenz. Aequat. Math. 23 (1981), 223 - 232.

61. S.S. Sane: Constructions of some (t,4)-PH-planes of type 4. To appear.

62. S.S. Sane and N.M. Singhi: On the structure of a finite projective Klingenberg plane. To appear.

63. J. Singer: A theorem in finite projective geometry and some applications to number theory. Trans. Amer. Math. Soc. 43 (1938), 377 - 385.

64. L.A. Skornyakov: Rings chain-like from the left (Russian). In: In memoriam N.G. Cebotarev. Izdat. Kazan Univ., Kazan (1964), pp. 75 - 88.

65. A.P. Sprague: Translation nets. Mitt. Math. Sem. Gießen 157 (1982),

46 - 88.

66. D.R. Stinson: A short proof of the non-existence of a pair of orthogonal Latin Squares of order six. J. Comb. Th. (A) 36 (1984), 373 - 376.

67. E. Study: Geometrie der Dynamen. Leipzig (1903).

68. G. Törner: Eine Klassifizierung von Hjelmslev-Ringen und Hjelmslev-Ebenen. Mitt. Math. Sem. Gießen 107 (1974).

69. G. Törner: Über den Stufenaufbau von Hjelmslev-Ebenen. Mitt. Math. Sem. Gießen 126 (1977).

70. G. Görner: (r^{n-1},r)-Hjelmslev-Ebenen des Typs n. Math. Z. 154 (1977), 189 - 201.

D. A. Drake
Dept. of Mathematics
University of Florida
Gainesville, Fla. 32611
USA

D. Jungnickel
Mathematisches Institut
Justus-Liebig-Universität
Gießen
6300 Gießen
F.R. Germany

Part III

Geometries over Alternative Rings

GENERALIZING THE MOUFANG PLANE

John R. Faulkner
University of Virginia
Charlottesville, Virginia

and

Joseph C. Ferrar
Ohio State University
Columbus, Ohio

ABSTRACT: In this article we trace the development of ideas, both geometric and algebraic, which lead from the real projective plane in its concrete, inhomogeneous coordinate realization, to the Moufang-Veldkamp plane realized as a homogeneous space for a group defined from a Jordan pair. Emphasis has been placed on the study of Moufang planes and Moufang-Veldkamp planes, both constructed from reduced exceptional simple Jordan algebras. For these planes we verify explicitly the axioms of a Moufang-Veldkamp plane and, by reworking the standard construction, obtain a method of constructing a Moufang-Veldkamp plane with coordinate ring any alternative ring of stable range 2.

R. Kaya et al. (eds.), Rings and Geometry, 235–288.

§1. Inhomogeneous and homogeneous coordinates.

The projection ϕ of a plane Π_1, in Euclidean 3-space to another plane Π_2 from a point O in neither plane is defined by $\phi(P) = OP \cap \Pi_2$ (Figure 1)

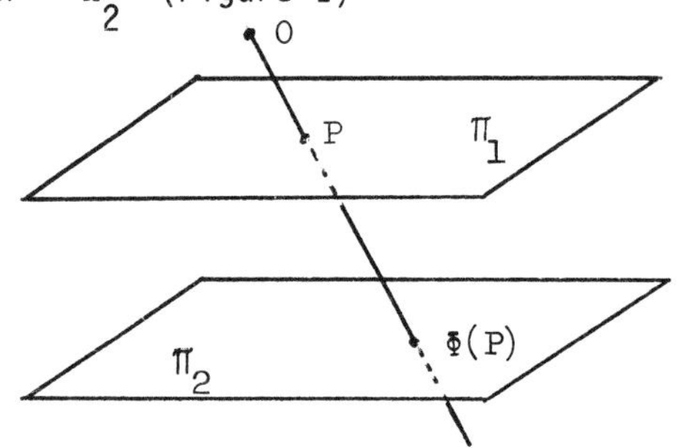

(Figure 1)

Unfortunately, ϕ is not defined for all $P \in \Pi_1$ since OP might be parallel with Π_2. Also, ϕ is not surjective since OQ might be parallel to Π_1 for $Q \in \Pi_2$. These defects can be overcome by extending the planes Π_i to "projective planes" by adding points and lines "at ∞". Specifically, one extends a plane Π to $\mathcal{P}$ by adding a point Ω for each family of parallel lines in Π with Ω lying (by definition) on each member of the family, and a line ∞ containing all points Ω.

If we now take $O\Omega$ to be the line through O parallel to the family it is clear that ϕ extends to a bijection between the extended planes preserving collinearity. Also we can match up points of either $\mathcal{P}_1$ or $\mathcal{P}_2$ with lines through O and lines of $\mathcal{P}_1$ or $\mathcal{P}_2$ with planes through O.

The two views of projective planes, as extended Euclidean planes or as lines and planes through a given point 0, give rise to two methods of coordinatizing the planes, i.e. labelling the points and lines. For inhomogeneous coordinates (suggested by the extended plane), we label the points of $\mathbb{P}$ by the usual Cartesian coordinates (x,y) and let (m) denote the point corresponding to the family of lines of slope m ($m = \infty$ is permissible). For lines, let $[a,b]$, a,b not both 0, denote the line with equation $xa + yb + 1 = 0$; $[m]$ the line through $(0,0)$ of slope m ($m = \infty$ is permissible); and $[0,0]$ the added line at ∞. For homogeneous coordinates (suggested by the second view of the plane) one assumes that 0 is the origin of a standard coordinate system for 3-space. Then the lines through 0 correspond to nonzero vectors (x,y,z) (with $(\lambda x,\lambda y,\lambda z)$, $\lambda \neq 0$, giving the same line) while the plane through 0 with equation $xa + yb + zc = 0$ is given by $(a,b,c)^t$ (up to scalar multiple). Note that $P = (x,y,z)$ lies on $\ell = (a,b,c)^t$ if $P\ell = 0$. If one takes Π to be the plane $z = 1$, it is easy to see that the points and lines with homogeneous coordinates (x,y), (m) , (∞) , $[a,b]$, $[m]$, $[\infty]$, correspond to $(x,y,1)$, $(1,m,0)$, $(0,1,0)$, $(a,b,1)^t$, $(-m,1,0)^t$, and $(1,0,0)^t$ respectively.

Before leaving inhomogeneous coordinates, we note the following simple relation of multiplication to collinearity. In the triangle with sides the x-axis (0) , the y-axis (∞) , and the line at ∞ $[0,0]$, let $(x_0,0)$, $(0,y_0)$ and $(\bar{m})$ be collinear (Figure 2). Since the line through (m) and $(0,y_0)$ has equation $y = xm + y_0$, we see $y_0 = -x_0 m$.

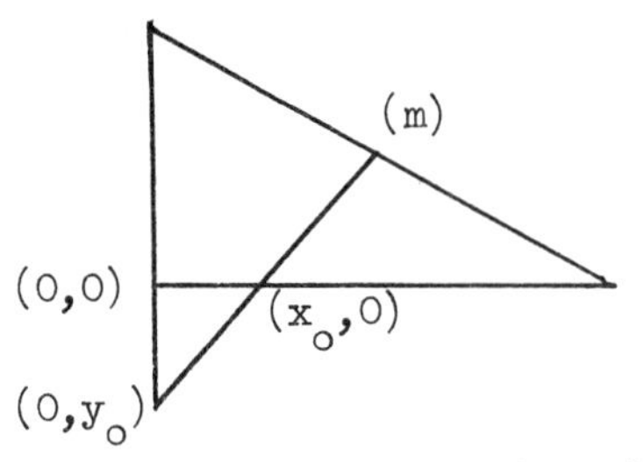

(Figure 2)

§2. Collineations of real projective planes.

The projection ϕ of §1 could be extended to a collineation between the extended (projective) planes $\mathcal{P}_1$ and $\mathcal{P}_2$; i.e. a bijection the points of the two planes preserving collinearity. Equivalently, this is a bijection of points with points and lines with lines preserving incidence. Clearly the set of all collineations of a plane to itself forms a group, which we wish to describe algebraically. Using homogeneous coordinates, we may define ϕ_A , for an invertible 3×3 matrix A, by $\phi_A(P) = PA^{-1}$, $\phi_A(\ell) = A\ell$. Clearly $P\ell = 0$ if and only if $(PA^{-1})(A\ell) = 0$ so ϕ_A is a collineation. The Fundamental Theorem of real projective planes states that every collineation is obtained in this fashion. Moreover, the kernel of $A \to \phi_A$ is the set of all non-zero scalar matrices so the collineation group is isomorphic is isomorphic to $PGL(3,\mathbb{R}) = GL(3,\mathbb{R})/\mathbb{R}^*I$.

The collineation of the The Euclidian plane Π given by $\sigma : (x,y) \to (x + u,y + v)$ is a translation. Since the translate of any line is parallel to the line, this collineation extends to a collineation $\tilde{\sigma}$ of $\mathcal{P}$ which fixes every point on the line at infinity. Since for any point (x,y) the slope of the line through (x,y) and $(x + u,y + v)$ is $m = \frac{v}{u}$ ($m = \infty$ allowed) if $\sigma \neq id$, every line incident to (m) is fixed by $\tilde{\sigma}$. In general, a collineation of $\mathcal{P}$ fixing all points on some line ℓ (the <u>axis</u>) and all lines through some point P (the <u>center</u>) is called a <u>transvection</u>. In homogeneous coordinates, we see $\tilde{\sigma}(x,y,1)=(x+u,y+v,1)$ and $\sigma(x,y,0) = (x,y,0)$. Thus $\sigma = \phi_A$ with $A^{-1} = \begin{pmatrix} 1 & 0 & 0 \\ 0 & 1 & 0 \\ u & v & 1 \end{pmatrix}$ so $A = \begin{pmatrix} 1 & 0 & 0 \\ 0 & 1 & 0 \\ -u & -v & 1 \end{pmatrix}$. In general, it is easy to see that every elementary matrix $E = E^t_{ij} = I + te_{ij}$, $i \neq j$, induces a transvection ϕ_E of the projective plane with axis v_i^t , center v_j, where $v_1 = (1,0,0)$, $v_2 = (0,1,0)$ and $v_3 = (0,0,1)$. If $F = E^s_{jk}$, $j \neq k \neq i$, then a matrix computation shows that the commutator $(E,F) = E^{-1}F^{-1}EF = E^{ts}_{ik}$. From this it is easy to see that the group $E(3,\mathbb{R})$ generated by all elementary matrices is also generated by the subgroups $\left\{\begin{pmatrix} 1 & 0 & 0 \\ 0 & 1 & 0 \\ * & * & 1 \end{pmatrix}\right\}$ and $\left\{\begin{pmatrix} 1 & 0 & * \\ 0 & 1 & * \\ 0 & 0 & 1 \end{pmatrix}\right\}$. Also, it is known that $E(3,\mathbb{R}) = SL(3,\mathbb{R}) \cong PSL(3,\mathbb{R})$. Both of these facts depend on special properties of $\mathbb{R}$ and are not in general true when $\mathbb{R}$ is replaced by an arbitrary ring. Interpreting these results for the plane $\mathcal{P}$, we see that the group G of collineations induced by $SL(3,\mathbb{R})$ is generated by U^+ and U^- , where U^+ is the abelian group of transvections with axis [0,0] (the translations of Π) which is isomorphic with $(V^+,+)$ where V^+ is the vector space of 2-tuples (u,v) and U^- is the abelian group of all transvections with center $(0,0)$ (sending lines $[a,b] \to [a+u, b+v]$) which is isomorphic with $(V^-,+)$ where V^- is the vector space of all $(u,v)^t$.

The group G, which is clearly generated by transvections, in fact contains all transvections of $\mathcal{P}$. To establish this, we note first, that if (x,y,z), (x',y',z') represent distinct points, we can find $A \in SL(3,\mathbb{R})$ with $A = \begin{pmatrix} x & y & z \\ x' & y' & z' \\ * & * & * \end{pmatrix}$. The associated ϕ_A maps (x,y,z) to $(1,0,0)$ and (x',y',z') to $(0,1,0)$. We thus see that G is transitive on pairs of points and hence on pairs (P,ℓ) with P on ℓ. Also, if θ is any collineation and ϕ is a transvection with axis ℓ and center P then $\theta\phi\theta^{-1}$ is clearly a transvection with axis $\theta(\ell)$ and center $\theta(P)$. Since we can take $\theta \in G$ to yield $\theta\phi\theta^{-1} \in G$, we see that $\phi \in G$ for any transvection ϕ.

We have shown

2.1. Proposition. The group G generated by all transvections is in fact generated by U^+ and U^- and is isomorphic with $PSL(3,\mathbb{R})$. Moreover, G is transitive on pairs (P,ℓ) with P not on ℓ.

§3. The real projective plane $\mathcal{P}(\mathbb{R})$ as a homogeneous space.

We have already seen that $G \cong PSL(3,\mathbb{R})$ is transitive on points in $\mathcal{P}(\mathbb{R})$. Moreover, we can easily compute the stabilizer of $(0,0,1)$ (the origin in Π) to be induced by matrices of the form $\begin{pmatrix} * & * & * \\ * & * & * \\ 0 & 0 & * \end{pmatrix}$.

We denote the stabilizer by P^- (since $P^- \supset U^-$ and P^- is, in fact, a parabolic subgroup of G). Similarly the stabilizer P^+ of $(0,0,1)^t$ is induced by matrices of the form $\begin{pmatrix} * & * & 0 \\ * & * & 0 \\ * & * & * \end{pmatrix}$ and contains U^+. Now $\phi_1 P^- \to \phi_1(0,0,1)$ and $\phi_2 P^+ \to \phi_2(0,0,1)^t$ are bijections of the cosets $G/_{P^-}$ with the points, and the cosets $G/_{P^+}$ with the lines of $\mathcal{P}(\mathbb{R})$. Since $(0,0,1)$ is not on $(0,0,1)^t$ and G is transitive on pairs (P,ℓ) with P not on ℓ by (2.1), we conclude that $\phi_1 P^-$ does not lie on $\phi_2 P^+$ if and only if for some $\phi \in G$, $\phi_1 P^- = \phi P^-$ and $\phi P_2^+ = \phi P^+$, i.e. $\phi \in \phi P_1^- \cap \phi P_2^+$.

We thus have

3.1. Proposition. The real projective plane is isomorphic to the geometry with point set $G/_{P^-}$, line set $G/_{P^+}$ and incidence defined by: $\phi_1 P^-$ is incident to $\phi_2 P^+$ if $\phi_1 P^- \cap \phi_2 P^+$ is the empty set.

An interesting fact about the representation of the points and lines in this "homogeneous space" version of the projective plane is that each point can be written in form $\phi_- \phi_+ P^-$ with $\phi_- \in U^-$, $\phi_+ \in U^+$ (a similar statement holds for lines). Recall that

$\phi_+(x,y,z) = (x + za, y + zb, z)$ for some $u = (a,b) \in V^+$ while

$\phi_-(x,y,z) = (x,y,xc + yd + z)$ for some $v^t = (c,d)^t \in V^-$. For any $(x,y,z) \neq (0,0,0)$, we can choose ϕ_+,ϕ_- so that $\phi_-^{-1}(x,y,z) = (x,y,z')$ with $z' \neq 0$ and $\phi_+^{-1}(x,y,z') = (0,0,z')$. Thus, up to scalar multiple $(x,y,z) = \phi_- \phi_+(0,0,1)$ as desired. (In general, this argument works for any geometry $\mathcal{P}(\mathcal{A})$ constructed from a ring $\mathcal{A}$ of stable range two, as $\mathcal{P}(\mathcal{A})$ is constructed from $\mathbb{R}$, via homogeneous coordinates, where stable range two means that $xa + y$ invertible implies $x + yb$ invertible for some b.) In this representation, suppose a point is represented in two different ways, i.e. $\phi_-\phi_+(0,0,1) = \tilde{\phi}_-\tilde{\phi}_+(0,0,1)$ where $\phi_\pm, \tilde{\phi}_\pm$ correspond respectively to $u, v^t, \tilde{u}, \tilde{v}^t$. If $\tilde{v} = (c,d) = (0,0)$, then $\phi_-\phi_+(0,0,1) = (a,b,1 + ac + bd) = (a,b,1 + uv^t)$ is a nonzero scalar multiple of $(\tilde{a},\tilde{b},1)$, i.e. $1 + uv^t$ is invertible and $\tilde{u} = (1 + uv^t)^{-1}u$. In general, $(\tilde{\phi}_-^{-1}\phi_-)\phi_+(0,0,1)$ and $\tilde{\phi}_+(0,0,1)$ are the same point so $1 + u(v^t - \tilde{v}^t)$ is invertible and $\tilde{u} = (1 + u(v^t - \tilde{v}^t))^{-1}u$.

3.2. <u>Proposition</u>. The points of the real projective plane correspond bijectively with the equivalence classes of pairs (u,v) with $u = (a,b)$, $v = (c,d)^t$ where $(u,v) \sim (\tilde{u},\tilde{v})$ provided $1 + u(v^t - \tilde{v}^t)$ is invertible with $\tilde{u} = (1 + u(v^t - \tilde{v}^t))^{-1}u$.
(A similar statement can be made for the generalized planes discussed briefly in §12, where the pairs (u,v) are interpreted as elements of a Jordan pair and the above condition translates to $(u, v - \tilde{v})$ is quasi-invertible with quasi-inverse $\tilde{u}$.)

§4. Abstract projective planes

Turning from the special case of the real projective plane, we now consider any projective plane $\mathcal{P}$; i.e. a set of <u>points</u>, a set of <u>lines</u> and an <u>incidence relation</u> $P|\ell$ with

(1) for points $P \neq Q$ there is a unique line ℓ (written $P \vee Q$) with $P|\ell$ and $Q|\ell$

(2) For lines $\ell \neq m$, there is a unique point P (written $\ell \wedge m$) with $P|\ell$ and $P|m$

(3) There are four points, no three of which are incident to the same line.

A bijection of points to points and lines to lines preserving incidence is a <u>collineation</u>. In view of the central role played by transvections in the real projective plane, it is natural to study them in $\mathcal{P}$. It is easy to see that a collineation ϕ fixing all points on a line ℓ and all lines through a point $P|\ell$ is uniquely determined by the image $\phi(Q)$ of a point Q not on ℓ. We call such a collineation a <u>transvection</u> with <u>axis</u> ℓ and <u>center</u> P. Moreover, we say that $\mathcal{P}$ is a <u>transvection plane</u> if for each line ℓ and points Q, Q' not on ℓ there is a transvection ϕ with axis ℓ such that $\phi(Q) = Q'$. One can also phrase the condition of being a transvection plane in terms of a <u>configuration condition</u> as follows. Let ℓ be a line, X_1, X_2, X_3, X_4 points not on ℓ such that no three are collinear. The six points $Y_{ij} = (X_i \vee X_j) \wedge \ell$ are called a <u>quadrangle section</u>. The "quadrangle section condition" says that if five of the corresponding pairs of points in two quadrangle sections are equal, then the sixth pair is also equal. This condition is too strong for our purposes, being in fact equivalent to Desarques condition for projective planes [10]. We shall, instead, use the following special case known as the <u>little quadrangle section condition:</u> If $\{Y_{ij}\}$ and $\{Y'_{ij}\}$ are quadrangle sections on a line ℓ with $Y_{14} = Y_{23}$ and $Y'_{14} = Y'_{23}$ and $Y_{ij} = Y'_{ij}$ for $\{i,j\} \neq \{1,3\}$, then $Y_{1,3} = Y'_{1,3}$.

4.1 Proposition [10]. $\mathcal{P}$ is a transvection plane if and only if $\mathcal{P}$ satisfies the little quadrangle section condition (for every line ℓ in $\mathcal{P}$).

Proof: If $\mathcal{P}$ is a transvection plane and τ is the transvection with axis $X_1 \vee X_4 = X_1 \vee Y_{14}$, center Y_{14} such that $\tau(Y_{24}) = Y_{34}$ (Figure 3), it is easy to see that $\tau(X_2) = X_3$ and $\tau(Y_{12}) = Y_{13}$. Similarly, τ' with axis $X'_1 \vee Y'_{14}$ sending Y'_{24} to Y'_{34} has $\tau'(Y'_{12}) = Y'_{13}$. If σ is the transvection with axis ℓ and $\sigma(X_1) = X'_1$, then $\sigma\tau\sigma^{-1}$ is a transvection with axis

(Figure 3)

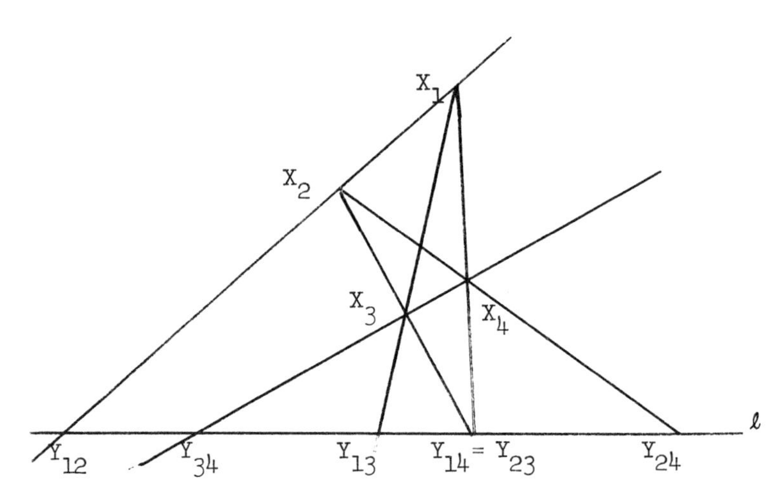

$\sigma(X_1 \vee Y_{14}) = X'_1 \vee Y_{14} = X'_1 \vee Y'_{14}$ sending $\sigma(Y_{24}) = Y_{24} = Y'_{24}$ to Y'_{34}. Thus $\tau' = \sigma\tau\sigma^{-1}$ and $Y'_{13} = \tau'(Y'_{12}) = \sigma\tau\sigma^{-1}(Y_{12}) = Y_{13}$.

Conversely, suppose $\mathcal{P}$ satisifes the little quadrangle section condition. First, we claim that if $\{P,P,P_1,P_2,Q_1,Q_2\}$ and $\{P,P,Q_1,Q_2,R_1,R_2\}$ are quadrangle sections with $P_1 \neq R_1$, then $\{P,P,P_1,P_2,R_1,R_2\}$ is also a quadrangle section

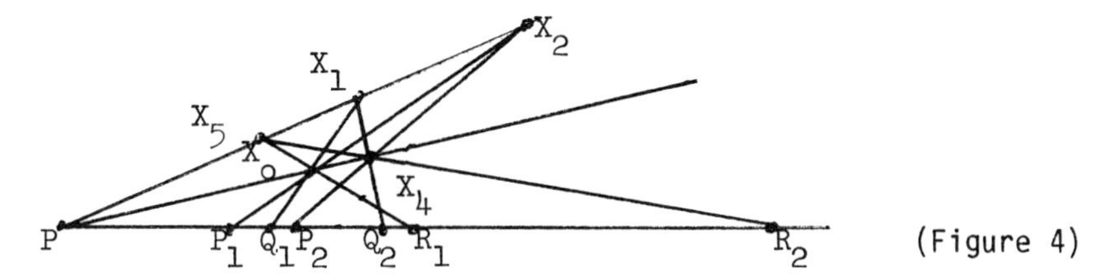

(Figure 4)

For this, let X_1, X_2, X_3, X_4 give the quadrangle section $\{P,P,P_1,P_2,Q_1,Q_2\}$ (Figure 4) and let $X_5 = (X_1 \vee X_2) \wedge (R_1 \vee X_3)$. Now X_5, X_1, X_3, X_4 give the quadrangle section $\{P,P,Q_1,Q_2,R_1,R_2\}$ so X_5, X_2, X_3, X_4 give the remaining section. Now let $Q \neq Q'$ be points not on a line ℓ. We shall find a transvection ϕ with axis ℓ so that $\phi(Q) = Q'$. Let $P = \ell \wedge (Q \vee Q')$. Define ϕ on $Q \vee Q'$ by $\phi(P) = P$, $\phi(Q) = Q'$ and $\phi(R) = R'$ if $\{P,P,Q,Q',R,R'\}$ is a quadrangle section. From the above result, we see that $\{P,P,R,\phi(R),S,\phi(S)\}$ is a quadrangle section of $R \vee S$. Define ϕ on lines by $\phi(m) = m$ for $m|P$ and $\phi(m) = \phi(m \wedge (Q \vee Q')) \vee (m \wedge \ell)$ for $m \nmid P$. Fix $T \neq P$, $T|\ell$ and define ϕ_T on points by $\phi_T(X) = X$ for $X|\ell$ and $\phi_T(X) = \phi(X \vee T) \wedge (X \vee P)$. Clearly $\phi_T = \phi$ on $Q \vee Q'$.

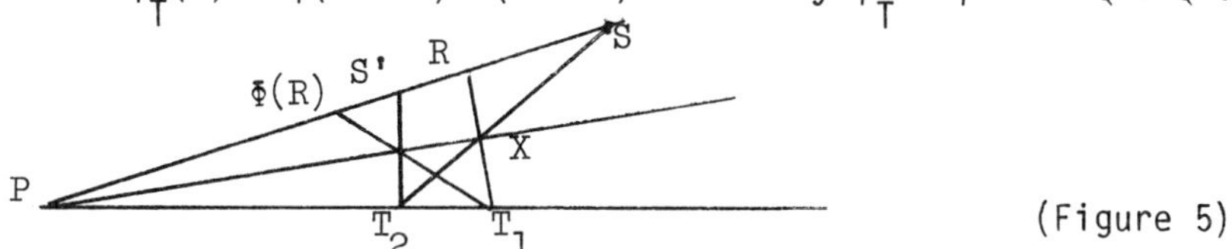

(Figure 5)

Since every line incident to X is of the form $X \vee T$ for some $T|\ell$ we see that $\phi = \phi_T$ is a collineation if $\phi_{T_1} = \phi_{T_2}$ for any $T_1 \neq T_2$ on ℓ. Using $T_1, T_2, X, \phi_{T_1}(X)$, we see that $\{P,P,R,\phi(R),S,S'\}$ is a quadrangle section (Figure 5) where $R = (Q \vee Q') \wedge (X \vee T_1)$, $S = (Q \vee Q') \wedge (X \vee T_2)$ and $S' = (Q \vee Q') \wedge (\phi_{T_1}(X) \vee T_2)$. Since $S' = \phi(S)$ we see $\phi_{T_1}(X) = \phi_{T_2}(X)$ as desired.

Our principal aim in this section will be to coordinatize planes in which the little quadrangle section condition holds, approaching the coordinatization in a somewhat nonstandard way which has the virtue of generalizing almost immediately to the geometries discussed in §12.

4.2. Proposition. If $\mathcal{P}$ is a transvection plane then

a) transvections with the same axis commute

b) if σ_i is a (P_i,ℓ_i) transvection, $i = 1, 2$, with $P_2|\ell_1$ then $\tau = (\sigma_1,\sigma_2) = \sigma_1^{-1}\sigma_2^{-1}\sigma_1\sigma_2$ is a (P_2,ℓ_1) transvection and $\tau(Q) = (\sigma_1(Q) \vee \sigma_2(P_1)) \wedge \ell_2$ for $Q|\ell_2$, $Q \neq P_2$.

Proof: Since σ_1 fixes points on ℓ_1 and σ_2 permutes them, we see that τ fixes points on ℓ_1 and, similarly, lines through P_2. Now $\tau(Q) = \sigma_1^{-1}\sigma_2^{-1}\sigma_1(Q) = \sigma_1^{-1}(T) = (\sigma_1(Q) \vee \sigma_2(P_1)) \wedge \ell_2$ where $T = (\sigma_1(Q) \vee P_2) \wedge (((\sigma_1(Q) \vee \sigma_2(P_1)) \wedge \ell_2) \vee P_1)$ (Figure 6).

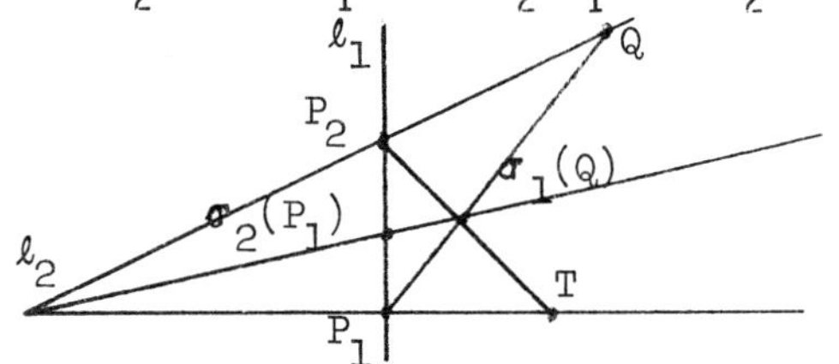

(Figure 6)

Thus (b) holds. If σ_1, σ_2 have axis ℓ and centers P_i then $\tau = (\sigma_1,\sigma_2)$ has axis ℓ and center P_i. Since any transvection with two centers must fix all points not on the axis, we see that $\tau = id$ if $P_1 \neq P_2$. If $P = P_1 = P_2$, we choose $m|P$, $m \neq \ell$, and $Q|\ell$ $Q \neq P$. By (b), $\sigma_1 = (\tau_1,\tau_2)$ where τ_1 is some (Q,ℓ)-transvection and τ_2 is some (P,m)-transvection. Since σ_2 commutes with τ_1 (by the above argument) and τ_2 by the dual argument it commutes with σ_1.

To introduce coordinates in a transvection plane $\mathcal{P}$, we fix a reference triangle P_1, P_2, P_3 (Figure 7). We would like to have P_3 as the origin of a coordinate system with $P_3 \vee P_1$ and $P_3 \vee P_2$ the coordinate axes and $P_2 \vee P_1$ the line at ∞. In this situation a point $X \neq P_1$ on $P_3 \vee P_1$ would have coordinates of form $(X_0, 0)$,

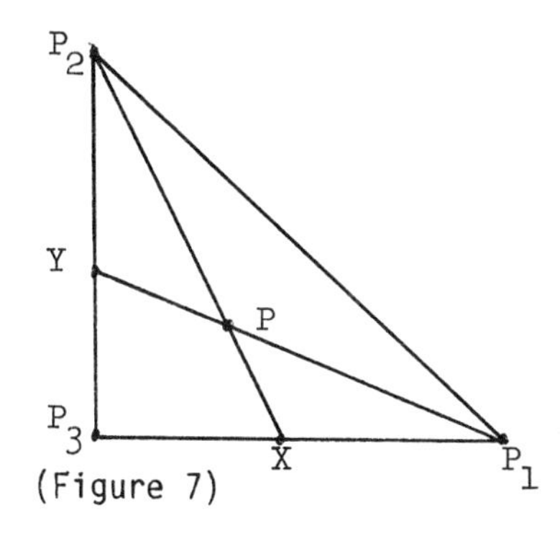

(Figure 7)

so the set of points $\{X : X | P_3 \vee P_1$ and $X \neq P_1\}$ would match up with the coordinate set. We shall endow this set with a(n) (alternative) ring structure, which will provide inhomogeneous coordinates for the entire plane.

Rather than working directly with given coordinate set, which leads to very complicated configurations in the verification of the assorted ring properties, we begin by viewing the sides of the reference triangle (each with a distinguished vertex) as six different coordinate sets. We denote by $\mathcal{a}_{ij}$ the set $\{X : X | P_i \vee P_j, X \neq P_j\}$ for $i \neq j$. If P is a point not on the line at $\infty(P_1 \vee P_2)$, we can project P to $\mathcal{a}_{31}$ and $\mathcal{a}_{32}$ via P_2 and P_1 respectively to get $X = (P_2 \vee P) \wedge (P_3 \vee P_1)$ and $Y = (P_1 \vee P) \wedge (P_3 \vee P_2)$. Clearly $P = (P_2 \vee X) \wedge (P_1 \vee Y)$ is uniquely determined by X and Y. We assign P "coordinates" (X,Y) with $X \in \mathcal{a}_{31})$ (the "x-axis") and $Y \in \mathcal{a}_{32}$ (the "y-axis"). Later, we will identify the two "axes" via selection of a suitably chosen fourth point U.

Since $\mathcal{P}$ is a transvection plane, the set V^+ of points P

with $P \nmid P_1 \vee P_2$ corresponds bijectively with the abelian group of transvections with axis $P_1 \vee P_2$ (see (4.2)) via $\sigma \to \sigma(P_3)$. Clearly this gives V^+ the structure of an abelian group in which $\mathcal{A}_{31}$ is the subgroup corresponding to the transvections with center P_1. We denote the induced operation on the points of V^+ by $P + Q$. If, for $X \in \mathcal{A}_{31}$, $Y \in \mathcal{A}_{32}$ we have $X = \sigma_1(P_3)$, $Y = \sigma_2(P_3)$, we see easily that $X + Y = \sigma_1(Y) = (X,Y)$. From this it is clear that $V^+ = \mathcal{A}_{31} \oplus \mathcal{A}_{32}$.

If $X \in \mathcal{A}_{31}$ and $M \in \mathcal{A}_{12}$, then $Y = (X \vee M) \wedge (P_3 \vee P_2) \in \mathcal{A}_{32}$ (Figure 8). In view of the calculation at the end of §1, it is natural to view y as a product $X * M$, so $\mathcal{A}_{31} * \mathcal{A}_{12} \subseteq \mathcal{A}_{32}$. Similarly, we have products $\mathcal{A}_{32} * \mathcal{A}_{21} \subseteq \mathcal{A}_{31}$, etc.

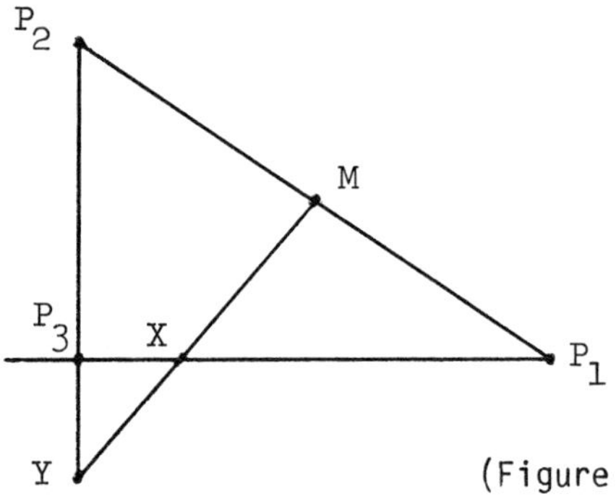

(Figure 8)

If σ_1 is the $(P_1, P_1 \vee P_2)$-transvection with $\sigma_1(P_3) = X$ and if σ_2 is the $(P_2, P_3 \vee P_2)$-transvection with $\sigma_2(P_1) = M$ then Proposition (4.2),b), shows $\tau = (\sigma_1, \sigma_2)$ is the $(P_2, P_1 \vee P_2)$-transvection with $\tau(P_3) = (X \vee M) \wedge (P_3 \vee P_2) = Y = X * M$. If $\tilde{\sigma}_1$ is another $(P_1, P_1 \vee P_2)$-transvection with $\tilde{\sigma}_1(P_3) = \tilde{X}$, then

$$(\sigma_1\tilde{\sigma}_1, \sigma_2) = \tilde{\sigma}_1^{-1}\sigma_1^{-1}\sigma_2^{-1}\sigma_1\tilde{\sigma}_1\sigma_2 = \tilde{\sigma}_1^{-1}(\sigma_1,\sigma_2)\sigma_2^{-1}\tilde{\sigma}_1\sigma_2 = (\sigma_1,\sigma_2)(\tilde{\sigma}_1,\sigma_2) .$$

Thus $(X + \tilde{X}) * M = X * M + \tilde{X} * M$, and the product is "right distributive".

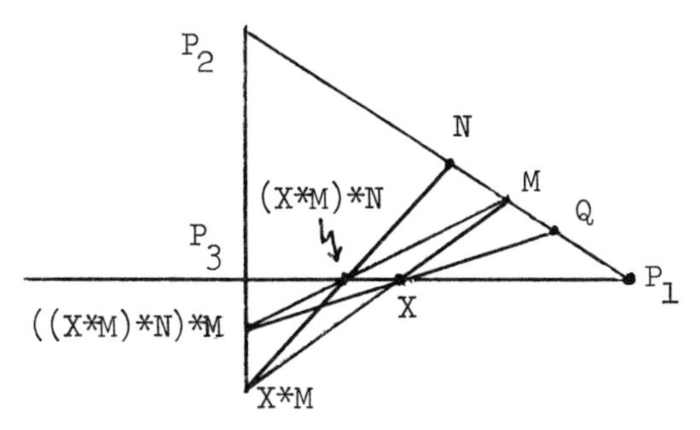

(Figure 9)

4.3. Proposition. If $M \in \mathfrak{a}_{12}$, $N \in \mathfrak{a}_{21}$, then there is $Q \in \mathfrak{a}_{12}$ with $X * Q = ((X * M) * N) * M$ for all $X \in \mathfrak{a}_{31}$.

Proof. If $M \neq P_1$ and $N \neq P_2$, take Q so that $\{M, M, P_1, P_2, N, Q\}$ is a quadrangle section (Figure 9). Otherwise let $Q = P_1$. If $X \neq P_3$, $M \neq P_1$, and $N \neq P_2$, then using the quadrangle X, $X * M$, $((X * M) * N) * M$, $(X * M) * N$ we get the desired equality. Otherwise, both sides equal P_3.

We are now prepared to restrict our attention to the single "axis" $\mathfrak{a}_{32}$. We pick a point U not on the sides of triangle P_1, P_2, P_3 and write $U = (U_2, U_1)$ with $U_1 \in \mathfrak{a}_{32} \cap \mathfrak{a}_{23}$, $U_2 \in \mathfrak{a}_{31} \cap \mathfrak{a}_{13}$ as before for $i = 1, 2$.

Set $U_3 = U_1 * U_2 \in \mathfrak{a}_{21} \cap \mathfrak{a}_{12}$. On $\mathfrak{a}_{32}$ we define a product $XZ = (X * U_3) * (U_2 * Z)$. Clearly this product is right distributive since (every) $*$ is. Also, it is easy to check that U_1 is a multiplicative identity element. Let $X = Z * U_3$, $M = U_2 * Y$, and $N = U_3$ in (4.3) to get

$(Z * U_3) * Q = (((Z * U_3) * (U_2 * Y)) * U_3) * (U_2 * Y) = (ZY)Y.$

In particular, for $Z = U_1$ we get $U_2 * Q = YY$ or $Q = U_2 * (YY)$. Thus $(ZY)Y = (Z * U_3) * (U_2 * (YY)) = Z(YY)$.

4.4 Proposition. If $\mathcal{P}$ is a transvection plane then $\mathcal{P}$ can be coordinated with inhomogeneous coordinates over an alternative division ring $\mathcal{A}$, i.e. $(YY)Z = Y(YZ)$ and $Z(YY) = (ZY)Y$ for $Y,Z \in \mathcal{A}$.

Proof. We have already shown that $\mathcal{A} = \mathcal{A}_{32}$ has a product which is right distributive and right alternative (i.e $Z(YY) = (ZY)Y$). For $Y \in \mathcal{A}_{32} \cap \mathcal{A}_{23}$, we can form $Z_1 = U_2 * ((Y * U_3) * U_1)$ and $Z_2 = (U_1 * (U_2 * Y)) * U_3$, $Z_i \in \mathcal{A}_{23} \cap \mathcal{A}_{32}$ and check that $Z_2 Y = YZ_1 = U_1$ so $\mathcal{A}_{32}$ is a division ring.

To see that the left distributive and alternative properties hold, we consider the dual plane. There $\ell_1 = P_2 \vee P_3$, $\ell_2 = P_3 \vee P_1$ and $\ell_3 = P_1 \vee P_2$ form a triangle and $\pi : T \to T \vee P_i$ is a bijection of $\mathcal{A}_{jk}$ with $\hat{\mathcal{A}}_{kj} = \{\ell : \ell | \ell_k \wedge \ell_j, \ell \neq \ell_j\}$ for $\{i,j,k\} = \{1,2,3\}$. Moreover, one can show $-\pi(X*Y) = \pi(Y) * \pi(X)$, the latter $*$ computed in the dual plane. The right alternative and distributive properties in the dual plane (which is itself a transvection plane) are carried by π to their left analogues for $\mathcal{P}$.

One can use the bijection $Y \to Y * U_3$ to identify $\mathcal{A}$ with $\mathcal{A}_{31}$. The coordinatization can now be completed just as the real projective plane was coordinatized in §1.

Just as in the real projective case, the group G generated by all transvections is generated by U^+, U^- where U^+ consists of transvections with axis $P_1 \vee P_2$ and is isomorphic with $V^+ = \{(X,Y) : X,Y \in \mathcal{A}\}$ while U^- consists of the transvections with center P_3 and is isomorphic with $V^- = \{(X,Y)^t : X,Y \in \mathcal{O}\}$. We would like to identify G explicitly as was done for real projective planes by the fundamental theorem of §2. The problem is that we do not have, in general, a model of the projective plane using homogeneous coordinates over $\mathcal{O}$. Indeed, since $\mathcal{O}$ may not be associative, the set of scalar multiples of (x,y,z) may not be closed under scalar multiplication. On the other hand, the model of inhomogeneous coordinates does not lend itself to a nice description of G. To get out of this quandry, we need another model of projective planes, which shall look at in the next section.

§5. A Jordan algebra construction of projective planes

Before turning to a new construction of a transvection plane, we return to the homogeneous coordinate construction of the real projective plane $\mathcal{P}$ (§1), so $v = (x,y,z)$ represents a point P, $w^t = (r,s,t)^t$ represents a line ℓ, and $P|\ell$ if and only if $vw^t = 0$. We may associate with P the real symmetric 3×3 matrix $A = v^t v$ (up to scalar multiple) and with ℓ the matrix $B = w^t w$. Clearly both of these symmetric matrices are of rank one. Since, as linear transformations in $\mathbb{R}^3$, A has image $\mathbb{R}v$ and B has image $\mathbb{R}w$, and $vw^t = 0$ if and only if $AB = 0$, we have an algebraic description of $\mathcal{P}$ as follows:

$$\text{points} \leftrightarrow \{A_* : A \text{ is of rank } 1\}$$

$$\text{lines} \leftrightarrow \{A^* : A \text{ is of rank } 1\}$$

$$A_* | B^* \leftrightarrow AB = 0$$

where A_* (resp. A^*) is the set of all scalar multiples of A.

At first glance, this construction appears to be even more specialized than the homogeneous coordinate construction which yields a projective plane when the reals are replaced by any associative division ring. In the above construction, $v^t v$ and $(v^t\lambda)(\lambda v)$ are to represent the same point for any λ in the coordinate ring $\mathcal{A}$, which causes difficulty unless $\mathcal{A}$ is commutative. With some restriction on $\mathcal{A}$ (much weaker than commutativity) we can introduce yet another variant of the homogeneous coordinate construction which reduces to the above construction when $\mathcal{A} = \mathbb{R}$. We assume that $\mathcal{A}$ is an associative division ring with involution $a \to \bar{a}$ (so $\bar{\bar{a}} = a$ and $\overline{ab} = \bar{b}\bar{a}$) and suppose that the set $k = \{a \in \mathcal{A} : \bar{a} = a\}$ is central in k. Now $(\overline{\lambda v})^t(\lambda v) = \bar{v}^t(\bar{\lambda}\lambda)v = \bar{\lambda}\lambda\bar{v}^t v$ is a "scalar multiple" of $\bar{v}^t v$. Thus we may, without ambiguity, represent points and lines by rank 1 Hermitian matrices $(\bar{A}^t = A)$ up to scalar multiple from k and define incidence by $A_* | B^*$ if $AB = 0$. In this view of the projective plane, the point with inhomogeneous coordinates (x,y) and homogeneous coordinates $(x,y,1)$ has matrix coordinates

$$\begin{pmatrix} \bar{x}x & \bar{x}y & \bar{x} \\ \bar{y}x & \bar{y}y & \bar{y} \\ x & y & 1 \end{pmatrix}.$$

Although this construction is still more specialized for associative rings than the homogeneous coordinate construction (since an involution is necessary), the following important theorem about alternative division rings suggests that this may often be a feasible approach for constructing transvection planes.

5.1. <u>Proposition</u> [2]. If $\mathcal{A}$ is an alternative division ring then either

(a) $\mathcal{A}$ is associative or

(b) $\mathcal{A}$ is an octonion (Cayley) algebra over its center (and hence is 8-dimensional over its center k and admits an involution $a \to \bar{a}$ with $k = \{a | a = \bar{a}\}$).

We denote henceforth in any octonion algebra:
$t(a) = a + \bar{a} \in k, n(a) = \bar{a}a \in k$ and note that an essential property of the quadratic form $n(a)$ is the composition property $n(ab) = n(a)n(b)$. Note also that by definition of "center" in $\mathcal{A}$, $\lambda(ab) = (\lambda a)b$ for $\lambda \in k$, a, $b \in \mathcal{A}$.

Now, if we wish to construct a projective plane with coordinate ring $\mathcal{A}$, $\mathcal{A}$ alternative, we can choose a construction according as $\mathcal{A}$ is of type (a) or (b). In case (a) we may use the standard homogeneous coordinate construction. In case (b), we make a Hermitian matrix construction. In subsequent sections we will show that, for suitable definition of <u>rank one</u> and of <u>incidence</u>, this construction does indeed yield a transvection plane with coordinate ring $\mathcal{A}$.

To begin to analyze the "plane" constructed in case (b), we denote by $H(\mathbb{O}_3)$ the set of Hermitian matrices

(1)
$$A = \begin{pmatrix} \alpha_1 & a_3 & \bar{a}_2 \\ \bar{a}_3 & \alpha_2 & a_1 \\ a_2 & \bar{a}_1 & \alpha_3 \end{pmatrix}$$

where $a_i \in \mathbb{O}$, $\mathbb{O}$ an octonion division algebra. We can no longer interpret A as a linear transformation of some vector space, so we need some substitutes for "A of rank 1" and "$AB = 0$".

The proper setting for studying $H(\mathbb{O}_3)$ is that of Jordan algebras. These algebras were introduced by physicists [6] as generalizations of the algebra of complex Hermitian matrices (actually, the algebra of self-adjoint operators on a Hilbert space). They noted that such matrices are not closed under the usual matrix multiplication but are closed under the product $A \cdot B = \frac{1}{2}(AB + BA)$. This product, while not associative, is easily seen to satisfy

(J1) $A \cdot B = B \cdot A$

(J2) $(A^2 \cdot B) \cdot A = A^2 \cdot (B \cdot A)$ where $A^2 = A \cdot A$

Any algebra satisfying (J1) and (J2) is called a (linear) _Jordan algebra_ (note that char. $\neq 2$ was essential in defining $A \cdot B$ in our original example). The structure theory of linear Jordan algebras in char. $\neq 2$ is well developed [5]. Playing an important role in that development is a certain quadratic operator U_A defined by

$$U_A(B) = 2A \cdot (A \cdot B) - A^2 \cdot B \tag{2}$$

(for complex Hermitian matrices $U_A B = ABA$). In extending the theory of Jordan algebras to characteristic 2, McCrimmon identified certain properties of the operator U_A:

(QJ1) $U_1 = \mathrm{id}$

(QJ2) $U_A V_{B,A} = V_{A,B} U_A$

(QJ3) $U_A U_B U_A = U_{U_A(B)}$

(QJ4) all linearizations of (QJ1) - (QJ3) hold

which he then used as the defining axioms for <u>quadratic Jordan algebras.</u> Here

$$V_{A,B}(C) = U_{A,C}(B) = (U_{A+C} - U_A - U_C)(B) \ . \tag{3}$$

Just as quadratic form theory is equivalent to the theory of symmetric bilinear forms for char. $k \neq 2$ and provides the "right" format in char. 2, the theory of quadratic Jordan algebras provides a unified approach to Jordan algebras in all characteristics. This is particularly important in applications to geometry where characteristic 2 plays no special role.

Since our particular interest is in $H(\mathbb{O}_n)$, we are interested in defining an operator U_A for this space, which satisfies (QJ1) - (QJ4). For char(k) $\neq 2$ and any n, we can form $A \cdot B = \frac{1}{2}(AB + BA)$. Since $\mathbb{O}$ is not associative, (J1) and (J2) fail for this product in general. Surprisingly, they hold for $n \leq 3$. (In some sense, this is the algebraic version of the geometric fact that a projective plane which can be embedded in a higher dimensional projective space necessarily satisfies Desargues' condition). Thus for char.(k) $\neq 2$, $H(\mathbb{O}_3)$ is a linear Jordan algebra with corresponding operators U_A. To give a characteristic free structure of quadratic Jordan algebra to $H(\mathbb{O}_3)$, we must express U_A without using $\frac{1}{2}$.

For A as in (1), define $a(A)=[a_1,a_2,a_3] \in \mathfrak{O}$ where, in general, $[x,y,z] = (xy)z - x(yz)$ is the <u>associator</u> in a nonassociative ring. Using the standard matrix product in $H(\mathfrak{O}_3)$ one computes

$$[A,A,A] = 2a(A)I \ .$$

Using the obvious modification of the notation (1) for $B \in H(\mathfrak{O}_3)$, we linearize the above to get

$$[B,A,A] + [A,B,A] + [A,A,B] = 2a(A,B)I \tag{4}$$

where $a(A,B) = [b_1,a_2,a_3] + [a_1,b_2,a_3] + [a_1,a_2,b_3]$. If $\frac{1}{2} \in k$, we use (2) and (4) to write

$$\begin{aligned} U_A B &= \frac{1}{2} (A(AB + BA) + (AB + BA)A - A^2B - BA^2) \\ &= \frac{1}{2} ((AB)A + A(BA) + [B,A,A] - [A,A,B]) \\ &= \frac{1}{2} (2(AB)A - [A,B,A] + [B,A,A] - [A,A,B]) \\ &= (AB)A + [B,A,A] - a(A,B)I \ . \end{aligned}$$

Thus, it seems appropriate to define

$$U_A(B) = (AB)A + [B,A,A] - a(A,B)I$$

for all fields k . It can be verified that $H(\mathfrak{O}_3)$ is a quadratic Jordan algebra [8].

The analogy between $H(\mathfrak{O}_3)$ and the algebra of complex Hermitian matrices goes beyond the existence of a quadratic structure. If, for A as in (1), we define a "determinant" function

(5) $\det A = \alpha_1\alpha_2\alpha_3 - \alpha_1 a_1\bar{a}_1 - \alpha_2 a_2\bar{a}_2 - \alpha_3 a_3\bar{a}_3 + (a_1a_2)a_3 + \bar{a}_3(\bar{a}_2\bar{a}_1) \in k$

and let $p(\lambda) = \det(\lambda I - A)$, there is a Cayley-Hamilton type theorem stating that $p(A) = 0$ (where $A^3 = U_A A, A^2 = U_A I$) [8]. Writing $p(\lambda) = \lambda^3 - T(A)\lambda^2 + S(A)\lambda - (\det A)I$, one finds that

(6) $$T(A) = \alpha_1 + \alpha_2 + \alpha_3 .$$

In our description of a projective plane related to $H(\mathfrak{O}_3)$ we will make use of the adjoint $A^\#$ of A defined by

$$A^\# = A^2 - T(A)A + S(A)I$$

It is not difficult to compute

(7) $$A^\# = \begin{pmatrix} \gamma_1 & c_3 & \bar{c}_2 \\ \bar{c}_3 & \gamma_2 & c_1 \\ c_2 & \bar{c}_1 & \gamma_3 \end{pmatrix} \text{ where}$$

$$\gamma_i = \alpha_j\alpha_k - a_i\bar{a}_i \ ; \ c_i = (\bar{a_ja_k} - \alpha_i a_i)$$

for (i,j,k) a cyclic permutation of $\{1,2,3\}$. We will often make use of the linearized version of the adjoint

(8) $$A \times B = (A + B)^\# - A^\# - B^\#$$

as well as a bilinear trace (suggested by $T(A \cdot B)$ if $\frac{1}{2} \in k$)

(9) $$T(A,B) = \alpha_1\beta_1 + \alpha_2\beta_2 + \alpha_3\beta_3 + n(a_1,b_1) + n(a_2,b_2) + n(a_3,b_3)$$

where $n(a,b) = n(a+b) - n(a) - n(b) = a\bar{b} + b\bar{a} \in k$. The form $T(A,B)$ is a nondegenerate bilinear form on $H(\mathfrak{O}_3)$ [8].

For the convenience of the reader, we list explicitly

(10) $e_i \times e_i = 0$; $e_i \times e_j = e_k$

$i, j, k \neq$; $e_i \times a_{jk} = (\delta_{ij} + \delta_{ik} - 1)a_{jk}$;

$a_{ij} \times b_{ij} = -n(a,b)e_k$ $i, j, k \neq$; $a_{ij} \times b_{jk} = (ab)_{ik}$

$i, j, k \neq$

where, for e_{ij} denoting the usual matrix units, $e_i = e_{ii}$ and $a_{ij} = ae_{ij} + \bar{a}e_{ji}$ for $a \in \mathfrak{G}$.

We return now to the definition of a projective plane, which we shall denote by $\mathcal{P}(\mathfrak{G})$. We define $\pi(\mathfrak{G}) = \{A \in H(\mathfrak{G}_3) : A^{\#} = 0\}$ (the "rank one" matrices). The <u>point set</u> of $\mathcal{P}(\mathfrak{G})$ will be the set of all A_*, $A \in \pi(\mathfrak{G})$, where $A_* = \{\lambda A | \lambda \in k^*\}$. Similarly, the <u>line</u> set is $\{A^* | A \in \pi(\mathfrak{G})\}$. (Note that when $\alpha_3 = 1$, (7) shows that $A \in \pi(\mathfrak{G})$ has form $A = \begin{pmatrix} n(a_2) & \overline{a_1 a_2} & \bar{a}_2 \\ a_1 a_2 & n(a_1) & \bar{a}_1 \\ a_2 & a_1 & 1 \end{pmatrix}$, the same form noted for associative matrix coordinates at the beginning of this section.)

Turning now to the incidence relation, we recall that for associative coordinates we took $A_* | B^*$ if and only if $AB = 0$. It is not difficult to verify that $AB = 0$, in that case, is equivalent to either trace $(A \cdot B) = 0$ or $V_{A,B} = 0$, these latter conditions being equivalent. Since these conditions have quadratic Jordan analogues we define in $\mathcal{P}(\mathfrak{G})$, $A_* | B^*$ <u>if and only if</u> $T(A,B) = 0$. Equivalently (recall that $\mathfrak{G}$ is a division algebra) $A_* | B^*$ if and only if $V_{A,B} = 0$.

Having thus defined the ingredients of a projective plane, we have

5.2. Proposition [11]. $\mathfrak{P}(\mathfrak{O})$, as constructed above, is a transvection plane with coordinate ring $\mathfrak{O}$.

Rather than proving 5.2 at this time, we defer the proof to §7 and §8, where it is a consequence of more general results (8.4).

§6. The Hjelmslev-Moufang plane

From an algebraic point of view, the construction of $\mathfrak{P}(\mathfrak{O})$ is quite specialized--"most" octonion algebras are not division algebras. In this section we investigate the geometric structure which arises when we relax the condition that $\mathfrak{O}$ be a division algebra in the construction of $\mathfrak{P}(\mathfrak{O})$. We cannot expect the resulting structure to be always a projective plane, since its "coordinate ring" need not be a division ring. Nevertheless, it bears a strong resemblance to a projective plane and, in a natural way, can be considered as a transvection plane. These "planes" were first introduced by Springer and Veldkamp [12] in char. ≠ 2, the extension to char. 2 being handled in [4]. We follow the latter reference in our development.

As in §5, we begin with an octonion algebra $\mathfrak{O}$, noting that either $\mathfrak{O}$ is a division algebra (in which case $n(a) \neq 0$ for all $0 \neq a \in \mathfrak{O}$) or a split algebra (in which case there are nonzero elements a with $n(a) = 0$). The space $H(\mathfrak{O}_3)$, with operators U_A defined in §5, is a quadratic Jordan algebra in both cases and all results from §5 carry over with one exception--the conditions

$T(A,B) = 0$ and $V_{A,B} = 0$ used to define collinearity are no longer equivalent. In particular, if $a \in \mathfrak{G}$ has $n(a) = 0$, $a_{12} \in \pi(\mathfrak{G})$ and $T(e_1, a_{12}) = 0$ while $V_{e_1,a_{12}} \neq 0$. We will use <u>both</u> conditions in our definition of the <u>Hjelmslev-Moufang</u> plane $\mathfrak{P}(\mathfrak{G})$.

To construct the plane, we define $\pi(\mathfrak{G})$ as before to be the set of rank one elements in $H(\mathfrak{G}_3)$. The <u>points</u> of $\mathfrak{P}(\mathfrak{G})$ are $\{A_*: A \in \pi(\mathfrak{G})\}$, the <u>lines</u> $\{A^*: A \in \pi(\mathfrak{G})\}$. It is quite easy to check that neither condition $T(A,B) = 0$ nor $V_{A,B} = 0$ defines an incidence relation on $\mathfrak{P}(\mathfrak{G})$ for which $\mathfrak{P}(\mathfrak{G})$ becomes a projective plane if $\mathfrak{G}$ is split. For example, we consider $P = e_{1*}$ and $Q = a_{12*}$ where $0 \neq a \in \mathfrak{G}$ has $n(a) = 0$. There is $0 \neq b \in \mathfrak{G}$ such that $n(b) = 0$ and $ab = 0$. Since

$T(e_1, e_3) = T(e_1, b_{23}) = 0 = T(a_{12}, e_3) = T(a_{12}, b_{23})$ and

$V_{e_1,e_3} = V_{e_1,b_{23}} = 0 = V_{a_{12},e_3} = V_{a_{12},b_{23}}$, use of either condition to define incidence would result in two lines, $\ell = e_3^*$ and $m = b_{23}^*$ incident to the two distinct points P and Q. By analogy with [4] or [12], we define two relations between points and lines

incident: $A_* | B^*$ if and only if $V_{A,B} = 0$

neighbor: $A_* \approx B^*$ if and only if $T(A,B) = 0$.

It is convenient to define two further relations, which we again call neighbor relations and denote by $\approx$:

$A_* \approx B_*$ if and only if $A_* \approx C^*$ for all C^* with $B_* | C^*$

$A^* \approx B^*$ if and only if $C_* \approx A^*$ for all C_* with $C_* | B^*$

Note: In both [4] and [12] the relation $\sim$ is called connected.

Our study of the geometric properties of $\mathcal{P}(\mathbb{O})$ will be carried out, for the most part, via explicit calculations in $H(\mathbb{O}_3)$. To simplify matters, we list several properties of this Jordan algebra, all of which can be verified directly [4].

(11)

$$\text{i) } V_{A,B}(C) = T(C,B)A + T(A,B)C - (A \times C) \times B$$

$$\text{ii) } T(V_{A,B}(C),D) = T(C,V_{B,A}(D))$$

$$\text{iii) } T(U_A(B),C) = T(B,U_A(C))$$

$$\text{iv) } V_{A,B}V_{C,D} - V_{C,D}V_{A,B} = V_{\{ABC\},D} - V_{C,\{BAD\}}$$

where $\{ABC\} = V_{A,B}(C) = U_{A,C}(B)$.

We begin our study of $\mathcal{P}(\mathbb{O})$ by noting that the elements e_i , a_{ij} (for $n(a) = 0$) , and $a[i,j] = e_i + a_{ij} + n(a)e_j$ are all elements of $\pi(\mathbb{O})$, as is $[\alpha,\beta,a]_{ij} = \alpha e_i + \beta e_j + a_{ij}$ if $n(a) = \alpha\beta$ (by (7)). This notation makes simple the description of the points incident to the line e_i^* or lines incident to e_{i*} . We treat one case, the others are entirely analogous.

6.1 Proposition. $e_{3*}|\ell$ if and only if $\ell = [\alpha,\beta,a]_{12}^*$ for $a \in \mathbb{O}$.

Proof. If $A = [\alpha,\beta,a]_{12}$, $T(e_3,A) = 0$ and $(e_3 \times C) \times A = T(C,A)e_3$ for any $C \in H(\mathbb{O}_3)$ (by explicit computation) so $V_{e_3,A} = 0$ by 11i) so $e_{3*}|\ell$ if $\ell = A^*$.

Conversely, $V_{e_3,A}(C) = 0$ implies

$0 = T(V_{e_3,A}(C),e_3) = T(C,V_{A,e_3}(e_3)) = T(C,A+T(A,e_3)e_3-(Axe)xe_3)$ for all $C \in H(\mathfrak{G}_3)$. Thus $A = (Axe_3)xe_3 - T(A,e_3)e_3$. Since $T((Axe_3)xe_3,e_3) = 0$ we see $T(A,e_3) = 0$ so $A = (Axe_3)xe_3$ and $A = [\alpha,\beta,a]_{12}$ for some α,β and a.

The task of verifying geometric properties of $\mathcal{P}(\mathfrak{G})$ is simplified by the observation that $\mathcal{P}(\mathfrak{G})$ is self-dual. Indeed, since $T(A,B) = T(B,A)$ (see (9)) and $V_{A,B} = 0$ if and only if $V_{B,A} = 0$ (by (11),ii) and nondegeneracy of $T(,))$ we have

6.2 <u>Lemma</u>. $A_* \to A^*$ is a polarity of $\mathcal{P}(\mathfrak{G})$

where by <u>polarity</u> we understand a mapping ρ of points to lines and lines to points, preserving incidence and neighboring and satisfying $\rho^2 = id$.

§7. Algebraic transvections in $\mathcal{P}(\mathfrak{G})$

By analogy with the situation for projective planes, we define a collineation of $\mathcal{P}(\mathfrak{G})$ to be a bijective mapping of points to points, lines to lines, preserving incidence and neighboring. A <u>transvection</u> with center P and axis ℓ is then defined exactly as in §2. Just as the group $SL(3,\mathbb{R})$ gave rise to collineations of the real projective plane in the homogeneous coordinate realization, collineations of $\mathcal{P}(\mathbb{O})$ can be induced by a(n) (algebraic) group of transformations of $H(\mathbb{O}_3)$. We denote by $\mathcal{G}$ <u>the group of k-linear transformations S of $H(\mathbb{O})_3$ satisfying $\det S(A) = \det A$ for all</u> $A \in H(\mathbb{O}_3)$. For such an S we denote by S^* the transformation satisfying $T(S(A),B) = T(A,S^*(B))$ and by $\hat{S}$ the transformation $\hat{S} = S^{*-1}$.

It is not difficult (see [4]) to verify that $\hat{S} \in \mathcal{G}$ if $S \in \mathcal{G}$ and that

$$\text{i) } SU_A S^* = U_{S(A)}$$

$$\text{(12)} \qquad \text{ii) } S(A)^\# = \hat{S}(A^\#)$$

$$\text{iii) } SV_{A,B}S^{-1} = V_{S(A),\hat{S}(B)} \quad \text{for all } A, B \in H(\mathbb{O}_3).$$

It is immediate that $S(A)$, $\hat{S}(A)$ are in $\pi(\mathbb{O})$ if A is and the induced mapping σ of $\mathcal{P}(\mathbb{O})$ defined by $\sigma(A_*) = S(A)_*$; $\sigma(A^*) = \hat{S}(A)^*$ is a collineation.

For the real projective plane, it was easy to produce matrices which induced transvections in $\mathcal{P}(\mathbb{R})$ with certain axes and centers. In particular, the transvections with center e_{j*}, axis e_i^*, were induced by the action of the elementary matrices E_{ij}^a, $a \in \mathbb{R}$. We define analogous elementary transformations S_{ij}^a, $a \in \mathbb{O}$, as follows: $S_{ij}^a = I + V_{a_{ij},e_i} + U_{a_{ij}}U_{e_i}$. (In passing we note that this is a well known operator in Jordan theory). It is not too difficult (see [4], p.17) to verify that $S_{ij}^a \in \mathcal{G}$ for all i, j, a. Moreover, one computes explicitly that for $S = S_{ji}^a$:

$$S(e_i) = e_i \qquad S(b_{kj}) = b_{kj} + (ba)_{ki}$$

$$S(e_j) = a[ji] \qquad S(b_{ik}) = b_{ik}$$

$$S(e_k) = e_k \qquad S(b_{ji}) = b_{ji} + n(b,a)e_i \qquad i, j, k \neq . \tag{13}$$

$$\hat{S} = S_{ij}^{-\bar{a}}$$

The last assertion can be easily verified as a consequence of the facts that $(S_{ij}^{a})^{-1} = S_{ij}^{-a}$, that

$(V_{a_{ij},e_i} + U_{a_{ij}} U_{e_i})^* = V_{e_i,a_{ij}} + U_{e_i} U_{a_{ij}}$ (by (11)), and that

$V_{e_i,a_{ij}} = V_{\bar{a}_{ji},e_j}$ and $U_{a_{ij}} U_{e_i} = U_{e_j} U_{\bar{a}_{ji}}$.

We denote by σ_{ij}^{a} the collineation of $\mathcal{P}(\mathfrak{G})$ induced by S_{ij}^{a} . (14) and (6.1) combine to show that σ_{ij}^{a} fixes every point on e_i^* and, since $\hat{S}_{ij}^{a} = S_{ji}^{-\bar{a}}$, every line incident to e_j^* . We have thus

7.1 <u>Lemma</u> σ_{ij}^{a} is a transvection with axis e_i^* , center e_j^* .

Another simple computation shows

$$\sigma_{ij}^{a} \sigma_{ij}^{b} = \sigma_{ij}^{a+b} . \tag{14}$$

Of particular geometric interest in the study of $\mathcal{P}(\mathfrak{G})$ is the group of collineations generated by transvections (in the real projective plane we saw this was precisely the group induced by $SL(3,\mathbb{R})$) . The analogous result is also true in $\mathcal{P}(\mathfrak{G})$ -- this is the group induced by $\mathfrak{g}$ -- but the proof is somewhat beyond the scope of these lectures. We will content ourselves here with a study of the transitivity properties of the group and with a proof that the group is induced by a subgroup of $\mathfrak{S}$. Our investigation of transitivity has two-fold purpose. On the one hand, it provides the information necessary to prove that $\mathfrak{P}(\mathfrak{G})$ admits all possible transvections. On the other hand, it provides us with a means of simplifying the computations necessary to verify the geometric properties of the plane. We denote by U_{ij} , $1 \leq i,j \leq 3$, the group of all σ_{ij}^{a} , $a \in \mathfrak{G}$, by U^+ (resp. U^-) the group generated by $U_{31} \cup U_{32}$ (resp. $U_{13} \cup U_{23}$) , and by Σ the group generated by U^+ and U^- .

7.2 Lemma. U^+ acts transitively on $\{P : P \neq e_3^*\}$

Proof. Let $P = A_*$, $A = \begin{pmatrix} * & * & \bar{a} \\ * & * & \bar{b} \\ a & b & \gamma \end{pmatrix}$ where we may assume $\gamma=1$ since $P \neq e_3^*$. After applying $\sigma_{32}^{-b}\,\sigma_{31}^{-a}$ we may assume $A = \begin{pmatrix} * & * & 0 \\ * & * & 0 \\ 0 & 0 & 1 \end{pmatrix}$, whence $A^{\#} = 0$ implies $A = e_3$. Transitivity follows.

7.3 Proposition. Σ acts transitively on points (dually on lines).

Proof. It suffices to show that any point can be translated to e_3^* by suitably chosen transvections. Let $P = A_*$ with A as in (1). If $\alpha_3 \neq 0$, we may apply (7.2). If $\alpha_3 = 0$, then the coefficient of e_3 in $S(A)$ for $S = S_{13}^{b}$ (resp. S_{23}^{b}) is $\alpha_1 n(b) + n(a_2, b)$ (resp. $\alpha_2 n(b) + n(a_1, b)$). Thus we may reduce to the case $\alpha_3 \neq 0$ unless $\alpha_1 = \alpha_2 = 0$ and $a_1 = a_2 = 0$. In this last case, we may apply σ_{23}^{1} to reduce to the case $a_1 \neq 0$.

For the dual, we need only act on $\ell = A^*$, noting that if S induces σ in $U^{\pm}$, $\hat{S}$ induces a collineation in $U^{\mp}$.

7.4 Corollary (to proof). The stabilizer of e_1^* and e_{1*} in Σ acts transitively on points of e_1^* (and dually).

Proof. Clearly U_{32} and U_{23} are contained in the stabilizer. If $P|e_1^*$, $P = [\alpha,\beta,a]_{23}$ by (6.1). The proof of (7.3) shows that P can be moved to e_{3*} using only elements of U_{23} and U_{32}.

We note also an interesting geometric corollary.

7.5 Proposition. $P|\ell \Rightarrow P \approx \ell$.

Proof: By (7.3) we may assume $P = e_{3*}$, so by (6.1) $\ell = ([\alpha,\beta,a]_{12})^*$. Clearly $T(e_3,[\alpha,\beta,a]_{12}) = 0$ so $P \approx \ell$. Returning to transitivity questions we have

7.6 Proposition. Σ acts transitively on $\{(P,\ell)\colon P \not| \ell\}$.

Proof. By (7.3) (dual) we may assume $\ell = e_3^*$, $P \not| e_3^*$. By (7.2) we may map P to e_{3*} while leaving e_3^* invariant. Thus (P,ℓ) is conjugate to (e_{3*}, e_3^*) and the result follows.

7.7 Proposition. Σ acts transitively on $\{(P,Q)\colon P \not\approx Q\}$ (and dually).

Proof. If $P \not\approx Q$, there is by definition a line ℓ with $Q|\ell$, $P \not| \ell$. By (7.6) we may assume $P = e_{1*}$, $Q|\ell = e_1^*$. By (7.4) we may map P to e_{1*}, Q to $e_{3*}|e_1^*$. Transitivity follows.

7.8 <u>Proposition</u>. Let $A, B \in \pi(\mathfrak{G})$. Then

i) $A_* \approx B_*$ ($A^* \approx B^*$) if and only if $A \times B = 0$

ii) If $A_* \not\approx B_*$ (resp. $A^* \not\approx B^*$) then $(A \times B)^*$ (resp. $(A \times B)_*$) is the unique line (resp. point) incident to both.

<u>Proof</u>: By duality it suffices to restrict ourselves to A_* and B_*. Suppose first that $A_* \not\approx B_*$. By (7.7) there is $S \in \mathfrak{S}$ with $S(A) = \lambda e_2$, $S(B) = \mu e_3$. By (12,ii) linearized, we have $\hat{S}(A \times B) = S(A) \times S(B) = \lambda e_2 \times \mu e_3 = \lambda\mu e_1 \neq 0$ so $A \times B \neq 0$. Moreover, since $\sigma((A \times B)^*) = \hat{S}(A \times B)^* = e_1^*$ and $e_{i*}|e_1^*$ for $i = 2, 3$, it follows that $A_*|(A \times B)^*$ and $B_*|(A \times B)^*$.

A close look at the analog of (6.1) for e_{2*} and e_{3*} shows that e_1^* is the only line incident to both. The uniqueness of the line incident to A_* and B_*, hence ii), follows.

Finally, if $A_* \approx B_*$, we may assume by (7.3) that $S(A) = \lambda e_1$ for some $S \in \mathfrak{S}$. Since $S(B)_* \approx \ell$ for all ℓ incident to e_{1*}, $T(S(B),C) = 0$ for all $[\alpha,\beta,a]_{12}$ by (6.1). An easy argument then shows $0 = e_1 \times S(B) = S(A) \times S(B) = \hat{S}(A \times B)$ so $A \times B = 0$ and i) is proved.

Since $A \times B = B \times A$ we have

7.9 <u>Corollary</u>: $A_* \approx B_*$ (resp. $A^* \approx B^*$) if and only if $B_* \approx A_*$ (resp. $B^* \approx A^*$).

We call a triangle P_1, P_2, P_3 in $\mathfrak{P}(\mathfrak{G})$ <u>regular</u> if $P_1 \not\approx P_2$ and $P_3 \not| P_1 \vee P_2$ (it is an immediate consequence of the definitions that now no vertices of the triangle neighbor, no sides neighbor, and no vertex neighbors a side to which it is not incident). Regular triangles, of course, play an important role in the coordinzation of projective planes. We have

7.10 Theorem. Σ acts transitively on regular triangles.

Proof: It suffices to show for any regular triangle P_1, P_2, P_3, there is $\sigma \in \Sigma$ with $\sigma(P_i) = e_{i*}$. By (7.7), we may assume $P_1 = e_{1*}$, $P_2 = e_{2*}$ and $P_3 = A_*$. By assumption, $A_* \not\approx e_{1*} \vee e_{2*} = e_3^*$. By (7.2) we can map A_* to e_{3*} while fixing e_{1*} and e_{2*}. Transitivity follows.

§8. Axiomatization and coordinatization of $\mathcal{P}(\mathfrak{G})$.

In §7, while studying transvections, we made several forays into the purely geometric. The properties noted there bear a striking resemblance to those introduced by Barbilian [1] and refined by Veldkamp [13]. We recall these properties:

- P1: $P|\ell$ implies $P \approx \ell$
- P2: If $P \not\approx Q$ there is a unique line (notation: $P \vee Q$) incident to both
- P2': If $\ell \not\approx m$ there is a unique point (notation: $\ell \wedge m$) incident to both
- P3: If $\ell \not\approx m$, $P|\ell$ and $\ell \wedge m \not\approx P$, then $P \not\approx m$
- P4: Every line is incident to at least one point
- P5: For every pair of points, there is a line which neighbors neither
- P6: There is a line

We call a plane satisfying P1-P6 a Barbilian plane.

8.1 Theorem. $\mathcal{P}(\mathfrak{O})$ is a Barbilian plane.

Proof: We have already proved P1 (7.5), P2 and P2' (7.8), P4 (6.1 and 7.3) and P6 .

For P3, we may assume, by (7.7) that $\ell = e_1^*$, $m = e_2^*$ and hence (by 7.8) that $\ell \wedge m = e_{3*}$. Now if $P = A_*$, $A = [\alpha,\beta,a]_{23}$ by (6.1) where $\alpha \neq 0$ since $P \not\sim e_{3*}$ implies $0 \neq A \times e_3 = \alpha e_1$ by (7.7) . Thus $\alpha = T(A,e_2) \neq 0$ so $P \not\sim m$.

For P5, if $P \not\sim Q$ we may assume by (7.7) that $P = e_{1*}$, $Q = e_{2*}$. Then $\ell = A^*$ with $A = [1,1,1]_{12}$ neighbors neither P nor Q . If $P \approx Q$, we may assume $P = e_{1*}$, $Q = B_*$ where

$$e_1 \times B = 0 \text{ , so } B = \begin{pmatrix} \alpha & a & \bar{b} \\ \bar{a} & 0 & 0 \\ b & 0 & 0 \end{pmatrix} .$$ If $a \neq 0$ there is $c \in \mathfrak{O}$ with $n(c) = 0$, $n(c,a) \neq -\alpha$. For $C = [1,0,c]_{12}$, one computes easily that $T(e_1,c) \neq 0 \neq T(B,C)$ so $C^* \not\sim P$, $C^* \not\sim Q$. A similar argument handles the case $a = 0$, $b \neq 0$.

A straightforward argument due to Veldkamp [13], which we will not reproduce here, yields

8.2 Proposition. In any Barbilian plane (hence in $\mathcal{P}(\mathfrak{O})$) , there is at most one transvection σ with axis ℓ , center P , such that $\sigma(P_1) = P_2$ where $P_1 \not\sim \ell \not\sim P_2$.

We will see that $\mathcal{P}(\mathfrak{O})$ can be coordinatized in a natural manner by the algebra $\mathfrak{O}$ which is alternative, but not associative.

In the study of abstract projective planes, the occurrence of such a coordinate ring suggests that the plane is a transvection plane or, equivalently, that some configurational condition such as the little quadrangle section theorem must hold. Following Faulkner [3] we consider the two conditions

P7: If $P \nmid \ell \nmid Q$ and $R|\ell$ with $Q|P \vee R$, there is a transvection σ with axis ℓ, center R such that $\sigma(P) = Q$.

P8: If P_1, P_2, P_3 is a regular triangle, $M \in \mathcal{A}_{12}$ and $N \in \mathcal{A}_{21}$, then there is $Q \in \mathcal{A}_{12}$ such that $X*Q = ((X*M)*N)*M$ for all $X \in \mathcal{A}_{31}$ (and analogously in the dual plane).

Here $\mathcal{A}_{ij} = \{P : P|P_i \vee P_j, P \nmid P_j\}$ and $U * V = (U \vee V) \wedge (P_i \vee P_k) \mathcal{A}_{ik}$ for $U \in \mathcal{A}_{ij}$, $V \in \mathcal{A}_{jk}$, $i, j, k \neq$. Note that P8 is proved for projective transvection planes as (4.3).

8.3 Proposition: P7 and P8 hold in $\mathcal{P}(\mathfrak{G})$.

Proof: For P8, we apply (7.10) to assume $P_i = e_{i*}$. Letting $X = A_*$, $M = B_*$, $N = C_*$, we see from the hypotheses that $A = a[3,1]$, $B = b[1,2]$, $C = c[2,1]$. If $((X*M)*N)*M = D_*$, $D = (((((A \times B) \times e_1) \times C) \times e_2) \times B) \times e_1 = d[3,2]$. A straightforward computation yields $d = -((ab)c)b$. Since $\mathfrak{G}$ is alternative, $d = -a(bcb)$ so $Q = (bcb)[1,2]$ satisfies the condition. The second condition is handled analogously.

For P7, we suppose P, Q, R and ℓ are as given and note, using (7.3), (7.4) and (6.1), that there is $S|\ell$ with $S \neq R$. By (7.10), there is $\tau \in \Sigma$ such that $\tau(R) = e_{1*}$, $\tau(S) = e_{2*}$ (hence $\tau(\ell) = e_3^*$)

and $\tau(P) = e_{3*}$. It follows from (7.8) that $\tau(Q)|e_2^*$ and $\tau(Q)\nmid e_3^*$. From (6.1) follows then that $\tau(Q) = A^*$, $A = a[3,1]$. By (7.1), $\sigma = \tau^{-1}\sigma_{31}^{a}\tau$ is thus a transvection with axis ℓ, center R, such that $\sigma(R) = Q$ as required.

In the special case that $\mathfrak{G}$ is an octonion division algebra, we obtain the result asserted in (5.2).

8.4 Corollary. If $\mathfrak{G}$ is octonian division algebra, $\mathfrak{P}(\mathfrak{G})$ is a projective transvection plane.

Proof. We show first that if $\mathfrak{G}$ is a division algebra, $P \approx \ell$ implies $P|\ell$. Thus, we assume $P\approx\ell$ and, by (7.3), that $P = e_{1*}$. $P\approx\ell$ implies $\ell=B^*$, $B = \begin{pmatrix} 0 & * & * \\ * & * & * \\ * & * & * \end{pmatrix}$. $B^\# = 0$ implies, by (7), that

$n(b_2) = n(b_3) = 0$, so $b_2 = b_3 = 0$. If $b_1 \neq 0$, $b_1\overline{b}_1 \neq 0$ so $\beta_2 \neq 0 \neq \beta_3$. If $b_1 = 0$, either $\beta_1 = 0$ or $\beta_2 = 0$. It follows that $B = [\alpha,\beta,b_1]_{23}$ so $P|\ell$ by (6.1). P6, P2, and P2′, together with the fact that e_{1*}, e_{2*}, e_{3*} and $\begin{pmatrix} 1 & 1 & 1 \\ 1 & 1 & 1 \\ 1 & 1 & 1 \end{pmatrix}$ are a quadruple of pairwise noncollinear points, imply that $\mathfrak{P}(\mathfrak{G})$ is indeed a projective plane. In this setting, P7 is precisely the requisite condition for $\mathfrak{P}(\mathfrak{G})$ to be a transvection plane.

We prove, for later use,

8.5. Theorem: Σ is the group generated by all transvections of $\mathfrak{P}(\mathfrak{G})$.

Proof. Clearly Σ, generated by transvections σ_{ij}^{a}, is contained in the group generated by all transvections. Conversely, if $\sigma\neq id$ is a transvection with axis ℓ , center $R|\ell$, we may pick $P\nmid\ell$ and set $Q=\sigma(P)$ (Q necessarily incident to $R \vee P$). For τ as in the proof of (8.3), $\tau\sigma\tau^{-1}$ is a transvection with center e_{1*} , axis e_3^* which carries e_{3*} to a[3,1]. Since σ_{31}^a has the same properties,

$\tau\sigma\tau^{-1} = \sigma_{31}^a$ by (8.2), hence $\sigma = \tau^{-1}\sigma_{31}^a\tau \in \Sigma$.

We now turn our attention to the problem of finding a coordinate ring for a plane satisfying P1-P8, as usual led by the analogous procedure in projective transvection planes. In the projective case one selects a line ℓ to play the role of the line at ∞ and then coordinatizes the affine points, i.e. points not incident to ℓ . A close comparision of P2, P2', and P7 with the analogous axioms for a projective plane suggests that the relevant _affine subplane_ in this setting should be $\{P:P \not\approx \ell\}$.

Suppose we now, with this one modification, attempt to mimic the coordinatization procedure used in §4, choosing as reference triangle e_{1*} , e_{2*} , e_{3*} , selecting e_{3*} to play the role of origin, $e_3^* = e_{1*} \vee e_{2*}$ to be the line at ∞ , $e_2^* = e_{3*} \vee e_{1*}$ to be the "x-axis". The coordinate ring $\mathcal{O}$ is identified with the affine points of e_2^* , i.e. with $\{P:P|e_2^* , P\neq e_{1*}\}$. By (6.1) this is just $\{a[31]:a \in \mathfrak{G}\}$, which corresponds bijectively to $\mathfrak{G}$.

We define addition in $\mathcal{O}$ by $\sigma_{31}^{P+Q} = \sigma_{31}^P\sigma_{31}^Q$ where σ_{31}^X is the unique transvection with axis e_3^*, center e_{1*}, mapping e_{3*} to X. If $P = A_*$, $Q = B_*$ where $A = a[31]$ and $B = b[31]$, direct computation using (14) yields $P + Q = C_*$ where $C = (a + b)[12]$.

To multiply in $\mathcal{A}$, we choose $U_1 = 1[23]_*$, $U_2 = 1[31]_*$ and $U_3 = (-1)[12]_* = U_1 * U_2$. Then $PQ = (P * U_3) * (U_1 * Q)$. For $P = A*$, $Q = B*$ as above,
$PQ = ((((A \times (-1)[12]) \times e_1) \times ((1[32] \times B) \times e_3)) \times e_2)_* = ab[31]$.
It follows that

8.6 Proposition. The coordinate ring of $\mathbb{P}(\mathbb{O})$, relative to the regular triangle e_{1*} , e_{2*} , e_{3*} is isomorphic (via $a \to a[31]_*$) to $\mathbb{O}$. The affine subplane in $\mathbb{P}(\mathbb{O})$ relative to the line at ∞, e_1^*, can be identified with

$$\{(a,b) \mid a,b \in \mathbb{O}\} \quad \text{via} \quad (a,b) \to \begin{pmatrix} n(a) & \bar{a}b & \bar{a} \\ \bar{b}\,a & n(b) & \bar{b} \\ a & b & 1 \end{pmatrix}.$$

§9. $\mathbb{P}(\mathbb{O})$ as homogeneous space

In our discussion of the similarities between the real projective plane and $\mathbb{P}(\mathbb{O})$, we have followed a somewhat contorted path of constructions of projective planes, beginning with the simple homogeneous coordinate view of $\mathbb{P}(\mathbb{R})$, ending with $\mathbb{P}(\mathbb{O})$ constructed with the help of the Jordan algebra $H(\mathbb{O}_3)$. Once we move into the realm of planes with non-division algebra coordinates, as we have with the Hjelmslev-Moufang plane, we are tempted to move yet another step along the path to obtain "planes" with arbitrary alternative rings as coordinates. The construction via Hermitian matrices will no longer suffice for our needs, since alternative rings need not admit involutions. Our development of the real projective plane suggests an alternative--the view of the plane as a homogeneous space for a certain group. In this section we shall investigate $\mathcal{P}(\mathfrak{G})$ as a homogeneous space, basing our considerations still on the realization of $\mathcal{P}(\mathfrak{G})$ in Jordan terms. In §10 we will free ourselves from the Jordan algebra, thus opening the way to generalization.

Following the example of the projective plane, we seek to describe the points and lines of $\mathcal{P}(\mathfrak{G})$ in terms of cosets of the group Σ generated by transvections, and to give group theoretic definitions of incident and neighbor. We denote by P^- the stabilizer in Σ of e_{3*}, by P^+ the stabilizer of e_3^*. We know, by (7.3) that Σ acts transitively on both points and lines, hence $G/_{P^-}$ (resp. $G/_{P^+}$) corresponds bijectively to the set of all points (resp. lines) of $\mathcal{P}(\mathfrak{G})$, via the correspondence $\phi P^- \leftrightarrow \phi(e_{3*})$, $\phi P^+ \leftrightarrow \phi(e_3^*)$. (7.6) implies that $P \not\approx \ell$ if and only if $P = \phi(e_{3*})$, $\ell = \phi(e_3^*)$ for some $\phi \in \Sigma$. If $P \leftrightarrow \Phi_1 P^-$, $\ell \leftrightarrow \phi_2 P^+$ this is equivalent to $\phi_1 P^- \cap \phi_2 P^+$ nonempty (note the analogy with incidence in §2). Combining (7.3) and (7.4) we see that $P|\ell$ if and only if $P = \phi(e_{3*})$, $\ell = \phi(e_2^*)$, if and only if $P = \phi(e_{3*})$, $\ell = \phi w(e_3^*)$ where $w = \sigma_{23}^{1}\sigma_{32}^{-1}\sigma_{23}^{1}$ (w was chosen to satisfy $w(e_3{*}) = e_2^*$). Thus for P, ℓ as above, $P|\ell$ if and only if for some $\phi \in \Sigma$, $\phi \in \phi P_1^-$, and $\phi w \in \phi P_2^+$.

The above considerations lead us, given any group G, subgroups P^-, P^+, and distinguished element w, to define a geometry $\mathcal{P}(G,P^-,P^+,w)$ as follows:

point set: $G/_{P^-}$

line set: $G/_{P^+}$

(15) incident: $\phi_1 P^- | \phi_2 P^+$ if and only if $\phi_1 P^- \cap \phi_2 w^{-1}(wP^+ w^{-1})$ is nonempty

neighbor: $\phi_1 P^- \approx \phi_2 P^+$ if and only if $\phi_1 P^- \cap \phi_2 P^+$ is empty

We now have

9.1 <u>Proposition</u>: $\mathbb{P}(\mathbb{O})$ is isomorphic to $\mathbb{P}(\Sigma, P^-, P^+, w)$ via the isomorphism $\phi_1 P^- \to \phi_1(e_{3*})$; $\phi_1 P^+ \to \phi_1(e_3^*)$.

By (9.1) and the definitions of Σ , P^- , P^+ , w , we can now reconstruct the plane $\mathbb{P}(\mathbb{O})$ directly from its affine coordinate ring (see (8.6)) by constructing $H(\mathbb{O}_3)$, defining $\mathcal{S}$ algebraically acting on $H(\mathbb{O}_3)$ and noting that only those elements of $\mathcal{S}$ acting as scalars on $H(\mathbb{O}_3)$ act trivially on $\mathbb{P}(\mathbb{O})$ so Σ can be defined algebraically as $\mathcal{S}/_Z$, Z the set of scalar transformations. Once Σ is defined algebraically, P^- , P^+ and w can also be easily described, and the geometry abstractly constructed. If $\mathbb{O}$ were alternative, but admitted no involution, this procedure would fail for lack of the representation space $H(\mathbb{O}_3)$ for $\mathbb{O}$. In §10 we propose an alternate definition for Σ which is more general than that using $H(\mathbb{O}_3)$. Before doing so, we look closely at the generation of Σ , P^+ and P^- in terms of certain (geometrically) distinguished subgroups.

In §7 we have seen that the groups U^+ and U^- generated by transvections σ_{ij}^a with axis e_3^* (resp. center e_{3*}), combine to generate the group Σ . Clearly $U^{\pm} \subset P^{\pm}$. We denote by L the stabilizer of e_{3*} and e_3^* in Σ .

9.2 Proposition. $P^{\pm} = U^{\pm}L$

Proof. We consider only the case for P^+, that for P^- is analogous. If $\sigma \in P^+$, $e_3^* = \sigma(e_3^*)$. If $Q = \sigma(e_{3*})$, $Q \neq e_3^*$ so by (7.2) there is $\tau \in U^+$ with $\tau(Q) = e_{3*}$. Thus $\tau\sigma \in L$ so $\sigma \in \tau^{-1}L \subseteq U^+L$.

§10. Another realization of Σ

In our search for a non-Jordan construction of $\mathcal{P}(\mathbb{O})$ we seek first another description of Σ . One such description is suggested by the fact that Σ is an algebraic group of adjoint type [11], thus acts faithfully on its Lie algebra which is known to be isomorphic to the "structure algebra" of $H(\mathbb{O}_3)$[9]. We shall later construct the relevant Lie algebra in a manner free of $H(\mathbb{O}_3)$. In this section, however, we introduce it as $\mathcal{L} = \text{span}\,\{V_{A,B} : A,B \in H(\mathbb{O}_3)\}$. It is not difficult to see, using (12), that $\mathcal{L}$ is closed under the product $[w_1,w_2] = w_1w_2 - w_2w_1$ and that for $S \in \mathcal{S}$, $w \in \mathcal{L}$, $SwS^{-1} \in \mathcal{L}$. $\mathcal{S}$ thus acts on $\mathcal{L}$ via Lie algebra automorphisms with the kernel of the action being the set of scalar transformations in $\mathcal{S}$. This action induces a faithful action of $\Sigma = \mathcal{S}/Z$ on $\mathcal{L}$. We denote the group of induced automorphisms by $\tilde{\Sigma}$.

We wish to consider a particular decomposition of $\mathcal{L}$ into subspaces on which the action of the transvections σ_{ij}^{a} is quite easy to describe explicitly. Following Jordan algebra notation (this is the Pierce 1-space for e_3) , we set

$\mathcal{J}_1 = \{a_{32} + b_{31} | a,b \in \mathbb{O}\}$ and define $N^+ = \{V_{A,e_3} : A \in \mathcal{J}_1\}$, $N^- = \{V_{e_3,A} : A \in \mathcal{J}_1\}$. Further, we define $D_{A,B} = [V_{A,e_3}, V_{e_3,B}]$ (where $[XY] = XY - YX$) and denote by $\mathcal{D}$ the subspace of $\mathcal{L}$ spanned by all $D_{A,B}$, A, $B \in \mathcal{J}_1$. It can be checked (see e.g. [9]) that $\mathcal{L} = N^+ \oplus \mathcal{D} \oplus N^-$. Straightforward, if tedious, computation yields the explicit products

(16)

$$\text{i)}\quad [N^+, N^+] = 0 = [N^-, N^-]$$

$$\text{ii)}\quad D_{A,B} = V_{A,B} - T(A,B)V_{e_3,e_3}$$

$$\text{iii)}\quad [D_{B,C}, V_{A,e_3}] = V_{\{BCA\},e_3}$$

$$\text{iv)}\quad [D_{B,C}, V_{e_3,A}] = -V_{e_3,\{CBA\}}$$

$$\text{v)}\quad [D_{A,B}, D_{C,E}] = D_{\{ABC\},E} - D_{C,\{BAE\}}$$

It follows from (16) that $N^+ \oplus N^-$ generates $\mathcal{L}$. Since $\mathcal{S}$, acting on $\mathcal{L}$ by conjugation, acts by Lie algebra isomorphisms, the action of each element of $\mathcal{S}$ is completely determined by its action on N^+ and N^-. More lengthy computation using (12,iii) leads to the following table showing the action of the transformations inducing the groups U_{ij}.

Group	conjugation by	maps	to
U_{12}	S^a_{12}	V_{A,e_3}	$V_{A+\{a_{12}e_1A\}, e_3}$
		$V_{e_3,A}$	$V_{e_3,A-\{a_{12}e_2A\}}$
U_{21}	S^a_{21}	V_{A,e_3}	$V_{A-\{a_{21}e_2A\},e_3}$
		$V_{e_3,A}$	$V_{e_3,A+\{a_{21}e_1A\}}$
(17) U_{13}	S^a_{13}	V_{A,e_3}	$V_{A,e_3} - D_{A,a_{13}} - V_{e_3, U_{a_{13}}(A)}$
		$V_{e_3,A}$	$V_{e_3,A}$
U_{31}	S^a_{31}	V_{A,e_3}	V_{A,e_3}
		$V_{e_3,A}$	$V_{e_3,A} + D_{a_{31},A} - V_{U_{a_{31}}(A),e_3}$
U_{23}	S^a_{23}	V_{A,e_3}	$V_{A,e_3} - D_{A,a_{23}} - V_{e_3, U_{a_{23}}(A)}$
		$V_{e_3,A}$	$V_{e_3,A}$
U_{32}	S^a_{32}	V_{A,e_3}	V_{A,e_3}
		$V_{e_3,A}$	$V_{e_3,A} + D_{a_{32},A} - V_{U_{a_{32}}(A),e_3}$

We have noted previously that $\tilde{\Sigma}$ is (naturally) isomorphic to Σ, hence is generated by the subgroups $\tilde{U}_{ij}$ induced by the elements S^a_{ij}. We need also to describe the groups $\tilde{P}^-$ and $\tilde{P}^+$ in purely Lie algebra terms. Following (9.2) we need to describe an analog of L. Clearly $L \subseteq \Sigma$ is induced by transformations $S \in \mathcal{S}$ such that $S(e_3) = \lambda e_3$, $\hat{S}(e_3) = \mu e_3$, $\lambda\mu \neq 0$. It is clear from (12,iii) that conjugation by any such S leaves invariant both N^+ and N^-. Led by this observation, we <u>define</u> $\tilde{L} \subseteq \tilde{\Sigma}$ to be the set of all elements of $\tilde{\Sigma}$ stabilizing N^+ and N^-. Clearly $\tilde{L} \supseteq \tilde{U}_{12} \cup \tilde{U}_{21}$ as expected (by (17)). We define further $\tilde{P}^- = \tilde{U}^-\tilde{L}$, $\tilde{P}^+ = \tilde{U}^+\tilde{L}$ where $\tilde{U}^+$ (resp. $\tilde{U}^-$) is generated by $\tilde{U}_{32} \cup \tilde{U}_{31}$ (resp. $\tilde{U}_{23} \cup \tilde{U}_{13}$). To complete the requirements for the construction of a geometry from $\tilde{\Sigma}$ as in (15), we need an element $\tilde{w} \in \tilde{\Sigma}$. We define $\tilde{w}$ to be induced by $S^1_{23}\, S^{-1}_{32}\, S^1_{23}$. It is indeed the case, though it is not easy to show, that the natural isomorphism $\Sigma \to \tilde{\Sigma}$ maps L onto $\tilde{L}$. It then follows easily that

(10.1) <u>Proposition</u>. The geometry $\mathcal{P}(\Sigma, P^-, P^+, w)$ of §9 is isomorphic to the geometry $\mathcal{P}(\tilde{\Sigma}, \tilde{P}^-, \tilde{P}^+, \tilde{w})$ (and both are isomorphic to $\mathcal{P}(\mathbb{O})$).

§11. Jordan pairs--a final look at the Hjelmslev-Moufang plane

At first glance, it may appear that in (10.1) we have succeeded in freeing ourselves from the Jordan algebra $H(\mathbb{O}_3)$ in defining a geometry isomorphic to $\mathcal{P}(\mathfrak{O})$ as a homogeneous space for a group of automorphisms of a Lie algebra. This is purely illusory, however, for our Lie algebra itself is defined as transformations acting on $H(\mathfrak{O}_3)$ in a manner prescribed by the Jordan structure. We seek here an intrinsic definition of a Lie algebra isomorphic to $\mathfrak{L}$, depending only on the structure of $\mathfrak{O}$ (which in turn is determined by the geometry of $\mathcal{P}(\mathfrak{O})$, (8.6)).

To achieve this we consider the pair (V^+,V^-) of vector spaces $V^+ = \{(a,b):a,b \in \mathfrak{O}\}$, $V^- = \{(a,b)^t:a,b \in \mathfrak{O}\}$ and define operators $Q_A^\sigma:V^{-\sigma} \to V^\sigma$, $\sigma = \pm 1$, $A \in V^\sigma$, by $Q_A^+(B) = A(BA)$; $Q_A^-(B) = (AB)A$, where the product of vectors is the usual matrix product. With these operations (since $\mathfrak{O}$ is alternative) $V = (V^+,V^-)$ becomes an algebraic structure known as a <u>Jordan pair</u> [7]. That is, for $\{A,B,C\} = Q_{A+C}^\sigma(B) - Q_A^\sigma(B) - Q_C^\sigma(B)$, the following hold:

JP1 $\quad \{A,B,Q_A^\sigma(C)\} = Q_A^\sigma(\{B,A,C\})$

JP2 $\quad \{Q_A^\sigma(B),B,C\} = \{A,Q_B^{-\sigma}(A),C\}$

JP3 $\quad Q^\sigma_{Q_A^\sigma(B)} = Q_A^\sigma Q_B^{-\sigma} Q_A^\sigma$

A second example of a Jordan pair will also interest us here--the pair $(\mathcal{J}_1,\mathcal{J}_1)$, $\mathcal{J}_1$ as in §10, with $Q_A^\sigma = U_A$. One can check that (V^+,V^-) is isomorphic to the pair $(\mathcal{J}_1,\mathcal{J}_1)$ via the "mapping" $\phi = (\phi^+,\phi^-)$ where $\phi^+((a,b)) = a_{31} + b_{32}$, $\phi^-((c,d)^t) = \bar{c}_{31} + \bar{d}_{32}$.

In any Jordan pair (V^+,V^-), one defines transformations $\delta^\sigma_{A,B}:V^\sigma \to V^\sigma$ for $A \in V^\sigma$, $B \in V^{-\sigma}$, by $\delta^\sigma_{A,B}(C) = \{ABC\}$ (in a Jordan pair such as $(\mathcal{J}_1,\mathcal{J}_1)$ induced from a Jordan algebra, the product $\{ABC\}$ is precisely that defined in the Jordan algebra and $\delta^\sigma_{A,B} = V_{A,B}$). For each $A \in V^+$, $B \in V^-$, we define a mapping of the pair V (i.e. a pair of mappings from $V^\pm$ to $V^\pm$) by $\delta_{A,B} = (\delta^+_{A,B},-\delta^-_{B,A})$. $\delta_{A,B}$ is in a natural sense a derivation of the pair V. The set spanned by $\{\delta_{A,B}:A \in V^+,B \in V^-\}$ is usually denoted by Inder (V). If we now form the vector space $\mathcal{L}(V) = V^+ \oplus \text{Inder } V \oplus V^-$, $\mathcal{L}(V)$ becomes a Lie algebra (because of the properties JP1, JP2, JP3) if we define:

i) $[V^\sigma,V^\sigma] = 0$

(18) ii) $[A,B] = \delta_{A,B}$ if $A \in V^+$, $B \in V^-$

iii) $[DD'] = DD' - D'D$ (as transformation of V) for D, $D' \in \text{Inder}(V)$

iv) $[D,A] = \delta^\sigma(A)$ for $D = (\delta^+,\delta^-)$, $A \in V^\sigma$.

That the product in iii) falls once again in Inder(V) follows from the Jordan pair identity:

$$[\delta^\sigma_{A,B},\delta^\sigma_{C,E}] = \delta^\sigma_{\{ABC\},E} - \delta^\sigma_{C,\{BAE\}} . \tag{19}$$

Comparing (16) and (18), in the light of the isomorphism $\phi:(V^+,V^-) \to (\mathcal{J}_1,\mathcal{J}_1)$ we see

11.1 Proposition. The Lie algebra $\mathcal{L}(V)$ is isomorphic to $\mathcal{L}$ via the mapping

$$\tilde{\phi} : \begin{cases} A \to V_{\phi^+(A),e_3} & , A \in V^+ \\ A \to V_{e_3,\phi^-(A)} & , A \in V^- \\ \delta_{A,B} \to D_{\phi^+(A),\phi^-(B)} & , A \in V^+ , B \in V^- \end{cases}$$

We have thus succeeded in constructing our Lie algebra in terms only of $\mathfrak{G}$. We need still to identify the analogues of $\tilde{\Sigma}$, $\tilde{P}^{\pm}$, $\tilde{U}_{ij}$, $\tilde{w}$ within Aut $\mathcal{L}(V)$. Of course, these analogues can be obtained very simply with the help of $\tilde{\phi}$. We take respectively $\tilde{\phi}^{-1}\tilde{\Sigma}\tilde{\phi}$, $\tilde{\phi}^{-1}\tilde{P}^{\pm}\tilde{\phi}$, etc. This still brings the structure of $H(\mathfrak{G}_3)$ into play, since it is involved in the definition of $\tilde{\Sigma}$, $\tilde{P}^{\pm}$, etc. so we seek intrinsic definitions of $\tilde{\phi}^{-1}\tilde{U}_{ij}\tilde{\phi}$, $\tilde{\phi}^{-1}\tilde{L}\tilde{\phi}$, and $\tilde{\phi}^{-1}\tilde{w}\phi$ in Aut $\mathcal{L}(V)$. The descriptions of all requisite groups follows from this. For convenience we denote $\tilde{\phi}^{-1}\tilde{U}_{ij}\tilde{\phi}$ by $\check{U}_{ij}$, similarly for other groups and elements.

Clearly, by definition of $\tilde{L}$, $\check{L}$ is the stabilizer of both V^+ and $V^- \subseteq (\mathcal{L}(V))$ in $\check{\Sigma}$ and $\check{\Sigma}$ is generated by $\{\check{U}_{ij}\}$. Now combining the explicit descriptions of $\tilde{U}_{ij}$ in (17) with the explicit form of $\tilde{\phi}$ we have the following table (where $(S^a_{ij})^{\vee}$ has the obvious meaning in $\Sigma^{\vee}$):

(20)

Group	Generated by	maps	to
U_{12}^{v}	$(S_{12}^{a})^{v}$	$A \in V^{+}$	$A+\{(0,a),\binom{1}{0},A\}$
		$A \in V^{-}$	$A-\{\binom{1}{0},(0,a),A\}$
U_{21}^{v}	$(S_{21}^{a})^{v}$	$A \in V^{+}$	$A-\{(1,0),\binom{0}{a},A\}$
		$A \in V^{-}$	$A+\{\binom{0}{a},(1,0),A\}$
U_{13}^{v}	$(S_{13}^{a})^{v}$	$A \in V^{+}$	$A-D_{A,\binom{a}{0}}-Q^{-}_{\binom{a}{0}}(A)$
		$A \in V^{-}$	A
U_{31}^{v}	$(S_{31}^{a})^{v}$	$A \in V^{+}$	A
		$A \in V^{-}$	$A+D_{(a,0),A}-Q_{(a,0)}(A)$
U_{23}^{v}	$(S_{23}^{a})^{v}$	$A \in V^{+}$	$A-D_{A,\binom{0}{a}}-\bar{Q}_{\binom{0}{a}}(A)$
		$A \in V^{-}$	A
U_{32}^{v}	$(S_{32}^{a})^{v}$	$A \in V^{+}$	A
		$A \in V^{-}$	$A+D_{(0,a)A}-Q_{(0,a)}(A)$

All results in this table are immediate with the exception of U_{12}^{v} and U_{21}^{v}. In these cases one must simply compute the given expressions in (17) and (20) and compare results. To complete our construction, we take w^{v} to be the product $w^{v} = (S_{23}^{1})^{v}(S_{32}^{-1})^{v}(S_{23}^{1})^{v}$ which is clearly intrinsically defined via (20). It is thus clear that

11.2 Theorem. $\mathcal{P}(\mathfrak{G})$ is isomorphic to the geometry $\mathcal{P}(\Sigma^V,(P^-)^V,(P^+)^V,w^V)$ constructed as in (15), Σ^V, $(P^\pm)^V$, w^V constructed as above.

§12. Abstract Moufang-Veldkamp planes

The Hjelmslev-Moufang plane provides (by (8.3)) one example of what has been called [3] a Moufang-Veldkamp plane--a geometry consisting of points, lines, an incidence relation and a neighbor relation satisfying P1-P8 of §8. Two natural questions arise: Are there other Moufang-Veldkamp planes? If so, what do they look like?

The answer to the first question is yes. Indeed, the construction given in §11, when carried out step by step beginning with the Jordan pair $V = (V^+,V^-)$ with $V^+ = \{(a,b):a,b \in \mathcal{A}\}$, $V^- = \{\binom{a}{b}:a,b \in \mathcal{A}\}$ yields a Moufang-Veldkamp plane with coordinate ring $\mathcal{A}$ whenever $\mathcal{A}$ is an alternative ring of stable range 2 ($ax + y$ left invertible implies $x + by$ is left invertible for some b) [3].

The answer to the second question is--they are all obtained by the procedure of §11 from some alternative ring of stable range 2. We close by sketching the method of proof of this result, following [3]:

Beginning with a Moufang-Veldkamp plane $\mathcal{P}$, one selects a regular triangle (one exists!) P_1, P_2, P_3. The procedure of coordinatization leading to (8.6) for $\mathcal{P}(\mathfrak{G})$ yields a coordinate ring $\mathcal{A}$ (in 1-1 correspondence with $\mathcal{A}_{31}$ such that $\mathcal{A} \times \mathcal{A}$ coordinatizes the "affine plane" of all points not neighboring $P_2 \vee P_1$. An argument very much like the proof of (4.4) shows that $\mathcal{A}$ (as a consequence of P7 and P8) is an alternative ring.

An analysis of the group Σ generated by transvections shows it to be generated by subgroups U_{ij} of transvections with axis $P_j \vee P_k$, center P_j, for $i, j, k \neq$.

Abstractly beginning with $\mathcal{A}$, one forms the Jordan pair from $\mathcal{A}$ as above and the corresponding Lie algebra $\mathcal{L}(V)$. The groups $\overset{\vee}{U}_{ij}$ defined in §11 are isomorphic to each other and to the U_{ij}.

Now two geometries arise in this context as homogeneous spaces: the original plane $\mathcal{P} \cong \mathcal{P}(\Sigma, P^-, P^+, w)$ for suitably chosen w, where P^- is the stabilizer of P_3 and P^+ is the stabilizer of $P_2 \vee P_1$; and the plane $\mathcal{P}(\Sigma^\vee, (P^-)^\vee, (P^+)^\vee, w^\vee)$ defined as in §11.

The isomorphism of the two planes is established by finding an appropriate isomorphism of Σ with $\Sigma^\vee$. It is at this point that axiom P5 for $\mathcal{P}$, on the geometric side, and stable range two for $\mathcal{A}$ on the algebraic side, enter crucially into the considerations. First, on the geometric side, if P is any point of $\mathcal{P}$, by P5 there is a line ℓ with $P \not\in \ell \not\ni P_3$. Thus, there are $\phi_+ \in U^+ = U_{31}U_{23}$ (a transvection with axis $P_1 \vee P_2$) and $\phi_- \in U^- = U_{13}U_{23}$ (a transvection with center P_3) such that $\phi_-^{-1}(\ell) = P_1 \vee P_2$ and $\phi_+^{-1}\phi_-^{-1}(P) = P_3$, i.e. $P = \phi_- \phi_+(P_3)$. Since $U^\pm$ is bijective with $V^\pm$, we see that every point $P \in \mathcal{P}$ is determined by a pair $(u,v) \in V^+ \times V^-$ for the Jordan pair $V = (V^+, V^-)$. A complicated geometric argument shows that (u,v) and (u_1, v_1) give the same point P if and only if

(21) $(u,v-v_1)$ is quasi-invertible in V with quasi-inverse u_1.

On the algebraic side, stable range two implies $\Sigma^\vee = (U^-)^\vee (U^+)^\vee (U^-)^\vee L^\vee$ so the cosets of $(P^-)^\vee = (U^-)^\vee L^\vee$ are also determined by pairs $(u,v) \in V^+ \times V^-$. Moreover, (u_1,v_1) gives the same coset as (u,v) precisely if (21) holds. Thus, both Σ and $\Sigma^\vee$ act faithfully on the set X of equivalence classes in $V^+ \times V^-$ defined by (21). One shows that the natural isomorphisms of $U^\pm$ with $(U^\pm)^\vee$ now extend to an isomorphism of Σ with $\Sigma^\vee$. One then gets the desired isomorphism of the planes.

REFERENCES

1. Barbilian, D., Zur Axiomatik der projektiven ebenen Ringgeometrien, I. Jahresbericht D.M.V. 50 (1940), 179-229; II. ibid. 51 (1941), 34-76.

2. Bruck, R.H., and Kleinfeld, E., The structure of alternative division rings, Proc. Amer. Math. Soc. 2(1951) 878-890.

3. Faulkner, J.R., Coordinatization of Moufang-Veldkamp Planes, Geom. Ded. 14 (1983), 189-201.

4. ________ , Octonion planes defined by quadratic Jordan algebras, Memoirs A.M.S. 104, 1970.

5. Jacobson, N., Structure and Representations of Jordan algebras, Amer. Math. Soc. Colloq. Publ., vol. 39, Amer. Math. Soc., Providence, R.I., 1969.

6. Jordan, P., von Neumann, J. and Wigner, E., On an algebraic generalization of the quantum mechanical formalism, Ann. of Math. (2), 36 (1934) 29-64.

7. Loos, O; Jordan Pairs, Lecture Notes in Math., 460; Springer-Verlag, 1975.

8. McCrimmon, K., The Freudenthal-Springer-Tits construction of exceptional Jordan algebras, Trans. Amer. Math. Soc., 127 (1967) 527-551.

9. Meyberg, K, Zur Konstruktion von Lie-Algebren aus Jordan-Tripelsystemen, Manuscr. Math. 3(1970) 115-132

10. Pickert, G., Projektive Ebene, Springer-Verlag, Berlin, 1955.

11. Springer, T.A., On the geometric algebra of the octave planes, Indag. Math. 24 (1962) 451-468.

12. ________ , and Veldkamp, F.D., On Hjelmslev-Moufang Planes, Math. Z. 107 (1968), 249-263.

13. Veldkamp, F.D., Projective planes over rings of stable rank 2, Geom. Dedicata 11 (1981) 285-308.

PROJECTIVE RING PLANES AND THEIR HOMOMORPHISMS

Ferdinand D. Veldkamp
Mathematical Institute, University of Utrecht
Budapestlaan 6
3508 TA Utrecht, The Netherlands.

ABSTRACT. This article surveys the theory of projective planes over rings of stable rank 2. Such a plane is described as a structure of points and lines together with an incidence relation and a neighbor relation and which has to satisfy two groups of axioms. The axioms in the first group express elementary relations between points and lines such as, e.g., the existence of a unique line joining any two non-neighboring points, and define what is called a Barbilian plane. In the second group of axioms the existence of sufficiently many transvections, dilatations, and generalizations of the latter, the affine dilatations and their duals, is required. Additional geometric properties of planes over special types of rings are then discussed. The paper ends with the treatment of homomorphisms between ring planes, i.e., of (not necessarily bijective) mappings which preserve incidence.

INTRODUCTION.

In this paper a projective plane over an associative ring R is the collection of free submodules having a free complement in a 3-dimensional free module over R. A most satisfactory situation is reached if we assume that R has stable rank 2. The notion of stable rank comes from algebraic K-theory. The results on stable rank which we need here are all scattered over the literature, with a firm amount of confusion in definitions and notations. For convenience of the reader we have therefore collected these results in §2, complete with proofs. That section, together with §1 where basic information about free modules is given, makes up part A of this paper.

Part B deals with the projective planes over rings of stable rank 2. In §3 we introduce them algebraically, working in a 3-dimensional free module R^3 over a ring R of stable rank 2. Points and lines are free submodules of dimension 1 resp. 2 having a (free) complement. Incidence is defined by inclusion. A point and a line are called distant if they span the whole R^3, and neighboring otherwise. Basic proporties of this geometry are derived in §3. In §4 we start off with the axiomatic description by introducing the first group of axioms, which

R. Kaya et al. (eds.), Rings and Geometry, 289–350.

express elementary relations between points and lines; see 4.1. A structure satisfying these axioms is called a Barbilian plane, after D. Barbilian, who first has studied projective planes over a large class of rings in [4]. In the remainder of §4 several proporties of Barbilian planes are derived. In §5 we introduce collineations of these planes, in particular transvections and dilatations. From a Barbilian plane we obtain an affine plane by deleting one line ℓ and all points neighboring ℓ. Then, affine collineations are introduced which, roughly speaking, are incidence preserving mappings of an affine plane. In particular, affine dilatations are needed for the further axiomatization of projective ring planes, as well as the duals of these. In §6 we more closely consider Barbilian planes in which all possible transvections exist. The collineation group of such a plane is shown to have the transitivity properties one expects from the point of view of classical projective geometry. In §7 we add the existence of dilatations, and of affine dilatations and their duals, and then in §8 the coordinatization of such an axiomatically given plane by a ring of stable rank 2 is described.

Most of the results in §§3-8 are from our paper [35]. The presentation here is sometimes different, notably in §§6,7 and 8.

In § 9 we give an overview of properties of projective planes over special types of rings such as rings without zero divisors, Bezout domains, local rings, Hjelmslev rings, valuation rings. These results come from our paper [36].

Part C, finally, presents the theory of incidence homomorphisms, i.e., mappings with preserve incidence. This is based mainly on joint work with J. Ferrar [17], completed in papers [37,38]. The presentation in this survey article is quite different from the original one as far as the main lines are concerned. The proofs of the individual results in this part can usually be taken over from the original papers and will mostly be omitted. The general theory is treated in §§10-13. The description of homomorphisms rests on a recoordinatization of the plane one starts from by a subring of the original coordinate ring, called an admissible subring. In §14 we deal with the purely algebraic question, to characterize admissible subrings of a ring R. This general characterization is considerably simplified in special cases of the ring R, as there are finite rings, Bezout domains, valuation rings, and other ones.

CONTENTS.

A. ALGEBRAIC PRELIMINARIES.

§1. FREE MODULES AND THEIR SUBSPACES.

All rings in this paper are assumed to be associative with unit element 1. By R^* we denote the group of units (invertible elements) in a ring R. We denote the free right (left) module generated by n elements over a ring R by R^n (${}^nR$, respectively). Between ${}^nR$ and R^n there is the dual pairing

$$\langle\ |\ \rangle : {}^nR \times R^n \to R,$$
$$\langle \ell | x\rangle = \langle \lambda_1,\ldots,\lambda_n | \xi_1,\ldots,\xi_n\rangle = \sum_{i=1} \lambda_i \xi_i .$$
$$\text{for } \ell = (\lambda_1,\ldots,\lambda_n) \in {}^nR,\ x = (\xi_1,\ldots,\xi_n) \in R^n.$$

This nondegenerate bilinear form makes ${}^nR$ the dual of R^n and vice versa.

$R°$ is the *opposite ring* of R, i.e., R with the same addition but the new multiplacation $\alpha * \beta = \beta\alpha$. One can identify R^n with ${}^nR°$ and ${}^nR$ with $R^{\circ n}$. Definitions and results given for R^n therefore also apply to ${}^nR$.

By a *subspace* of R^n we understand a free direct summand L, i.e., $L \cong R^p$ for some p, and $R^n = L \oplus M$ for some submodule M. xR is a subspace of R^n if and only if $\langle \ell | x\rangle = 1$ for some $\ell \in {}^nR$; then $R^n = xR \oplus \ker(\ell)$. An $x \in R^n$ with the property that $\langle \ell | x\rangle = 1$ for some $\ell \in {}^nR$ is called *unimodular*. Notice that $x = (\xi_1,\ldots,\xi_n)$ is unimodular if and only if $R = R\xi_1 + \ldots + R\xi_n$ (sum of left ideals).

The group $GL_n(R)$ of invertible n×n-matrices acts on R^n on the left and on ${}^nR$ on the right, and $\langle \ell | Ax\rangle = \langle \ell A | x\rangle$ for $A \in GL_n(R)$, $\ell \in {}^nR$, $x \in R^n$. Let E_{ij} denote the matrix unit with 1 on the i,j-place and zeros elsewhere. Then $E_{ij}(\alpha) = I+\alpha E_{ij}$, with $i \neq j$ and $\alpha \in R$, is an invertible matrix, called an *elementary matrix*. The subgroup of $GL_n(R)$ generated by all elementary matrices is the *elementary subgroup* $E_n(R)$. It is evident that $GL_n(R)$ transforms subspaces of R^n into subspaces.

A ring R is said to have *invariant basis number* if the rank of a free R-module is unique, i.e., $R^s \cong R^t$ implies $s \cong t$; see [11] for an exhaustive discussion of this notion.

Tentatively, we define the n-dimensional projective space over a ring R, denoted by $P_n(R)$, as the set of subspaces of R^{n+1}, with the incidence relation $L|M$ defined by $L \subseteq M$ or $L \supseteq M$, for any two subspaces L and M. To get "good" geometries we begin by imposing the following conditions (for all n):

1.1. R *has invariant basis number.*

1.2_n. *If* $R^n = L \oplus M$ *with* L *free (i.e.,* L *is a subspace), then* M *is free (so a subspace too);*

1.3_n. $E_n(R)$ *acts transitively on the set of* 1*-dimensional subspaces of* R^n.

Condition 1.2_n is equivalent to

$1.2'_n$. *Every unimodular vector is part of a basis of* R^n.

It immediate that 1.2_n implies $1.2'_n$. Conversely, assume $1.2'_n$. It suffices to prove 1.2_n for the case $L = aR$, a unimodular. Let $a = e_1$, $e_2,\dots,e_n$ be a basis of R. Write each e_j as $e_j = e_1\alpha_j+f_j$ with $\alpha_j \in R$, $f_j \in M$ $(j = 2,\dots,n)$, then $f_2,\dots,f_n$ form a basis of M.

Since condition 1.3_n of course implies $1.2'_n$, we see that 1.2_n is a consequence of 1.3_n.

§2. STABLE RANK OF A RING.

The following condition comes from K-theory and will turn out to be very useful in our geometry.

SR_n ($n \geq 2$): *For each unimodular* $(\xi_1,\dots,\xi_n) \in R^n$ *there exist* $\alpha_1,\dots,\alpha_{n-1} \in R$ *such that* $(\xi_1+\alpha_1\xi_n,\dots,\xi_{n-1}+\alpha_{n-1}\xi_n)$ *is unimodular in* R^{n-1}.

The *stable rank* of R, denoted by sr(R), is the least $n \geq 2$ such that SR_n holds (∞ if no such n exists). Notice that we follow the convention of [23] which agrees more or less with [5], whereas, e.g., in [34] the stable rank is 1 lower than ours.

Since we shall need several properties of rings of stable rank 2, and also to get a general idea about stable rank and in particular about rings of stable rank 2, we shall give a systematic account of these matters in the present section. For convenience of the reader we add complete proofs since these are scattered over the literature. The original proofs can be found in [5,6,23,29,33,34], except for 2.6, 2.14 and 2.16. The result of 2.14 is a generalization of [35], Prop. (1.3), whereas 2.16 is folklore.

2.1. SR_n *implies* SR_{n+1} *for all* $n \geq 2$.

Proof. Let $(\xi_1,\dots,\xi_{n+1})$ be unimodular in R^{n+1}, say $\sum_{i=1}^{n+1} \lambda_i\xi_i = 1$. Then $(\xi_1,\dots,\xi_{n-1},\lambda_n\xi_n+\lambda_{n+1}\xi_{n+1})$ is unimodular in R^n, hence by SR_n we can find $\alpha_1,\dots,\alpha_{n-1} \in R$ such that

$$(\xi_1+\alpha_1(\lambda_n\xi_n+\lambda_{n+1}\xi_{n+1}),\dots\dots,\xi_{n-1}+\alpha_{n-1}(\lambda_n\xi_n+\lambda_{n+1}\xi_{n+1}))$$

is unimodular. But then the vector

$$(\xi_1+\alpha_1\lambda_{n+1}\xi_{n+1},\dots\dots,\xi_{n-1}+\alpha_{n-1}\lambda_{n+1}\xi_{n+1},\xi_n)$$

is unimodular too.

2.2. $sr(R) = sr(R^\circ)$.

Proof. Since $(R^\circ)^\circ = R$, it suffices to show: if SR_n holds for R, then so it does for R°. So consider a unimodular vector $(\lambda_1,\dots,\lambda_n) \in R^{\circ n} = {}^nR$. For suitable $\xi_1,\dots,\xi_n \in R$ we have $\sum_{i=1}^{n} \lambda_i\xi_i = 1$. The vector $(\xi_1,\dots,\xi_{n-1},\lambda_n\xi_n)$ is unimodular in R^n. Since SR_n holds for R, there exist $\alpha_1,\dots,\alpha_{n-1} \in R$ such that $(\xi_1+\alpha_1\lambda_n\xi_n,\dots,\xi_{n-1}+\alpha_{n-1}\lambda_n\xi_n)$ is unimodular in R^{n-1}, so we can find $\mu_1,\dots,\mu_{n-1} \in R$ such that

$$\sum_{i=1}^{n-1}\mu_i(\xi_i+\alpha_i\lambda_n\xi_n) = 1.$$

Now take the matrix

$$A = \prod_{i=1}^{n-1} E_{ni}(-\lambda_n \xi_n \mu_i) . \prod_{i=1}^{n-1} E_{in}(\alpha_i) . \prod_{i=1}^{n-1} E_{ni}(-\lambda_i) . \prod_{i=1}^{n-1} E_{in}(\xi_i),$$

which is in $E_n(R)$. A straightforward computation yields that $(0,\dots,0,-1)A$ is of the form

$$(\lambda_1+\lambda_n\beta_1,\dots,\lambda_{n-1}+\lambda_n\beta_{n-1},0),$$

which shows that the vector $(\lambda_1+\lambda_n\beta_1,\dots,\lambda_{n-1}+\lambda_n\beta_{n-1})$ is unimodular. Thus, SR_n holds for $R°$.

The importance of the above result is, of course, that we can interchange left and right modules over R in the context of stable rank. This will be of great value for the geometry (duality!).

By rad R we denote the *Jacobson radical* of the ring R. We recall that rad R is a two-sided ideal, which is the intersection of all maximal left ideals in R, also of all maximal right ideals, and that $\alpha \in$ rad R if and only if $1+\alpha\xi$ ($1+\xi\alpha$, respectively) is invertible for all $\xi \in R$. See e.g., [1,12,13,14,20,21,22].

The following result says that factoring out the radical does not change the stable rank of a ring.

2.3. sr(R) = sr(R/rad R).

Proof. We must show that the condition SR_n is valid for R if and only if it is so for R/rad R. The truth of this is an immediate consequence of the following

LEMMA. *Let* $\xi \mapsto \bar{\xi}$ *denote the projection* R → R/rad R. *For any* n, $(\xi_1,\dots,\xi_n)$ *is unimodular in* R^n *if and only if* $(\bar{\xi}_1,\dots,\bar{\xi}_n)$ *is unimodular in* $(R/\text{rad } R)^n$.

Proof. Let $(\bar{\xi}_1,\dots,\bar{\xi}_n)$ be unimodular, then there are $\lambda_1,\dots,\lambda_n \in R$ such that

$$\sum_{i=1}^{n} \bar{\lambda}_i\bar{\xi}_i = \bar{1}.$$

$$\sum_{i=1}^{n} \lambda_i\xi_i = 1+\alpha \text{ with } \alpha \in \text{rad } R.$$

Since $1+\alpha$ is invertible,

$$\sum_{i=1}^{n} (1+\alpha)^{-1}\lambda_i\xi_i = 1,$$

which shows that $(\xi_1,\ldots,\xi_n)$ is unimodular. The converse is immediate.

More generally, one may ask what happens with the stable rank of a ring if a homomorphism is applied. It may change: $\mathbb{Z}$ has stable rank 3 (see 2.7), $\mathbb{Z}/(p)$, for p a prime, has stable rank 2 since it is a field. The general result is:

2.4. *If* $\varphi : R \to S$ *is a surjective homomorphism, then* $sr(S) \leq sr(R)$.
Proof. Assume SR_n holds in R. Consider a unimodular vector $(\varphi(\xi_1),\ldots,\varphi(\xi_n))$ in S^n. There are $\lambda_1,\ldots,\lambda_n$ in R such that

$$\lambda_1\xi_1+\ldots+\lambda_n\xi_n = 1+\mu \text{ with } \mu \in \ker \varphi.$$

This means that $(\xi_1,\ldots,\xi_{n-1},\lambda_n\xi_n-\mu)$ is unimodular in R^n. By SR_n, we can find $\alpha_i \in R$ such that

$$(\xi_1+\alpha_1(\lambda_n\xi_n-\mu),\ldots,\xi_{n-1}+\alpha_{n-1}(\lambda_n\xi_n-\mu))$$

is unimodular. Applying φ on the coordinates of this vector, we get a unimodular vector

$$(\varphi(\xi_1)+\varphi(\alpha_1\lambda_n)\varphi(\xi_n),\ldots,\varphi(\xi_{n-1})+\varphi(\alpha_{n-1}\lambda_n)\varphi(\xi_n)).$$

Thus we find that SR_n holds in S.

2.5. *The stable rank of the direct product of (any number of) rings* R_α *is the maximum of all* $sr(R_\alpha)$.
Proof. Clear.

2.6. $M_n(D)$, *the ring of* $n\times n$*-matrices over the skew field* D, *has stable rank* 2.
Proof. We let the elements of $M_n(D)$ act as linear transformations on the right vector space D^n. Let (A,B) be unimodular in $M_n(D)^2$. The relation $XA+YB = I$ implies that $\ker(A) \cap \ker(B) = 0$. We can therefore split $D^n = L \oplus \ker(A)$ with $L \supseteq \ker(B)$. There exist an invertible $U \in M_n(D)$, and a $V \in M_n(D)$ such that UA is the projection on L and VB the projection on ker(A). Thus $UA+VB = I$, i.e., $A+U^{-1}VB$ is invertible.

2.7. *If* R *is semiprimary (i.e.,* R/rad R *satisfies the minimum conditions for left and right ideals), then* $sr(R) = 2$.

Proof. This follows from 2.3, 2.5 and 2.6, since R/rad R is a direct sum of full matrix rings.

2.8. $sr(\mathbb{Z}) = 3$.
Proof. (5,7) is unimodular, but no element 5+7n is invertible, hence SR_2 does not hold in $\mathbb{Z}$. Let (a_1,a_2,a_3) be unimodular in $\mathbb{Z}^3$. If $a_1 = 0$, then (a_3,a_2) is unimodular, so now assume $a_1 \neq 0$. The ring $R = \mathbb{Z}/a_1\mathbb{Z}$ is semiprimary, hence it has stable rank 2. Since $(a_2 \bmod a_1, a_3 \bmod a_1)$ is unimodular in R^2, there exists $b \in R$ such that a_2+ba_3 is invertible mod a_1, which means that (a_1,a_2+ba_3) is unimodular in $\mathbb{Z}^2$.

2.9. *If* K *is a (commutative) field, then the polynomial ring* $K[X_1,\ldots,X_n]$ *has stable rank* r *with* $3 \leqq r \leqq n+2$.
Proof. That $r > 2$ is shown by a similar argument as for $\mathbb{Z}$: $(1+X_1,X_1^2)$ is unimodular, but no polynomial $1+X_1+fX_1^2$ is invertible. We will not prove $r \leqq n+2$; this follows from a theorem of H. Bass ([5], Th. V,(3.5) or [33], Th. 2.5).

In a matrix ring over a skew field an element has a left inverse if and only if it has a right inverse, hence the same is true in, e.g., any semiprimary ring. A nice result of Kaplansky and Lenstra says that all rings of stable rank 2 have this property:

2.10. *If* $sr(R) = 2$, *then* $\xi\eta = 1$ *implies* $\eta\xi = 1$ *in* R.
Proof. $(\xi, 1-\eta\xi)$ being unimodular, we can find $\alpha,\beta \in R$ with $\alpha(\xi+\beta(1-\eta\xi))=$ $= 1$ by SR_2. Take $\zeta = \xi+\beta(1-\eta\xi)$. Then $\xi\eta = 1$ implies $\zeta\eta = 1$. So now $\xi\eta = \zeta\eta$ $= \alpha\zeta = 1$. Thus ζ is invertible, η its inverse and ξ the inverse of η.

We shall now prove that SR_n implies condition 1.3_n. In fact, the following stronger result is true.

2.11. $E_n(R)$ *is transitive on unimodular vectors in* R^n *for all* $n \geqq sr(R)$.
Proof. Let $x = (\xi_1,\ldots,\xi_n)$ be unimodular in R^n. Pick $\alpha_1,\ldots,\alpha_{n-1}$ in R such that $(\xi_1+\alpha_1\xi_n,\ldots,\xi_{n-1}+\alpha_{n-1}\xi_n)$ is unimodular, say, $\sum_{i=1}^{n-1}\mu_j(\xi_j+\alpha_j\xi_n) = 1$. The transformation

$$\prod_{i=1}^{n-1} E_{n,i}(\mu_i) \cdot \prod_{i=1}^{n-1} E_{i,n}(\alpha_i)$$

transforms x into $(\xi_1+\alpha_1\xi_n,\ldots,\xi_{n-1}+\alpha_{n-1}\xi_n,1)$, which is mapped upon $(0,\ldots,0,1)$ by

$$\prod_{i=1}^{n-1} E_{i,n}(-\xi_i-\alpha_i\xi_n).$$

We finally want to prove that the condition SR_n implies that R has invariant basis number, i.e., condition 1.1. Before we can prove this, we need a theorem of Vasershtein which gives an equivalent formulation of SR_n. We start with some definitions and lemmas.

2.12. We use the notation I_n for the n×n identity matrix. An n×k-matrix A is called *unimodular* provided there exists a k×n-matrix B such that $BA = I_k$.

The following lemmas are readily verified.

LEMMA A. *If* A *is an* $(n-1)\times(k-1)$ *matrix and* a *a column vector of length* $n-1$, *then the* n×k-*matrix*

$$\begin{pmatrix} 1 & 0 \\ a & A \end{pmatrix}$$

is unimodular if and only if A *is unimodular.*

LEMMA B. *If* A *is an* n×k-*matrix and* $C \in GL_k(R)$, *then* A *is unimodular if and only if* AC *is unimodular.*

We denote by $F_n = F_n(R)$ the subgroup of $E_n(R)$ generated by all matrices $E_{in}(\alpha)$ with $i = 1,\ldots,n-1$ and $\alpha \in R$, i.e., F_n consists of the matrices

$$\begin{pmatrix} I_{n-1} & a \\ 0 & 1 \end{pmatrix}$$

with column vectors a of length n-1. $F_n(R)$ is a commutative subgroup of $E_n(R)$, which is normalized in $GL_n(R)$ by all matrices of the form

$$\begin{pmatrix} A & 0 \\ 0 & 1 \end{pmatrix}$$

with $A \in GL_{n-1}(R)$.

For $n\times k$-matrices B_1, B_2 the notation $B_2 \in F_n B_1$ means that there exists $C \in F_n(R)$ such that $B_1 = CB_2$; in that case B_1 and B_2 have the same last rows. We can now formulate a generalization of the condition SR_n to matrices, viz., for $n \geqq 2$ and $k \geqq 1$,

SR_n^k: *For each unimodular* $(n+k-1)\times k$*-matrix* B *there exists a unimodular* $(n+k-2)\times k$*-matrix* B' *such that*

$$\begin{pmatrix} B' \\ u \end{pmatrix} \in F_{n+k-1}B, \text{ if } u \text{ is the last row of } B.$$

Notice that SR_n^1 is precisely the same as SR_n. The following theorem of L.N. Vasershtein [34] says that SR_n^k is equivalent to SR_n.

2.13. THEOREM. *For every* $k \geqq 1$ *and every* $n \geqq 2$, *a ring* R *satisfies* SR_n *if and only if it satisfies* SR_n^k.

Proof. We proceed by induction on k. For k = 1 it is clear, therefore we assume that $k \geqq 2$ and that the equivalence of SR_n and SR_n^{k-1} has been settled for all n. Assuming SR_n we shall prove SR_n^k now. Consider a unimodular $(n+k-1)\times k$-matrix B. We say that SR_n^k holds for B if we can find $C \in F_{n+k-1}(R)$ such that $CB = \binom{B'}{u}$ with unimodular $(n+k-2)\times k$-matrix B'.

Denote the first column of B by $b = (\beta_1,\ldots,\beta_{n+k-1})^T$ where T indicates transposition of a matrix. Since SR_n implies SR_{n+k-1}, there are $\alpha_1,\ldots,\alpha_{n+k-2} \in R$ such that

$$b' = (\beta_1+\alpha_1\beta_{n+k-1},\ldots,\beta_{n+k-2}+\alpha_{n+k-2}\beta_{n-k+1})^T$$

is unimodular. By 2.11 we can find $A \in E_{n+k-2}(R)$ with $Ab' = (1,0,\ldots,0)^T$. Denoting $(\alpha_1,\ldots,\alpha_{n+k-2})^T$ by a, we have

$$B_1 = \begin{pmatrix} A & 0 \\ 0 & 1 \end{pmatrix}\begin{pmatrix} I_{n+k-2} & a \\ 0 & 1 \end{pmatrix} B = \begin{pmatrix} 1 & v \\ 0 & \\ \vdots & \\ 0 & D_1 \\ \beta_{n+k-1} & \end{pmatrix}$$

where v is a row of length k-1 and D_1 an $(n+k-2)\times(k-1)$-matrix. Since the matrices

$$\begin{pmatrix} A & 0 \\ 0 & 1 \end{pmatrix} \text{ with } A \in GL_{n+k-2}(R)$$

normalize F_{n+k-1}, SR_n^k holds for B if and only if it holds for B_1. Take (with v as in B_1 above)

$$B_2 = B_1 \begin{pmatrix} 1 & -v \\ 0 & I_{k-1} \end{pmatrix} = \begin{pmatrix} 1 & v \\ 0 & \\ 0 & \\ \beta_{n+k-1} & D_2 \end{pmatrix}.$$

Again it is readily verified that SR_n^k holds for B_1 if and only if it holds for B_2. Since B_2 is unimodular, so is D_2 by Lemma A. Applying SR_{n+k-2}^{k-1} to D_2 we find $C' \in F_{n+k-2}$ with the property that

$$C'D_2 = \begin{pmatrix} D' \\ u' \end{pmatrix},$$

where D' is a unimodular $(n+k-3)\times(k-1)$-matrix. Then

$$\begin{pmatrix} 1 & 0 \\ 0 & C' \end{pmatrix} B_2 = \begin{pmatrix} 1 & 0 \\ * & D' \\ \beta_{n+k-1} & u' \end{pmatrix}$$

and the matrix

$$\begin{pmatrix} 1 & 0 \\ * & D' \end{pmatrix}$$

is unimodular by Lemma A. Thus, SR_n^k holds for B_2, and we are done.

To derive SR_n from SR_n^k, just apply the latter condition to an $(n+k-1)\times k$-matrix of the form

$$\begin{pmatrix} I_{k-1} & 0 \\ 0 & b \end{pmatrix},$$

where b is any unimodular column vector of length n.

2.14. PROPOSITION. *A ring* R *of stable rank* $n < \infty$ *has invariant basis number*.

Proof. Suppose R not to have invariant basis number, i.e., $R^s \cong R^{s+t}$ for some $s > 0$, $t > 0$. Then $R^s \cong R^{s+kt}$ for all $k \in \mathbb{N}$, so we may as well assume $t \geq n-1$, thus, SR_{t+1} holds. $R^s \cong R^{s+t}$ means that there exist an $s\times(s+t)$-matrix A and an $(s+t)\times s$-matrix B such that $AB = I_s$, $BA = I_{s+t}$.

By SR^s_{t+1} there exists a column vector v of length s+t−1 such that

$$\begin{pmatrix} I_{s+t-1} & v \\ 0 & 1 \end{pmatrix} B = \begin{pmatrix} B' \\ u \end{pmatrix},$$

with unimodular $(s+t-1)\times s$-matrix B', i.e., $CB' = I_s$ for some $s\times(s+t-1)$-matrix C. Then

$$\begin{pmatrix} I_{s+t-1} & 0 \\ -uC & 1 \end{pmatrix} \begin{pmatrix} B' \\ u \end{pmatrix} = \begin{pmatrix} B' \\ 0 \end{pmatrix}.$$

Thus we have an invertible $(s+t)\times(s+t)$-matrix U such that

$$UB = \begin{pmatrix} B' \\ 0 \end{pmatrix},$$

from which we infer

$$U = UBA = \begin{pmatrix} B' \\ u \end{pmatrix} A = \begin{pmatrix} A' \\ 0 \end{pmatrix}.$$

The last matrix not being invertible, we have arrived at a contradiction.

2.15. A unimodular column vector in $M_n(R)^s$ is nothing but a unimodular $ns\times n$-matrix. Application of Theorem 2.13 easily yields that

$$sr(M_n(R)) = 2 +]\frac{sr(R)-2}{n}[,$$

if $]x[$ denotes the least integer $\geq x$. This yields in particular that $M_n(R)$ has stable rank 2 if R has stable rank 2, which generalizes the result given in 2.6.

2.16. *Let* R *be a ring of stable rank* 2 *and* $\varphi : R \to S$ *a surjective homomorphism. If* $\alpha'\beta' = 1$ *in* S, *there exist* $\alpha, \beta \in R$ *with* $\alpha\beta = 1$, $\varphi(\alpha) = \alpha'$, $\varphi(\beta) = \beta'$.

Proof. Pick $\alpha_1, \beta_1 \in R$ with $\varphi(\alpha_1) = \alpha'$, $\varphi(\beta_1) = \beta'$. Then $\alpha_1\beta_1 = 1+\varepsilon$ with $\varepsilon \in \ker\varphi$. The pair (β_1, ε) is therefore unimodular in R^2, hence $\beta = \beta_1 + \gamma\varepsilon$ is invertible in R for suitable γ. Take $\alpha = \beta^{-1}$, then $\varphi(\beta) = \beta'$ and $\varphi(\alpha) = \alpha'$.

B. PROJECTIVE RING PLANES.

§3. THE PROJECTIVE PLANE OVER A RING OF STABLE RANK 2.

In this section R always denotes a ring with stable rank 2. We recall that R then satisfies the conditions 1.1, 1.2_n ($\Leftrightarrow 1.2'_n$) and 1.3_n. We define a projective plane over the ring R as in [9].

3.1. DEFINITION. The projective plane $P = P_2(R)$ is defined by:

points of P are the 1-dimensional subspaces of R^3;

lines of P are the 2-dimensional subspaces of R^3;

incidence between a point and a line is defined by inclusion; notation $|$;

two points x and y are *distant* if they span a 2-dimensional subspace of R^3, and *neighboring* (or *neighbor points*) otherwise; notation for neighbor points: $x \approx y$;

two lines ℓ and m (or a point x and a line ℓ) are *distant* if they span R^3, and *neighboring* otherwise; notation for neighbor lines (or neighboring point and line); $\ell \approx m$ ($x \approx \ell$, respectively).

If $a = (\alpha_0, \alpha_1, \alpha_2)$ is unimodular in R^3, we denote the 1-dimensional subspace aR, considered as a point of $P_2(R)$, by $\ulcorner a \urcorner$ or $\ulcorner \alpha_0, \alpha_1, \alpha_2 \urcorner$. A line of $P_2(R)$, i.e., a 2-dimensional subspace of R^3, is the nullspace of a unimodular $\ell \in {}^3R$ and will be denoted by $\llcorner \ell \lrcorner$ or $\llcorner \lambda_0, \lambda_1, \lambda_2 \lrcorner$ if $\ell = (\lambda_0, \lambda_1, \lambda_2)$. The incidence relation $\ulcorner a \urcorner | \llcorner \ell \lrcorner$ means $\langle \ell | a \rangle = 0$. Upon identifying ${}^3R$ with $R^{\circ 3}$ (R° is the opposite ring of R), we can interpret a line $\llcorner \ell \lrcorner$ of $P_2(R)$ as a point of $P_2(R^\circ)$, and similarly a point of $P_2(R)$ as a line of $P_2(R^\circ)$. We call $P_2(R^\circ)$ with such an identification the *dual plane* of $P_2(R)$, denoted by $P_2^*(R)$.

Distant points x and y have, by definition, a unique line in common; we denote that by $x \vee y$. If two lines ℓ and m are distant when considered as points of $P_2^*(R)$, they have a unique point of $P_2(R)$ in common, which we denote by $\ell \wedge m$. We shall see in 3.3 that $\ell \approx m$ as lines in $P_2(R)$ is equivalent to $\ell \approx m$ as points in $P_2^*(R)$, and dually for points. As a consequence, if a statement S about points, lines, incidence and the neighbor relations holds in all planes $P_2(R)$, it also holds in all $P_2^*(R) = P_2(R^\circ)$, i.e., we can interchange the words *point* and *line* in S and thus get a true statement S*, called the *dual* of S.

For $a \in R^3$ and $\ell \in {}^3R$, $\ulcorner a \urcorner \not\approx \llcorner \ell \lrcorner$ if and only if there is a basis $a = e_1, e_2, e_3$ with e_2 and e_3 in the nullspace of ℓ. Then $\ell = (\lambda,0,0)$ with invertible λ, hence

$$\ulcorner a \urcorner \not\approx \llcorner \ell \lrcorner \Leftrightarrow \langle \ell | a \rangle \text{ is a unit.}$$

Points $x_1,\ldots,x_s$ and lines $\ell_1,\ldots,\ell_t$ are said to be *in general position* if $x_i \not\approx x_j$, $x_i \vee x_j \not\approx x_k$, $\ell_i \not\approx \ell_j$, $\ell_i \wedge \ell_j \not\approx \ell_k$, $x_i \not\approx \ell_i$, $x_i \not\approx \ell_j$ for all distinct i,j,k.

We shall now derive a number of properties of $P_2(R)$. Observe that the action of $GL_3(R)$ on R^3 defines an action on $P_2(R)$ which preserves incidence and the neighbor relations.

3.2. PROPOSITION. (i) *Two points* x *and* y *are distant if and only if there exists a line* $\ell | y$ *such that* $x \not\approx \ell$.
(ii) $\ulcorner 1,0,0 \urcorner \not\approx \ulcorner \eta_0,\eta_1,\eta_2 \urcorner$ *if and only if* (η_1,η_2) *is unimodular in* R^2.
(iii) $E_3(R)$ *is transitive on triples of points in general position.*
Proof. (i) It is easily seen that x and y are distant if and only if there is a basis e_0, e_1, e_2 of R^3 such that $x = \ulcorner e_0 \urcorner$ and $y = \ulcorner e_1 \urcorner$. The latter condition is equivalent to the existence of a line $\ell | y$ with $x \not\approx \ell$.
(ii) By (i), $\ulcorner \eta_0,\eta_1,\eta_2 \urcorner \not\approx \ulcorner 1,0,0 \urcorner$ if and only if there exists a line $\llcorner 0,\lambda_1,\lambda_2 \lrcorner$ which is $\not\approx \ulcorner \eta_0,\eta_1,\eta_2 \urcorner$, i.e., $\lambda_1\eta_1+\lambda_2\eta_2$ is invertible. This means that (η_1,η_2) is unimodular.
(iii) Let x,y en z be in general position. By 2.10 we may assume that $x = \ulcorner 1,0,0 \urcorner$, and so $y = \ulcorner \eta_0,\eta_1,\eta_2 \urcorner$ with unimodular (η_1,η_2). By SR_2 we can find $\alpha \in R$ such that $\eta_1+\alpha\eta_2$ is invertible. The product of elementary matrices.

$$E_{12}(-\eta_0(\eta_1+\alpha\eta_2)^{-1})E_{32}(-\eta_2(\eta_1+\alpha\eta_2)^{-1})E_{23}(\alpha)$$

leaves $\ulcorner 1,0,0 \urcorner$ fixed and maps y on $\ulcorner 0,\eta_1+\alpha\eta_2,0 \urcorner = \ulcorner 0,1,0 \urcorner$. Then z is transformed into $\ulcorner \zeta_0,\zeta_1,\zeta_2 \urcorner$ with invertible ζ_2. The transformation

$$E_{13}(-\zeta_0\zeta_2^{-1})E_{23}(-\zeta_1\zeta_2^{-1})$$

brings the latter point to $\ulcorner 0,0,1 \urcorner$ whereas it leaves the other two points fixed.

3.3. PROPOSITION. *Let* ℓ *and* m *be lines in* $P_2(R)$. *Then* $\ell \approx m$ *as lines in* $P_2(R)$ *if and only if* $\ell \approx m$ *as points in the dual plane* $P_2^*(R)$.
Proof. Assume $\ell \not\approx m$ in $P_2^*(R)$. By the dual of the previous proposition we may assume that $\ell = \llcorner 1,0,0 \lrcorner$ and $m = \llcorner 0,1,0 \lrcorner$. These two subspaces span R^3 so $\ell \not\approx m$ in $P_2(R)$. Conversely, let $\ell \not\approx m$ in $P_2(R)$. We may assume $\ell = \llcorner 1,0,0 \lrcorner$. Since $(1,0,0)$ belongs to the span of the subspaces ℓ and m, there must exist a point $x = \ulcorner 1,\alpha_1,\alpha_2 \urcorner | m$. Clearly, $x \not\approx \ell$, so by the dual of 3.2 (i), $\ell \not\approx m$ as points in $P_2^*(R)$.

From (3.2) (i) and its dual, in combination with 3.3 we infer that the neighbor relation between a point and a line determines the other two neighbor relations (that between points, and that between lines). This will be used in the axiomatic description which starts in the next section. In fact, any of the three neighbor relations determines the other two but we will not use that here; for details, see [35], (2.6).

The following proposition gives two properties of $P_2(R)$ of a technical nature, which shall play a role in the axiomatic setup in the following section.

3.4. PROPOSITION. (i) *If* x *is a point and* ℓ *and* m *are lines in* $P_2(R)$ *such that* $\ell \not\approx m$, $x | \ell$, $x \not\approx \ell \wedge m$, *then* $x \not\approx m$.
(ii) *For any two points* x *and* y *there exists a line* ℓ *in* $P_2(R)$ *with* $\ell \not\approx x$, $\ell \not\approx y$.
Proof. (i) The subspaces ℓ and m span R^3, and x and $\ell \wedge m$ span ℓ, so x and m span R^3.
(ii) We may assume $x = \ulcorner 1,0,0 \urcorner$; let $y = \ulcorner \eta_0,\eta_1,\eta_2 \urcorner$. Since (η_0,η_1,η_2) is unimodular, there exist $\alpha_i \in R$ such that $(\eta_0+\alpha_0\eta_2, \eta_1+\alpha_1\eta_2)$ is unimodular by SR_3. Applying SR_2 to the latter vector yields the existence of $\beta,\gamma \in R$ such that

$$\gamma(\eta_0+\alpha_0\eta_2+\beta(\eta_1+\alpha_1\eta_2)) = 1.$$

The line $\ell = \llcorner \gamma,\gamma\beta,\gamma\alpha_0+\gamma\beta\alpha_1 \lrcorner$ indeed satisfies $\ell \not\approx x$, $\ell \not\approx y$.

Much can be said about planes $P_2(R)$ over special classes of rings R, e.g., local rings, Bezout domains, and so on. We shall postpone the discussion of the geometric properties of such planes till §9, and turn our attention first to an axiomatic description of the general case. One of the questions to be dealt with in §9 is the following. The neighbor relation $\approx$ between points (or between lines) is apperently reflexive and symmetric, but need it be transitive? The answer is: in general not. In fact, $\approx$ is transitive if and only if R is a (not necessarily commutative) local ring; see 9.7. Another important question is whether the neighbor relations are already determined by the incidence relation, e.g., if for points x and y it is true that $x \not\approx y$ if and only if x and y have a unique line in common. We shall see that this too is not true in general but it is true for quite a large class of rings, viz., if every non-invertible element is a zero divisor; this will be proved in 9.1.

§4. BARBILIAN PLANES.

We shall now write down a set of basic axioms which are satisfied by projective rings of stable rank 2. They can be compared to the basic axioms by which not necessarily Desarguesian ordinary projective planes are defined. We call the geometric structure they define a *Barbilian plane*, since D. Barbilian was the first who has studied projective planes over what we now call rings of stable rank 2 in his papers [4]. For the characterization of projective planes over rings of stable rank 2 we need, besides the axioms for a Barbilian plane, a suitable substitute for Desargues' axiom. Such a substitute will be the existence of sufficiently many transvections and dilatations, which will be formulated in two axioms that we shall give in the next section.

4.1. DEFINITION. A *(projective) Barbilian plane* $P = (P_*, P^*, |, \approx)$ consists of a set of *points* P_* and a set of *lines* P^* together with two symmetric relations $x | \ell$ (*incidence*) and $x \approx \ell$ (*neighbor*) between P_* and P^*, subject to seven axioms given below. We will generally denote by x, y, z, u, v points and by ℓ, m, n, p lines in P. One defines relations in P_* (resp. P^*) by

$x \approx y$ (resp. $\ell \approx m$) if and only if $x \approx n$ for all $n | y$ (resp. $\ell \approx z$ for all $z|m$);

these are also called *neighbor relations*. The axioms read as follows:

1. $x|\ell$ *implies* $x \approx \ell$.
2. *If* $x \not\approx y$, *there is a unique* ℓ, *denoted* $x \vee y$ *and called the join of* x *and* y, *with* $\ell|x$ *and* $\ell|y$.

2'. *If* $\ell \not\approx m$, *there is a unique* x, *denotes* $\ell \wedge m$ *and called the meet of* ℓ *and* m, *with* $x|\ell$ *and* $x|m$.

3. *If* $\ell \not\approx m$, $x|\ell$, $\ell \wedge m \not\approx x$, *then* $x \not\approx m$.
4. *For all* ℓ *there is* x *with* $x|\ell$.
5. *For all* x, y *there is* ℓ *with* $\ell \not\approx x$, $\ell \not\approx m$.
6. *There exists a line.*

A point and a line (or two points, or two lines) are often called *distant* if they are not neighbors, i.e., distant means: $\not\approx$.

It is clear from the preceding section that projective planes over rings of stable rank 2, with incidence and neighbor relation as defined there, satisfy the axioms for a Barbilian plane. Examples of Barbilian planes which can not be coordinatized by associative rings of stable rank 2 are the Moufang-Veldkamp planes studied by J.R. Faulkner in [15,16], which papers form the link between our paper [35] and the article [32] by T.A. Springer and the present author. We shall not include these planes in the present contribution.

If the words *point* and *line* are interchanged in a statement, we get its *dual statement*. Axiom 2', e.g., is the dual of axiom 2. In fact, the dual of each of the above axioms holds for all Barbilian planes, and consequently we have the *principle of duality*: If any statement is true for all Barbilian planes, then so is the dual statement. Axioms 1 and 3 are selfdual, and 2 and 2' are each others duals. The following proposition gives the duals of the remaining axioms, together with some other results. One of these is the symmetry of the neighbor relation $\approx$ for two points, and dually for two lines; this symmetry was not explicitly required a priori. Notations for points and lines are as before.

4.2. PROPOSITION. (i) *For all* x *there is* ℓ *with* $\ell|x$ (*dual of* 4).

(ii) *For all* x *and* ℓ *there is* y *such that* $x \not\approx y \not\approx \ell$.

(iii) *For all* x, y *and* ℓ *there is* $z|\ell$ *with* $y \not\approx z \not\approx x$.

(iv) *The relation* $\approx$ *for points is reflexive and symmetric; dually for lines.*

(v) *For all* ℓ *and* m *there is* x *with* $x \not\approx \ell$ *and* $x \not\approx m$ *(dual of* 5).

(vi) *There exists a point (dual of* 6).

(vii) *If* $x \not\approx \ell$, $y|\ell$, $x \not\approx y$, *then there is* $z|\ell$ *such that* $x \approx z \not\approx y$.

(viii) *If* $x \not\approx \ell$, $\ell \not\approx m$, $x \not\approx m$, *then there is* $z|\ell$ *such that* $x \approx z \not\approx m$.

Proof. This is found in [35], (3.2). It should be pointed out that in the proof of (vii) the principle of duality is used. This is allowed at that point, since in (i) - (vi) all the duals of the axioms have already been proved.

General position of points and lines can be defined in Barbilian planes precisely as in §3. From the axioms and the above proposition we infer the existence of a *general quadrangle*, i.e., a set of four points in general position, which we shall need for the coordinatization. The statements of the proposition and their duals will be frequently used in the rest of this paper, often without explicitly saying that; usually, when points and/or lines in a given position are chosen, their existence is guaranteed by 4.2.

§5. COLLINEATIONS AND AFFINE COLLINEATIONS.

In this section we shall deal with collineations in Barbilian planes, in particular with transvections and dilatations. We shall, moreover, need affine collineations, which are only defined in an affine plane, i.e., on the points and lines not neighboring some fixed line.

$P = (P_*, P^*, |, \approx)$ is a Barbilian plane, as is $P' = (P'_*, P'^*, |, \approx)$. We continue to denote points of P by $a,b,c,\dots,x,y,z,u,v,\dots$, and lines of P by $\ell,m,n,p,\dots$.

5.1. DEFINITION. A *collineation* $\psi : P \to P'$ between Barbilian planes P, P' consists of a pair of bijective mappings $\psi_* : P_* \to P'_*$, $\psi^* : P^* \to P'^*$

such that

$x|\ell \Leftrightarrow \psi_* x|\psi^*\ell$, and $x \approx \ell \Leftrightarrow \psi_* x \approx \psi^*\ell$ for all $x \in P^*$ and $\ell \in P^*$.

Usually, we shall just write ψ instead of ψ_* and ψ^* if there is no danger of confusion.

We call two Barbilian planes P and P' *isomorphic* if there exists a collineation $P \to P'$.

It is immediate that the neighbor relation between points and that between lines are also preserved:

$x \approx y \Leftrightarrow \psi_* x \approx \psi_* y$ for all $x,y \in P_*$,

$\ell \approx m \Leftrightarrow \psi^*\ell \approx \psi^* m$ for all $\ell,m \in P^*$.

ψ_* already determines ψ^* and vice versa, since any $\ell \in P^*$ can be written as $\ell = a\vee b$ for distant $a,b \in P_*$.

The collineations of a plane onto itself form a group. In $P_2(R)$, the projective plane over a ring R of stable rank 2, bijective semilinear transformations induce collineations; in fact, all collineations in $P_2(R)$ are thus obtained: see Theorem 8.1.

The following lemma ensures uniqueness of a collineation for which certain images are prescribed.

5.2. LEMMA. *A collineation* $\psi : P \to P'$ *between Barbilian planes is uniquely determined (provided it exists) by any of the following three sets of data:*

(i) *The images* ψu *of all points* $u \not\approx \ell$ *for some fixed line* $\ell \in P^*$.

(ii) *The images* ψa, ψb *and* ψu *for all* $u|\ell$ *if the points* a, b *and the line* ℓ *are given in general position in* P;

(iii) *The images* ψa, ψu *for all* $u|\ell$ *and* ψm *for a point* a *and lines* ℓ, m *such that* a, ℓ *and* m *are in general position in P.*

Proof. (i) Consider any point $x \in P_*$. Pick a line $m_1|x$, $m_1 \not\approx \ell$, and another line $m_2|x$, $m_2 \not\approx \ell$, $m_2 \not\approx m_1$. On each m_i next choose a point $y_i \not\approx \ell\wedge m_i$ and another point $z_i \not\approx \ell\wedge m_i$, $z_i \not\approx y_i$. Then the points y_i and z_i are $\not\approx \ell$ whence their images under ψ are known. $x = m_1\wedge m_2$ and each

$m_i = y_i \vee z_i$, thus $\psi x_1 = (\psi y_1 \vee \psi z_1) \wedge (\psi y_2 \vee \psi z_2)$.

(ii) Consider any point $x \not\approx a \vee b$. Take $u = (a \vee x) \wedge \ell$ and $v = (b \vee x) \wedge \ell$, then $x = (a \vee u) \wedge (b \vee v)$ and its image is therefore determined by $\psi x = (\psi a \vee \psi u) \wedge \wedge (\psi b \vee \psi v)$. By case (i) above, the images ψx for all $x \not\approx a \vee b$ completely determine ψ.

(iii) Pick a point $b | m$, $b \not\approx c = \ell \wedge m$, then $b \not\approx \ell$, $b \not\approx a \vee c$. Take $v = = (a \vee b) \wedge \ell$. Now ψb is determined by $\psi b = (\psi a \vee \psi v) \wedge \psi m$. Since a, b and ℓ are in general position, ψ is determined by ψa, ψb and ψu for all $u | \ell$.

As in the classical theory of projective planes, an important role is played by two special kinds of collineations: transvections and dilatations. Here are definitions of such collineations in Barbilian planes.

5.3. DEFINITION. Let P be a Barbilian plane, $c \in P_*$ and $\ell \in P^*$. For $c \not\approx \ell$, a *dilatation* with *center* c and *axis* ℓ, or a (c,ℓ)-*dilatation*, is any collineation which leaves c and all points on ℓ fixed.

For $c | \ell$, a *central transvection* with *center* c and *axis* ℓ, or a (c,ℓ)-*transvection*, is any collineation which leaves all points on ℓ and all lines through c fixed (i.e., a collineation ψ such that $\psi_* x = x$ for all $x | \ell$ and $\psi^* m = m$ for all $m | c$).

We speak of *central* transvections in the above definition because further on we shall introduce a more general kind of transvections which need not have a center; see 5.6. for an example and Def. 6.2.

The notions of dilatation and central transvection are selfdual. Center and axis of these need not be uniquely determined; see also the remark after Def. 6.2.

One might consider (c,ℓ)-collineations with possibly $c \approx \ell$, $c \not| \ell$; since these are not relevant for our purpose, we shall not pursue this matter.

5.4. PROPOSITION. *Let* a, b, $c \in P_*$ *and* $\ell \in P^*$.

(i) *If* a, c *and* ℓ *are in general position*, $b | a \vee c$, $\ell \not\approx b \not\approx c$, *there exists at most one* (c,ℓ)*-dilatation which maps* a *on* b. *We denote this by* $T_{c,\ell:a,b}$.

(ii) *If* $c|\ell$, $a \not\approx \ell$, $b|a\vee c$, $b \not\approx c$, *there exists at most one* (c,ℓ)*-transvection which maps* a *on* b. *We denote this too by* $T_{c,\ell:a,b}$.

Proof. These statements immediately follow from 5.2, (ii) and (iii), respectively.

5.5. In a plane $P_2(R)$ with R of stable rank 2 all $T_{c,\ell:a,b}$ exist if a, b, c and ℓ are in the position as in (i) and (ii) of the above proposition. In case $c \not\approx \ell$ we may take $c = \ulcorner 1,0,0 \urcorner$, $\ell = \llcorner 1,0,0 \lrcorner$, $a = \ulcorner 1,1,0 \urcorner$, $b = \ulcorner 1,\alpha,0 \urcorner$ where α has to be invertible to ensure $b \not\approx c$. The linear transformation $\mathrm{diag}(\alpha^{-1},1,1)$ induces the dilatation $T_{c,\ell:a,b}$. If $c|\ell$, we may assume $c = \ulcorner 0,1,0 \urcorner$, $\ell = \llcorner 1,0,0 \lrcorner$, $a = \ulcorner 1,0,0 \urcorner$, $b = \ulcorner 1,\alpha,0 \urcorner$ (with arbitrary $\alpha \in R$). The linear transformation $E_{21}(\alpha)$ then induces the translation $T_{c,\ell:a,b}$.

5.6. The fact that all points on some line (called "axis") remain fixed under a collineation ψ does not imply the existence of a point c (called "center") such that all lines through c are fixed under ψ. Consider, e.g., the plane $P_2(R)$ over a local ring R without zero divisors and whose maximal ideal contains elements α and β with $\alpha \notin R\beta$ and $\beta \notin R\alpha$. Then the collineation induced by $E_{21}(\alpha)E_{31}(\beta)$ has $\llcorner 1,0,0 \lrcorner$ as an axis but it has no center.

5.7. We consider again the dilatation $T_{c,\ell:a,b}$ in $P_2(R)$ induced by $\mathrm{diag}(\alpha^{-1},1,1)$ as in 5.5. It acts on the points $\not\approx \ell = \llcorner 1,0,0 \lrcorner$ by

$$\ulcorner 1,\eta_1,\eta_2 \urcorner \mapsto \ulcorner \alpha^{-1},\eta_1,\eta_2 \urcorner = \ulcorner 1,\eta_1\alpha,\eta_2\alpha \urcorner,$$

whereas it fixes the points on ℓ. Here α is required to be invertible in order to get a decent dilatation. However, for the points $\not\approx \ell$ or on ℓ the above makes sense for all $\alpha \in R$. The same is true for ℓ and all lines $m \not\approx \ell$: if $m = \llcorner \mu_0,\mu_1,\mu_2 \lrcorner$ with unimodular $(\mu_1,\mu_2) \in {}^2R$, which means $m \not\approx \ell$, we can map this on $\llcorner \mu_0\alpha,\mu_1,\mu_2 \lrcorner$. The mapping thus defined preserves incidence and the neighbor relation. It is something like an "affine dilatation", viz., multiplication of the "affine plane" by (a not necessarily invertible) α. Let us give formal definitions now.

5.8. DEFINITION. Let $P = (P_*, P^*, |, \approx)$ be a Barbilian plane. For $\ell \in P^*$ the *affine (Barbilian) plane* $P^\ell = (P_*^\ell, P^{*\ell}, |, \approx)$ is defined by

$$P_*^\ell = \{x \in P_* | x \not\approx \ell \text{ or } x | \ell\},$$
$$P^{\ell *} = \{m \in P^* | m \not\approx \ell \text{ or } m = \ell\},$$

whereas | and $\approx$ are as in P.

For $a \in P_*$ the *dual affine plane* P^a consists of the set of points $\not\approx a$ together with a itself, and the set of lines $\not\approx a$ or |a, again with | and $\approx$ as in P.

An *ℓ-affine collineation* ψ of P, for $\ell \in P^*$, is a pair of mappings

$$\psi_* : P_*^\ell \to P_*^\ell, \ \psi^* : P^{\ell *} \to P^{\ell *},$$

having the following properties:

$$\psi^*\ell = \ell, \ x \not\approx \ell \Rightarrow \psi_* x \approx \ell,$$
$$x|m \Rightarrow \psi_* x | \psi^* m, \ x \approx m \Rightarrow \psi_* x \approx \psi^* m,$$
ψ_* is bijective on the set of points $|\ell$,
for $x, y | \ell$, $x \not\approx y \Rightarrow \psi_* x \not\approx \psi_* y$.

A *dual a-affine collineation* is defined similarly in P^a, for $a \in P_*$.

Usually we just write ψ for ψ_* and ψ^*.

If c is a point $\not\approx \ell$, an *ℓ-affine dilatation* with center c, or *affine (c,ℓ)-dilatation*, is an ℓ-affine collineation which leaves c and all points of ℓ fixed. Dually, we define a *dual affine (c,ℓ)-dilatation* in P^c.

The following lemma gives a simple characterization of a class of affine collineations, which covers all cases that will be relevant.

5.9. LEMMA. *Let P^ℓ be an affine plane. If a pair of mappings*

$$\psi_* : P_*^\ell \to P_*^\ell, \ \psi^* : P^{\ell *} \to P^{\ell *}$$

satisfies

$$\psi_* x = x \text{ for all } x|\ell,$$
$$x \not\approx \ell \Rightarrow \psi_* x \not\approx \ell,$$
$$x|m \Rightarrow \psi_* x | \psi^* m \text{ for all } x \in P_*^\ell, \ m \in P^{\ell *},$$

then it is an ℓ-affine collineation.

Proof. We have to show that $x \approx m$ implies $\psi_* x \approx \psi^* m$. This being obvious for $x|\ell$, we assume $x \not\approx \ell$, $m \not\approx \ell$ and $x \approx m$. Pick a point $y|m$, $y \not\approx \ell$, $y \not\approx x$. Then $x \vee y \approx m$. For $p = (x \vee y) \wedge \ell$ and $q = m \wedge \ell$ we have $p \approx q$. Since $\psi_* y \not\approx \ell$, we find $\psi_* y \vee p \approx \psi_* y \vee q$. This implies $\psi_* x \approx \psi^* m$.

Next we prove an elementary property of ℓ-affine collineations.

5.10. LEMMA. *If* ψ *is an* ℓ*-affine collineation, then*

$$x \approx y \Rightarrow \psi x \approx \psi y \text{ for } x,y \in P_*^{\ell}.$$

Proof. Assume first that e.g., $y | \ell$. Then $x \approx y$ implies $x \approx \ell$, whence $x | \ell$ since $x \in P_*^{\ell}$. Taking $m | y$, $m \not\approx \ell$ we see that $x \approx m$, hence $\psi x \approx \psi m$ and therefore $\psi x \approx \psi y$, as $\psi m \not\approx \ell$.

It remains to consider the case that x and y are both $\not\approx \ell$. For any $n | \psi y$ we have $n = \psi y \vee (n \wedge \ell)$, since $\psi y \not\approx \ell$ and therefore $n \not\approx \ell$. It follows that $n = \psi m$ with $m = y \vee \psi^{-1}(n \wedge \ell)$ (remember that ψ is bijective on the points of ℓ). From $m | y$ it follows that $x \approx m$ and so $\psi x \approx \psi m = n$. This means that $\psi x \approx \psi y$.

REMARK. In the above lemma one would also expect the property $m \approx n \Rightarrow$ $\Rightarrow \psi m \approx \psi n$. We have no proof for this and, in fact, we do not need it in the sequel.

5.11. If an ℓ-affine collineation ψ can be extended to a collineation of the whole Barbilian plane P, the extension is unique by 5.2 (i); we shall then identify ψ and its extension.

The following lemma is an analogue for ℓ-affine collineations of Lemma 5.2.

5.12. LEMMA. *An* ℓ*-affine collineation of a Barbilian plane* P *is uniquely determined (provided it exists) by any of the following three sets of data:*

(i) *the images* ψu *for all points* $u \in P_*^{\ell}$ *such that* $u \not\approx m$ *for a given line* $m \not\approx \ell$.

(ii) *the images* ψa, ψb *and* ψu *for all* $u | \ell$ *where* $a, b \in P_*$ *are given such that* a, b *and* ℓ *are in general position;*

(iii) *the images* ψa, ψu *for all* $u | \ell$, *and* ψm *for a point* a *and a line* m *such that* a, ℓ *and* m *are in general position.*

Proof. (i) Consider first a point $x \not\approx \ell$. Pick points y_1, y_2 on ℓ with $y_i \not\approx m$, $y_1 \not\approx y_2$. Then $x \vee y_i \not\approx \ell$, m. Take on each $x \vee y_i$ a point z_i with $z_i \not\approx \ell$, m. Then $x = (y_1 \vee z_1) \wedge (y_2 \vee z_2)$ and its image under ψ is determined by $\psi x = (\psi y_1 \vee \psi z_1) \wedge (\psi y_2 \vee \psi z_2)$.

If $x | \ell$ we can write it as $x = \ell \wedge (y \vee z)$ with y, z chosen so as to satisfy y, $z \not\approx \ell$, $z | x \vee y$, $y \not\approx z$. Then $\psi x = \ell \wedge (\psi y \vee \psi z)$.

(ii) Consider a point $x \not\approx \ell$, $a \vee b$. Define $p = (x \vee a) \wedge \ell$ and $q = (x \vee b) \wedge \ell$. Then $p \not\approx q$, whence $\psi p \not\approx \psi q$. Since $x = (p \vee a) \wedge (q \vee b)$, we find its image by $\psi x = (\psi p \vee \psi a) \wedge (\psi q \vee \psi b)$. Now the results follows from (i).

(iii) Let $c = \ell \wedge m$. Take a point $b | m$, $b \not\approx c$, then $b \not\approx \ell$. With $d = (a \vee b) \wedge \ell$. we have $c \not\approx d$, hence $\psi c \not\approx \psi d$ and therefore $\psi d \not\approx \psi m$. We find that $\psi b = (\psi a \vee \psi d) \wedge \psi m$. Now we can apply (ii) to get the result.

As a special case of part (ii) of the above lemma we find:

5.13. PROPOSITION. *An affine dilatation is uniquely determined by its center* c, *its axis* ℓ *(with* $\ell \not\approx c$*) and the image* b *of any point* $a \not\approx \ell$, $a \not\approx c$, *where of course* b *must satisfy* $b \not\approx \ell$, $b | a \vee c$. *This affine dilatation is denoted by* $\hat{T}_{c,\ell:a,b}$. *If* b *satisfies, moreover,* $b \not\approx c$, *then* $\hat{T}_{c,\ell:a,b}$ *is identified with* $T_{c,\ell:a,b}$ *(if existing).*

5.14. DEFINITION. Let P be a Barbilian plane, $c \in P_*$, $\ell \in P^*$.

If $c | \ell$, the plane P is said to be (c,ℓ)-*transitive* if $T_{c,\ell:a,b}$ exists for all $a \not\approx \ell$, $b | a \vee c$, $b \not\approx \ell$.

If $c \not\approx \ell$, then P is said to be (c,ℓ)-*transitive* if $T_{c,\ell:a,b}$ exists for all $a \not\approx \ell$, $a \not\approx c$ and $b | a \vee c$, $b \not\approx \ell$, $b \not\approx c$ and if, in addition, $\hat{T}_{c,\ell:a,b}$ exists in P^{ℓ} for all $a \not\approx \ell$, $a \not\approx c$ and $b | a \vee c$, $b \not\approx \ell$.

P is called *dually* (c,ℓ)-*transitive* for $c \not\approx \ell$, if for all $a \not\approx \ell$, $a \not\approx c$ and $b | a \vee c$, $b \not\approx \ell$, the dilatation $T_{c,\ell:a,b}$ exists if b satisfies, moreover, $b \not\approx c$, whereas in all other cases the dual affine (c,ℓ)-dilatation mapping a on b has to exist in the dual affine plane P^c.

Notice that in the above definition it suffices to require the existence of all $T_{c,\ell:a,b}$ and $\hat{T}_{c,\ell:a,b}$ for only one point a satisfying the conditions. The notion of (c,ℓ)-transitivity is selfdual in case $c|\ell$, whereas for $c \not\sim \ell$ the dual notion of (c,ℓ)-transitivity is dual (c,ℓ)-transitivity and vice versa.

§6. BARBILIAN TRANSVECTION PLANES.

After the previous section it is clear that in order to characterize projective planes over rings of stable rank 2 we have to require in a Barbilian plane (c,ℓ)-transitivity for all points c and lines ℓ, where either $c|\ell$ or $c \not\sim \ell$ (or something equivalent to this). It has certain advantages to do this separately for central transvections and for (affine) dilatations. Thus we now introduce one more axiom requiring the existence of transvections.

6.1. DEFINITION. A Barbilian plane P is said to be a *Barbilian transvection plane* if, in addition to the axioms of 4.1, it satifies the following axiom:

7. P *is* (c,ℓ)*-transitive for all* $c \in P_*$ *and* $\ell \in P^*$ *such that* $c|\ell$.

The (c,ℓ)-transvections for given c and ℓ with $c|\ell$ form a group. However, the central transvections with given axis ℓ need not form a group, as we see from the example in 5.6. This defect can be easily corrected, of course, by calling transvections arbitrary products of central transvections with the same axis. However, we shall also need that in certain cases the product of central transvections is again central. For that purpose we introduce proper transvections.

6.2. DEFINITION. A *transvection* with *axis* ℓ ($\ell \in P^*$) is an arbitrary (finite) product of (c_i,ℓ)-transvections with centers $c_i|\ell$.

A (c,ℓ)-transvection T is callled *proper* whenever $Ta \not\sim a$ for some point $a \not\sim \ell$.

The notion of proper central transvection is easily seen to be selfdual. Proper central transvections have a unique center, and by duality a unique axis. The following lemma tells us that the choice of the point a in the above definition is not of importance.

6.3. LEMMA. *If* T *is a proper* (c,ℓ)*-transvection,* $Tx \not\approx x$ *for all* $x \not\approx \ell$.
Proof. Let $Ta \not\approx a$ with $a \not\approx \ell$. If $x \not\approx \ell$, $x \not\approx a \vee c$, let $u = (a \vee x) \wedge \ell$. Then $Ta \vee u \not\approx a \vee u$, and $Tx = (c \vee x) \wedge (Ta \vee u)$ implies $Tx \not\approx x$. For $y | c \vee a$, $y \not\approx \ell$, one gets $Ty \not\approx y$ using an x as above. For arbitrary $z \not\approx \ell$, finally, we deduce $Tz \not\approx z$ by means of a point $d | \ell$, $d \not\approx c$, and $y = (d \vee z) \wedge (a \vee c)$.

We now want to show that in a Barbilian transvection plane the group of all transvections with a given axis ℓ is commutative. This result will be derived in a series of lemmas, whose proofs are to a large extent the same as those of (4.19)-(4.21) in [35]. The main difference with the original presentation lies in the insertion of Lemma 6.7, which in [35] came later, and which we need to correct a mistake in the proof of (4.21) in [35]. In that proof we claim on line 3 of p. 299 that T_4 is proper; however, this is only true if $T_1 a \not\approx a$, i.e., if T_1 is proper. The formulation of Lemma 6.6 is also different from that in (4.21) for the same reason.

6.4. LEMMA. *Any* (c_1,ℓ)*-transvection and* (c_2,ℓ)*-transvection with* $c_i | \ell$, $c_1 \not\approx c_2$, *commute.*
Proof. An obvious modification of the usual proof for ordinary projective planes works (cf., [2] or [19]).

6.5. LEMMA. *If the* (c_i,ℓ)*-transvections* T_i *are proper* (i = 1,2) *and* $c_1 \not\approx c_2$, *then* $T_1 T_2$ *is a proper* (c_3,ℓ)*-transvection for some* $c_3 \not\approx c_1, c_2$

The proof is found in (4.20) of [35], using Lemma 5.2 (ii) of the present paper instead of (4.2) (ii) of [35].

6.6. LEMMA. *A* (c_1,ℓ)*-transvection* T_1 *and a proper* (c_2,ℓ)*-transvection* T_2 *always commute.*
Proof. The proof of (4.21) of [35] works if the word "proper" is deleted on line 3 of p. 299, which was incorrect as we remarked before 6.4. The

conclusion of the last sentence of that proof is valid because T_2 is assumed to be proper now.

6.7. LEMMA. *Every central transvection wiht axis ℓ is a product of two proper central transvections with the same axis ℓ.*

Proof. Let T be a (c,ℓ)-transvection. Choose points a and u with $a \not\approx \ell$, $u \not\approx \ell$, $u \not\approx a \vee c$. Take $c_1 = (a \vee u) \wedge \ell$ and $c_2 = (Ta \vee u) \wedge \ell$. Let T_1 be the proper (c_1,ℓ)-transvection which carries a to u, and T_2 the proper (c_2,ℓ)-transvection which carries u to Ta. Then $T_2 T_1 a = Ta$, and $T_2 T_1 u = T_1 T_2 u$ (since T_1 and T_2 commute by the previous lemma), the latter point being equal to $T_1 Ta = Tu$. Thus we are in a position that we can apply 5.2 (ii) to reach the conclusion that $T_2 T_1 = T$.

6.8. PROPOSITION. *In a Barbilian transvection plane the group of transvections with a given axis ℓ is commutative. Its action on the points $\not\approx \ell$ is sharply transitive.*

Proof. The first statement is immediate from the last two lemmas. As for the second statement, consider points a and b, both $\not\approx \ell$. Pick points c and d on ℓ with $c \not\approx d$. Then $c \vee a \not\approx d \vee b$, and for $u = (c \vee a) \wedge (d \vee b)$ we see that $u \not\approx \ell$. Now we can find a (c,ℓ)-transvection carrying a to u and a (d,ℓ)-transvection carrying u to b. This proves transitivity.

Finally, consider any transvection T with axis ℓ which leaves some point $a \not\approx \ell$ fixed. Let T_0 be any proper central transvection with axis ℓ. Since T and T_0 commute, $T(T_0 a) = T_0 Ta = T_0 a$. Thus, T leaves a, $T_0 a$ and al points on ℓ fixed, which implies that T is the identity according to 5.2 (ii). This completes the proof.

We now shift our attention to the group generated by all transvections.

6.9. DEFINITION. The group generated by all transvections in a Barbilian plane P is called the *little projective group* of P, denoted by LPG or LPG(P).

The transitivity property of the transvections with a fixed axis that we gave in 6.8 has its consequences, of course, for LPG.

6.10. PROPOSITION. *If* P *is a Barbilian plane with transvections, then* LPG(P) *is transitive on general triangles in* P, *and also on ordered triples of collinear points* x_1, x_2, x_3 *such that* $x_i \not\approx x_j$ *for* $i \neq j$.
Proof. The first statement is proved in the usual way with the aid of the transitivity property in 6.8. Now consider two triples of collinear points, x_1, x_2, x_3 respectively y_1, y_2, y_3, such that $x_i \not\approx x_j$ and $y_i \not\approx y_j$ for $i \neq j$. From the transitivity of LPG on general triangles we see that we can carry y_1 to x_1, y_2 to x_2, so we may assume $x_1 = y_1$, $x_2 = y_2$. Call $x_1 \vee x_2 = \ell$. Take a point $a \not\approx \ell$ and a line m such that $m | x_3$, $m \not\approx \ell$, $x_1 \vee a$. Then $m \not\approx x_3 \vee a$, $y_3 \vee a$. Call $m \wedge (x_3 \vee a) = b$ and $m \wedge (y_3 \vee a) = c$. Both b and c are $\not\approx \ell$, and the lines $b \vee x_2$ and $c \vee x_2$ are $\not\approx a, x_3, y_3$. The product of central transvections

$$T_{b, b\vee x_2 : a, x_3} T_{c, c\vee x_2 : y_3, a}$$

carries y_3 to x_3 while leaving x_1 and x_2 fixed. This proves the second transitivity statement.

§7. PROJECTIVE RING PLANES.

As we already announced in the introduction to the previous section, the characterization of projective planes over rings of stable rank 2 will be completed by requiring for a Barbilian plane that it admits all possible transvections (axiom 7 in 6.1) and all possible (affine) dilatations and their duals (axiom 8 below).

7.1. DEFINITION. A *(projective) ring plane* is a Barbilian transvection plane P which, moreover, satisfies the axiom

8. P *is* (c,ℓ)*-transitive and dually* (c,ℓ)*-transitive for some* $c \in P_*$ *and* $\ell \in P^*$ *such that* $c \not\approx \ell$.

From 6.10 we infer that the little projective group in P is transitive on distant point-line pairs. Hence it suffices in axiom 8 to require

(c,ℓ)-transitivity and dual (c,ℓ)-transitivity for just one pair (c,ℓ) with $c \not\approx \ell$ to conclude the same for all such pairs. Thus,

7.2. PROPOSITION. *A projective ring plane* P *is* (c,ℓ)*-transitive and dually* (c,ℓ)*-transitive for all* $c \in P_*$, $\ell \in P^*$ *such that* $c \not\approx \ell$.

7.3. DEFINITION. The *(full) projective group* of a Barbilian plane P is the group generated by all transvections and dilatations in P. Notation: PG, or PG(P).

LPG is a normal subgroup of PG. Since LPG is already transitive on general triangles, one shows in the wellknown manner:

7.4. PROPOSITION. *The full projective group of a ring plane* P *is transitive on general quadrangles in* P.

§8. COORDINATIZATION OF PROJECTIVE RING PLANES.

The axiomatic framework is now complete; a projective ring plane P is indeed what we want it to be: coordinates can be introduced which form a ring R of stable rank 2, and P is isomorphic with $P_2(R)$. Let us briefly indicate how the coordinatization is performed. We begin by introducing coordinates in an affine plane, using transvections and affine dilatations to define addition and multiplication, respectively. The extension of coordinates to the whole projective plane then requires some toilsome shuffling with certain types of collineations and affine collineations. The result we eventually obtain is:

8.1. THEOREM. *Every projective ring plane is isomorphic to a projective plane over a ring* R *of stable rank* 2. R *is unique up to isomorphism. Collineations are induced by bijective semilinear transformations in* R^3. *A collineation belongs to the full projective group if and only if it is induced by a bijective linear transformation, and it belongs to the little projective group if and only if it is induced by an element of* $E_3(R)$.

We shall now give a more detailed description of the coordinatization procedure, ommitting most of the proofs, which can be found in §5 of our paper [35].

$P = (P_*, P^*, |, \approx)$ is a projective ring plane.

8.2. Starting point of the introduction of coordinates is the choice of a general quadrangle o, p_1, p_2, e in P. The line $\ell_\infty = p_1 \vee p_2$ is taken as *line at infinity*, the affine ring plane P^{ℓ_∞} is called the *affine plane*. The points $\not\approx \ell_\infty$ are called *affine points*, the lines $\not\approx \ell_\infty$ *affine lines*. Furthermore, we call ℓ_∞-affine collineations (dilatations) *affine* collineations (dilatations, respectively), and (central) transvections with ℓ_∞ as axis: (central) *translations*.

Further notation: $\ell_i = o \vee p_i$, $1 = \ell_1 \wedge (e \vee p_2)$, $1' = \ell_2 \wedge (e \vee p_1)$. The set of coordinates R consists of the points on ℓ_1 which are $\not\approx \ell_\infty$; its elements are in general denoted by lower case Greek characters except for such specific elements as $0 = o$ and 1.

T_a denotes the translation which carries o to $a \not\approx \ell_\infty$; if $a = \alpha \in R$, we may also write T_α. *Addition* of affine points is defined by

$$a+b = T_a b.$$

This makes the set V of affine points a commutative group isomorphic to the group of translations, and R a subgroup of V.

For $\alpha \in R$, D_α denotes the affine dilatation with center o which maps 1 on α. Such a D_α is invertible, with a D_β as inverse, if and only if $\alpha \not\approx o$ and so D_α is a dilatation. For $x \in V$ and $\alpha \in R$ we define

$$x\alpha = D_\alpha x.$$

If α and $\beta \in R$, then $\beta\alpha \in R$ and $D_\alpha D_\beta = D_{\beta\alpha}$; further, $D_1 = 1$. Hence α is invertible if and only if $\alpha \not\approx 0$, and these units form a group anti-isomorphic to the group of $(0,\ell_\infty)$-dilatations.

It can now be shown that with the two operations we have defined, R is an associative ring with 1 as its identity element, and V is a free right module over R with 1 and 1' as basis elements. For the proof we refer to [35], (5.1).

8.3. We denote the points of V by their coordinates (α,β) with respect to the basis 1, 1'. Instead of (α,β) we will often write $\ulcorner 1,\alpha,\beta \urcorner$ or $\ulcorner \lambda,\alpha\lambda,\beta\lambda \urcorner$ with arbitrary invertible $\lambda \in R$, i.e., we coordinatize the points of V by triples $\ulcorner \xi_0,\xi_1,\xi_2 \urcorner$ with invertible ξ_0 under identification of right proportional triples.

In these coordinates a translation is described by

$$T_{(\alpha,\beta)} : \ulcorner \xi_0,\xi_1,\xi_2 \urcorner \mapsto \ulcorner \xi_0,\alpha\xi_0+\xi_1,\beta\xi_0+\xi_2 \urcorner,$$

i.e., it is induced by the linear transformation $E_{21}(\alpha)E_{31}(\beta)$ in R^3.

Besides translations we need another class of transvections which transform V into itself, viz., the shears:

8.4. DEFINITION. A *shear* is an $(\ell\wedge\ell_\infty,\ell)$-transvection with $\ell \not\approx \ell_\infty$.

8.5. LEMMA. *If* U *is a shear with fixes* o *(i.e., its axis passes through* o*), then*

$$U(a\lambda+b\mu) = (Ua)\lambda+(Ub)\mu \text{ for } a,b \in V, \lambda,\mu \in R.$$

Proof. We have to show that $U(a+b) = Ua+Ub$ and $U(a\lambda) = (Ua)\lambda$. The first statement can be translated into $UT_a = T_{Ua}U$, and this follows from 5.12 (iii) by observing that UT_a and $T_{Ua}U$ have the same action on o, on the points of ℓ_∞ and on the lines through $(o\vee a)\wedge\ell_\infty$ (the center of T_a).

$U(a\lambda) = (Ua)\lambda$ means $UD_\lambda = D_\lambda U$. This is again proved with the aid of 5.12 (iii), since UD_λ and $D_\lambda U$ have the same action on o, on the points of ℓ_∞ and on the lines through the center of U.

REMARK. In [35], Lemma (5.3), the above result was proved for a more general class of affine collineations U. In the statement of that lemma one has to add the condition: $Ux \not\approx Uy$ for $x,y|\ell_\infty$ with $x \not\approx y$.

8.6. The shear with axis ℓ_2 which maps (1,1) on $(1,\alpha+1)$ in V is given by the matrix $E_{21}(\alpha)$, and the shear with axis ℓ_1 which maps (1,1) on $(\alpha+1,1)$ by $E_{12}(\alpha)$. Call GS the group generated by all shears with axis through o. It is already generated by the shears with ℓ_1 or ℓ_2 as an axis, so it is isomorphic to the elementary group $E_3(R)$.

The group GST generated by all translations and shears consists of all products of an element of GS and a translation. Hence it consists of the transformations which act on the points $\ulcorner \xi_0, \xi_1, \xi_2 \urcorner$ (with invertible ξ_0) as a linear transformation

$$\begin{pmatrix} 1 & 0 & 0 \\ \alpha_{21} & \alpha_{22} & \alpha_{23} \\ \alpha_{31} & \alpha_{32} & \alpha_{33} \end{pmatrix} \quad \text{with } \alpha_{21}, \alpha_{31} \in R \text{ and } \begin{pmatrix} \alpha_{22} & \alpha_{23} \\ \alpha_{32} & \alpha_{33} \end{pmatrix} \in E_2(R).$$

GST is transitive on pairs of lines m_1, m_2 with $m_1 \not\approx m_2$ and $m_1 \wedge m_2 \not\approx \ell_\infty$.

8.7. Now we are ready to introduce coordinates for the affine lines. The line ℓ_1 has as affine points all $\ulcorner \xi_0, \xi_1, \xi_2 \urcorner$ with $\xi_2 = 0$ (and ξ_0 invertible); it will be denoted by $\llcorner 0,0,1 \lrcorner$. An arbitrary line $\not\approx \ell_\infty$ can be transformed into ℓ_1 by some $A \in$ GST, and therefore it is described by a linear equation

$$\langle \ell | x \rangle = 0 \text{ with } \ell = (\lambda_0, \lambda_1, \lambda_2) = (0,0,1)A \in {}^3R.$$

It is clear that the coefficients of the equation for ℓ_1 are unique up to an invertible scalar on the left; hence the same holds for all lines $\not\approx \ell_\infty$. From the form of the matrix A we infer that (λ_1, λ_2) is unimodular in ${}^2R$.

Each line $\ell \not\approx \ell_\infty$ has a unique point of intersection with ℓ_∞. We want to represent this by coordinates $\ulcorner 0, \eta_1, \eta_2 \urcorner$ (unique up to right multiplication by an invertible element of R) with unimodular $(\eta_1, \eta_2) \in R^2$ satisfying the equation of ℓ: $\langle \ell | x \rangle = 0$. For $\ell = \ell_1 = \llcorner 0,0,1 \lrcorner$ the solution is $\llcorner 0,1,0 \lrcorner$. Application of the group GST shows that there is a unique solution for each $\ell \not\approx \ell_\infty$. The line ℓ_∞ itself is thus represented by the equation $\xi_0 = 0$, i.e., by $\llcorner 1,0,0 \lrcorner$.

For an affine point on $\ell_1 = \ulcorner 0,0,1 \urcorner$ the distant-relation to ℓ_2 is given by

$$\ulcorner 1,\alpha,0 \urcorner \not\approx \llcorner 0,1,0 \lrcorner \Leftrightarrow \ulcorner 1,\alpha,0 \urcorner \not\approx o$$
$$\Leftrightarrow \alpha \text{ invertible (see 8.1).}$$

Now $\alpha = \langle 0,1,0 | 1,\alpha,0 \rangle$. Using the action of the group GST again we find for all affine points and lines:

$$x \not\approx \ell \Leftrightarrow \langle \ell | x \rangle \text{ invertible.}$$

This result is easily extended to the points of ℓ_∞; see [35], (5.6).

8.8. The rest of the points of P is coordinatized by considering them as the intersection of two distant affine lines, which is always possible, and afterwards the remaining lines get their coordinates. For further details we again refer to [35], (5.8)-(5.11). Here we shall only show how the stable rank 2 of R comes in from the geometry.

8.9. PROPOSITION. R *has stable rank* 2.
Proof. Let (α_1,α_2) be unimodular in R^2. Consider the points $a = \ulcorner 0,\alpha_1,\alpha_2 \urcorner$ and $b = \ulcorner 0,1,0 \urcorner$ on ℓ_∞. By axiom 5 there exists a line $\ell \not\approx a, b$. Since $\ell \not\approx b$, we may represent it in the form $\ell = \llcorner \lambda_0, 1, \lambda_2 \lrcorner$. Then $\ell \not\approx a$ means that $\alpha_1 + \lambda_2\alpha_2$ is invertible, which proves SR_2.

§9. PROJECTIVE PLANES OVER SPECIAL TYPES OF RINGS.

In this section we will display a number of relations between algebraic properties of the coordinate ring and geometric properties of the projective plane over that ring. The rings will always be assumed to have stable rank 2 unless the contrary is explicitely stated. We shall omit proofs in most cases; they can be found in [35,36].

In many cases the neighbor relations are already determined by the incidence relation in the sense that, e.g., two points which have precisely one line in common are necessarily distant. This is, in fact, the way the neighbor relation is defined in Hjelmslev planes; see [24,25,26]. We shall come back to Hjelmslev planes in 9.10. More generally, we have

9.1. PROPOSITION. *The following conditions are equivalent.*
(a) *For points* x,y *(lines* ℓ, m*) in* $P_2(R)$, $x \approx y$ $(\ell \approx m)$ *if and only if* x *and* y *(*ℓ *and* m*) have either no or at least two lines (points, respectively) in common.*
(b) *If* $\xi \in R$ *is not invertible, there exists* $\eta \neq 0$ *in* R *such that* $\eta\xi = 0$ $(\xi\eta = 0$, *respectively)*

For a proof, see [35], (2.7).

Regarding condition (b) in the above proposition, we recall the following result of R. Baer [7]. Remember that an Artin ring has stable rank 2 (see 2.7), and that by 2.10 an element in such a ring is left invertible if and only if it is right invertible.

PROPOSITION. *Let* R *be an Artin ring, i.e., the minimum conditions hold for left and right ideals. The following are equivalent for any* $\alpha \in R$:
(i) α *is not a left zero divisor, i.e.,* $\alpha\beta = 0$ *implies* $\beta = 0$.
(ii) α *is not a right zero divisor, i.e.,* $\beta\alpha = 0$ *implies* $\beta = 0$.
(iii) α *is a unit.*

The possibility of more than one line through two distinct points corresponds to the presence of zero divisors in the coordinate ring:

9.2. PROPOSITION. *Any two distinct points in* $P_2(R)$ *have at most one line in common if and only if* R *has no zero divisors.*

This is proved in (2.10) of [35].

A Barbilian plane is said to be *linearly connected* if every two points have at least one line in common.

For affine planes over commutative rings the property of being linearly connected was examined in [8], in which paper condition (c) below already appears. For planes over rings of stable rank 2 the following result was proved in [36], §3.

9.3. PROPOSITION. *The three conditions below are equivalent:*
(a) *Any two points in* $P_2(R)$ *have at least one line in common.*
(b) *For any two elements* η_1, η_2 *of* R *there exists* $\alpha \in R$ *such that*

$$R\eta_1+R\eta_2 = R(\eta_2+\alpha\eta_1).$$

(c) *Each vector* $(\eta_1,\eta_2) \in R^2$ *is right proportional to a unimodular vector* $(\gamma_1,\gamma_2) \in R^2$:

$$(\eta_1,\eta_2) = (\gamma_1,\gamma_2)\delta \text{ for some } \delta \in R.$$

Actually, for arbitrary associative rings R *with* 1 (*not a priori assumed to have stable rank* 2), *condition* (b) *is equivalent to condition* (c) *together with* SR_2.

By combining 9.2 and 9.3 we get a characterization of those rings R such that any two distinct points (or lines) have a unique line (point, respectively) in common. These are the Bezout domains. We recall that a *left* (or *right*) *Bezout domain* is an integral domain (i.e., a ring without zero divisors) in which any two elements generate a principal left (resp. right) ideal. Affine planes over Bezout domains were studied already in [30].

In a left Bezout domain R the *left Ore condition* holds. This says that $R_0\xi \cap R_0\eta \neq \emptyset$ for any $\xi, \eta \in R_0$, where $R_0 = R\setminus\{0\}$. To prove this, take δ such that $R\xi+R\eta = R\delta$. Then we can find $\lambda, \mu, \alpha, \beta \in R$ such that $\xi = \lambda\delta$, $\eta = \mu\delta$, $\delta = \alpha\xi+\beta\eta$. Since $\delta \neq 0$, at least one of α and β is nonzero, say $\alpha \neq 0$. Now $\mu\alpha\xi = (1-\mu\beta)\eta$, and since $\mu\alpha \neq 0$, also $1-\mu\beta \neq 0$. Thus we have found an element in $R_0\xi \cap R_0\eta$.

If R is an integral domain in which the left Ore condition holds, it has a *left quotient field* $K = \{\alpha^{-1}\beta \mid \alpha,\beta \in R,\ \alpha \neq 0\}$ (cf. [12], p.23. or [20], Ch.6. §1). If R satifies both the left and the right Ore condition, its left quotient field and its right quotient field are naturally isomorphic and will therefore be identified with each other.

In a left Bezout domain any two elements ξ, η have a *highest common right factor* (HCRF). For consider δ with $R\xi+R\eta = R\delta$, so $\xi = \lambda\delta$, $\eta = \mu\delta$ and $\delta = \alpha\xi+\beta\eta$. This δ is a common right factor of ξ and η, whereas $\xi = \lambda'\delta'$, $\eta = \mu'\delta'$ imply that $\delta = (\alpha\lambda'+\beta\mu')\delta'$. So, indeed, δ is an HCRF. Moreover, $\alpha\lambda+\beta\mu = 1$ (as R is free of zero divisors), i.e., (λ,μ) is unimodular in R^2. By induction, any $\xi_1,\ldots,\xi_n$ in R have an HCRF δ, and if $\xi_i = \lambda_i\delta$ for $1 \leqq i \leqq n$, then $(\lambda_1,\ldots,\lambda_n)$ is unimodular in R^n. Applying this for n = 3, we see that each point $\ulcorner\xi_0,\xi_1,\xi_2\urcorner$ in $P_2(K)$ can be identified with a unique point $\ulcorner\alpha_0,\alpha_1,\alpha_2\urcorner$ in $P_2(R)$. If R is both a left and a right Bezout domain, $P_2(R)$ is just the ordinary projective plane $P_2(K)$ if we forget about the neighbor relation.

Precisely stating the results:

9.4. PROPOSITION. *Let* R *have stable rank* 2. *Any two distinct points (or lines) have exactly one line (resp. point) in common if and only if* R *is a left (resp. right) Bezout domain. If* R *is a left and right Bezout domain, the points and lines of* $P_2(R)$ *with the incidence relation form the projective plane* $P_2(K)$ *over the quotient field* K *of* R.

In view of the preceding we should, of course, also ask: what is the place of classical Desarguesian projective planes in the context of the ring planes? Quite simply this: the neighbor relations are trivial. To put it in exact terms:

9.5. PROPOSITION. *For a projective ring plane* $P_2(R) = (P_*, P^*, |, \approx)$ *the following properties are equivalent.*

(a) $x \approx y$ *if and only if* $x = y$ *for* $x, y \in P_*$.

(b) $\ell \approx m$ *if and only if* $\ell = m$ *for* $\ell, m \in P^*$.

(c) $x \approx \ell$ *if and only if* $x | \ell$ *for* $x \in P_*$, $\ell \in P^*$.

(d) R *is a skew field.*

Proof. One can either argue directly from the axioms, which in the case of trivial neighbor relations are those of classical Desarguesian planes, or use the following observation. Consider $x = \ulcorner 1,0,0 \urcorner$ and $y = \ulcorner 1,\eta,0 \urcorner$. The statement $x \neq y \Leftrightarrow x \not\approx y$ amounts to saying $\eta \neq 0 \Leftrightarrow \eta$ invertible. This proves the equivalence of (a) and (d). The other cases are similar.

In the rest of §9 we focus attention on planes over local rings, and over special types of local rings: Hjelmslev rings and valuatian rings. First local rings in general. Recall all local rings have stable rank 2.

9.6. PROPOSITION. *In* $P_2(R)$ *the neighbor relation* $\approx$ *between points (or equivalently, between lines) is transitive if and only if* R *is a local ring, i.e.,* R/rad R *is a skew field.*

For a proof, see [35], (2.10).

The planes over local rings are precisely the projective planes with homomorphism of W. Klingenberg [27]. To make this remark clear, take in 10.2 for R a local ring and for φ the projection of R on R/rad R. We now come back to linearly connected planes, which we considered in 9.3, in the case of local rings. First a definition.

9.7 DEFINITION. The *left* (or *right*) *chain condition* is said to hold in a ring R provided for any two elements α, β in R either $\alpha \in R\beta$ or $\beta \in R\alpha$ (respectively, either $\alpha \in \beta R$ or $\beta \in \alpha R$).

A ring which satisfies either of the chain conditions is local. The geometric meaning of the chain conditions is given in the following proposition.

9.8. PROPOSITION. *In* $P_2(R)$ *the neighbor relation* $\approx$ *between points (or between lines) is transitive and any two points have at least one line in common, if and only if the left chain condition holds in* R. *Similarly for lines in* $P_2(R)$ *and the right chain condition in* R.

The proof is found in [36], §7.

We recall that a *Hjelmslev ring* or *H-ring* is a ring in which the left and right chain condition hold (hence it is a local ring) and whose maximal ideal contains only zero divisors. The planes $P_2(R)$ over a Hjelmslev ring R, the *Hjelmslev planes*, have been studied first by W. Klingenberg in [24,25,26]. In the set-up of projective ring planes the following characterization of projective Hjelmslev planes is an immediate consequence of propositions 9.1 and 9.8.

9.9. PROPOSITION. R *is a Hjelmslev ring if and only if* $P_2(R)$ *has the following properties: the neighbor relation* $\approx$ *is transitive for points (or, equivalently, for lines)*; $x \approx y$ *if and only if the points* x *and* y *have at least two lines in common*; $\ell \approx m$ *if and only if the lines* ℓ *and* m *have at least two points in common.*

A (not necessarily commutative) *valuation ring* is a ring without zero divisors in which the left and right chain condition hold; cf. [10,27]. Such a ring is always local. From 9.2 and 9.8 we infer:

9.10. PROPOSITION. R *is a valuation ring if and only if in* $P_2(R)$ *the relation* $\approx$ *is transitive for points (or, equivalently, for lines), any two distinct points have a unique line in common, and any two distinct lines meet in a unique point.*

By 9.4, a valuation ring can also be characterized as a local ring (whence it automatically has stable rank 2) which is at the same time a Bezout domain; it is not hard, by the way, to give a direct algebraic proof of this fact, avoiding the geometry. It is also immediate from the chain conditions that the two Ore conditions hold in R. So R has a quotient field K, and the chain conditions mean that for any $x \in K$ either x or x^{-1} belongs to R. Conversely, any ring R in a skew field K which has the latter property is easily seen to be a valuation ring.

To the plane $P_2(R)$ over a valution ring R there are related two ordinary Desarguesian projective planes, viz., the plane $P_2(K)$ over the quotient field K of R as in 9.4, which is just $P_2(R)$ with the neighbor relation made trivial, and the plane $P_2(k)$ over the residue class field $k = R/\text{rad}\, R$. The projection $R \to k = R/\text{rad}\, R$ has to do with a homomorphism between the projective planes $P_2(K)$ and $P_2(k)$ in the sense of W. Klingenberg [27]; cf. also [10]. More about this in 14.9.

C. HOMOMORPHISMS OF PROJECTIVE RING PLANES.

§10. HOMOMORPHISMS OF BARBILIAN PLANES.

Homomorphisms between ordinary Desarguesian projective planes and geometries have first been studied by W. Klingenberg [27]. For skew fields K and K', a homomorphism $\psi : P_2(K) \to P_2(K')$ consists of a pair of mappings, one from points to points and one from lines to lines (not necessarily injective or surjective), such that a point and a line which are incident have incident images. If the image of ψ contains a general quadrangle (i.e., ψ is full in the terminology of [10]), then ψ can, after appropriate choice of basic quadrangles in the planes, be described as follows: there is a valuation ring S in K and a nonzero ring homomorphism $\varphi : S \to K'$ such that each point $\ulcorner\xi_0,\xi_1,\xi_2\urcorner$ in $P_2(K)$ can be assumed to have all its coordinates ξ_i in S, not all in rad S, and that $\ulcorner\xi_0,\xi_1,\xi_2\urcorner$ is mapped by ψ upon $\ulcorner\varphi(\xi_0),\varphi(\xi_1),\varphi(\xi_2)\urcorner$, and similarly for lines.

In the above description valuation rings have made their entrance on the stage. Since these are local rings, hence of stable rank 2, it is natural to ask for a generalization to homomorphisms between projective planes over rings of stable rank 2; this generalization has been carried out in three papers [17,37,38]. Such a homomorphism should, of course, preserve incidence. Less obvious is, what else it should do: preserve the neighbor relation, or just its opposite, the distant relation? It turns out that either possibility makes sense and that descriptions in terms of ring homomorphisms can be given, even in the case that neither the neighbor relation nor the distant relation but only incidence is preserved. Crucial in all this is an appropriate generalization of fullness. More information in 10.2, 10.4-10.6, after we have given a number of definitions. For generality we begin working in Barbilian planes, to restrict the discussion to ring planes later on.

10.1. DEFINITION. Let $P = (P_*,P^*,|,\approx)$ and $P' = (P'_*,P'^*,|,\approx)$ be Barbilian planes. An *incidence homomorphism*, or *homomorphism* for short, $\psi : P \to P'$ consists of a pair of mappings $\psi_* : P_* \to P'_*$, $\psi^* : P^* \to P'^*$ such that

a) $x|\ell \Rightarrow \psi_*x|\psi^*\ell$ *for* $x \in P_*$, $\ell \in P^*$.

ψ is called *neighbor-preserving*, abbreviated as *n-p*, if moreover

b) $x \approx \ell \Rightarrow \psi_*x \approx \psi^*\ell$ *for* $x \in P_*$, $\ell \in P^*$.

On the other hand, a homormorphism ψ is said to be *distant-preserving*, abbreviated as *d-p*, if in addition to a) it satisfies

c) $x \not\approx \ell \Rightarrow \psi_*x \not\approx \psi^*\ell$ *for* $x \in P_*$, $\ell \in P^*$.

Finally, a homomorphism ψ is *full* provided

d) *For all* $x, y \in P_*$ *there exists* $\ell \in P^*$ *such that* $\ell \not\approx x$, $\ell \not\approx y$, $\psi^*\ell \not\approx \psi_*x$, $\psi^*\ell \not\approx \psi_*y$.

Usually, we shall simply write ψ for ψ_* and ψ^*.

Notice that *a d-p homomorphism is always full*, as is immediate from axiom 5 for Barbilian planes (see 4.1) and condition c).

For an n-p homomorphism condition d) reduces, in view of b), to the condition

e) *for all* x, y $\in P_*$ *there exists* $\ell \in P^*$ *such that* $\psi^*\ell \approx \psi_* x$, $\psi^*\ell \not\approx \psi_* y$.

From a) and d) one easily derives the dual of d), i.e.,

d') *for all* ℓ, m $\in P^*$ *there exists* x $\in P_*$ *such that* $x \not\approx \ell$, $x \not\approx m$, $\psi_* x \not\approx \psi^*\ell$, $\psi_* x \not\approx \psi^* m$.

The proof is a straightforward adaption of the proof of condition c') in [17], 1.3. Condition d) is needed, e.g., to guarantee that the image of a full homomorphism is again a Barbilian plane.

An n-p homomorphism ψ with surjective ψ^* is obviously full; notice that ψ^* is surjective if and only if ψ_* is so. For general incidence homomorphisms it is not clear, unfortunately, if ψ is automatically full if ψ_* and ψ^* are surjective.

If P and P' are ordinary projective planes, i.e., if $x \approx \ell$ is defined by $x|\ell$, then a homomorphism ψ is always n-p, and it is full if and only if Im ψ_* contains a general quadrangle (as was the definition of full homomorphism in the classical case, cf. [10]); for a proof we refer to [17], 1.3, Remark.

In the rest of this paper we shall only deal with full homomorphisms. It is an open question what the possibilities are for a homomorphism, if it is not full. Only in case of ordinary Désarguesian projective planes the answer is known; see [10].

10.2. The simplest example of a homomorphism between projective ring planes is obtained by the action of a ring homomorphism on the coordinates of points and lines. Thus, let R and R' be rings of stable rank 2, $\varphi : R \to R'$ a homomorphism with $\varphi(1) = 1$. If (ξ_0, ξ_1, ξ_2) is unimodular in R^3, then so is $(\varphi(\xi_0), \varphi(\xi_1), \varphi(\xi_2))$ in R'^3; similarly in ${}^3R$ and ${}^3R'$. Clearly,

$$\langle\varphi(\lambda_0), \varphi(\lambda_1), \varphi(\lambda_2) | \varphi(\xi_0), \varphi(\xi_1), \varphi(\xi_2)\rangle = \varphi(\langle\lambda_0, \lambda_1, \lambda_2 | \xi_0, \xi_1, \xi_2\rangle).$$

Therefore, φ defines mappings

$$\tilde{\varphi}_* : \ulcorner \xi_0, \xi_1, \xi_2 \urcorner \mapsto \ulcorner \varphi(\xi_0), \varphi(\xi_1), \varphi(\xi_2) \urcorner,$$
$$\tilde{\varphi}^* : \llcorner \lambda_0, \lambda_1, \lambda_2 \lrcorner \mapsto \llcorner \varphi(\lambda_0), \varphi(\lambda_1), \varphi(\lambda_2) \lrcorner .$$

This pair of mappings preserves incidence and the distant relation between points and lines:

$$x \mid \ell \Rightarrow \tilde{\varphi}_* x \mid \tilde{\varphi}^* \ell, \quad x \not\approx \ell \Rightarrow \tilde{\varphi}_* x \not\approx \tilde{\varphi}^* \ell.$$

The latter implication follows since φ maps units in R to units in R'. Thus we have found a d-p homomorphism $\tilde{\varphi} : P_2(R) \to P_2(R')$. We say that $\tilde{\varphi}$ is *induced* by φ.

We shall prove in §11 that all d-p homomorphisms between projective ring planes are of the above kind, i.e., induced by a ring homomorphism carrying 1 to 1; see Theorem 11.9. Distant-preserving homomorphisms have already been considered in the context of affine planes over commutative rings in [28]; the methods used there are completely different from ours.

As we recalled in the introduction of §10, the algebraic characterization of full homomorphisms between skew field planes given by W. Klingenberg depends on a recoordinatization of the original plane by a valuation ring in the coordinate skew field. Exploiting this idea in the context of ring planes, we must first recoordinatize a plane $P_2(R)$, R a ring of stable rank 2, by a suitable subring S of R. This leads to the notion of admissible subring which we shall now introduce.

10.3. DEFINITION. By an *admissible subring* of a ring R of stable rank 2 we understand a subring S (with the same 1) which is also of stable rank 2 and which has the property: for each unimodular $x \in R^3$ or $\in {}^3R$ there exists $\tau \in R$ such that $x\tau$ (resp. τx) is unimodular in S^3 (resp. ${}^3S$).

Notice that τ as in this definition must be a unit in R. If (ξ_0, ξ_1, ξ_2) is unimodular in R^3, and $y = (\xi_0\rho, \xi_1\rho, \xi_2\rho)$ as well as $z = (\xi_0\tau, \xi_1\tau, \xi_2\tau)$ is unimodular in S^3, then $z = y\rho^{-1}\tau$ with $\rho^{-1}\tau \in S$, for assume $\sum_i \lambda_i\xi_i\rho = 1$ with $\lambda_0, \lambda_1, \lambda_2 \in S$, then

$$\rho^{-1}\tau = (\sum_i \lambda_i\xi_i\rho)\rho^{-1}\tau = \sum_i \lambda_i\xi_i\tau \in S.$$

This means that the point $\ulcorner \xi_0, \xi_1, \xi_2 \urcorner$ in $P_2(R)$ determines a unique point $\ulcorner \xi_0\tau, \xi_1\tau, \xi_2\tau \urcorner$ in $P_2(S)$; we shall identify the two points from now on. Similarly for lines. Of course, incidence is the same in the two planes.

However, the neighbor relations may be different. For let $\approx$ denote the neighbor relation in $P_2(R)$ and $\sim$ that in $P_2(S)$. A nonunit in R is certainly a nonunit in S but not conversely, so $x \approx \ell$ in $P_2(R)$ implies $x \sim \ell$ in $P_2(S)$ but not conversely. We say that $\sim$ is a coarser neighbor relation than $\approx$. Generally, if $P = (P_*, P^*, \mathrm{I}, \approx)$ is a Barbilian plane, and $P_1 = (P_*, P^*, \mathrm{I}, \sim)$ a Barbilian plane which has the same points, lines and incidence relation, then $\sim$ is said to be *coarser* then $\approx$ if $x \approx \ell$ implies $x \sim \ell$ for $x \in P_*$, $\ell \in P^*$. By abuse of language we shall call $\sim$ a coarser neighbor relation on P rather than distinguishing between P and P_1. Notice that $x \approx y$ implies $x \sim y$ for $x, y \in P_*$, and similarly for any two lines. Concluding, we see that $P_2(S)$ is $P_2(R)$ with a coarser neighbor relation, if S is an admissible subring of R. This means that the identity map from $P_2(R)$ onto $P_2(S)$ is an n-p homomorphism.

EXAMPLES. Every ring of stable rank 2 is an admissible subring of itself; a valuation ring (not necessartily commutative) is an admissible subring of its quotient field (see 9.8); a subring which contains an admissible subring is itself admissible.

In §14 we shall give an intrinsic characterization of admissible subrings; see, in particular, Theorem 14.3.

We are now in a position to carry out the generalization we announced at the end of 10.2.

10.4. Let R and R' be rings of stable rank 2, S an admissible subring of R, and $\varphi : S \to R'$ a homomorphism of rings which carries 1 to 1. Then the homomorphism $\tilde{\varphi} : P_2(S) \to P_2(R')$ defines a full incidence homomorphism $\tilde{\varphi}_S : P_2(R) \to P_2(R')$. To prove condition d) of 10.1, just take, for given points x and y, a line ℓ with $\ell \not\sim x$, $\ell \not\sim y$, where $\sim$ denotes the neighbor relation in $P_2(S)$. This $\tilde{\varphi}_S$ is d-p if S = R, but in general it is neither necessary that $\tilde{\varphi}_S$ is d-p nor that it is n-p. It is our aim to show that all full incidence homomorphisms are of the type described here; see Theorem 12.1.

10.5. Under what conditions is $\tilde{\varphi}_S$ n-p? The identity mapping $P_2(R) \to P_2(S)$ is n-p as we remarked in 10.3, so it suffices to require that $\tilde{\varphi} : P_2(S) \to P_2(R')$ be n-p. Now, if x is a point and ℓ a line in $P_2(S)$, then $x \approx \ell$ means that $\langle\ell|x\rangle$ is not a unit, and $\langle\tilde{\varphi}\ell|\tilde{\varphi}x\rangle = \varphi(\langle\ell|x\rangle)$. Thus, $\tilde{\varphi}$ preserves the neighbor relation if and only if $\tilde{\varphi} : S \to R'$ carries nonunits to nonunits. The following lemma will give simpler conditions equivalent to this property of φ. Anyhow, if φ is such a homomorphism, then $\tilde{\varphi}_S$ is a full n-p homomorphism. Here again, we can announce the result that we have, in fact, described all full n-p homomorphisms; see Theorem 13.1.

10.6. LEMMA. *Let* $\varphi : R \to R'$ *be a homomorphism of associative rings with 1 (not necessarily of stable rank 2),* $\varphi(1) = 1$. *The following properties are equivalent.*

i) φ *maps nonunits in* R *on nonunits in* R'.

ii) $\ker \varphi \subseteq \text{rad } R$ *and* $\varphi(R)^* = R'^* \cap \varphi(R)$.

Proof. See [17], 1.5.

REMARK. In contrast with the situation for finite projective planes where every full homomorphism is injective by a theorem of Hughes [18], or [31], one can easily construct non-injective full n-p homomorphisms between finite projective ring planes using, for example, $\tilde{\varphi}$ induced by the natural homomorphism $\varphi : R \to R/\text{rad } R$ for R any finite ring having nonzero radical.

Before tackling the general theory of incidence homomorphisms, we further analyze the example given in 10.4; we stick to the notations used there. If S' is a subring of R containing S such that φ can be extended to the homomorphism $\varphi' : S' \to R'$, then we call $(S',\varphi') \geqq (S,\varphi)$. In this situation, $\tilde{\varphi}'_{S'} = \tilde{\varphi}_S$. By Zorn's lemma, any pair (S,φ) is contained in a maximal such pair, so we may as well assume that (S,φ) is already *maximal*, i.e., that φ cannot be extended as a homomorphism to a larger subring of R.

10.7. DEFINITION. If R and R' are arbitrary associative rings, S a subring of R and $\varphi : S \to R'$ a homomorphism, then we call S *maximal for* φ if there is no subring S' of R with $S' \supset S$, $S' \neq S$ such that φ can be extended to a homomorphism $\varphi' : S' \to R'$.

For the situation in this definition we have a lemma which will turn out to have an important geometric consequence.

10.8. LEMMA. *Let* R *and* R' *be associative rings with* 1, S *a subring of* R *with* $1 \in S$, *and* $\varphi : S \to R'$ *a homomorphism with* $\varphi(1) = 1$ *such that S is maximal for* φ. *Then* $\xi \in S$ *is a unit in* S *if and only if* ξ *is a unit in* R *and* $\varphi(\xi)$ *a unit in* R'.

Proof. The "only if"-part being clear, we assume that ξ is a unit in R and $\varphi(\xi)$ a unit in R'. Assume ξ^{-1} satisfies a relation

$$\sum_{i=0}^{n} \alpha_i (\xi^{-1})^i = 0 \text{ with } \alpha_i \in S.$$

Then, after multiplication by ξ^n,

$$\sum_{i=0}^{n} \alpha_i \xi^{n-i} = 0,$$

whence

$$\sum_{i=0}^{n} \varphi(\alpha_i)\varphi(\xi)^{n-i} = 0.$$

Dividing by $\varphi(\xi)^n$ we get

$$\sum_{i=0}^{n} \varphi(\alpha_i)(\varphi(\xi)^{-1})^i = 0.$$

It follows that we can extend φ to a homomorphism from $S[\xi^{-1}]$ to R' by defining $\varphi(\xi^{-1}) = \varphi(\xi)^{-1}$. Since, however, φ cannot be extended any more, $\xi^{-1} \in S$, which was to be proved.

Now recall from 3.1 that in the projective plane $P_2(R)$ over the ring R, $\ulcorner a \urcorner \not\approx \llcorner \ell \lrcorner$ if and only if $\langle \ell | a \rangle$ is invertible. Thus, the above lemma can be translated into geometric terms:

10.9. LEMMA. *Let* R *and* R' *be rings of stable rank* 2, S *an admissible subring of* R, *and* $\varphi : S \to R'$ *a homomorphism carrying* 1 *to* 1 *such that* S *is maximal for* φ. *Then for any point* x *and line* ℓ *in* $P_2(R)$,

$$x \not\approx \ell \text{ in } P_2(S) \Leftrightarrow x \not\approx \ell \text{ in } P_2(R) \text{ and } \tilde{\varphi}_S x \not\approx \tilde{\varphi}_S \ell \text{ in } P_2(R').$$

So in the above situation the neighbor relation in $P_2(S)$ is determined by the neighbor relations in $P_2(R)$ and $P_2(R')$ together with the incidence homomorphism $\tilde{\varphi}_S$. Now, starting at last the general study of an incidence homomorphism $\psi : P \to P'$ between Barbilian planes, we convert this result into the definition of a coarser neighbor relation $\sim$ in P in the following proposition, where we use the notations of 10.1 again.

10.10. PROPOSITION. *Let* $\psi : P \to P'$ *be a full incidence homomorphism of Barbilian planes. Then* ${}^{\psi}P = (P_*, P^*, \mathrm{I}, \sim)$ *with* $\sim$ *defined by*

$$x \sim \ell \Leftrightarrow x \approx \ell \text{ or } \psi x \approx \psi\ell, \quad \text{for } x \in P_*,\ \ell \in P^*$$

is a Barbilian plane.

The proof simply consists of verifying the axioms for a Barbilian plane as stated in 4.1; see [38], 1.2 for details. Notice that condition d) of 10.1 (fullness) is just axiom 5 for ψP !

The statements of Lemma 5.2 carry over to the case of a full incidence homomorphism provided we replace the neighbor relation $\approx$ in P by the coarser neighbor relation $\sim$ of ψP. Precisely stated we have the following lemma, whose proof consists of an adaption of the arguments used in the proof of 5.2. to the present situation.

10.11. LEMMA. *Let* $\psi : P \to P'$ *be a full incidence homomorphism between Barbilian planes,* ${}^{\psi}P = (P_*, P^*, \mathrm{I}, \sim)$ *as defined in* 10.10. *Then* ψ *is uniquely determined by any of the following three sets of data:*

(i) *the images* ψu *of all points* $u \not\sim \ell$ *for some fixed line* $\ell \in P^*$;

(ii) *the images* ψa, ψb *and* ψu *for all* $u \mathrm{I} \ell$, *if the points* a, b *and the line* ℓ *are in general position in* ${}^{\psi}P$;

(iii) *the images* ψa, ψu *for all* $u \mathrm{I} \ell$ *and* ψm *for a point* a *and lines* ℓ, m *such that* a, ℓ *and* m *are in general position in* ${}^{\psi}P$.

The definition of the neighbor relation $\sim$ in ${}^{\psi}P$ is such that the homomorphism ψ is split in a full n-p homomorphism $P \to {}^{\psi}P$, viz., the identity, and a d-p homomorphism ${}^{\psi}P \to P'$. The next important step is to show that ${}^{\psi}P$ is a ring plane if P and P' are ring planes. Formulated in a different way: from the existence of transvections, dilatations and affine dilatations in P and P' one has to derive the existence of similar devices in ${}^{\psi}P$. This will be an easy consequence of a series of "commutation rules" for the homomorphism ψ and certain transvections etc. in P and P'. For convenience of notations we adopt the convention to denote the image of $x \in P_*$ by x'; often, $x' \in P'_*$ is given and then x denotes some or a specified point in P such that $\psi x = x'$. The same notations for lines.

10.12. LEMMA. *Assume that* P *and* P' *are projective ring planes. Let* a, b, $c \in P_*$ *and* $\ell \in P^*$ *be such that* $a \not\sim \ell$, $a \not\sim c$, $b \not\sim \ell$ *and* $b | a \vee c$, *with* $\sim$ *as in* 10.10.

(i) *If* $c | \ell$, *then*

$$\psi \circ T_{c,\ell:a,b} = T_{c',\ell':a',b'} \circ \psi,$$

and $T_{c,\ell:a,b}$ *acts as a* (c,ℓ)*-transvection in* ${}^{\psi}P$.

(ii) *If* $c \not\sim \ell$, $b \not\sim c$, *then again*

$$\psi \circ T_{c,\ell:a,b} = T_{c',\ell':a',b'} \circ \psi,$$

and $T_{c,\ell:a,b}$ *acts as a* (c,ℓ)*-dilatation in* ${}^{\psi}P$.

(iii) *If* $c \not\sim \ell$, *then in the affine plane* ${}^{\psi}P^{\ell}$,

$$\psi \circ \hat{T}_{c,\ell:a,b} = \hat{T}_{c',\ell':a',b'} \circ \psi,$$

and $\hat{T}_{c,\ell:a,b}$ *acts as an affine* (c,ℓ)*-dilatation in* ${}^{\psi}P^{\ell}$.

Proof. (i) Notice that a, b, c and ℓ satisfy the conditions $c|\ell$, $a \not\approx \ell$, $b \not\approx \ell$, $b | a \vee c$, so $T_{c,\ell:a,b}$ exists in P. For similar reasons, $T_{c',\ell':a',b'}$ exists in P'. Now the "commutation rule"

$$\psi \circ T_{c,\ell:a,b} = T_{c',\ell':a',b'} \circ \psi$$

immediately follows from the previous lemma by the observation that the left hand side and the right hand side act in the same way on a, on the points of ℓ, and on the lines through c.

To prove that $T_{c,\ell:a,b}$ is also a collineation in ${}^{\psi}P$ we only have to show that it preserves the neighbor relation $\sim$. So consider x and m with $x \sim m$. Either $x \approx m$ or $\psi x \approx \psi m$. If $x \approx m$, then $T_{c,\ell:a,b}x \approx T_{c,\ell:a,b}m$, hence $T_{c,\ell:a,b}x \sim T_{c,\ell:a,b}m$. If, on the other hand, $\psi x \approx \psi m$, then $T_{c',\ell':a',b'} \circ \psi x \approx T_{c',\ell':a',b'} \circ \psi m$, so by the "commutation rule" we just proved, $\psi \circ T_{c,\ell:a,b}x \approx \psi \circ T_{c,\ell:a,b}m$. This proves that $T_{c,\ell:a,b}x \sim T_{c,\ell:a,b}m$, and we are finished. The proof of case (ii) is similar.

(iii) From the definition of the neighbor relation $\sim$ in ${}^{\psi}P$ it is immediate that ${}^{\psi}P^{\ell}$ is contained in P^{ℓ}, and that ψ maps ${}^{\psi}P^{\ell}$ into P'^{ℓ}. Since ${}^{\psi}P$ is a Barbilian plane, the affine dilatation $\hat{T}_{c,\ell:a,b}$ (defined in P^{ℓ}) maps ${}^{\psi}P^{\ell}$ into itself. Thus the commutation relation in (iii) makes sense. For its proof one can use the arguments given in the proof of (i) and (ii) of Lemma 5.12, provided they are adapted to the present situation by arguing with ${}^{\psi}P$ and its neighbor relation $\sim$ instead of with P and $\approx$.

Lemma 5.9, finally, yields that $T_{c,\ell:a,b}$ acts as an affine dilatation in ${}^{\psi}P^{\ell}$.

From 10.10 and 10.12 we immediately infer:

10.13. COROLLARY. *If* P *and* P' *are projective ring planes and* $\psi : P \to P'$ *is a full homomorphism, then* ${}^{\psi}P$ *as defined in* 10.10 *is a projective ring plane.*

As we remarked already after Proposition 10.10, ψ is a d-p homomorphism from ${}^{\psi}P$ to P'. Thus we are lead to the study of d-p homomorphisms, which we shall enter upon in the next section.

§11. DISTANT-PRESERVING HOMOMORPHISMS.

In this section we shall prove the result we announced in 10.2, viz., that every d-p homomorphism between projective ring planes is induced by a homomorphism between the coordinate rings. The key to this is a study of the image under a d-p homomorphism.

11.1. To get an idea of the situation, we have a closer look at the example given in 10.2. So let $\varphi : R \to R'$ be a homomorphism between rings of stable rank 2 with $\varphi(1) = 1$, and $\tilde{\varphi} : P_2(R) \to P_2(R')$ the d-p homomorphisme induced by φ. Set $U = \varphi(R)$, a subring of R' containing 1, which has stable rank 2 by 2.4.

A unimodular triple in U^3 is unimodular in R'^3, and if two such triples are right proportional with a proportionality factor $\rho \in R'^*$, then in fact $\rho \in U^*$ as one easily deduces from unimodularity. This allows us to identify $P_2(U)_*$ with a subset of $P_2(R')_*$. Similarly for lines. Notice that $P_2(U)_*$ is not the subset of points of $P_2(R')$ all of whose coordinates are in U, since a triple in U^3 may well be unimodular in R'^3 without being proportional to a unimodular triple in U ; for an example of this, see [37], 1.3, Remark.

Using the conditions SR_3 and SR_2 in U together with 2.16, one easily shows that $P_2(U)_* = \tilde{\varphi}P_2(R)_*$ and $P_2(U)^* = \tilde{\varphi}P_2(R)_*$; see [37], 1.4. The incidence relation in $P_2(U)$ is, of course, that of $P_2(R')$. On the other hand, the neighbor relation in $P_2(U)$, which we denote by $\sim$, is in general not induced by the neighbor relation $\approx$ in $P_2(R')$, since a nonunit in U may well be a unit in R'. It is easy, however, to describe $\sim$ by the neighbor relation in $P_2(R)$ and $\tilde{\varphi}$. The following two equivalences are proved in [37], 1.5, using the result of 2.16 again:

For $x' \in P_2(U)_*$, $\ell' \in P_2(U)^*$,

$$x' \sim \ell' \text{ in } P_2(U) \Leftrightarrow x \approx \ell \text{ for all } x \in P_2(R)_*,\ \ell \in P_2(R)^* \text{ such that } \tilde{\varphi}x = x',\ \tilde{\varphi}\ell = \ell'. \quad \text{(a)}$$

$$x' \sim \ell' \text{ in } P_2(U) \Leftrightarrow \exists x \in P_2(R)_* \text{ with } \tilde{\varphi}x = x' \text{ such that } x \approx \ell \text{ for all } \ell \in P_2(R)^* \text{ with } \tilde{\varphi}\ell = \ell'. \quad \text{(b)}$$

We now know enough about the d-p homomorphism $\tilde{\varphi}$ induced by a ring homomorphism φ to deal with the general d-p homomorphisms.

11.2. DEFINITION. For a d-p homomorphism $\psi : P \to P'$ of Barbilian planes, its *image* is $\psi P = (\psi P_*, \psi P^*, \mathrm{I}, \sim)$ with for all $x' \in \psi P_*$, $\ell' \in \psi P^*$:

$$x' \mathrm{I} \ell' \text{ in } \psi P \Leftrightarrow x' \mathrm{I} \ell' \text{ in } P',$$

$$x' \sim \ell' \text{ in } \psi P \Leftrightarrow x \approx \ell \text{ in } P \text{ for all } x \in P_*,\ \ell \in P^* \text{ such that } \psi x = x',\ \psi\ell = \ell'.$$

The definition of $\sim$ in ψP agrees with (a) in 11.1. Before showing that ψP is a Barbilian plane we must derive some ancillary results, which are to serve in several reasonings. For two of these we have to impose on the planes P and P' that they are Barbilian transvection planes as defined in 6.1.

11.3. LEMMA. *Notations as in* 11.2.
(i) *If* $\ell \in P^*$, $x' \in \psi P_*$ *with* $x'|\ell' = \psi\ell$, *then* $x \in P_*$ *exists such that* $\psi x = x'$ *and* $x|\ell$.

For the following two properties assume, moreover, that P *and* P' *are Barbilian transvection planes.*
(ii) *If* $x' \in \psi P_*$, $\ell' \in \psi P^*$, $x' \not\sim \ell'$, *then for each* $x \in P_*$ *with* $\psi x = x'$ *there exists* $\ell \in P^*$ *such that* $\psi\ell = \ell'$ *and* $x \not\approx \ell$.
(iii) *If* $x \in P_*$, $\ell \in P^*$, $x|\ell$, *and* $y' \in \psi P_*$, $y'|\ell' = \psi\ell$, $y' \not\sim x' = \psi x$, *then there exists* $y \in P_*$ *satisfying* $\psi y = y'$, $y|\ell$, $y \not\approx x$.
Proof. See [37], 2.3. For the proof of (ii), Lemma 2.1 of [37] is needed; this is a special case of (i) of Lemma 10.12 in the present paper, where we only need that P and P' have transvections.

From (i) of the above lemma and its dual, one derives that in the situation of Definition 11.2:

11.4. COROLLARY. *For* $x', y' \in \psi P_*$, $\ell', m' \in \psi P^*$,

$$x' \sim y' \Leftrightarrow x \approx y \text{ for all } x, y \in P_* \text{ with } \psi x = x', \psi y = y',$$
$$\ell' \sim m' \Leftrightarrow \ell \approx m \text{ for all } \ell, m \in P^* \text{ with } \psi\ell = \ell', \psi m = m'.$$

Now one can prove as in 2.4 of [37]:

11.5. PROPOSITION. *If* P *and* P' *are Barbilian transvection planes and* $\psi : P \to P'$ *is a d-p homomorphism, then* ψP *is a Barbilian plane.*

11.6. Consider $\ell' \in \psi P^*$, $a', b', c' \in \psi P_*$ with $c'|\ell'$, $a' \not\sim \ell'$, $b' \not\sim \ell'$, $b'|a' \vee c'$. From 11.3 we infer the existence of $\ell \in P^*$, $a, b, c \in P_*$ which are mapped by ψ upon ℓ', a', b', c', respectively, and which satisfy $c|\ell$, $a \not\approx \ell$, $b \not\approx \ell$, $b|a \vee c$. Using (i) of Lemma 10.12 one sees that the transvection $T_{c',\ell':a',b'}$ in P' maps ψP_* and ψP^* upon themselves. Hence it induces

a collinearity-preserving bijection in ψP, which we call $T^{\psi P}_{c',\ell':a',b'}$.
Again using case (i) of 10.12 one easily shows that

$$x' \sim m' \Leftrightarrow T^{\psi P}_{c',\ell':a',b'}x' \sim T^{\psi P}_{c',\ell':a',b'}m' \text{ for } x' \in \psi P_*,\ m' \in \psi P^*$$

(see 3.1 of [37]). This shows that $T^{\psi P}_{c',\ell':a',b'}$ acts as a (c',ℓ')-transvection in ψP.

We call $T^{\psi P}_{c',\ell':a',b'}$ *induced by* $T_{c,\ell:a,b}$, or *lifted up to* $T_{c,\ell:a,b}$.

We conclude that ψP is a Barbilian transvection plane in the situation of 11.5. This allows us to apply the results of §6 to ψP. In particular, the little projective group of ψP has the transitivity properties of 6.10. Using that property, one can prove the following extension of 11.3 (see [37], Lemma 3.3):

11.7. LEMMA. *Assume* $\psi : P \to P'$ *is a d-p homomorphism of Barbilian transvection planes. Given* $x, y \in P_*$, $\ell \in P^*$, $x|\ell$, $y|\ell$, $x \not\sim y$, *with* ψ*-images* x', y', ℓ', *respectively, there exists for each* $z' \in \psi P_*$ *such that* $z'|\ell'$, $z' \not\sim x'$, $z' \not\sim y'$, *a point* $z|\ell$ *such that* $\psi z = z'$ *and* $z \not\sim x$, $z \not\sim y$.

Using the above lemma in combination with 11.3, one can prove: if P and P' are projective ring planes, then the image ψP is also a projective ring plane. The existence of dilatations $T^{\psi P}_{c',\ell':a',b'}$ and affine dilatations $\hat{T}^{\psi P}_{c',\ell':a',b'}$, which are just the restrictions to ψP of the analogous (affine) dilatations in P', is proved by similar techniques as in 11.6; for details, see [37], Lemma 3.5. The conclusion is:

11.8. PROPOSITION. *If* $\psi : P \to P'$ *is a d-p homomorphism of projective ring planes, then the image* ψP *is a projective ring plane.*

We are now ready for the algebraic characterization of d-p homomorphisms we announced in 10.2. Notice that a d-p homomorphism preserves general position, hence the image of a general quadrangle in P is a general quadrangle in ψP as well as in P'.

11.9. THEOREM. *Let* $\psi : P_2(R) \to P_2(R')$ *be a d-p homomorphism of projective planes over rings of stable rank* 2. *Let* $P_2(R)$ *be coordinatized with respect to a general quadrangle* p_0, p_1, p_2, e *and* $P_2(R')$ *with respect to the general quadrangle* p'_0, p'_1, p'_2, e' *with* $p'_i = \psi p_i$, $e' = \psi e$. *Then there exists a ring homomorphism* $\varphi : R \to R'$ *with* $\varphi(1) = 1$ *such that* $\psi = \tilde{\varphi}$, *the d-p homomorphism induced by* φ *(see* 10.2*). The ring* $U = \varphi(R)$ *is the coordinate ring of the plane* $\psi P_2(R)$ *with respect to* p'_0, p'_1, p'_2, e'.

Proof. We call $P_2(R) = P$, $P_2(R') = P'$ with the usual notations such as $\approx$ in P, P', $\sim$ in ψP, etc. Set $\ell_\infty = p_1 \vee p_2$, $\ell_1 = p_0 \vee p_1$, and similarly ℓ'_∞, ℓ'_1 in P'. From the definition of $\sim$ and Lemma 11.3 we see that ψ maps $\{x \in P_* | x | \ell_1, x \not\approx \ell_\infty\} = \{\ulcorner 1,\alpha,0 \urcorner | \alpha \in R\}$ onto $\{x' \in P'_* | x' | \ell'_1, x' \not\sim \ell'_\infty\}$. The latter is a subset of $\{x' \in P'_* | x' | \ell'_1, x' \not\approx \ell'_\infty\} = \{\ulcorner 1,\alpha,0 \urcorner | \alpha \in R'\}$. This allows to define $\varphi : R \to R'$ by

$$\ulcorner 1,\varphi(\alpha),0 \urcorner = \psi \ulcorner 1,\alpha,0 \urcorner \text{ for } \alpha \in R.$$

Set $\varphi(R) = U$.

For $\alpha \in R$, let T_α denote the (p_1,ℓ_∞)-translation which carries $\ulcorner 1,0,0 \urcorner$ to $\ulcorner 1,\alpha,0 \urcorner$, and D_α the (affine) (p_1,ℓ_∞)-dilatation which carries $\ulcorner 1,1,0 \urcorner$ to $\ulcorner 1,\alpha,0 \urcorner$. Similarly one has in $P_2(R')$ the translations T_α and the (affine) dilatations D_α for $\alpha \in R'$. If $\alpha' \in U$, then $T_{\alpha'}$ and $D_{\alpha'}$ induce in $\psi P_2(R)$ the translation $T^U_{\alpha'}$, resp. the (affine) dilatation $D^U_{\alpha'}$. Since T_α and D_α define addition and multiplication, we see in the first place that U is a subring of R' with the same identity element, which is clearly the coordinate ring of the image plane $\psi P_2(R)$. From the commutation rules in Lemma 10.12 we infer that

$$\psi \circ T_\alpha = T_{\varphi(\alpha)} \circ \psi, \quad \psi \circ D_\alpha = D_{\varphi(\alpha)} \circ \psi.$$

Translated into coordinates this means that φ is a ringhomomorphism. Since ψ maps the basic quadrangle p_0, p_1, p_2, e upon the quadrangle p'_0, p'_1, p'_2, e', it follows that $\varphi(1) = 1$.

The points in $P_2(R)$ which are $\not\approx \ell_\infty$ are precisely the points $[1,\alpha,\beta]$ with $\alpha, \beta \in R$. From

$$\ulcorner 1,\alpha,\beta \urcorner = T_{D_\alpha \ulcorner 1,1,0 \urcorner} D_\alpha \ulcorner 1,0,1 \urcorner$$

together with the analogous relation in $P_2(R')$, we infer, again using the commutation rules of Lemma 10.12, that

$$\ulcorner 1,\alpha,\beta \urcorner = \ulcorner 1,\varphi(\alpha),\varphi(\beta) \urcorner.$$

Thus we see that ψ has the same action on the points $\not\approx \ell_\infty$ in $P_2(R)$ as $\tilde{\varphi}$ has. By Lemma 10.11, we conclude that $\psi = \tilde{\varphi}$, which completes the proof.

§12. ALGEBRAIC CHARACTERIZATION OF FULL INCIDENCE HOMOMORPHISMS.

Now we have proved in the previous section that all d-p homomorphisms between projective ring planes are induced by ring homomorphisms, we can take up the line of §10 again and characterize full incidence homomorphisms.

12.1. THEOREM. *Let* $\psi : P_2(R) \to P_2(R')$ *be a full incidence homomorphism of projective planes over rings of stable rank* 2. *Let* $P_2(R)$ *be coordinatized with respect to a general quadrangle* p_0, p_1, p_2, e *which is also a general quadrangle in* ${}^{\psi}P_2(R)$ *(i.e., with respect to the neighbor relation* $\sim$ *as defined in* 10.10), *and* $P_2(R')$ *with respect to the general quadrangle* p_0', p_1', p_2', e' *with* $p_i' = \psi p_i$, $e' = \psi e$. *Then there are a unique admissible subring* S *of* R *and a unique homomorphism* $\varphi : S \to R'$ *with* S *maximal for* φ *such that* $\psi = \tilde{\varphi}_S$, *the homomorphism induced by* φ *(see* 10.4).

Proof. Take for S the admissible subring of R which coordinatizes the plane ${}^{\psi}P_2(R)$, with neighbor relation $\sim$ as defined in 10.10, with respect to the quadrangle p_0, p_1, p_2, e. From the definition of $\sim$ it is immediate that ψ defines a distant-preserving homomorphism $\psi' : P_2(S) \to P_2(R')$ such that $\psi = \psi' \circ \mathrm{id}$, where id denotes the identity map $P_2(R) \to P_2(S)$. Now Theorem 11.9 tells us that $\psi' = \tilde{\varphi}$ for some ring homomorphism $\varphi : S \to R'$ with $\varphi(1) = 1$, i.e., $\psi = \tilde{\varphi}_S$.

Let $\overline{S}$ be a maximal extension of S in R such that φ can be extended to a homomorphism $\overline{\varphi} : \overline{S} \to R'$. Then $\overline{\varphi}$ induces a homomorphism $P_2(\overline{S}) \to P_2(R')$ which is in fact ψ, for S is an admissible subring of $\overline{S}$ whence the action of $\tilde{\overline{\varphi}}$ on points and lines is precisely that of $\tilde{\varphi}$. By Lemma 10.9, the neighbor relation in $P_2(\overline{S})$ must therefore be the same as the neighbor relation $\sim$ in $P_2(S)$. From this we see that $\overline{S} = S$, i.e., S is already maximal for φ. The uniqueness of S and φ follow in the same way: the fact that S is maximal for φ implies that the neighbor relation in $P_2(S)$ cannot be anything else but that in ${}^{\psi}P$, i.e., $P_2(S)$ has to be ${}^{\psi}P$.

Properties of ψ can often be translated in properties of φ and vice versa. Here we'll see this in some cases; one more will follow in the next section. It should be observed that ψ_* is injective or surjective if and only if ψ^* is injective resp. surjective; we simply call ψ *injective* or *surjective* in these cases, respectively.

12.2. COROLLARY. *With the notations of the theorem above we have:*
(i) ψ *is distant-preserving if and only if* $S = R$.
(ii) ψ *is injective if and only if* φ *is injective.*
(iii) ψ *is surjective if and only if* $\varphi(S)$ *is an admissible subring of* R'.
(iv) φ *is surjective if and only if* ψ *is surjective and* $x' \approx \ell'$ *in* $P_2(R')$ *if* $x \sim \ell$ *for all* x *with* $\psi x = x'$ *and all* ℓ *with* $\psi\ell = \ell'$.
Proof. Obviously, ψ is d-p if and only if the neighbor relation $\sim$ of $^{\psi}P_2(R) = P_2(S)$ is the same as the neighbor relation $\approx$ of $P_2(R)$. By the coordinatization theorem of §8 the latter is the case precisely if $S = R$.

For the proof of (ii), (iii) and (iv), we start with the observation that $\mathrm{id} : P_2(R) \to P_2(S)$ is bijective. Thus, it suffices to consider $\tilde{\varphi} : P_2(S) \to P_2(R')$ instead of ψ.
(ii) If $\tilde{\varphi}$ is injective, so is φ since $\tilde{\varphi}\ulcorner 1,\alpha,0\urcorner = \ulcorner 1,\varphi(\alpha),0\urcorner$. Conversely, let φ be injective. Assume $\tilde{\varphi}\ulcorner\alpha_0,\alpha_1,\alpha_2\urcorner = \tilde{\varphi}\ulcorner\beta_0,\beta_1,\beta_2\urcorner$, then there is $\gamma \in R'$ such that $\varphi(\beta_i) = \varphi(\alpha_i)\gamma$ for $i = 0,1,2$. Pick $\lambda_0,\lambda_1,\lambda_2 \in S$ such that $\Sigma\lambda_i\alpha_i = 1$. Then $\gamma = \varphi(\delta)$ if we take $\delta = \Sigma\lambda_i\beta_i$. So $\varphi(\beta_i) = \varphi(\alpha_i\delta)$ and, by the injectivity of φ, we conclude that $\beta_i = \alpha_i\delta$. Thus $\ulcorner\alpha_0,\alpha_1,\alpha_2\urcorner = \ulcorner\beta_0,\beta_1,\beta_2\urcorner$, which proves injectivity of $\tilde{\varphi}$.
(iii) $\tilde{\varphi}$ is surjective means that the image plane $\psi P_2(S)$, which is coordinatized by $\varphi(S)$, has the same points and lines as $P_2(R')$ but a coarser neighbor relation; see 11.2 and 11.9. This means that $\varphi(S)$ is an admissible subring of R'; see 10.2.
(iv) φ surjective means $\varphi(S) = R'$, which is equivalent to the conditions: $\varphi(S)$ is admissible in R' and $R'^*\cap\varphi(S) = \varphi(S)^*$ as will follow from 14.3. By (iii), $\varphi(S)$ is admissible in R' if and only if $\tilde{\varphi}$ is surjective. Since $\varphi(S)^* = \varphi(S^*)$ by 2.16, the condition $R'^*\cap\varphi(S) = \varphi(S)^*$ means that every $\eta \in \varphi(S)$ which is a unit in R' is the image of a unit in S : $\eta = \varphi(\xi)$ for some $\xi \in S^*$. The geometric translation of this property is : if $x' \not\approx \ell'$ in $P_2(R')$, then there exist x and ℓ in $P_2(S)$ with $\tilde{\varphi}x = x'$, $\tilde{\varphi}\ell = \ell'$ and $x' \not\sim \ell'$. This completes the proof.

§13. FULL NEIGHBOR-PRESERVING HOMOMORPHISMS.

With the results of the three previous sections in hand it is easy to characterize full n-p homomorphisms as a special instance of general incidence homomorphisms as described in Theorem 12.1.

13.1. THEOREM. *Let* $\psi : P_2(R) \to P_2(R')$ *be a full incidence homomorphism. Coordinatize* $P_2(R)$ *again with respect to a general quadrangle* $p_0, p_1, p_2,$ *e which is also a general quadrangle in* ${}^{\psi}P_2(R)$, *and* $P_2(R)$ *with respect to the general quadrangle* p_0', p_1', p_2', e' *with* $p_i' = \psi p_i$, $e' = \psi e$. *Let* S *be the admissible subring of* R *and* $\varphi : S \to R'$ *the homomorphism with* S *maximal for* φ *such that* $\psi = \tilde{\varphi}_S$. *Then* ψ *is neighbor-preserving if and only if* $\ker \varphi \subseteq \operatorname{rad} S$ *and* $\varphi(S)^* = R'^* \cap \varphi(S)$.

Proof. If ψ is n-p, the neighbor relation $\sim$ on ${}^{\psi}P_2(R) = P_2(S)$ is given by $x \sim \ell \Leftrightarrow \psi x \approx \psi \ell$, so $\tilde{\varphi} : P_2(S) \to P_2(R')$ is n-p. The converse is immediate since $\mathrm{id} : P_2(R) \to P_2(S)$ is n-p. We have seen in 10.2 that $\tilde{\varphi}$ is n-p if and only if $\varphi : S \to R'$ maps nonunits on nonunits, and by Lemma 10.6 the latter is equivalent to the conditions on φ given in the statement of the theorem.

Concerning the correlation between properties of ψ on the one hand and properties of S and φ on the other hand we can remark that (i) and (ii) of Corollary 12.2 need no further attention in the present case, but (iii) and (iv) of the same corollary can be taken together and simplified if ψ is n-p. Namely,

13.2. COROLLARY. *If* ψ *is a full* n-p *homomorphism, then with the notations of the above theorem,*

ψ is surjective if and only if φ is surjective.

Proof. This is immediate from 12.2 (iv) by observing that for n-p ψ we have: $x \sim \ell$ in ${}^{\psi}P_2(R) \Leftrightarrow \psi x \approx \psi \ell$ in $P_2(R')$.

§14. ADMISSIBLE SUBRINGS.

In 10.3 we introduced the notion of admissible subring of a ring of stable rank 2; the definition there was given in terms of free 3-dimensional modules. In the present section we shall give a more ring-theoretic characterization of admissible subrings, although the modules cannot be completely eliminated, in fact; for more details, see Theorem 14.3.

14.1. Let R be a ring of stable rank 2, and S an admissible subring. For any $\xi \in R$, $(1,\xi,0)$ is unimodular in R^3, hence by Definition 10.3 there exists $\tau \in R^*$ such that $(\tau,\xi\tau,0)$ is unimodular in S^3. This means that $\tau \in R^* \cap S$, $\alpha = \xi\tau \in S$ and that (τ,α) is unimodular in S^2.

Call $R^* \cap S = T$. This is clearly a multiplicative subsemigroup containing S^* and whose elements are not zerodivisors in S. We have seen that each $\xi \in R$ can be written as $\alpha\tau^{-1}$ with $\alpha \in S$, $\tau \in T$, i.e., $R = ST^{-1}$. So T is a right denominator set and henceforth the right Ore condition must hold:

$$\sigma T \cap \tau S \neq \emptyset \text{ for all } \sigma \in S \text{ and } \tau \in T.$$

Similar considerations about the left modules ${}^3R$ and ${}^3S$ show that T is at the same time a left denominator set in S and $R = T^{-1}S$.

A further analysis shows: any finite number of elements $\xi_1,\ldots,\xi_n$ in R can be written as $\xi_i = \sigma_i\tau^{-1}$ with $\sigma_1,\ldots,\sigma_n \in S$ and a common denominator $\tau \in T$, and also $\xi_i = \bar{\tau}^{-1}\bar{\sigma}_i$ with $\bar{\sigma}_1,\ldots,\bar{\sigma}_n \in S$ and $\bar{\tau} \in T$. See [17], 5.3 for a proof of this.

Now consider any unimodular $x = (\xi_1,\xi_2,\xi_3)$ in R^3. Write $\xi_i = \sigma_i\tau^{-1}$ with $\sigma_i \in S$ and a common denominator $\tau \in T$. Unimodularity of x can be expressed by the requirement that there are $\lambda_i \in S$, $\rho \in T$ such that

$$\sum_i \rho^{-1}\lambda_i\sigma_i\tau^{-1} = 1,$$

i.e.,

$$\sum_i \lambda_i\sigma_i = \rho\tau.$$

Thus, the admissibility of S in R means: for any triple $\sigma_1,\sigma_2,\sigma_3$ with all $\sigma_i \in S$ such that $\sum_i \lambda_i\sigma_i \in T$ for some $\lambda_i \in S$ there exist $\beta \in R^*$ and $\mu_i \in S$ such that

$$\sigma_i' = \sigma_i\beta \in S \ (i = 1,2,3) \text{ and } \sum_i \mu_i\sigma_i' = 1.$$

Take $\alpha = \sum_i \mu_i\sigma_i$, then $\alpha \in S$ and $\alpha\beta = 1$, so $\alpha \in R^* \cap S = T$ and $\beta = \alpha^{-1}$. Thus,

$$\sigma_i = \sigma_i'\alpha \text{ with } \sigma_i' \in S,\ \alpha \in T, \text{ and}$$

$(\sigma'_1, \sigma'_2, \sigma'_3)$ is unimodular in S^3.

A similar condition with left and right interchanged must hold.

The properties of S and T we have found lead to the following

14.2. DEFINITION. A subset R of an associative ring S with 1 is a *planar denominator set* (PDS for short) provided:

i) T is a left and right denominator set in S, i.e., a multiplicative semigroup not containing zero divisors and satisfying the left and right Ore conditions:

$T\sigma \cap S\tau \neq \emptyset$ resp. $\sigma T \cap \tau S \neq \emptyset$ for all $\sigma \in S, \tau \in T$.

ii) $T \supseteq S^*$, the set of units in S.

iii) If $\sigma_1, \sigma_2, \sigma_3 \in S$ satisfy $\Sigma\lambda_i\sigma_i \in T$ for certain $\lambda_i \in S$, then there is $\alpha \in T$ such that $\sigma_i = \sigma'_i\alpha$ with all $\sigma'_i \in S$ and $(\sigma'_1, \sigma'_2, \sigma'_3)$ unimodular in S^3.

iii') If $\lambda_1, \lambda_2, \lambda_3 \in S$ satisfy: $\Sigma\lambda_i\sigma_i \in T$ for certain $\sigma_i \in S$, then there is $\alpha \in T$ such that $\lambda_i = \alpha\lambda'_i$ with all $\lambda'_i \in S$ and $(\lambda'_1, \lambda'_2, \lambda'_3)$ unimodular in ${}^3S$.

In the situation of a ring S with a PDS T, the right quotient ring $R = ST^{-1}$ exists, as well as the left quotient ring $T^{-1}S$. We will consider S as a subring of R, thus $T \subseteq S \subseteq R$, and then $T^{-1}S \subseteq R$. It is immediate that $T^{-1}S = R$.

We have thus seen that if S is an admissible subring of a ring R of stable rank 2, then S contains a PDS T such that $R = ST^{-1} = T^{-1}S$. The converse is also true, i.e., the following theorem holds (see [17], 5.10):

14.3. THEOREM. *Let* R *be an associative ring with* 1, *and* S *a subring with* $1 \in S$. *The following are equivalent:*

(i) R *has stable rank* 2 *and* S *is an admissible subring of* R.

(ii) S *has stable rank* 2 *and contains a* PDS T *such that* $R = ST^{-1} = T^{-1}S$.

Moreover, if (ii) holds, then necessarily $T = R^* \cap S$.

REMARK. If T is a PDS in a ring S and $R = ST^{-1}$ had stable rank 2, then S need not have stable rank 2 as one sees by taking $S = \mathbb{Z}$, $T = \mathbb{Z}\setminus\{0\}$, so $R = \mathbb{Q}$.

In the remainder of this section we will explore the relation between a ring S with a PDS T and its quotient ring $R = ST^{-1}$ in the particular cases of rings we have dealt with in §9. We do not assume a priori the stable rank of R or S to be 2, unless we explicitly say so. In several cases it will be possible to greatly simplify the definition of a PDS.

14.4. PROPOSITION. *If every nonunit* α *in* S *is a right zero divisor, i.e., there exists* $\beta \in S$ *with* $\beta\alpha = 0$, $\beta \neq 0$, *then* $R = S$. *Similarly if every nonunit in* S *is a left zero divisor.*

Proof. If T is a PDS, then $T \supseteq S^*$ by definition. Since T is not allowed to contain zero divisors, we must have $T = S^*$ in the present situation, which implies $R = S$.

If, conversely, every nonunit in R is a right (left, respectively) zero divisor, the same need not hold in S, as one sees with $R = \mathbb{Q}$ and $S = \mathbb{Z}_{(p)}$ (the p-adic integers). The condition in the proposition is satisfied if S is a left and right Artin ring (cf. 9.1), so in particular in finite rings. Hence we find (cf. [17], 6.6):

COROLLARY. *If* R *is a finite ring and* S *an admissible subring, then* $R = S$.

14.5. PROPOSITION. *Assume that* R *and* S *have stable rank* 2, *so* S *is an admissible subring of* R. *Then* R *satisfies the equivalent conditions* (b) *and* (c) *of* 9.3 *if and only if* S *does so.*

Proof. Assume that R satisfies (c). Take $a_1, a_2 \in S$. Then by (c) we can write $(\alpha_1,\alpha_2) = (\xi_1,\xi_2)\eta$ with unimodular (ξ_1,ξ_2) in R^2 and $\eta \in R$. Since S is admissible in R, $(\xi_1,\xi_2) = (\beta_1,\beta_2)\zeta$ with unimodular (β_1,β_2) in S^2 and $\zeta \in R$, so $(\alpha_1,\alpha_2) = (\beta_1,\beta_2)\zeta\eta$. Let $\lambda_1, \lambda_2 \in S$ be such that $\lambda_1\beta_1+\lambda_2\beta_2 = 1$, then $\zeta\eta = \lambda_1\alpha_1+\lambda_2\alpha_2 \in S$. This proves that (c) holds in S. Let, conversely, S satisfy (c). For $\xi_1, \xi_2 \in R$ write $(\xi_1,\xi_2) = (\sigma_1,\sigma_2)\tau^{-1}$ with $\sigma_i \in S$, $\tau \in T$ (cf. 14.1). Writing $(\sigma_1,\sigma_2) = (\alpha_1,\alpha_2)\delta$ with unimodular $(\alpha_1,\alpha_2) \in S^2$, we see that $(\xi_1,\xi_2) = (\alpha_1,\alpha_2)\delta\tau^{-1}$. Thus, (c) holds in R.

Another way of proving this proposition is, of course, by observing that (b) and (c) are equivalent to the geometric condition (a) of 9.3 and that $P_2(S)$ is just $P_2(R)$ with a coarser neighbor relation.

14.6. PROPOSITION. (i) R *is a left (resp. right) Bezout domain if and only if* S *is so.*
(ii) *In an arbitrary left and right Bezout domain* S *a subset* T *is a* PDS *if and only if* T *is a left and right denominator set in* S *with* $1 \in T$ *which is saturated, i.e., for* ξ *and* $\eta \in S$, $\xi\eta \in T$ *implies that both* ξ *and* $\eta \in T$.
(iii) *If* S *is an arbitrary associative ring with* 1 *not containing zero divisors, and* X *a nonempty left and right denominator set in* S, *then its saturation* $\overline{X}$ *which is defined by*

$$\overline{X} = \{\lambda \in S \mid \exists \mu \in S \text{ such that } \lambda\mu \in X\}$$

is a saturated left and right denominator set containing 1.
The proofs of these three satements are given in 6.1, 6.2 and 6.3 of [17]. For the notion of *saturated subset* and *saturation* in commutative rings we refer to [3], p. 44, ex. 7 and 8. For S to be admissible in R we need the extra condition that S has stable rank 2.

We recall that a *strongly prime ideal* in an arbitrary associative ring R with 1 is a twosided ideal $P \neq R$ such that $\alpha\beta \in P$ implies $\alpha \in P$ or $\beta \in P$, i.e., that $R \setminus P$ is a multiplicative subsemigroup. See [12], p. 132, ex. 4, and p. 254; in [13], p. 445, ex. 7, and in [14], p. 60, the term *completely prime ideal* is used. Strongly prime ideals are prime but not conversely; see e.g., [21], Ch. VIII, §2.

As we remarked already after Proposition 9.10, a valuation ring can also be characterized as a Bezout domain which is at the same time a local ring. Now a subring S and a PDS T which yield a valuation ring $R = ST^{-1}$ are characterized in the following result.

14.7. PROPOSITION. R *is a valuation ring if and only if* S *is a left and right Bezout domain and* $T = S \setminus P$ *for a strongly prime ideal* P *in* S.
The proof can be found in [17], 6.4. Notice that here again we need the

extra condition that S has stable rank 2 if we want it to be an admissible subring of R.

One would like to have a similar result for the case of a local ring R in general but, unfortunately, we do not have that. From the proof of 14.7 one sees that T must still be the complement of a strongly prime ideal if R is local, but that condition is not sufficient.

The following result is an immediate consequence of 14.7.

14.8. COROLLARY. R *is a skew field if and only if* S *is a Bezout domain and* $T = S\setminus\{0\}$.

The classical result of W. Klingenberg [27] about homomorphisms between projective planes over skew fields is an easy consequence of Theorem 12.1 or 13.1, and the above corollary; see [17], 6.5 for a proof.

14.9. PROPOSITION (Klingenberg). *Let* $\psi : P_2(K) \to P_2(L)$ *be a full incidence homomorphism between projective planes over skew fields* K *and* L. *respectively. Then there is a valuation ring* S *in* K *and a homomorphism* $\varphi : S \to L$ *with* $\ker \varphi = \operatorname{rad} S$ *such that*

(i) *every point (resp. line) in* $P_2(K)$ *can be represented as* $\ulcorner\alpha_0,\alpha_1,\alpha_2\urcorner$ *(resp.* $\llcorner\alpha_0,\alpha_1,\alpha_2\lrcorner$ *with all* $\alpha_i \in S$, *not all in* rad S.

(ii) $\psi\ulcorner\alpha_0,\alpha_1,\alpha_2\urcorner = \ulcorner\varphi(\alpha_0),\varphi(\alpha_1),\varphi(\alpha_2)\urcorner$ *and similary for lines, if the* α_i *are as in i), i.e.,* $\psi = \tilde{\varphi}_S$.

Conversely, a valuation ring S *in* K *and a homomorphism* $\varphi : S \to L$ *with* $\ker \varphi = \operatorname{rad} S$ *determine a full neighbor preserving homomorphism* $\tilde{\varphi}_S : P_2(K) \to P_2(L)$.

BIBLIOGRAPHY.

1. Adamson, I.T.: *Rings, Modules and Algebras.* Oliver and Boyd, Edinburgh, 1971.
2. Artin, E.: *Geometric Algebra.* Interscience Publishers, New York, London, 1957.

3. Atiyah, M.F. and MacDonald, I.G.: *Introduction to Commutative Algebra*. Addison Wesley, Reading, MA, 1969.
4. Barbilian, D.: 'Zur Axiomatik der projektiven ebenen Ringgeometrien'. I, *Jahresbericht D.M.V.* 50 (1940), 179-229; II, *Jahresbericht D.M.V.* 51 (1941), 34-76.
5. Bass, H.: *Algebraic K-theory*. Benjamin, New York, Amsterdam, 1968.
6. Bass, H.: *Introduction to some Methods of Algebraic K-theory*. CBMS regional conference series in mathematics, No. 20. Amer. Math. Soc., Providence, R.I., 1974.
7. Baer, R.: 'Inverses and Zero Divisors'. *Bulletin Amer. Math. Soc.* 48 (1942), 630-638.
8. Benz, W.: 'Ebene Geometrie über einem Ring'. *Math. Nachr.* 59 (1974), 163-193.
9. Bingen, F.: 'Géométrie projective sur un anneau semiprimaire'. *Acad. Roy. Belg., Bull. Cl. Sci.* (5) 52 (1966), 13-24.
10. Carter, D.S. and Vogt, A.: *Collinearity-preserving Functions between Desarguesian Planes*. Memoir Amer. Math. Soc. 27 (1980), No. 235. Providence, R.I., 1980.
11. Cohn, P.M.: 'Some Remarks on the Invariant Basis Property'. *Topology* 5 (1966), 215-228.
12. Cohn, P.M.: *Free Rings and their Relations*. Academic Press, London, New York, 1971.
13. Cohn, P.M.: *Algebra*, Vol. 2. Wiley and Sons, London, New York, Sydney, Toronto, 1977.
14. Divinsky, N.J.: *Rings and Radicals*. Allen and Unwin, London, 1965.
15. Faulkner, J.R.: 'Stable Range and Linear Groups for Alternative Rings'. *Geometriae Dedicata* 14 (1983), 177-188.
16. Faulkner, J.R.: 'Coordinatization of Moufang-Veldkamp Planes'. *Geometriae Dedicata* 14 (1983), 189-201.
17. Ferrar, J.C., and Veldkamp, F.D.: 'Neighbor-preserving Homomorphisms between Projective Ring Planes'. *Geometriae Dedicata* 18 (1985), 11-33.
18. Hughes, D.R.: 'On Homomorphisms of Projective Planes'. *Proc. Symp. Appl. Math.* 10 (1958), 45-52.
19. Hughes, D.R. and Piper, F.C.: *Projective Planes*. Springer-Verlag,

New York, Heidelberg, Berlin, 1973.

20. Jacobson, N.: 'The Radical and Semi-simplicity for Arbitrary Rings'. *Amer. J. Math.* 67 (1945), 300-320.
21. Jacobson, N.: *The Structure of Rings*. Coll. Publ. Vol. 37, Amer. Math. Soc. 1964 (2nd ed.).
22. Jacobson, N.: *Basic Algebra, Vol. II*. Freeman, San Fransisco, 1980.
23. Kallen, W. van der: 'Injective Stability for K_2'. In: *Algebraic K-theory, Evanston 1976*. Lecture Notes in Math., Vol. 551, Springer-Verlag, Berlin, Heidelberg, New York, 1976, pp. 77-154.
24. Klingenberg, W.: 'Projektive und affine Ebenen mit Nachbarelementen'. *Math. Z.* 60 (1954), 384-406.
25. Klingenberg, W.: 'Euklidische Ebenen mit Nachbarelementen'. *Math. Z.* 61 (1954), 1-25.
26. Klingenberg, W.: 'Desarguesche Ebenen mit Nachbarelementen'. *Abh. Math. Sem. Univ. Hamburg* 20 (1955), 97-111.
27. Klingenberg, W.: 'Projektive Ebenen mit Homomorphismus'. *Math. Ann.* 132 (1956), 180-200.
28. Knüppel, F. und Kunze, M.: 'Homomorphismen von geometrischen Strukturen über kommutativen Ringen'. *Unpublished Manuscript* 1978.
29. Krusemeyer, M.I.: 'Fundamental Groups, Algebraic K-theory and a Problem of Abhyankar'. *Invent. Math.* 19 (1973), 15-47.
30. Leiszner, W.: 'Affine Barbilian-Ebenen'. I, *J. Geometry* 6 (1975), 31-57; II, *ibid*, 105-129.
31. Mortimer, B.: 'A Geometric Proof of a Theorem of Hughes on Homomorphisms of Projective Planes'. *Bull. London Math. Soc.* 7 (1975), 167-168.
32. Springer, T.A. and Veldkamp, F.D.: 'On Hjelmslev-Moufang Planes'. *Math. Z.* 107 (1968), 249-263.
33. Vasershtein, L.N.: 'On the Stabilization of the General Linear Group over Rings'. *Math. USSR Sbornik* 8 (1969), 383-400.
34. Vasershtein, L.N.: 'Stable Rank of Rings and Dimensionality of Topological Spaces'. *Functional Anal. Appl.* 5 (1971), 102-110.
35. Veldkamp, F.D.: 'Projective Planes over Rings of Stable Rank 2'. *Geometriae Dedicata* 11 (1981), 285-308.

36. Veldkamp, F.D.: 'Projective Ring Planes: Some Special Cases'. In: *Atti del Convegno "Geometria Combinatoria e di incidenza: fondamenti e applicazioni", La Mendola 1982*. Rendiconti del Seminario Matematico di Brescia 7 (1984), 609-615.

37. Veldkamp, F.D.: 'Distant-preserving Homomorphisms between Projective Ring Planes'. *Preprint Nr. 336*. Dept. of Mathematics, Utrecht, 1984.

38. Veldkamp, F.D.: 'Incidence-preserving Mappings between Projective Ring Planes'. *Preprint Nr. 338*. Dept. of Mathematics, Utrecht, 1984.

Part IV

Metric Ring Geometries, Linear Groups over Rings and Coordinatization

TOPICS IN GEOMETRIC ALGEBRA OVER RINGS

Claudio Bartolone and Federico Bartolozzi
Istituto Matematico dell'Università
Via Archirafi 34
90123 Palermo (Italy)

E. ARTIN (1957), R. BAER (1952) and J.DIEUDONNE (1951) emphasized many times the structural identity between classical projective geometry and linear algebra over a division ring. Then Baer pointed out a possible extension of this structural identity to the case of a ring, generating intense research activity in the area of geometric algebra over rings. In this direction the most significant results appear: Ojanguren and Sridharan's article on the fundamental theorem of projective geometry over a commutative ring; some theorems due to Klingenberg, Bass and Suslin on the structure of the general linear group over appropriate rings and some valuable notes of O'Meara's on the automorphisms of linear groups over an integral domain.

We will try to give, without any pretension of completeness, some results and methods in this research area. Also we will discuss in detail the case of the projective line over a ring where we have obtained some new results.

When trying to extend the concepts of projective geometry over a division ring in order to achieve, for a given ring, similar results to

R. Kaya et al. (eds.), Rings and Geometry, 353–389.

the classical ones, the following question arises: what is a projective space over a ring? The most natural way to introduce the projective space of dimension n over the ring R is to consider the set of submodules of rank 1 of the free R-module R^{n+1}. Nevertheless this leads to some pathological situations (for example one point containing another point). In order to avoid degeneracy it is customary to assume as a (right) projective space of dimension n over R, $P_n(R)$, the set of R-free direct summands of rank 1 of the (right) R-module R^{n+1}: clearly if we put $\langle x_o,\dots,x_n\rangle = \{(x_o a,\dots,x_n a) \mid a \in R\}$, $x_i \in R$, $\langle x_o,\dots,x_n\rangle \in P_n(R)$ if and only if $(x_o,\dots,x_n)$ is unimodular, i.e. if and only if there exists a linear map $g: R^{n+1} \to R$ with $(x_o,\dots,x_n)^g = 1$ and this is equivalent to $\sum_{i=0}^{n} Rx_i = R$. Clearly, if two unimodular elements p and q generate the same point, then $p = qu$ where u is invertible in R on the left. If R is a ring containing an element invertible only on the left, then a further critical situation can occur: we shall consider only rings with 1 where $uv = 1$ is equivalent to $vu = 1$.

1. COLLINEATIONS BETWEEN PROJECTIVE SPACES

M. OJANGUREN and R. SRIDHARAN (1969) proved a theorem which generalizes to commutative rings the classical fundamental theorem of projective geometry:

THEOREM (Ojanguren-Sridharan): Let R and R' be commutative rings and $P_n(R)$ and $P_{n'}(R')$ projective spaces of dimension ≥ 2. If there exists a collineation between $P_n(R)$ and $P_{n'}(R')$, then $n = n'$ and $R \cong R'$. Moreover the group of collineations $\mathrm{Aut}P_n(R)$ of $P_n(R)$ is exactly the group $P\Gamma L_{n+1}(R)$ of collineations induced by the semilinear maps of R^{n+1}.

We have to go back to the definition of a collineation between projective spaces over rings: a collineation $k: P_n(R) \to P_{n'}(R')$ is a 1-1 map of $P_n(R)$ onto $P_{n'}(R')$ such that k and k^{-1} preserve the "collinearity of points" i.e., if U, V and W are points of $P_n(R)$, then $W \subseteq U+V$ if and only if $W^k \subseteq U^k+V^k$. It is well known that if R is a division ring and k preserves the collinearity of points, then k^{-1} also preserves the collinearity of points. However this is not the case if R and R' are arbitrary rings.

Example: Let F be a field, $A = F\langle x \rangle$ the ring of formal power series in x and B the quotient field of A. We can define a natural map $g: P_2(A) \to P_2(B)$ by putting $\langle x_o, x_1, x_2 \rangle_A^g = \langle x_o, x_1, x_2 \rangle_B$, where the indices A and B denote whether $\langle x_o, x_1, x_2 \rangle$ is regarded as a point of $P_2(A)$ or of $P_2(B)$. It is easy to check that g is a well defined bijective map which preserves the collinearity of points. On the contrary we have: $\langle 1,0,0 \rangle_B = \langle x,0,0 \rangle_B \subseteq \langle x,1,0 \rangle_B + \langle 0,1,0 \rangle_B$ while $\langle 1,0,0 \rangle_A \not\subseteq \langle x,1,0 \rangle_A + \langle 0,1,0 \rangle_A$ because x is not invertible in A. Thus g^{-1} does not preserve the collinearity of points.

In order to extend the Ojanguren-Sridharan theorem to non-commutative rings we have to overcome a big handicap: the bases of a free R-module do not necessarily have the same order if R is not commutative. This leads to the existence of collineations between projective spaces of different dimensions.

2. COLLINEATIONS BETWEEN LINES

2.1 - The case of a division ring.

The concept of a collineation between projective spaces is somewhat different in the case of lines (projective spaces of dimension 1). A collineation k between the lines $P_1(D)$ and $P_1(D')$, over the division rings D and D', is defined to be a bijective map $k:P_1(D) \to P_1(D')$ preserving harmonic quadruples. We wish to discuss these known arguments so that the generalization to the case of an arbitrary ring will appear rather natural. To this end we recall that four distinct points A_1,A_2,A_3,A_4 of the line $P_1(D)$ have <u>cross-ratio</u> $s \in D$ (we write $(A_1,A_2,A_3,A_4) = s$) if there exist generators p_i of A_i $(i = 1,\ldots,4)$ such that $p_3 = p_1 + p_2$ and $p_4 = p_1 + p_2 s$: of course s $(\neq 0,1)$ is uniquely determined up to conjugation. In case $(A_1,A_2,A_3,A_4) = -1$, the points $A_1,\ldots,A_4$ are said to form a <u>harmonic quadruple</u> (we have to admit $A_3 = A_4$ in order to define a harmonic quadruple also in characteristic two).

A map $\beta:R \to R'$ is called a <u>Jordan homomorphism</u> if, for $a,b \in R$, $(a+b)^\beta = a^\beta + b^\beta$ and $(a*b)^\beta = a^\beta * b^\beta$, where $x*y = xy + yx$. By a well known theorem of Hua's, a Jordan homomorphism β between two division rings D and D' is either a homomorphism or an anti-homomorphism of rings. If in addition β is bijective, then β induces, in a natural way, a bijective map $k(\beta):P_1(D) \to P_1(D')$ which maps harmonic quadruples of $P_1(D)$ into harmonic quadruples of $P_1(D')$, i.e. $k(\beta)$ is a collineation. Furthermore $k(\beta)$ maps $\langle 0,1\rangle \to \langle 0,1\rangle$, $\langle 1,0\rangle \to \langle 1,0\rangle$, $\langle 1,1\rangle \to \langle 1,1\rangle$ and, if $xy \neq 0$,

(1) $\langle x,y\rangle \to \langle x^{\beta},y^{\beta}\rangle$ if β is an isomorphism,

(2) $\langle x,y\rangle \to \langle y^{-\beta},x^{-\beta}\rangle$ if β is an anti-isomorphism.

If the characteristic is not two, it is known (the von Staudt-Hua theorem) that (1) and (2) fill up the examples of collineations between the lines $P_1(D)$ and $P_1(D')$ sending $\langle 1,0\rangle$ onto $\langle 1,0\rangle$, $\langle 0,1\rangle$ onto $\langle 0,1\rangle$ and $\langle 1,1\rangle$ onto $\langle 1,1\rangle$ (because $PGL_2(D)$ is transitive on the set of ordered triplets of distinct points of $P_1(D)$, we may assume (up to an element of $PGL_2(D)$) such a collineation).

Remark: Let $k:P_1(D) \to P_1(D')$ be a collineation and let D' be a field. If $k = k(\beta)$, then β is in any case an isomorphism: thus $P\Gamma L_2(D)$ is the full collineation group $\mathrm{Aut}P_1(D)$ in case $D = D'$. In the general case the above maps (1) and (2) can be combined into

$$\langle x,y\rangle \to \langle x^{\beta},y^{\beta}u\rangle$$

where $\beta:D \to D'$ is a Jordan isomorphism and u is an invertible element of D' satisfying the condition $(xy)^{\beta} = x^{\beta}y^{\beta}u$ (precisely $u = 1$ if β is an isomorphism or $xy = 0$ and $u = y^{-\beta}x^{-\beta}y^{\beta}x^{\beta}$ otherwise).

2.2 - The case of a ring.

Most of the results stated in the last section can be extended to a line over a ring R provided we replace the concept of a pair of distinct points by that of a pair of non-neighbouring points according to the following

Definition 1: Two points P and Q of the projective line $P_1(R)$ are said to be non-neighbouring if $R^2 = P \oplus Q$.

Clearly any pair of distinct points of $P_1(R)$ is non-neighbouring if and only if R is a division ring.

Proposition 1: The group $PGL_2(R)$ is transitive on the set T of ordered triplets (U,V,W) of non-neighbouring points of $P_1(R)$ (i.e. $R^2 = U \oplus V = U \oplus W = V \oplus W$).

PROOF: Let (A_1,A_2,A_3) and (B_1,B_2,B_3) be elements of T and set $A_i = \langle a_i \rangle$, $B_i = \langle b_i \rangle$ $(i = 1,2,3)$. As $\{a_1,a_2\}$ and $\{b_1,b_2\}$ are bases for R^2, there exist x_1,x_2,y_1,y_2 in R such that $a_3 = a_1x_1+a_2x_2$ and $b_3 = b_1y_1+b_2y_2$. At the same time $\{a_1,a_3\}$ and $\{a_2,a_3\}$ are bases whence x_1 and x_2 must be invertible in R (we recall that in our rings $uv = 1$ is equivalent to $vu = 1$). Likewise y_1 and y_2 are invertible. Thus if $a'_i = a_ix_i$ and $b'_i = b_iy_i$ $(i = 1,2)$, we have $A_i = \langle a'_i \rangle$, $B_i = \langle b'_i \rangle$, $a_3 = a'_1 + a'_2$ and $b_3 = b'_1 + b'_2$. Now the linear isomorphism g defined by $(a'_i)^g = b'_i$ induces the required element of $PGL_2(R)$.

Proposition 2: Suppose A_1,A_2,A_3,A_4 are non-neighbouring points in $P_1(R)$: then there exist generators p_i of A_i $(i = 1,\dots,4)$ and an element s in R such that $p_3 = p_1+p_2$ and $p_4 = p_1+p_2s$. Moreover the "cross-ratio" s is invertible in R and s is uniquely determined up to conjugation.

PROOF: By using arguments similar to the ones used in the proof of proposition 1, we can find invertible elements x_1,x_2,y_1,y_2 of R such that $a_3 = a_1x_1+a_2x_2$ and $a_4 = a_1y_1+a_2y_2$, where $\langle a_i \rangle = A_i$ $(i = 1,\dots,4)$. Put $p_1 = a_1y_1$, $p_2 = a_2x_2x_1^{-1}y_1$, $p_3 = a_3x_1^{-1}y_1$, $p_4 = a_4$, $s = y_1^{-1}x_1x_2^{-1}y_2$: the proposition follows now.

In view of proposition 2 we can define a harmonic quadruple of $P_1(R)$ as a quadruple of non-neighbouring points having cross-ratio -1. Clearly if -1 = 1 the concept is trivial: let us exclude this possibility assuming that 2 is invertible in our rings. At this point the following definition should appear rather natural by analogy with the classical case of a division ring:

D e f i n i t i o n 2: A bijective map $k:P_1(R) \to P_1(R')$ between the lines over the rings R and R' is a <u>collineation</u> if k and k^{-1} preserve harmonic quadruples.

<u>Remark 1</u>: In definition 2 we require k^{-1} to preserve harmonic quadruples: this is superfluous in the case of division rings. Nevertheless there exist bijective maps between lines over rings preserving harmonic quadruples but sending some non-harmonic quadruple onto a harmonic quadruple. As an example we can consider the rings A and B from the example in no. 1 and the bijective map g defined there. The "restriction" of g to $P_1(A)$ is a bijective map preserving harmonic quadruples, yet the quadruple $\langle x,1\rangle, \langle 0,1\rangle, \langle x,2\rangle, \langle 1,0\rangle$ is harmonic only in $P_1(B)$.

Now what one expects to obtain is the generalization of the von Staudt-Hua theorem to the line over an arbitrary ring R : in particular, in the commutative case, one expects to get that $P\Gamma L_2(R)$ is the full collineation group of $P_1(R)$. Unfortunately we cannot prove this since, for instance, the following theorem holds

THEOREM (Bartolone-Di Franco): Let $R = F[x]$, where F is a field of characteristic $\neq 2$. Then $P\Gamma L_2(R)$ is a proper subgroup of the full collineation group of $P_1(R)$.

PROOF: Consider the bijection $h:P_1(R) \to P_1(R)$ defined by

$$<f,g>^h = \begin{cases} <-f,g> \text{ if } fg \neq 0 \text{ and } \deg(f)+\deg(g) \text{ is odd,} \\ < f,g> \text{ otherwise.} \end{cases}$$

Since R is commutative the pair of points of $P_1(R)$ $<f,g>$, $<f',g'>$ is non-neighbouring if and only if the "determinant" $fg'-f'g$ is invertible, i.e. if and only if $fg'-f'g \in F^*$. This leads to $\deg(f)+\deg(g') = \deg(f')+\deg(g)$ in case none of the considered polynomials is zero. Therefore we have only two possibilities: either $<f,g>^h = <f,g>$ and $<f',g'>^h = <f',g'>$ or $<f,g>^h = <-f,g>$ and $<f',g'>^h = <-f',g'>$ (if one of the polynomials is zero, the previous statement is still true because only $<0,1>$ and $<1,0>$ have one null coordinate). Consider now a quadruple A_1,A_2,A_3,A_4 of non-neighbouring points and let $A_i = <f_i,g_i>$, $i = 1,\dots,4$. Then only two possibilities can occur: either $A_i^h = <f_i,g_i>$ or $A_i^h = <-f_i,g_i>$ (for each i). Consequently $(A_1,A_2,A_3,A_4) = -1$ if and only if $(A_1^h,A_2^h,A_3^h,A_4^h) = -1$, i.e. h is a collineation. However $h \notin P\Gamma L_2(R)$: indeed, if $s:R^2 \to R^2$ were a β-semilinear isomorphism, with $\beta \in \mathrm{Aut}R$, such that $<f,g>^h = <(f,g)^s>$ for any point $<f,g>$ in $P_1(R)$, then we should have $<1,0> = <1,0>^h = <(1,0)^s>$ and $<0,1> =<0,1>^h = <(0,1)^s>$. Hence $(1,0)^s = (u,0)$ and $(0,1)^s = (0,v)$ with u and v invertible. Moreover : $<1,1> = <1,1>^h = <(1,1)^s> = <u,v>$ so that $u = v$. Consequently, for any f in R , $<\varepsilon f,1> = <f,1>^h = <(f,1)^s> = <uf^\beta,u> = <f^\beta,1>$, with $\varepsilon = \pm 1$ depending on $\deg(f)$: therefore $f^\beta = \varepsilon f$ and β would not be an automorphism.

Remark 2: In spite of the previous theorem one can extend the von Staudt-Hua theorem to some special commutative rings: in particular $\mathrm{Aut}P_1(R) = P\Gamma L_2(R)$ if R is a commutative local or semilocal ring (N.B. LIMAYE (1971/72)), or if R is a commutative algebra of finite dimension over a field of order sufficiently large (H. SCHAEFFER (1974)), or if R is a commutative primitive ring, where this means that for every polynomial $f(x) \in R[x]$, whose entries generate R as an ideal (i.e. a primitive polynomial), there exists $a \in R$ such that $f(a)$ is a unit (B.R. McDONALD (1981)). Furthermore B.V. LIMAYE and N.B. LIMAYE (1977), by adopting a definition of collineation equivalent, in the commutative case, to definition 2, generalized the von Staudt-Hua theorem to non-commutative local rings.

2.3 - Geometric characterization of $P\Gamma L_2(R)$ in $\mathrm{Aut}P_1(R)$.

In view of the theorem in no. 2.2 it is natural to ask the following question: what should we require of a collineation of $P_1(R)$ in order for it to belong to $P\Gamma L_2(R)$? The next definition is given for this purpose:

Definition: A quadruple of points A_1, A_2, A_3, A_4 of $P_1(R)$ is said to be a <u>generalized harmonic quadruple</u> if there exist a generator p_i of A_i $(i = 1,\ldots,4)$ and elements $x_1,\ldots,x_4 \in R$ such that $p_1x_1+p_2x_2 = p_3x_3$ and $p_1x_1-p_2x_2 = p_4x_4$, where x_1 and x_2 are not both zero and x_3 (resp. x_4) is either zero or a unit in R.

<u>Remark 1</u>: In the case of division rings, a generalized harmonic quadruple is either a harmonic quadruple or at least three of its points

coincide ([1]): hence every collineation k preserves trivially generalized harmonic quadruples. This is still true for arbitrary rings provided k is induced by a semilinear map. Our goal is to prove that every collineation $k: P_1(R) \to P_1(R')$ preserving generalized harmonic quadruples is induced, in the commutative case, by a semilinear map.

We start with the general case of non-necessarily commutative rings. As $PGL_2(R)$ is transitive on ordered triplets of non-neighbouring points and its elements preserve generalized harmonic quadruples, we may assume (up to an element in $PGL_2(R)$) $\langle 0,1\rangle^k = \langle 0,1\rangle$, $\langle 1,0\rangle^k = \langle 1,0\rangle$, $\langle 1,1\rangle^k = \langle 1,1\rangle$. If $a \in R$, then $\langle a,1\rangle^k$ is a point of $P_1(R')$, say $\langle a',b'\rangle$. Clearly, $\langle a,1\rangle$ and $\langle 1,0\rangle$ are non-neighbouring points, whence $\langle a,1\rangle$, $\langle 1,0\rangle$, $\langle a+1,1\rangle$, $\langle a-1,1\rangle$ is a quadruple of non-neighbouring points which is harmonic. Since k is a collineation, the image under k of this quadruple is a harmonic quadruple of $P_1(R')$. In particular, $\langle a',b'\rangle = \langle a,1\rangle^k$ and $\langle 1,0\rangle = \langle 1,0\rangle^k$ are non-neighbouring points. Therefore b' is invertible because $\{(a',b'), (1,0)\}$ is a basis for R'^2. Thus we can put $\langle a,1\rangle^k = \langle a^{\beta_1},1\rangle$, where $a^{\beta_1} = a'b'^{-1}$ and $\beta_1 : R \to R'$ is a well defined map such that $0^{\beta_1} = 0$, $1^{\beta_1} = 1$. Moreover, if $a' \in R'$ and $\langle a',1\rangle = \langle a,b\rangle^k$ with $a,b \in R$, by using k^{-1} we see that b is invertible. Hence β_1 is surjective. Since the injectivity of β_1 follows from the injectivity of k, we have that β_1 is bijective. Now we can similarly define a bijection $\beta_2 : R \to R'$ such that $\langle 1,b\rangle^k = \langle 1,b^{\beta_2}\rangle$, $0^{\beta_2} = 0$, $1^{\beta_2} = 1$.

---

(1) Therefore the concept of generalized harmonic quadruple extends the one of "harmonic relation" defined for division rings by F. BUEKENHOUT (1965).

Let us consider the generalized harmonic quadruple of $P_1(R)$ $\langle 0,1\rangle$, $\langle 1,0\rangle$, $\langle a,1\rangle$, $\langle -a,1\rangle$, with $a \in R$. Then $\langle 0,1\rangle$, $\langle 1,0\rangle$, $\langle a^{\beta_1},1\rangle$, $\langle (-a)^{\beta_1},1\rangle$ is a generalized harmonic quadruple of $P_1(R')$. Hence there exist $x_1,x_2,x_3,x_4 \in R'$ such that $(0,1)x_1+(1,0)x_2 = (a^{\beta_1},1)x_3$ and $(0,1)x_1-(1,0)x_2 = ((-a^{\beta_1}),1)x_4$. Thus $x_1 = x_3 = x_4 \neq 0$ must be a unit by definition and we find

$$(-a)^{\beta_1} = -a^{\beta_1} \quad \text{for any} \quad a \in R \,.$$

Moreover, for $a,b \in R$, $\langle a^{\beta_1},1\rangle$, $\langle 1,0\rangle$, $\langle (a+b)^{\beta_1},1\rangle$, $\langle (a-b)^{\beta_1},1\rangle$ is a generalized harmonic quadruple of $P_1(R')$, since it is the image under k of the generalized harmonic quadruple of $P_1(R)$ $\langle a,1\rangle$, $\langle 1,0\rangle$, $\langle a+b,1\rangle$, $\langle a-b,1\rangle$. Therefore $(a^{\beta_1},1)x_1+(1,0)x_2 = ((a+b)^{\beta_1},1)x_3$ and $(a^{\beta_1},1)x_1-(1,0)x_2 = ((a-b)^{\beta_1},1)x_4$ for appropriate x_1,x_2,x_3,x_4 in R'. We obtain that $x_1 = x_3 = x_4$ must be a unit whence $(a+b)^{\beta_1}-a^{\beta_1} = a^{\beta_1}-(a-b)^{\beta_1}$, i.e. $(a+b)^{\beta_1}+(a-b)^{\beta_1} = 2a^{\beta_1}$. By interchanging a and b, we have $(a+b)^{\beta_1}+(b-a)^{\beta_1} = 2b^{\beta_1}$ whence, as $(b-a)^{\beta_1} = (-(a-b))^{\beta_1} = -(a-b)^{\beta_1}$,

$$(a+b)^{\beta_1} = a^{\beta_1}+b^{\beta_1} \quad \text{for any} \quad a,b \in R \,.$$

Likewise we can show that β_2 is additive. Let us prove now $\beta_1 = \beta_2$. If $a \in R$, $\langle 1,1\rangle$, $\langle 1,-1\rangle$, $\langle 2a-1,1\rangle$, $\langle 1,2a-1\rangle$ is a generalized harmonic quadruple and so $\langle 1,1\rangle$, $\langle 1,-1\rangle$, $\langle 2a^{\beta_1}-1,1\rangle$, $\langle 1,2a^{\beta_2}-1\rangle$ is a generalized harmonic quadruple in $P_1(R')$. Therefore there exist x_1,x_2,x_3,x_4 in R', x_3 and x_4 units, such that $(1,1)x_1+(1,-1)x_2 = (2a^{\beta_1}-1,1)x_3$ and $(1,1)x_1-(1,-1)x_2 = (1,2a^{\beta_2}-1)x_4$. Hence $x_3 = x_4$ and $a^{\beta_1} = a^{\beta_2}$ for any $a \in R$.

Let us put $\beta = \beta_1 = \beta_2$. Again let $a \in R$. Since $(a-1,1)(a+1)$ +

$(-a,1)a = (-1,2a+1)$ and $(a-1,1)(a+1)-(-a,1)a = (2a^2-1,1)$, then $\langle a^\beta-1,1\rangle$, $\langle -a^\beta,1\rangle$, $\langle -1,2a^\beta+1\rangle$, $\langle 2(a^2)^\beta-1,1\rangle$ is a generalized harmonic quadruple in $P_1(R')$. Thus for suitable $x_1,x_2,x_3,x_4 \in R'$ we can write the identities:

(1) $-x_3 = (a^\beta-1)x_1-a^\beta x_2 = a^\beta(x_1-x_2)-x_1$;

(2) $x_1+x_2 = (2a^\beta+1)x_3$;

(3) $(2(a^2)^\beta-1)x_4 = (a^\beta-1)x_1+a^\beta x_2 = a^\beta(x_1+x_2)-x_1$;

(4) $x_1-x_2 = x_4$.

By (4) and (1) we have $x_1 = a^\beta x_4+x_3$ and $x_2 = a^\beta x_4+x_3-x_4$, i.e. $x_1+x_2 = 2(a^\beta x_4+x_3)-x_4$; also by (2) $2(a^\beta x_4+x_3)-x_4 = (2a^\beta+1)x_3$, i.e. $x_3-x_4 = 2a^\beta(x_3-x_4)$, and by (3) $(2(a^2)^\beta-1)x_4 = a^\beta(2(a^\beta x_4+x_3)-x_4)-(a^\beta x_4+x_3)$, i.e. $2((a^2)^\beta-(a^\beta)^2)x_4-2a^\beta(x_3-x_4) = x_4-x_3$. Hence $2((a^2)^\beta-(a^\beta)^2)x_4 = 0$. If $x_4 \neq 0$, then x_4 is a unit by definition and we get $(a^2)^\beta = (a^\beta)^2$. Otherwise x_3 must be a unit and we obtain $x_1 = x_2 = x_3$ by (4) and (1). Thus $2a^\beta = 1$ by (2). Hence $(a^2)^\beta = (a^\beta)^2$ for every a such that $2a^\beta \neq 1$. If $2a^\beta = 1$, then $2(2a^\beta) = 2(2a)^\beta \neq 1$ whence $((2a)^2)^\beta = ((2a)^\beta)^2$ i.e. $(a^2)^\beta = (a^\beta)^2$ for every $a \in R$. Therefore we obtain, for any $a,b \in R$, $((a+b)^2)^\beta = ((a+b)^\beta)^2$ and consequently β is a Jordan isomorphism.

Finally, if $\langle a,b\rangle \in P_1(R)$, $\langle 0,1\rangle \neq \langle a,b\rangle \neq \langle 1,0\rangle$, we have $(a,1)(1+b)/2+(-a,1)(1-b)/2 = (ab,1)$ and $(a,1)(1+b)/2-(-a,1)(1-b)/2 = (a,b)$. Therefore $\langle a^\beta,1\rangle$, $\langle -a^\beta,1\rangle$, $\langle (ab)^\beta,1\rangle$, $\langle a',b'\rangle$ is a generalized harmonic quadruple in $P_1(R')$, where we put $\langle a',b'\rangle = \langle a,b\rangle^k$. Consequently we can write, for suitable $x_1,x_2,x_3,x_4 \in R'$.

(5) $(ab)^\beta x_3 = a^\beta(x_1-x_2)$; (6) $x_3 = x_1+x_2$; (7) $a'x_4 = a^\beta(x_1+x_2)$;

(8) $b'x_4 = x_1 - x_2$. If $x_4 = 0$, then $x_1 = x_2$ by (8) and $2x_1 = x_3$ by (6). Thus x_1 is a unit because x_3 is a unit by definition and 2 is invertible. So $a^ß = 0$ by (7) i.e. $a = 0$, a contradiction to $<a,b> \neq <0,1>$. Therefore x_4 must be a unit and by multiplying by x_4^{-1} we may assume $x_4 = 1$. Now by (7) and (6) we get $a' = a^ß x_3$. As $<a,b> \neq <0,1>$ it follows that $<a',b'> \neq <0,1>$ too, i.e. $a' \neq 0$ and x_3 is a unit by definition. By choosing another suitable generalized harmonic quadruple we can get that $b' = b^ß y_3$, where y_3 is another unit. Hence $<a,b>^k = <a^ß x_3, b^ß y_3> = <a^ß, b^ß u>$, where $u = y_3 x_3^{-1}$. Moreover by (5) and (8) we have $a^ß b' = (ab)^ß x_3$ i.e. $(ab)^ß = a^ß b^ß u$. Summing up we have

THEOREM: Let R and R' be rings with 1 in which 2 is invertible and $uv = 1$ is equivalent to $vu = 1$. If $k: P_1(R) \to P_1(R')$ is a collineation preserving generalized harmonic quadruples, then k is the product of an element of $PGL_2(R)$ and of a collineation of the form

$$<x,y> \to <x^ß, y^ß u>,$$

where $ß: R \to R'$ is a Jordan isomorphism with $1^ß = 1$ and u is an invertible element in R' such that $(xy)^ß = x^ß y^ß u$.

Corollary (Bartolone-Di Franco): If R is a commutative ring with 1 and 2 invertible, then $P\Gamma L_2(R)$ is the group of collineations of $P_1(R)$ which preserve generalized harmonic quadruples.

PROOF: We have only to prove that every collineation of the form $<x,y> \to <x^ß, y^ß u>$, where $ß \in \mathrm{Aut}R$ and u is an invertible element satisfying $(x\,y)^ß = x^ß y^ß u$, is in $P\Gamma L_2(R)$. Since (x,y) is unimodular

(x^β, y^β) is unimodular too: hence there exist a' and b' in R' such that $a'x^\beta + b'y^\beta = 1$. If we set $c = a'x^\beta + b'y^\beta u$, by using $x^\beta y^\beta = x^\beta y^\beta u$, we get $x^\beta c = x^\beta$ and $y^\beta c = y^\beta u$. Hence $\langle x^\beta, y^\beta u\rangle = \langle x^\beta c, y^\beta c\rangle = \langle x^\beta, y^\beta\rangle$ and the considered collineation is in $P\Gamma L_2(R)$.

Remark 2: In view of remark 1 and of remark in no. 2.1, the above theorem generalizes the von Staudt-Hua theorem. Unfortunately, for each Jordan isomorphism $\beta: R \to R'$ satisfying the conditions of the above theorem, $\langle x,y\rangle \to \langle x^\beta, y^\beta u\rangle$ does not always define a map, least of all a collineation. Nevertheless $\langle x,y\rangle \to \langle x^\beta, y^\beta u\rangle$ defines a collineation preserving generalized harmonic quadruples if β belongs to the following class of Jordan isomorphisms:

Let $R = \prod_{i\in I} R_i$ and $R' = \prod_{i\in I} R'_i$ be direct products of the rings R_i and R'_i respectively. Also let $\beta_i: R_i \to R'_i$ be an isomorphism or an anti-isomorphism, provided R_i is a division ring if β_i is an anti-isomorphism. If we put $(a_i)^\beta_{i\in I} = (a_i^{\beta_i})_{i\in I}$, then we define a Jordan isomorphism $\beta: R \to R'$ satisfying $(xy)^\beta = x^\beta y^\beta u$, where $u_i = 1$ if β_i is an isomorphism or $x_i y_i = 0$ and $u_i = y_i^{-\beta_i} x_i^{-\beta_i} y_i^{\beta_i} x_i^{\beta_i}$ otherwise. Also $\langle x,y\rangle \in P_1(R)$ (resp. $P_1(R')$) if and only if $\langle x_i y_i\rangle \in P_1(R_i)$ (resp. $P_1(R'_i)$) for any i: thus $\langle x,y\rangle \to \langle x^\beta, y^\beta u\rangle$ defines a collineation preserving generalized harmonic quadruples because $\langle x_i, y_i\rangle \to \langle x_i^{\beta_i}, y_i^{\beta_i} u_i\rangle$ behaves so for any $i \in I$.

We note that the above class of Jordan isomorphisms includes all known Jordan isomorphisms defining collineations, precisely: the isomorphisms for arbitrary rings; the anti-isomorphisms for arbitrary division rings. Thus the following question arises naturally: is every Jordan isomorphism defining a collineation which preserves generalized harmonic quadruples of this type? The answer is in the affirmative for the special class of artinian semisimple rings. Let $\beta: R \to R'$ be a Jordan isomorphism satisfying $(xy)^{\beta} = x^{\beta} y^{\beta} u$ for each unimodular element (x,y) of R^2 (u a unit). Assume that R' is an artinian semisimple ring. By the structure theory of artinian semisimple rings $R' \simeq D^{o}_{n_o} \oplus \ldots \oplus D^{r}_{n_r}$, where $D^{i}_{n_i}$ is the ring of $n_i x n_i$ matrices over an appropriate division ring D^i. Let p_i be the projection of R' onto $D^{i}_{n_i}$ and put $\beta_i = \beta p_i$. By a well known theorem due to I.N. HERSTEIN (1956), $\beta_i : R \to D^{i}_{n_i}$ is a homomorphism or an anti-homomorphism. Let us define now for every i a ring R'_i by putting: $R'_i = D^{i}_{n_i}$ in case β_i is a homomorphism, $R'_i = (D^{i}_{n_i})^{0}$ otherwise, where $(D^{i}_{n_i})^{0}$ denotes the opposite ring of $D^{i}_{n_i}$ (i.e. xy in $(D^{i}_{n_i})^{0}$ is equivalent to yx in $D^{i}_{n_i}$). Since R'_i and $D^{i}_{n_i}$ coincide as sets, the map which defines β_i defines at the same time (in any case) a homomorphism of R onto R'_i. Therefore, since $a^{\beta} = \sum_{i=o}^{r} a^{\beta_i}$ for $a \in R$, by the same map defining β one defines an isomorphism $\bar{\beta}$ between the rings R and $\bigoplus_{i=o}^{r} R'_i$. Thus if we put $R_i = (R'_i)^{\bar{\beta}-1}$, then $R = \bigoplus_{i=o}^{r} R_i$ and $\beta_i : R_i \to D^{i}_{n_i}$ is an isomorphism or an anti-isomorphism. Let us assume β_i is an anti-isomorphism and consider the matrices $(a_{ij}), (t_{ij}), (v_{ij}) \in D^{i}_{n_i}$, where $s_{ij} = t_{ij} = v_{ij} = 0$ except for $s_{11} = t_{n_i n_i} = v_{i,n_i-i+1} = 1$. We have $(s_{ij})(v_{ij})(t_{ij}) \neq 0$. As $((t_{ij})^{\beta_i^{-1}}, (v_{ij})^{\beta_i^{-1}})$ is unimodular in R^2, there exists an invertible matrix (u_{ij}) such that $(t_{ij})(v_{ij}) = (v_{ij})(t_{ij})(u_{ij})$ because β_i is an anti-isomorphism. Hence $(t_{ij}) = (v_{ij})(t_{ij})(u_{ij})(v_{ij})^{-1}$. Assume $n_i > 1$, then $(s_{ij})(t_{ij}) = 0$. This implies $(s_{ij})(v_{ij})(t_{ij})(u_{ij})(v_{ij})^{-1} = 0$, i.e. $(s_{ij})(v_{ij})(t_{ij}) = 0$, a contradiction. Therefore $n_i = 1$ and ß belongs to the class of Jordan isomorphism described in remark 2.

3. NON-INJECTIVE MAPS WHICH PRESERVE GENERALIZED HARMONIC QUADRUPLES

W. KLINGENBERG (1956) introduced the idea of "non-injective collineation" between projective spaces in two and three dimensions. His work was later extended by A. DRESS (1964), J. ANDRE (1969), F. RADÒ (1969/70) and D.S. CARTER and A. VOGT (1980). The first article on non-injective collineations between lines was due to F. BUEKENHOUT (1965); afterward D.G. JAMES (1982) got the same result as Buekenhout. As Buekenhout and James work only with division rings, the problem of studying the general case of arbitrary rings arises. In this section we will report on Buekenhout's result and on our attempt to extend this result to arbitrary rings.

Let $k:P_1(R) \to P_1(R')$ be a (not necessarily bijective) map which preserves generalized harmonic quadruples. Our goal is to describe k algebraically. We begin by giving two examples of such maps which are not collineations:

(1) Let $ß:R \to R'$ be a ring homomorphism, then $\langle a,b\rangle \to \langle a^ß,b^ß\rangle$ defines a map from $P_1(R)$ into $P_1(R')$ which preserve generalized harmonic quadruples. This map is a collineation if and only if ß is an isomorphism.

(2) (Buekenhout): Let D and D' be division rings and add to D' (as a set) an extra element ∞ . If $ß:D \to D'\cup\{\infty\}$ is a place or an

anti-place (according to Krull's valuation theory), $\langle 0,1\rangle \to \langle 0,1\rangle$, $\langle 1,a\rangle \to \langle 1,a^{\beta}\rangle$ if $a^{\beta} \neq \infty$, $\langle 1,a\rangle \to \langle 0,1\rangle$ if $a^{\beta} = \infty$ is a map from $P_1(D)$ into $P_1(D')$ which preserves generalized harmonic quadruples.

For division rings we have an essentially complete characterization of the maps preserving generalized harmonic quadruples:

THEOREM (Buekenhout): Let $k:P_1(D) \to P_1(D')$ be a map preserving generalized harmonic quadruples, where D and D' are division rings of characteristic $\neq 2$. If $\mathrm{Im}\, k$ contains at least three distinct points, then k is the product of collineations and of a map of type (2).

Let us pass to the general case where $k:P_1(R) \to P_1(R')$ is a map (preserving generalized harmonic quadruples) between lines over arbitrary rings. Assume that 2 is invertible and $\mathrm{Im}\, k$ contains at least three non-neighbouring points which are the images of three non-neighbouring points of $P_1(R)$. This hypothesis is a natural generalization of Buekenhout's hypothesis that the characteristic is not two and $\mathrm{Im}\, k$ contains at least three distinct points. As the groups $PGL_2(R)$ and $PGL_2(R')$ are transitive on ordered triplets of non-neighbouring points and their elements preserve generalized harmonic quadruples, we may assume (up to a collineation) that $\langle 0,1\rangle^k = \langle 0,1\rangle$, $\langle 1,0\rangle^k = \langle 1,0\rangle$, $\langle 1,1\rangle^k = \langle 1,1\rangle$. Let us define $Q_1(R)$ (resp. $Q_2(R)$) to be the subset of points of $P_1(R)$ of type $\langle a,1\rangle$ (resp. $\langle 1,a\rangle$) and also $Q_1(R')$ (resp. $Q_2(R')$) to be the analogous subset of $P_1(R')$. As

$$(a,1)+(1,0)(1-a) = (1,1) \quad\text{and}\quad (a,1)-(1,0)(1-a) = (2a-1,1) ,$$

the points $\langle a,1\rangle$, $\langle 1,0\rangle$, $\langle 1,1\rangle$, $\langle 2a-1,1\rangle$ form a generalized harmonic quadruple in $P_1(R)$. Hence, if we put $\langle a,1\rangle^k = \langle a',b'\rangle$, the points $\langle a',b'\rangle$, $\langle 1,0\rangle$, $\langle 1,1\rangle$, $\langle 2a-1,1\rangle^k$ form a generalized harmonic quadruple in $P_1(R')$. Assume that $\langle a,1\rangle^k \neq \langle 1,0\rangle$, then it is easy to check, by elementary computations, that b' must be a unit whence $\langle a,1\rangle^k \in Q_1(R')$. Likewise we can prove that $\langle 1,a\rangle^k \in Q_2(R')$ in case $\langle 1,a\rangle^k \neq \langle 0,1\rangle$. Then we have

$$(Q_1(R))^k \subseteq Q_1(R') \cup \{\langle 1,0\rangle\} \quad \text{and} \quad (Q_2(R))^k \subseteq Q_2(R') \cup \{\langle 0,1\rangle\} .$$

Let us consider now the set $\bar{R}'$ obtained by adding to R' an extra element ∞ . In view of the previous inclusions we can define two maps β_1 and β_2 from R into $\bar{R}'$ by putting:

$$x^{\beta_1} = \begin{cases} y \in R' & \text{if } \langle x,1\rangle^k = \langle y,1\rangle , \\ \infty & \text{if } \langle x,1\rangle^k = \langle 1,0\rangle , \end{cases}$$

$$x^{\beta_2} = \begin{cases} y \in R' & \text{if } \langle 1,x\rangle^k = \langle 1,y\rangle , \\ \infty & \text{if } \langle 1,x\rangle^k = \langle 0,1\rangle . \end{cases}$$

Clearly $0^{\beta_i} = 0$ and $1^{\beta_i} = 1$ for $i = 1,2$. Let $R_{\beta_i} = \{x \in R \mid x^{\beta_i} \neq \infty\}$. Then

THEOREM: R_{β_1} and R_{β_2} coincide as subalgebras of the Jordan algebra $R(+,*)$, where $x*y = (1/2)(xy+yx)$. Moreover the maps β_1 and β_2 define the same algebra homomorphism between the algebras $R_\beta = R_{\beta_1} = R_{\beta_2}$ and $R'(+,*)$. In particular, in the commutative case, R_β is an associative subring of R and $\beta: R_\beta \to R'$ is a ring homomorphism.

The proof of this theorem is quite similar to the one of the theorem in no. 2.3. However, here we shall use continually the following lemma on generalized harmonic quadruples.

L e m m a : If A_1,A_2,A_3,A_4 is a generalized harmonic quadruple then exactly one of the following holds:

(a) the points are all distinct;

(b) only the points A_3 and A_4 coincide;

(c) at least three points coincide.

PROOF: By definition, if A_i is generated by p_i $(i = 1,\dots,4)$, then for appropriate scalars x_1,x_2,x_3,x_4 we have $p_1x_1+p_2x_2 = p_3x_3$ and $p_1x_1-p_2x_2 = p_4x_4$. As A_2,A_1,A_3,A_4 and A_1,A_2,A_4,A_3 are generalized harmonic quadruples too, we need only to examine the following two cases:

$A_1 = A_2$. As x_1 and x_2 are not both zero, x_3 or x_4 must be invertible whence we have $A_3 = A_1$ or $A_4 = A_1$.

$A_1 = A_3$. We can write $p_2x_2 = p_1(x_3-x_1)$ and $p_4x_4 = p_1(2x_1-x_3)$. If x_4 is invertible then $A_4 = A_1$, otherwise $x_4 = 0$ and x_3 is invertible. In this case we have $x_3 = 2x_1$ and $p_2x_2 = p_1x_1$, i.e. $A_1 = A_2$.

We note that, in the case of division rings (of characteristic not two), only the cases (a) and (c) can occur (this can be shown for arbitrary domains too). Nevertheless the case (b) can actually occur in the general case: to this end consider $P_1(\mathbb{Z}_{15})$ and its points $A_1 = \langle 9,1\rangle$, $A_2 = \langle 1,3\rangle$, $A_3 = \langle 1,9\rangle$. As 2 and 7 are invertible in $\mathbb{Z}_{15}$ and the identities $(9,1)3+(1,3)10 = (1,9)7$ and $(9,1)3-(1,3)10 = (1,9)2$ hold, the points A_1,A_2,A_3,A_3 form a generalized harmonic quadruple.

Let us come back to the proof of the previous theorem by applying the above lemma to the generalized harmonic quadruples

(3) $\langle 0,1\rangle^k$, $\langle 1,0\rangle^k$, $\langle a,1\rangle^k$, $\langle -a,1\rangle^k$;

(4) $\langle a,1\rangle^k$, $\langle 1,0\rangle^k$, $\langle 2a,1\rangle^k$, $\langle 0,1\rangle^k$;

(5) $\langle 1,0\rangle^k$, $\langle a+b,1\rangle^k$, $\langle 2a,1\rangle^k$, $\langle 2b,1\rangle^k$.

(3) (resp. (4)) shows that $-a$ (resp. $2a$) $\in R_{\beta_1}$ if and only if $a \in R_{\beta_1}$. Consequently (5) says that R_{β_1} is an additive group. Likewise we can prove that R_{β_2} is an additive group. At this point we can show that β_i is additive on R_{β_i} as in the proof of the theorem in no. 2.3.

Suppose now that there exists $a \in R_{\beta_1}$ such that $a \notin R_{\beta_2}$: hence $(2a+1)^{\beta_1} \neq \infty$ and $(2a+1)^{\beta_2} = \infty$ since $1 \in R_{\beta_1} \cap R_{\beta_2}$. Then $\langle 1,1\rangle^k = \langle 1,1\rangle$, $\langle -1,1\rangle^k = \langle -1,1\rangle$, $\langle 1,2a+1\rangle^k = \langle 0,1\rangle$, $\langle 2a+1,1\rangle^k \neq \langle 1,0\rangle$ would be a generalized harmonic quadruple of $P_1(R')$ because $(1,2a+1) = (1,1)(a+1)+(-1,1)a$ and $(2a+1,1) = (1,1)(a+1)-(-1,1)a$. But it is easy to check that the previous four points of $P_1(R')$ do not form a generalized harmonic quadruple, hence $R_{\beta_1} \subseteq R_{\beta_2}$. Likewise we can prove that $R_{\beta_2} \subseteq R_{\beta_1}$. At this point we can prove that $\beta_1 = \beta_2 = \beta$ as in the proof of the theorem in no. 2.3.

As $(1,a-1)(a+1)+(1,-a)a = -(-2a-1,1)$ and $(1,a-1)(a+1)-(1,-a)a = (1,2a^2-1)$ for every $a \in R$, $\langle 1,a-1\rangle^k$, $\langle 1,-a\rangle^k$, $\langle -2a-1,1\rangle^k$, $\langle 1,2a^2-1\rangle^k$ is a generalized harmonic quadruple of $P_1(R')$. If $a^\beta = \infty$, then also $(a-1)^\beta = (-a)^\beta = (-2a-1)^\beta = \infty$, hence by applying the lemma to the previous generalized harmonic quadruple we get $\langle 1,2a^2-1\rangle^k = \langle 0,1\rangle$, i.e. $(a^2)^\beta = \infty$. Conversely, let us suppose that there exists $a \in R_\beta$ such that $a^2 \notin R_\beta$. Then also $2a^2-1 \notin R_\beta$ and by again using the previous generalized harmonic quadruple of $P_1(R')$ we obtain the identities

$$(1,a^{\beta}-1)x_1+(1,-a^{\beta})x_2 = (-2a^{\beta}-1,1)x_3 ,$$
$$(1,a^{\beta}-1)x_1-(1,-a^{\beta})x_2 = (0,1)x_4 ,$$

for suitable $x_1,x_2,x_3,x_4 \in R'$. Thus we must have $2a^{\beta} = 1$, hence, if $a \in R_{\beta}$ and $2a^{\beta} \neq 1$ it follows that $a^2 \in R_{\beta}$. If, on the contrary, $a \in R_{\beta}$ and $2a^{\beta} = 1$, then $(2a)^{\beta} = 1$ whence $2(2a)^{\beta} \neq 1$. Therefore $(2a)^2 = 4a^2 \in R_{\beta}$, i.e. $a^2 \in R_{\beta}$. So in any case we have $a \in R_{\beta}$ if and only if $a^2 \in R_{\beta}$. At this point we have that $a*b \in R_{\beta}$ if a and b are in R_{β} . In fact $(a+b)^2 = a^2+b^2+ab+ba$ and a^2,b^2, $a+b$, $(a+b)^2$ are all elements of R_{β} . Finally we can prove that β is an algebra homomorphism as in the proof of the theorem in no. 2.3. The last statement of the theorem is trivial.

Remark: If R is a local ring, then $P_1(R) = Q_1(R) \cup Q_2(R)$ since the non-units of R form a two sided ideal. Therefore in this case β describes completely the map β .

4. THE STRUCTURE OF $GL_n(R)$

So far we have considered linear groups only geometrically. Now we will consider them from an algebraic point of view. We will discuss the problem of locating and classifying the normal subgroups and automorphisms of linear groups over arbitrary rings. We shall try to report the most meaningful results without any pretence of completeness.

4.1 - The normal subgroups of $GL_n(R)$.

We may identify endomorphisms of R^n with the corresponding matrices. Let I_n denote the identity matrix and e_{ij} be the matrix with 1 in position (i,j) and zeros elsewhere. The matrix $I_n + e_{ij}a$ $(a \in R$, $i \neq j)$ is said to be J-elementary, where J is a two sided ideal in R, if $a \in J$. We denote by $E_n(J)$ the group generated by all the J-elementary matrices. If R is commutative we have the determinant and its kernel $SL_n(R)$: clearly $E_n(R) \subseteq SL_n(R)$. Also, if R is a field, $E_n(R) = SL_n(R)$, whence $E_n(R)$ is a normal subgroup of $GL_n(R)$. If R is an arbitrary division ring one can define the determinant using Dieudonnè's definition, and one finds again $E_n(R) = SL_n(R) \trianglelefteq GL_n(R)$. In the general case of arbitrary rings it is unknown if $E_n(R)$ is normal in $GL_n(R)$. Nevertheless, if R is commutative and $n \geq 3$, in spite of an example due to Bass where $E_n(R) \neq SL_n(R)$, one can again prove that $E_n(R)$ is normal in $GL_n(R)$ (A.A. SUSLIN (1977)) and even characteristic (W.C. WATERHOUSE (1980)). Actually Suslin's proof of the normality of $E_n(R)$ can be adapted to prove the normality of $E_n(J)$ for each two sided ideal J even using the weaker hypothesis that R is finitely generated as a module over its centre.

In the special case of division rings we have essentially complete results on the normal subgroups of GL_n due to Dieudonnè:

THEOREM (Dieudonnè): Let D be a division ring and suppose that either $n \geq 3$ or that $n = 2$ but $|D| \geq 4$. If H is a subgroup of $GL_n(D)$ which is not contained in the centre of $GL_n(D)$, then $E_n(D) \subseteq H$ if (and only if) H is normalized by $E_n(D)$.

Since $GL_n(D)/SL_n(D)$ is commutative, as an immediate consequence we have:

Corollary: Unless $n = 2$ and $|D| < 4$, the normal subgroups of $GL_n(D)$ are either the subgroups of the centre of $GL_n(D)$ or the subgroups containing $E_n(D)$.

Remark: If R is not a division ring, then for each two sided ideal J of R the canonical map $R \to R/J$ induces a homomorphism $h_j: GL_n(R) \to GL_n(R/J)$ and, if $0 \neq J \neq R$, $H = \text{kern}\, h_j$ is a non-central normal subgroup which does not contain $E_n(R)$. Thus the previous corollary fails if R is not a simple ring. Also we note that H is sandwiched between two normal subgroups. Indeed H satisfies

$$\bar{E}_n(J) \trianglelefteq H \trianglelefteq Z_n(J) \tag{1}$$

where $\bar{E}_n(J)$ denotes the normal closure of $E_n(J)$ in $GL_n(R)$ and $Z_n(J)$ is the inverse image of the centre of $GL_n(R/J)$ under the homomorphism h_j.

The most that may be expected is that each normal subgroup H of $GL_n(R)$ satisfies the inclusions (1) for a suitable ideal J. This has been proved for $n \geq 3$ when R is either a local ring (W. KLINGENBERG (1961/62)) or the ring of rational integers (J.L. MENNICKE (1965)). Moreover they proved a stronger result for these rings, generalizing Dieudonnè's theorem. Precisely:

THEOREM (Klingenberg-Mennicke): Let R be either a local ring or the ring of rational integers and assume that $n \geq 3$. A subgroup H of $GL_n(R)$ which is normalized by $E_n(R)$ determines a unique two sided ideal J of R such that

$$K_n(J) \trianglelefteq H \trianglelefteq Z_n(J) ,$$

where $K_n(J) = SL_n(R) \cap \text{kern}\, h_j$ (1) . Conversely, if H satisfies the previous condition for a suitable ideal J , then H is a normal subgroup of $GL_n(R)$ (2) .

That the previous theorem breaks down for normal subgroups of an arbitrary commutative ring (even for large n) follows from Bass's example of $E_n(R) \neq SL_n(R)$. On the contrary, J.S. WILSON (1972), I.Z. GOLUBČHIK (1973) and A.A. SUSLIN (1977), in subsequent steps, showed that the expected inclusions (1) hold for commutative rings provided $n \geq 3$:

THEOREM (Wilson-Golubčhik-Suslin): Let R be a commutative ring and let $n \geq 3$. Then a subgroup H of $GL_n(R)$ is normalized by $E_n(R)$ if and only if there exists a unique ideal J of R such that $E_n(J) \trianglelefteq H \trianglelefteq Z_n(J)$ (3) .

At an early date H. BASS (1964) proved the previous theorem under a different hypothesis on R (satisfied by both commutative and non-commutative rings). Bass's hypothesis is very technical, yet it is satisfied by a reasonably large class of rings. We will formulate an equivalent one due to L.N. VASERSTEIN (1971):

D e f i n i t i o n : The ring R has m in its stable range, write $sr(R) \leq m$ for short, if for any $n \geq m$ and $(b_o, \ldots, b_n)$ uni-

---

(1) Klingenberg defines $K_n(J)$ also for non-commutative local rings by means of a suitable definition of determinant.

(2) We can prove that $\bar{E}_n(J)=E_n(J)=K_n(J)$ under the hypotheses of the theorem.

(3) We note that $\bar{E}_n(R)=E_n(R)$ because $E_n(R)$ is normal under the hypotheses of the theorem.

modular in R^{n+1}, there exist $c_1,\ldots,c_n \in R$ such that $(b_1+c_1b_o,\ldots,b_n+c_nb_o)$ is unimodular in R^n.

Bass proves his theorem assuming that R (not necessarily commutative) satisfies $sr(R) \leq n-1$ $(n \geq 3)$.

Recently L.N. VASERSTEIN (1980) proved the theorem under discussion for a ring R such that for every maximal two sided ideal M of the centre Z(R) of R, there exists a multiplicative subset S contained in Z(R)-M such that $sr(S^{-1}R) \leq n-1$. Vaserstein's condition is satisfied in case R is commutative (take $S = Z(R)-M = R-M$) and in case $sr(R) \leq n-1$ (take $S = \{1\}$ for any M), hence Vaserstein generalizes Wilson's, Golubčhik's, Suslin's and Bass's theorems. Furthermore Vaserstein observes that his hypothesis is satisfied if R is finitely generated as a module over its centre.

When $n = 2$ the situation is quite different. J.L. MENNICKE (1965) showed that pathology occurs in $GL_2(\mathbb{Z})$. On the other hand W. KLINGENBERG (1961) was able to extend his theorem to $n = 2$ (limited to the commutative case). Probably this different behaviour depends on the size of the group of units of R. This is strengthened by B.R. McDONALD (1980) who extends Klingenberg's theorem to a commutative primitive ring, which is a ring having many units.

4.2 - The isomorphisms of linear groups.

The problem of classifying the isomorphisms of linear groups over rings dates to the solution for groups PSL_n $(n \geq 3)$ over fields by O. SCHREIER and B.L. VAN DER WAERDEN (1928). The origin of this problem was probably in determining when two simple groups are isomorphic. Later J. DIEUDONNÈ (1951) and C.E. RICKART (1950), by introducing the method of involutions, were able to describe the isomorphisms for groups GL_n over division rings, except for some open cases solved by L.A. HUA and C.H. WAN (1953). The classical result is that each isomorphism $GL_n(R) \to GL_{n'}(R')$, in the case of division rings, is the product of the following standard isomorphisms:

(1) the inner automorphism $x \to gxg^{-1}$, where $g \in GL_n(R)$;

(2) the isomorphism $x = (x_{ij}) \to x^{\bar{\beta}} = (x_{ij}^{\beta})$, where $\beta : R \to R'$ is a ring isomorphism;

(3) the contragredient automorphism $x = (x_{ij}) \to (x^{t\bar{\beta}})^{-1} = (x_{ji}^{\beta})^{-1}$, where β is an anti-automorphism of R;

(4) the radial automorphism $y \to y^{\beta}y$, where $\beta : GL_{n'}(R') \to$ centre $(GL_{n'}(R'))$ is a group homomorphism.

As an immediate consequence we have $n = n'$ and R is isomorphic or anti-isomorphic to R'. Clearly in this case the isomorphism theory is essentially an automorphism theory.

The following question arises now: if R is an arbitrary ring, is every automorphism of $GL_n(R)$ again product of standard automorphisms? The answer is yes in case R is commutative and $n \geq 4$ (V.M. PETECHUK (1983)) provided we make some natural modifications in (1) and (3): first, g in (1) can be taken as an element satisfying $gGL_n(R)g^{-1} \subseteq GL_n(R)$ in a "proper extension" $GL_n(S)$ of $GL_n(R)$ (for example $S = \prod_M R_M$, where the product is taken over all maximal ideals M of R and R_M is the localization); secondly, if the ring R splits into

$R_1 \oplus R_2$, then $GL_n(R) = GL_n(R_1) \oplus GL_n(R_2)$ and thus the contragredient automorphism (3) must change to allow for a contragredient automorphism on $GL_n(R_1)$ and the identity on $GL_n(R_2)$.

We refer to O.T. O'MEARA (1974) and Y.I. MERZLYAKOV (1980) for a detailed history of the steps leading to Petechuk's result. Here we wish to recall the most interesting ones: the first step was taken by L.K. HUA and I. REINER (1951) specifically for $GL_n(\mathbb{Z})$, $n \geq 2$. Subsequently, J. LANDIN and I. REINER (1957) and Z. WAN (1958) extended the result of Hua and Reiner to (not necessarily commutative) principal ideal domains for $n \geq 3$. The methods used in these articles are mainly based on the study of involutions. O.T. O'MEARA (1974) devised a completely new method, not using involutions (the so-called method of residual spaces), which enabled him to describe the automorphisms of GL_n $(n \geq 3)$ over integral domains. J. POMFRET and B.R. McDONALD (1972) used Kaplansky's theorem, which states that projective modules over local rings are free, to determine the automorphisms of GL_n $(n \geq 3)$ over a commutative local ring in which 2 is invertible. Finally, B.R. McDONALD (1978) and W.C. WATERHOUSE (1980) described the automorphisms of $GL_n(R)$ $(n \geq 3)$ for the case where R is commutative and 2 is a unit.

Remark: The hypothesis that 2 is a unit is fundamental in order to prove that every automorphism of $GL_3(R)$, R a commutative ring, is a product of standard automorphisms. In fact, if we assume that 2 is not a unit, further automorphisms can occur in addition to the standard ones, even if R is a local commutative ring (V.M. PETECHUK

(1982)): let $J(R)$ be the radical of R and assume that $R/J(R) \simeq \mathbb{Z}_2$. Also let $\beta : R \to R$ be a bijective map satisfying the following conditions: (a) $x-x^{\beta} \in J(R)$; (b) $(x+y)^{\beta} = x^{\beta}+y^{\beta}+xy(s-1)$; (c) $(xy)^{\beta} = x^{\beta}y^{\beta}$, where s is an element of R such that $(s-1)J(R) = 0$. (We note that such a map ($\neq 1$) actually exists: take $R = \mathbb{Z}_8$, $s = 5$ and put $2^{\beta} = 6$, $6^{\beta} = 2$, $3^{\beta} = 7$, $7^{\beta} = 3$ and $x^{\beta} = x$ otherwise.) Furthermore, let us define the following map k_s^{β} on the set of matrices

(5) $\{d_i(x) = I_3+e_{ii}(x-1) \mid x$ is invertible in $R\} \cup \{t_{ij}(x) = I_3+e_{ij}x \mid x \in R\}$,

where we have used the same symbolism of no. 4.1:

$$d_i(x)^{k_s^{\beta}} = d_i(x^{\beta}) \ , \ t_{ij}(x)^{k_s^{\beta}} = t_{ij}(x^{\beta})+h_{ij}x(s-1) \ ,$$

with $h_{12} = e_{22}+e_{33}$, $h_{21} = e_{22}+e_{23}+e_{31}+e_{33}$, $h_{13} = e_{11}+e_{22}+e_{23}$, $h_{31} = e_{21}+e_{22}+e_{32}+e_{33}$, $h_{23} = e_{11}+e_{12}+e_{33}$, $h_{32} = e_{11}+e_{22}$. Since the matrices (5) generate the group $GL_3(R)$ if R is a local ring (N.S. ROMANOVSKII (1971)), the map k_s^{β} can be extended to an automorphism of $GL_3(R)$ which is not a product of standard automorphisms (see Petechuk's article for more details). Moreover Petechuk proves that no further automorphisms of $GL_3(R)$ can occur if R is a local commutative ring, while the automorphisms are only standard in the semisimple commutative case.

What can we say about the isomorphisms between linear groups in the commutative case? Petechuk observes that his proof of the theorem on the automorphisms of $GL_n(R)$ $(n \geq 4)$ can be appropriately modified to prove that an isomorphism $GL_n(R) \to GL_{n'}(R')$ (n and $n' \geq 4$) is a product of standard isomorphisms. Therefore, in the commutative case, we have an extension of the classical result for fields that $n = n'$

and $R \simeq R'$. O.T. O'MEARA (1977) proves that an analogous result holds if R and R' are (not necessarily commutative) principal ideal domains or symmetric ideal domains (a symmetric ideal domain being a domain in which all ideals are two sided). Clearly in the non-commutative situation, in view of generalized contragredient isomorphisms, $R \simeq R'$ turns into $R \simeq \bar{R}'$, where $\bar{R}' = (R_1')^0 \oplus R_2'$ for a (possibly trivial) suitable decomposition $R_1' \oplus R_2'$ of R' ($(R_1')^0$ the opposite ring of R_1'). Nevertheless O'Meara's result cannot be extended to an arbitrary Ore domain:

Counterexample (O'Meara): Let F be the quadratic number field $\mathbb{Q}(\sqrt{3})$ and R be the ring of all algebraic integers in F , i.e. $R = \mathbb{Z} + \mathbb{Z}\sqrt{3}$. Let D be the usual quaternion algebra over F with defining basis $\{1,i,j,k\}$. Further, let us define

$$A = R(1+\sqrt{3}i/2)+Ri+Rj+R(k+\sqrt{3}j/2) ,$$

$$B = R(1+\sqrt{3})(1+i)/2+R(1-\sqrt{3})(1+j)/2+Rk+R(1+i+j+k/2) .$$

We have (we omit the proof): A and B are maximal orders in D such that A contains at least eight elements with square -1 , while B has only six. So A is neither isomorphic nor anti-isomorphic to B . Nevertheless $GL_n(A) \simeq GL_n(B)$ for an infinite number of n .

Vaserstein observed that the rings A and B in O'Meara's counterexample are Morita equivalent and asked: does $GL_n(R) \simeq GL_{n'}(R')$ imply that R is Morita equivalent to $\bar{R}'$ for some decomposition of R' ? We recall that two rings R and R' are Morita equivalent if the category $\mathcal{M}_R$ of right (unital) R-modules is equivalent to the category $\mathcal{M}_{R'}$ of right R'-modules. The relevancy of this concept to the present discussion is immediate: if $f:\mathcal{M}_R \to \mathcal{M}_{R'}$ is a category equivalence and $f(R^n) = (R')^{n'}$, then f restricts to an isomorphism of matrix rings $M_n(R) \to M_{n'}(R')$ which, in turn, restricts to an isomorphism $GL_n(R) \to GL_{n'}(R')$. This family of "Morita isomorphisms" includes every composition of a generalized inner automorphism by an isomorphism of type (2). On the other hand, in view of O'Meara's counterexample, there exist Morita isomorphisms which cannot be realized as a composition of standard isomorphisms. A.J. HAHN (1982) proves that every isomorphisms $GL_n(R) \to GL_{n'}(R')$, n and $n' \geq 3$, is a product of standard isomorphisms and of a Morita isomorphism in two special cases: (a) R and R' semisimple artinian rings; (b) R and R' Ore domains. Therefore Hahn answers positively to Vaserstein's question in these two rather diverse situations.

As Petechuk shows that in the commutative case an isomorphism $GL_n(R) \to GL_{n'}(R')$, n and $n' \geq 4$, is determined (up to a radial automorphism (4)) by its restriction to $E_n(R)$ (this uses among other things the normality of $E_n(R)$ in $GL_n(R)$), Petechuk's result characterizes the isomorphisms between groups E_n $(n \geq 4)$ over commutative rings too. In a non-published recent note M. BOLLA classifies the isomorphisms between groups E_n, $n \geq 3$, over (not necessarily commutative) rings subject only to the condition that 2 is invertible. Bolla proves that such an isomorphism is a product of standard isomorphisms and of a Morita isomorphism. Also Bolla observes that his result can be extended to isomorphisms $GL_n(R) \to GL_{n'}(R')$ provided $E_n(R)$ is normal in $GL_n(R)$.

We wish to recall that the isomorphisms between groups E_n over Ore domains has been treated for $n \geq 5$ by O.T. O'MEARA (1977) also.

The case $n = 2$ is very difficult. It was first solved for various special types of domains. In addition to the above mentioned result of Hua and Reiner for $GL_2(\mathbb{Z})$, J. LANDIN and I. REINER (1958) proved, in two different articles, that each automorphism of $GL_2(R)$ is a product of standard automorphisms if R is the ring of gaussian integers or if R is a commutative euclidean domain generated by its units. Yet I. REINER (1957) showed that this pattern could not be expected to hold in the general case. He analysed the automorphism group of $GL_2(F[x])$, for F a field, and found there a new type of automorphism which can be viewed as a generalization of an automorphism of type (2). This is defined as follows: let β be an automorphism of $F[x]$ as a F-vector space such that $1^\beta = 1$. Then the map $x = (x_{ij}) \to x^{\bar{\beta}} = (x_{ij}^\beta)$ defined on the set of matrices (5) (written for $n = 2$) induces an automorphism of $GL_2(F[x])$, because the matrices (5) generate $GL_2(F[x])$ in this special case. (Note that under this automorphism it is not the case that $x \to x^{\bar{\beta}}$ for any $x \in GL_2(F[x])$.)

P.M. COHN (1966) generalized this automorphism by introducing a special type of function on a ring, called a <u>U-homomorphism</u>, which is defined to preserve the additive structure and multiplication by units. He then showed that, for a certain class of commutative domains (containing the polynomial rings) such a map applied to the entries of the matrices in (5) induces a homomorphism of the group $GE_2(R)$ generated by those matrices ($GE_2(R)$ may not be all of $GL_2(R)$ in case R is not a polynomial ring). Cohn's result for his class of rings is that every automorphism of $GE_2(R)$ is obtained by taking the automorphism induced by a U-automorphism, followed by standard automorphisms. This result was extended by M.H. DULL (1974) to commutative domains having either $2 = 0$ or 2 a unit.

For rings having zero divisors there is a result due to B.R. McDONALD (1981). He proves for a commutative primitive ring R having only trivial idempotents the classical result that $\mathrm{AutGL}_2(R)$ is generated by the standard automorphisms.

Finally, we wish to mention the striking result, due to P.M. COHN (1966), that if A is any free associative algebra over a field F, on at most countably many free generators, then $GL_2(A)$ is isomorphic to $GL_2(F[x])$.

ADDED NOTE

(1) Recently, A.J. HAHN delivered at the International Group Theory Symposium at Beijin (China) some lectures on "Algebraic K-theory, Morita theory and the classical groups". The arguments treated by Hahn are a useful complement to section 4 of this article (see also "Homomorphisms of algebraic and classical groups: a survey" by A.J. HAHN, D.G. JAMES, B. WEISFEILER (to appear)).

(2) We wish also to mention a recent article due to B. SARATH and K. VARADARAJAN ("Fundamental theorem of projective geometry" Comm. in Algebra 12 (8) 937 - 952 (1984)), where the fundamental theorem of projective geometry in dimension ≥ 2 is proved for suitable regular rings (in the sense of Von-Neumann) and for semiprime right Goldie rings (see the open question at the end of section 1).

REFERENCES

ANDRÈ J. (1969): Über Homomorphismen projektiver Ebenen, Abh. Math. Sem. Univ. Hamburg 34, 98 - 114.

ARTIN E. (1957): Geometric Algebra, Wiley-Interscience. New York.

BAER R. (1952): Linear Algebra and Projective Geometry, Academic Press. New York.

BARTOLONE C. - DI FRANCO F. (1979): A remark on the projectivities of the projective line over a commutative ring, Math. Z. 169, 23 - 29.

BASS H. (1964): K-theory and stable algebra, Publ. Math. IHES 22, 5 - 60.

BOLLA M. (1984): Isomorphisms of general linear groups over rings, J. Algebra (to appear).

BUEKENHOUT F. (1965): Une généralisation du théorème de von Staudt-Hua, Acad. Roy. Bel. Bull. Cl. Sci. 51, 1282 - 1293.

CARTER D. S. - VOGT A. (1980): Collinearity preserving functions between desarguesian planes, Memoirs Amer. Math. Soc. 235.

COHN P.M. (1966): On the structure of the GL_2 of a ring, Publ. Math. IHES 30, 5 - 53.

DIEUDONNÈ J. (1951): La géométrie des groupes classiques, Springer-Verlag, Berlin Heidelberg New York.

DRESS A. (1964): Metrische Ebenen und projektive Homomorphismen, Math. Z. 85, 116 - 140.

DULL M.H. (1974): Automorphisms of the two-dimensional linear groups over integral domains, Amer. J. Math. 96, 1 - 40.

GOLUBČHIK I.Z. (1973): On the full linear group over an associative ring, Uspehi Mat. Nauk 28 (171), 179 - 180 (Russian).

HAHN A. J. (1982): Category equivalence and linear groups over rings, J. Algebra 77, 505 - 543.

HERSTEIN I.N. (1956): Jordan homomorphisms, Trans. A.M.S.81, 331 - 341.

HUA L.K. - REINER I. (1951): Automorphisms of the unimodular group, Trans. A.M.S. 71, 331 - 348.

HUA L.K. - WAN C.H. (1953): On the automorphisms and isomorphisms of linear groups, J. Chinese Math. Soc. 2, 1 - 32.

JAMES D.G. (1982): Projective homomorphisms and von Staudt's theorem, Geom. Ded. 13, 291 - 294.

KLINGENBERG W. (1956): Projective Geometrien mit Homomorphismus, Math. Ann. 132, 180 - 200.

KLINGENBERG W. (1961): Lineare Gruppe über localen Ringen, Amer. J. Math., 83, 137 - 153.

KLINGENBERG W. (1962): Die Struktur der linearen Gruppe über einem nicht kommutativen localen Ring, Arch. Math. 13, 73 - 81 .

LANDIN J. - REINER I. (1957): Automorphisms of the general linear group over a principal ideal domain, Ann. of Math. 65, 519 - 526.

LANDIN J. - REINER I. (1958): Automorphisms of the Gaussian modular group, Trans. A.M.S. 87, 76 - 89.

LANDIN J. - REINER I. (1958): Automorphisms of the two-dimensional general linear group over a euclidean ring. Proc. A.M.S. 9, 209 - 216.

LIMAYE N.B. (1971): Projectivities over local rings, Math. Z. 121, 175 - 180.

LIMAYE N.B. (1972): Cross-ratios and projectivities of a line, Math. Z. 129, 49 - 53.

LIMAYE B.V. - LIMAYE N.B. (1977): Fundamental theorem for the projective line over non-commutative local rings, Arch. Math. 28, 102 - 109.

McDONALD B.R. (1978): Automorphisms of $GL_n(R)$, Trans. A.M.S. 246, 155 - 171.

McDONALD B.R. (1980): GL_2 of rings with many units, Comm. in

Algebra 9 (2), 205 - 220.

McDONALD B.R. (1981): Projectivities over rings with many units, Comm. in Algebra 9 (2), 195 - 204.

McDONALD B.R. (1982): Aut(GL_2(R)) for rings with many units, Comm. in Algebra 9 (2), 205 - 220.

MENNICKE J.L. (1965): Finite factor groups of the unimodular group, Ann. of Math. 81, 31 - 37.

MERZLYAKOV Y.I. (1980): Linear groups, Itogi Nauki: Algebra, Topologia, Geometriya 16, VINITI, Moscow 1978, 35 - 89. English transl. in J. Soviet Math. 14 (1).

OJANGUREN M. - SRIDHARAN R. (1969): A note on the fundamental theorem of Projective Geometry, Comment. Math. Helv. 44, 310 - 315.

O'MEARA O.T. (1966): The automorphisms of the linear groups over any integral domain, J. reine angew. Math. 223, 56 - 100.

O'MEARA O.T. (1974): Lectures on linear groups, Conf. Board Math. Sci. Regional Conf. Ser. Math. 22, A.M.S., Providence, R.I.

O'MEARA O.T. (1977): A general isomorphism theory for linear groups, J. Algebra 44, 93 - 142.

PETECHUK V.M.(1982): Automorphisms of the groups SL_3(K) and GL_3(K), Math. Notes 31, 335 - 340.

PETECHUK V.M.(1983): Automorphisms of matrix groups over commutative rings, Math. USSR Sbornik 45, 527 - 542.

POMFRET J. - McDONALD B.R. (1972): Automorphisms of GL_n(R), R a local ring, Trans. A.M.S. 173, 379 - 388.

RADÒ F. (1969): Darstellung nicht-injektiver Kollineationen eines projectiven Raumes durch verallgemeinerte semilineare Abbildungen, Math. Z. 100, 153 - 170.

RADÒ F. (1970): Non-injective collineations on some sets in desarguesian projective planes and extensions of non-commutative valuations,

Aequationes Math. 4, 307 - 321.

REINER I. (1957): A new type of automorphism of the general linear group over a ring, Ann. of Math. 66, 461 - 466.

RICKART C. E. (1950): Isomorphic groups of linear transformations, Amer. J. Math. 72, 451 - 464.

ROMANOVSKII N.S. (1971): Generators and defining relations of the full linear group over a local ring. Sib. Mat. Zh. 12 (4), 922 - 925 (Russian).

SCHAEFFER H. (1974): Das von Staudtsche Theorem in der Geometrie der Algebren, J. reine angew. Math. 267, 133 - 142.

SCHREIER O. - VAN DER WAERDEN B.L. (1928): Die Automorphismen der projectiven Gruppen, Abh. Math. Sem. Univ. Hamburg 6, 303 - 322.

SUSLIN A.A. (1977): On the structure of the special linear group over polynomial rings, Izv. Akad. Nauk., ser. mat. 41 (2), 235 - 252.

VASERSTEIN L.N. (1971): The stable range of rings and the dimension of topological spaces, Funct. Anal. Appl. 5 (2), 102 - 110.

VASERSTEIN L.N. (1980): On the normal subgroups of GL_n over a ring, Algebraic K-theory Evanstone 1980. Springer-Verlag, 456 - 465.

WAN Z. (1958): On the automorphisms of linear groups over a non-commutative principal ideal domain of characteristic $\neq 2$, Sci. Sinica 7, 885 - 933.

WATERHOUSE W.C. (1980): Automorphisms of $GL_n(R)$, Proc. AMS 79, 347 - 351.

WILSON J.S. (1972): The normal and subnormal structure of general linear groups, Proc. Cambridge Philos. Soc. 71, 163 - 177.

ACKNOWLEDGEMENTS

The authors wish to thank Prof. B.R. McDONALD and Dr. M. BOLLA for their suggestions. Our thanks go also to Prof. I.R. FAULKNER, Prof. J.C. FERRAR, Prof. I.N. HERSTEIN and Prof. J.W. LORIMER for their help concerning style and wording.

METRIC GEOMETRY OVER LOCAL-GLOBAL COMMUTATIVE RINGS

Bernard R. McDonald
Division of Mathematical Sciences
National Science Foundation
Washington, D.C., U. S. A.

The research topics and directions of metric geometry in U. S. institutions were heavily influenced by two publications in the 1950's -- Dieudonné's American Mathematical Society Memoir #2, On the automorphisms of classical groups in 1950 and Artin's graduate text Geometric Algebra published by Wiley-Interscience in 1957.

These works determined the boundaries of the subject and gathered the scattered literature into a coherent presentation.

Within the context of a finite dimensional vector space V over either a field or a division ring, these surveys examined the general linear group $GL(V)$, the orthogonal group $O(V)$ relative to a symmetric bilinear form, the symplectic group $Sp(V)$, the unitary group $U(V)$, and their commutator subgroups. Basic questions were asked. What are the generators? What are the normal subgroups? What transitivity arises from the actions of the group on the underlying space? What are the automorphisms of the groups and do isomorphisms exist between the groups? The theory of the orthogonal group was inter-

R. Kaya et al. (eds.), Rings and Geometry, 391–415.

woven with the study of quadratic forms. The automorphism theory and normal subgroup theory were connected to and motivated by the classification of simple groups. The solution of each of the questions was related to the solution of the others.

Three parameters became apparent in the study of these questions. First, there was the group under consideration, i.e. GL, O, Sp, U, or their commutator subgroups. Second, there was the dimension of the space onwhich the groups lived. The groups often behaved differently in small dimensions (for example in dimension 2) than they did in higher dimensions. Third, often surprising results occurred when the field varied; for example, finite fields, real fields, complex fields, algebraically closed fields, and, more dramatically, when the field was replaced by a ring.

As the field came to be replaced by more general rings, two separate studies emerged and appear to be determined by the relative 'size' of the group of units within the ring. For example, if the field is replaced by the ring of integers Z (which has only 1 and -1 as units) then the above questions were difficult and often require deep results from algebraic number theory. On the other hand, if the field is replaced by a commutative local ring, then field theoretic arguments can often be mimicked, giving results that resemble the theory over a field. Here the maximal ideal of the local ring may be thought of as simply a large 'zero' and calculations and proofs are constructed so as to utilize the units of the local ring and avoid the maximal ideal. W. Klingenberg [18],

[19] , and [20] was the first to make extensive use of this in the early 1960's. Results were gradually extended in part to semilocal rings and von Neumann regular rings. These separate studies again merge to produce the theory over an arbitrary commutative ring.

In this paper, we discuss the evolution of this latter idea from the local scalar ring to what is currently called a Local-Global ring. Local-global commutative rings are described in Section I and include local, semilocal, von Neumann regular, and zero dimensional rings. The later sections sketch the results of several papers on metric geometry over local-global rings. Actually, for the most part these papers were not developed over local-global rings; however, their contents can be easily modified to this natural setting.

Relative to changing the dimension, generally different and often difficult arguments are needed for low dimensions. For larger finite dimensions, very sharp ´field-like´ results can be obtained for local-global rings. Finally, as dimension increases arbitrarily we often obtain a 'stable behavior', that is, behavior resembling the field case, for an arbitrary commutative (and sometimes even noncommutative) ring.

The general linear group GL is the best known and most widely studied of the groups listed above. Thus, its theory has been pushed quite far for both arbitrary commutative and noncommutative rings. For the other groups, the theory is less complete and follows the development of GL more slowly. Indeed, to obtain basic results we often must 'linearize' the context by assuming the existence of hyperbolic planes in order to introduce 'transvections' whose actions resemble the actions of the transvections of the general linear group.

Section I LG-Rings

Let R denote a commutative ring. A polynomial $F(X) = a_o + a_2X + \dots + a_nX^n$ is a primitive polynomial if the ideal generated by the coefficients of F is all of R. A primitive polynomial F(X) is said to represent a unit if there is an element a in R with F(a) invertible. The ring R is called primitive if every primitive polynomial in R[X] represents a unit. Clearly if a polynomial represents a unit then the polynomial must be primitive. A field is primitive if and only if it is infinite.

The local-global principle is said to hold for R and R is called a LG-ring if whenever a polynomial $F(X_1, X_2, \dots, X_t)$ represents a unit over the local ring R_m for each maximal ideal m in R, then F represents a unit over R. The idea of an LG-ring was introduced in [5] and [32] with the terminology "Local-Global ring" first occurring in [5].

The simplest example of a LG-ring is a field. In [5] and [32] it is shown that the following are equivalent:

(a) R is a primitive ring.

(b) (1) R is an LG-ring and

(2) Every residue field R/m is infinite.

Examples and constructions of LG-rings (See [5], [14], [32]) include the following:

(a) If J(R) denotes the Jacobson radical and R/J(R) is an LG-ring then R is an LG-ring.

(b) Products and direct limits of LG-rings are LG-rings.

(c) Local, semilocal, von Neumann regular, and zero dimensional rings are LG-rings.

(d) Integral extensions of LG-rings are LG-rings.

The above results also may be stated for primitive rings (provided in (c) that we require all residue class fields be infinite). In addition:

(a) Homomorphic images of primitive rings are primitive.

(b) If R is any commutative ring and S denotes the set { F in R[X] : F is a primitive polynomial }, then the ring of fractions $S^{-1}R[X]$ is a primitive ring. Thus every commutative ring can be embedded in a primitive ring.

This shows that an LG-ring can be very far from dimension zero. Indeed, since S is the set of primitive polynomials in R[X], then the maximal ideal space of $S^{-1}R[X]$ is identical with that of R.

(c) If R is primitive, then so is the formal power series ring $R[[X_1, \dots, X_t]]$.

(d) The algebraic closures of the integers Z or k[X] where

k is a finite field are primitive. The ring of all real algebraic integers is primitive.

Section II Linear Algebra

In this section we summarize certain results concerning modules and matrices over LG-rings. The proofs of these results can be found in [5] and [31]. The purpose of this section is to provide the reader with certain linear algebraic results that are well-known for local rings and extend to LG-rings.

We begin with a classification of projective modules over LG-rings. The proof of this result will be provided to illustrate the polynomial techniques.

Theorem. Let R be a LG-ring and P be a finitely generated projective R-module. Then P is a direct sum of cyclic modules of the form Re_i where the e_i are idempotents in R.

Proof. It is known (see [28]) that a finitely generated projective module P may be written as a direct sum of modules of the form e_iP where e_i is an idempotent in R and e_iP has constant local rank. Thus, we may assume that P is a finitely generated module of constant rank. Under this assumption we will show that such a P is free.

For some integer n we can write $P \oplus Q = R^n$. Let A be the matrix of the map $R^n \longrightarrow R^n$ that projects onto P with kernel Q. Let $X = [X_{st}]$ be the matrix of indeterminates and let $Ad_J(X)$ be the adjoint of X, so that $XAd_J(X) = (\det X)I$. Let m be the constant local rank of P, and let $g(X_{ts})$ be the polynomial which is the upper left $m \times m$ minor determinant of $XAAd_J(X)$. Set $f(X_{st}) = g(X_{st})\det(X_{st})$.

Let z be a maximal ideal of R. Over the local ring R_z, both of the modules P_z and Q_z are free, with P_z having dimension m. Select a basis of $(R_z)^n$ compatible with the decomposition $P_z \oplus Q_z$. In the basis the projection to P_z has a matrix with upper left block equal to I_m and other blocks equal to 0. Thus if $[c_{ts}]$ is the matrix over R_z giving the change of basis, then $g(c_{ts}) = (\det(c_{ts}))^m$. Hence $f(c_{ts})$ is a unit in R_z.

Since this is true for each maximal ideal z, there are values b_{st} in R with $f(b_{st})$ invertible. Then the matrix $[b_{st}]$ is invertible, and the upper left $m \times m$ submatrix of $D = [b_{st}]\, A\, [b_{st}]^{-1}$ is invertible.

The rest of the proof is standard. A change of basis shows that P is isomorphic to R^m.

The above implies that the Picard group of R is trivial and that a finitely generated projective module of constant rank is free.

Estes and Guralnick[5] have generalized the above, showing that if

R is an LG-ring, M and N are finitely presented modules with M_z isomorphic to N_z for every prime ideal z, then M is isomorphic to N.

Then also show the following:

(a) If R is an LG-ring, M and N are finitely presented R-modules, then

(1) If M^n is isomorphic to a summand of N^n, then M is isomorphic to a summand of N.

(2) If M^n is isomorphic to N^n, then M is isomorphic to N.

(b) If A and B are n x n matrices over an LG-ring R with A and B equivalent (similar, left equivalent, resp.) over R_z for each prime ideal z in R, then A and B are equivalent (similar, left euqivalent, resp.) over R.

Section III GL (2)

Let R denote a commutative ring and let GL(n) denote the group of invertible n x n matrices with coefficients in R. As noted elsewhere in these proceedings, Petechuk [35] has determined the automorphisms of GL (n) when n $\geq$ 4. If 2 is a unit and n $\geq$ 3, then the automorphisms of GL (n) were given in McDonald [24] /Waterhouse [38] . Also Petechuk [35] shows that, in the case n = 3, the assumption that 2 a unit is necessary for the automorphism theory of GL(n) for arbitrary commutative ring.

On the other hand, Golubchik [6] has determined the normal subgroups of GL(n) for $n \geq 3$.

Thus, one remaining difficult case is where the dimension n = 2. In general, this case remains open. In dimension n = 2 the structure of GL(2) appears to be a function of the size of the group of units R^* in R. For example, if R is a local ring, then expected results similar to the field case will occur; while, if R has few units then one may expect pathology. For example, in [36] Reiner discovered an exceptional automorphism for GL(2) where R = k [x] (k a field). Later this was examined in detail by Cohn [2] and Dull [4].

When R is a local ring the normal subgroups of GL(2) were described by Klingenberg [18]. Klingenberg's work suggests that an automorphism theory and normal subgroup theory for GL(2) over a LG-ring might be possible. Indeed, the three papers [25], [26], and [27] show that this is the case.

The results in [25], [26], and [27] were stated for primitive rings; however, they can easily be adapted to LG-rings in the following fashion. In order to obtain desired results, certain primitive polynomials were utilized to produce units. These polynomials were of low degree and over a local ring these polynomials would produce units under substitution provided the residue class field of the local ring had a sufficient number of elements. Thus, we must avoid certain small finite fields. Hence, for an LG-ring R we must require that R possess no residue class fields of a suitable specified 'smallness'.

Let R be an LG-ring having 2 a unit and no residue class fields of 5 or fewer elements. Then, [26] shows that Hua's theorem (Theorem II.1 [26]) is available for use in the automorphism theory and that The Fundamental Theorem of Projective Geometry is valid for the projective line, that is, von Staudt's Theorem. (For dimensions ≥ 3, the Fundamental Theorem of Projective Geometry was generalized to arbitrary commutative rings by Ojanguren and Sridharen[34]. In dimension 2, under stronger conditions, Bartolone and Di Franco obtained the Fundamental Theorem for the projective line over an arbitrary commutative ring).

In [25] and [27] the generators, normal subgroups, transitivity, commutators, and automorphisms of GL(2) were determined for commutative rings having 2 a unit and satisfying certain conditions on the representation of units by one or more primitive polynomials. These conditions can be replaced by the assumption that R is an LG-ring having 2 a unit and no 'small' residue class fields.

The arguments in the above papers ([25],[26],[27]) were modifications of arguments over local rings. Thus, after a careful reexamination of the original proofs, it is highly likely that the size of the residue class rings that must be avoided may be further reduced.

Section IV Inner Product Spaces and the Orthogonal Group

Throughout this section R will denote a commutative LG-ring and V will denote a free R-space of dimension n. We assume that V possesses a symmetric inner product (,) : V x V ---> R. We assume that 2 is a unit in R. The results summarized in this section occur in [9], [19], [29], [30], and [31]. (In Section 7 of [5], Estes and Guralnick describe the classification of quadratic lattices over a domain satisfying the primitive criteria. They show that this theory parallels the theory for LG-rings.)

In the first paper [29] it is shown that the above spaces have orthogonal bases and a split space is a direct sum of hyperbolic planes. Further, symmetric inner product spaces may be 'cancelled' (Witt Cancellation). Under the assumption that the hyperbolic rank of V is ≥ 1, generators and transitivity results are given for the orthogonal group O(V). In particular, O(V) is transitive on unimodular vectors of the same norm and is transitive on the set of hyperbolic planes of V. The generators (for R connected in [29], and later for a general R in [30]) are given in terms of Eichler-Siegel-Dickson transvections and orthogonal transformations of hyperbolic planes. At this point, symmetries do not enter the discussion in any significant fashion.

Quadratic primitive polynomials play an important and interesting role in this papers:

(a) The basic theory in [29] and the normal subgroup theory for

O(V) by Ishibashi in [9] requires that R be a ring for which any primitive quadratic polynomial

$$f(X) = aX^2 + bX + c$$

represents a unit.

(b) The normal subgroup theory for O(V) by McDonald-Kirkwood in [30] requires that two primitive quadratic polynomials

$$f_1(X) = a_1X^2 + b_1X + c_1$$
$$f_2(X) = a_2X^2 + b_2X + c_2$$

simultaneously represent units.

(c) The theory of Witt rings as given in [16] by Kirkwood-McDonald requires that three primitive quadratic polynomials of the form

$$f_i(X) = a_iX^2 + b_iX + c_i \qquad 1 \leq i \leq 3$$

simultaneously represent units.

The above indicates that the 'stable' theory of the orthogonal group is intimately connected to representation of units by quadratic polynomials.

To illustrate this use of quadratic polynomials we construct an orthogonal basis for a symmetric inner product space over an LG-ring. Let V be free of dimension n and let x be any unimodular vector of V. Since (,) is an inner product, there is a unimodular y in V with $(x,y) = 1$. For an element a in R, consider the vector $x + ay$. We have $(x + ay, x + ay) = (x,x) + 2a(x,y) + (y,y)a^2$. Thus, the polynomial $f(X) = (x,x) + 2(x,y)X + (y,y)X^2$ is primitive. Since 2 is a unit, the polynomial represents a unit over every local ring and thus represents a unit over R. That is, there is an element a such that the vector $z = x + ay$ has unit norm. Thus, Rz splits as a free orthogonal summand of V. Since Rz is free and R is an LG-ring, the orthogonal complement of Rz is also free. By an induction argument, we can produce an orthogonal basis for V.

Next we sketch the contents of the above papers. The theory in these papers is presented within the context of a commutative ring having the property that units may be simultaneously represented by one or more primitive quadratic polynomials. The arguments may easily be adapted to LG-rings utilizing assumptions on the sizes of the residue class fields. For the remainder of this section we assume that <u>the residue class fields of the LG-ring R are sufficiently large so that one or more primitive quadratic or higher degree polynomials will simultaneously represent units</u>.

McDonald-Kirkwood[30] and Ishibashi [9] provide two somewhat different characterizations of the normal subgroups of O(V). In both papers, when the hyperbolic rank of V is greater than or equal to one, then the normal subgroups are shown to be sandwiched between congruence subgroups. In Ishibashi's work it is assumed that, in addition to 2, both 3 and 5 are units and that the dimension of V is greater than or equal to 7. McDonald-Kirkwood's characterization also requires that 2, 3 and 5 be units and that the dimension of V be either 3 or greater than or equal to 5. Both papers analyze the generators of the congruence subgroups and describe the commutator subgroups. In [30] the transitivity properties of both the orthogonal group O (V) and the special orthogonal group SO(V) on the space V are examined. Also McDonald-Kirkwood examine the Clifford algebra of the space V, relating it to the special orthogonal group in the fashion analogous to the well-known theory over a field.

Both approaches to the characterization of the normal subgroups in the above papers are based on the characterization of the normal subgroups of O(V) over a local ring given by D. James in [12]. The proofs are patterned after arguments given by James. The earliest work on this subject over a local ring was that of Klingenberg in [19]. In these papers, the normal subgroup theory is developed in the so-called 'linearized' setting. By this we mean the assumption that the hyperbolic rank of V is greater than or equal to 1. This essential condition allows the use of the Eichler-Siegel-Dickson transvections, permits the description of generators, transitivity, and moves many of the vector calculations into a hyperbolic plane. Little is known for the case of no hyperbolic rank.

Another interesting aspect is the use of symmetries in the above papers - or perhaps, more accurately, the complete absence of the role of the symmetry. Although Klingenberg had shown in [19] that the orthogonal group over a local ring is generated by symmetries, it was not until a later paper by McDonald-Kirkwood [31], in which the existence of transversals for inner product spaces over LG-rings was established, was it possible to show that the orthogonal group O(V) over LG-rings was generated by symmetries. Suppose that V can be decomposed as an orthogonal direct sum of subspaces $E_1, E_2, \ldots, E_t$. A unimodular vector x is said to be a <u>transversal</u> relative to this decomposition if x can be written as $x = a_1 + a_2 + \ldots + a_t$ where each a_i is in E_i and the norm of a_i, that is (a_i, a_i), is a unit for $1 \leq i \leq t$. The key result is that over an LG-ring, if x is a unimodular vector, then there is a product # of symmetries such that #(x) is a transversal for the given decomposition. In itself, the existence of transversals does not seem to be significant; however, in [31] it is shown that this is a crucial step in showing that O(V) is generated by symmetries and that O(V) and SO(V) act transitively on sets of nonisotropic vectors of the same norm. Further, transversal theory is essential to a deeper understanding of Witt rings.

The automorphism theory for the orthogonal group over an LG-ring has not at this date been developed. The automorphisms of O(V) over a local ring were first determined by E. M. Keenan in his unpublished Ph.D. thesis at Massachusetts Institute of Technology in 1965. Portions of Keenan's work now appears in the monograph [15]. The automorphisms over a local ring are of the standard type. It is presumed that, given the current development of the structure of O(V) over an LG-ring, the automorphism theory for O(V) could be obtained and that the automorphisms should be of the standard type.

Section V Witt Rings

We continue the assumptions of the previous section that R is a commutative local-global ring with 2 a unit. We also assume that R has no residue class fields of 'small' size. Recall that the category of free symmetric inner product spaces BIL(R) over R has operations of orthogonal direct sum and tensor product. Let $K_o(BIL(R))$ be the Witt-Grothendieck ring of BIL(R). Let H(R) denote the ideal of $K_o(BIL(R))$ generated by differences $[V_1] - [V_2]$ where V_1 and V_2 are split spaces. Since split spaces are direct sums of hyperbolic planes, H(R) is the group ring Z[H] where H is a hyperbolic plane. The quotient ring W(R)= $K_o(BIL(R))/H(R)$ is called the <u>Witt ring of the free symmetric inner product spaces over R</u>.

The Witt ring W(R) is examined in detail in [16] and [31]. These two papers extend the theory of the Witt ring for semilocal rings developed by M. Knebusch, A. Rosenberg, and R. Ware in a series of papers in the 1970's to the setting of LG-rings. The central results of these papers are summarized below.

Let Q(R) denote the quotient group $R^*/(R^*)^2$ of the square classes of units R^* of R. The group Q(R) may be thought of as the group of isometry classes $\langle a \rangle$ of one-dimensional spaces over R with tensor product as multiplication. There is a natural map $\langle a \rangle \longrightarrow [\langle a \rangle]$ (denote $[\langle a \rangle]$ by $[a]$) from Q(R) to W(R) which is easily seen to be injective. Identifying Q(R) with its image in W(R), the group Q(R) my be regarded as a subgroup of the group of units $W(R)^*$ of $W(R)$ and it is easy to show that $W(R)$ is additively generated by $Q(R)$. Thus, there is a surjective ring morphism from the integral group ring $Z[Q(R)]$ to $W(R)$ which we denote by $\wedge: Z[Q(R)] \rightarrow W(R)$. Suppose that the square class $a(R^*)^2$ in $Q(R)$ is denoted by (a). Then $\wedge: (a) \rightarrow [a]$. Let K denote the kernel of $\wedge$. In [16] the additive generators of K are determined. Further, using the isomorphism between $W(R)$ and $Z[Q(R)]/K$ the generators and relations for $W(R)$ are explicitly given.

Due to the '2-connectability' of orthogonal bases in symmetric inner product spaces over LG-rings, the prime ideals of the $W(R)$ may be classified in the same fashion as was done by Knebusch, Rosenberg, and Ware([21],[22]) for semilocal rings. A <u>signature</u> # of R is a ring morphism from $W(R)$ to Z. There may exist no signatures of R. The ring R is called <u>real</u> if R has at least one signature; otherwise, R is called <u>nonreal</u>. It is easy to see that the signatures are related to prime ideals $P_\#$ where $P_\#$ denotes the kernel of #. If R is real, then the $P_\#$ are precisely the minimal prime ideals of $W(R)$.

The remainder of [16] concerns the torsion theory of the Witt ring of R and the relation of torsion or torsion-free to the representation of units as sums of squares. The nonreal case may be dismissed since R is an nonreal LG-ring if and only if for some positive integer t we have $2^t W(R) = 0$. Thus, if R is a nonreal LG-ring then all elements of $W(R)$ are torsion. Further, R is nonreal if and only if -1 is a sum of squares. Hence, assume that R is real and let $W(R)_t$ denote the torsion elements of $W(R)$. In [16] it is shown that $W(R)_t$ is generated by elements of the form $[\langle 1, -a\rangle]$ where a is a unit and is a sum of squares. Further, the Witt ring $W(R)$ is torsion free if and only if every unit which is a sum of squares is itself a square and -1 is not a square.

A powerful tool in the study of Witt rings is Pfister's theory of multiplicative forms, which was simplified by the concept of the round form, introduced by Witt and later generalized by Knebusch. Recall that a *n-fold Pfister form* over R is the symmetric inner product space $\langle\langle a_1, a_2,\ldots,a_n\rangle\rangle$ which represents the tensor product of the n two-dimensional spaces $\langle 1, a_1\rangle, \langle 1, a_2\rangle,\ldots, \langle 1, a_n\rangle$. Let $D(V) = \{(x,x)$ in $R^* : x$ is in the inner product space $V\}$. The inner product $(\, , \,)$ is called a *group form* if $D(V)$ is a subgroup of the group of units R^* of R. A unit z in R is called a *similarity norm of V* if V is isometric to the tensor product of $\langle z\rangle$ and V. Let $N(V)$ denote the similarity norms of V. The similarity norms $N(V)$ form a multiplicative subgroup of R^* and if 1 is in $D(V)$ then $N(V)$ is a subset of $D(V)$. The symmetric inner product space V is called *round* if $N(V) = D(V)$. In [31] the so-called *Main Theorem* of Pfister forms is proven for LG-rings R:

(a) $\langle\langle a_1, a_2,\ldots,a_n\rangle\rangle$ is a group form.

(b) If $\langle\langle a_1, a_2,\ldots,a_n\rangle\rangle$ is isotropic, then it is a hyperbolic form, i.e. a direct sum of hyperbolic planes.

This paper also proves the *Pure Subform Theorem*:

Let $V = \langle\langle a_1, a_2,\ldots,a_n\rangle\rangle$ and b be a unit. We have that 1 is in $D(V)$

and thus <1> splits V as a orthogonal direct summand with orthogonal complement, say W. Then b is in D(W) if and only if V is isometric to $\langle b, b_2, \ldots, b_n \rangle$ for suitable b_i in the ring R.

The paper [31] also extends the theory of signature spaces over semilocal rings developed by Rosenberg and Ware [37] to LG-rings. Precisely, let R be a LG-ring and X denote its space of signatures. Then the following are equivalent:

(a) X satisfies the Weak Approximation Property.

(b) X satisfies the Strong Approximation Property.

(c) R satisfies the Hasse-Minkowski Property.

(d) Any two n-Pfister spaces are stably linked.

(The definitions of the above terms can be found in [31] .)

The final result for Witt rings given in [31] is that M. Marshall´s [33] 'abstract theory of Witt rings' is satisfied by an LG-ring.

Section VI The Symplectic and Unitary Groups

The symplectic and the unitary groups have been studied over LG-rings in much same fashion as the orthogonal group.

The symplectic group is examined in papers by Kirkwood-McDonald [17] and Ishibashi [11]. Actually, these papers are developed with only the assumption that 2 is a unit in R and that R has stable range one, that is, if a and b are in R with (a,b) = R, then there is a z in R with a + zb = unit.

Let V be a free R-module of finite dimension possessing a nonsingular symplectic form (,) : V x V —> R. We assume that 2 is a unit in R and that the dimension of V is $\geq$ 2. Under these assumptions, it is shown in [17] that V is a direct sum of hyperbolic planes. Let Sp(V) denote the symplectic group on the space V. In [17] it is shown that each element in Sp(V) is a product of Eichler-Siegel-Dickson transvections (and we can bound the number of transvections that appear in the product), that Sp(V) is transitive on both unimodular vectors and on hyperbolic planes, and that cancellation of symplectic spaces is valid. The commutator subgroups are determined together with their transitivity. The approach is based on the work of Klingenberg [20] for symplectic spaces over local rings.

Ishibashi [11] continued this study. Assuming that both 2 and 3 are units in R, Ishibashi classified the normal subgroups of Sp(V) when either the dimension of V is infinite and R is an arbitrary commutative ring or when R has stable range one and the dimension of V is $\geq$ 4. (In the former case, Sp(V) is replaced by Sp'(V). An isometry # of Sp(V) is said to be of finite type if # moves only a finite number of basis elements. The group Sp'(V) consists of all the isometries # of finite type. If the dimension of V is finite then Sp'(V)

= Sp(V). If the dimension of V is infinite, then Sp'(V) is a proper subgroup of Sp(V). If R has stable range one then Sp'(V) is generated by the Eichler-Siegel-Dickson transvections.) The normal subgroups are shown by Ishibashi to be sandwiched between congruence subgroups.

The automorphisms of the symplectic group were determined when R is a local ring in 1974. Recently, Petechuk in 'Isomorphisms of symplectic groups over commutative rings', Algebra and Logic 22 (1983), 397-405 (Russian translation) described the automorphisms of Sp(V) for a commutative ring R and V of dimension 6 or greater. The automorphisms are of standard type.

The theory of unitary groups of Hermitian forms with hyperbolic rank of at least one over a local ring has been developed by D. G. James in [13]. Using James' techniques, Ishibashi extended these results in [10] and obtained generators, congruence subgroups, and the classification of the normal subgroups. Since, in these papers, the ring R need not be commutative, we sketch below the new ring theoretic setting.

Assume that R is a ring with involution * and that V is a left free R-module with ϵ-Hermitian form (,) : V x V —> R. Namely, ϵ is a fixed element in the center of R with $\epsilon\epsilon^* = 1$ and (,) is a sesquilinear form satisfying $(y,x) = \epsilon^*(x,y)^*$ for all x and y in V. The unitary group U(V) of V is the set of all R-linear isomorphisms $t : V \longrightarrow V$ satisfying $(t(x), t(y)) = (x,y)$ for all x and y in V.

An ideal A in R is called involutory if $A = A^*$. To achieve the geometric theory, ring R is assumed to have units 2, 3 and 5 and V is assumed to be nonsingular with hyperbolic rank at least one.

The ring R is said to be <u>full</u> if for each positive integer m and each set of 2 m + 1 elements $\{a, b_1, \ldots, b_m, c_1, \ldots, c_m\}$ in R with

$$aR + b_1R + \ldots + b_mR + c_1R + \ldots c_mR = R,$$

then there exist elements $w_1, \ldots, w_m$ in R such that $a + b_1w_1^* + \ldots + b_mw_m^* + c_1w_1w_1^* + \ldots + c_mw_mw_m^*$ is a <u>unit</u> in R. One observes that this definition is a 'quadratic' type of condition on R assuring the existence of units. Technical difficulties arise since R is not assumed to be commutative and has a involution. Further, we can no longer describe the construction of the units by the substitution morphism. Finally, localization is not available. Nevertheless the idea of the production of units having 'quadratic' form is analogous to the ideas giving rise to the theory of the earlier sections.

Under the above setting (with some modest assumptions on V), Ishibashi shows that the normal subgroups of U(V) are sandwiched between congruence subgroups determined by involutory ideals.

The automorphisms of U(V) for the above setting have not been determined.

References

[1] R. Baeza, Quadratic Forms Over Semilocal Rings, Lecture Notes In Mathematics 655, Springer-Verlag, New-York, 1978.

[2] P. M. Cohn, On the structure of GL(2) of a ring, Inst. Hautes Etud. Sci. 30 (1966), 365-413.

[3] I.G. Connell, Some ring theoretic Schroder-Bernstein theorems, Transactions of the Amer. Math. Soc. 132 (1968), 335-351.

[4] M.H. Dull, Automorphisms of the two-dimensional linear groups over integral domains, American J. Math. 96 (1974), 1-40.

[5] D.R. Estes and R. M. Guralnick, Module equivalences: Local to global when primitive polynomials represent units, J. of Algebra 77 (1982), 138-157.

[6] I.Z.Golubchik, On the general linear group over an associative ring, Uspekhi Mat. Nauk. 28 (1973), 179-180.

[7] K.R. Goodearl and R.B. Warfield, Jr., Algebras over zero-dimensional rings, Math. Ann. 223 (1976), 157-168.

[8] R.M. Guralnick, The genus of a module, J. of Algebra, 18 (1984), 169-177.

[9] H. Ishibashi, Structure of O(V) over full rings, J. of Algebra 75 (1982), 1-9.

[10] , Unitary groups with excellent S and entire E (u,L), J. of Algebra 76 (1982), 442-458.

[11] , Multiplicative and transitive symplectic groups, J. of Algebra 84 (1983), 115-127.

[12] D.J. James, On the structure of orthogonal groups over local rings, American J. Math. 95 (1973), 255-265.

[13] , Unitary groups over local rings, J. of Algebra 52 (1978), 354-363.

[14] W. van der Kallen, The K(2) of rings with many units, Ann. scient. Ec. Norm. Sup. 10 (1977), 473-515.

[15] E.M. Keenan, On the automorphisms of classical groups over local rings, Ph. D. Thesis, Massachusetts Institute of Technology (1965).

[16] B. Kirkwood and B. McDonald, The Witt ring of a full ring, J. of Algebra 64 (1980), 148-166.

[17] , The symplectic group over a ring with one in stable range, Pacific J. of Math. 92 (1981), 111-125.

[18] W. Klingenberg, Lineare Gruppen über lokalen Ringen, American J. Math. 83 (1961), 137-153.

[19] , Orthogonale Gruppen über lokalen Ringen, American J. Math. 83 (1961), 281-320.

[20] , Symplectic groups over local rings, American J. Math. 85 (1963), 232-240

[21] M. Knebusch, A. Rosenberg, and R. Ware, Structure of Witt rings and quotients of abelian group rings, American J. of Math. 94 (1972), 119-155.

[22] , Signatures on semilocal rings, J. of Algebra 26 (1973), 208-249.

[23] B. McDonald, Geometric Algebra Over Local Rings, Pure and Applied Math. # 36, Marcel Dekker, Inc., New York and Basel (1976).

[24] , Automorphisms of GL(n,R), Transactions of the Amer. Math. Soc. 246 (1978), 155-171.

[25] , GL(2) of rings with many units, Communications in Alg. 8 (1980), 869-888.

[26] , Projectivities over rings with many units, Communications in Alg. 9 (1981), 195-204.

[27] ______, Aut(GL(2,R)) for rings with many units, Communications in Alg. 9 (1981), 204-220.

[28] ______, Linear Algebra Over Commutative Rings, Pure and Applied Math. # 87, Marcel Dekker, Inc., New York and Basel (1984).

[29] B. McDonald and B. Hershberger, The orthogonal group over a full ring, J. of Algebra 51 (1978), 536-549.

[30] B. McDonald and B. Kirkwood, The orthogonal and special orthogonal groups over a full ring, J. of Algebra 68 (1981), 121-143.

[31] ______, Transversals and symmetric inner product spaces, to appear: J. of Algebra.

[32] B. McDonald and W. Waterhouse, Projective modules over rings with many units, Proceedings of Amer. Math. Soc. 83 (1981), 455-458.

[33] M. Marshall, Abstract Witt Rings, Queen's Papers in Math. # 57, Queen's University, Kingston, Canada (1980).

[34] M. Ojanguren and R. Sridharen, A note on the fundamental theorem of projective geometry, Commen. Math. Helv. 44 (1969), 310-315.

[35] V.M. Petechuk, Automorphisms of matrix groups over commutative rings, Math. USSR Sb. 45 (1983) 527-542.

[36] I. Reiner, A new type of automorphism of the general linear group over a ring, Ann. of Math. 66 (1957), 461-466.

[37] A. Rosenberg and R.Ware, Equivalent topological properties of the space of signatures of a semilocal ring, Pub. Mathematicae 23 (1977), 283-289.

[38] W.C. Waterhouse, Automorphisms of GL(n,R), Proceedings Amer. Math. Soc. 79 (1980), 347 - 351.

LINEAR MAPPINGS OF MATRIX RINGS PRESERVING INVARIANTS

Bernard R. McDonald
Division of Mathematical Sciences
National Science Foundation
Washington, D. C., U. S. A.

Section I Introduction

Let R denote a commutative ring and $(R)_n$ be the $n \times n$ matrix ring over R. For over 90 years much effort has been devoted to following question and its variations:

> Suppose that $\lambda(A)$ is an invariant defined on matrices A in $(R)_n$. Determine the set of R-linear mappings $T : (R)_n \to (R)_n$ that preserve the invariant λ.

For example, the invariant might be the determinant, that is, $\lambda(A) = \det(A)$, then we would seek all R-linear mappings T satisfying $\det(T(A)) = \det(A)$.

Other invariants might include the trace, various determinantal ideals, rank, etc. Also, similar questions may be asked for specific subrings or subgroups of $(R)_n$ or for mappings other than linear maps. Finally, certain questions may only be answerable for certain classes of scalar rings, that is, algebraically closed fields, domains, etc.

R. Kaya et al. (eds.), Rings and Geometry, 417–436.

The case of the determinant is perhaps best known and was first studied for the case R = $\mathbb{C}$ ($\mathbb{C}$ denotes the complex numbers) by Frobenius in 1897 [1]. Frobenius proved that if $T : (\mathbb{C})_n \to (\mathbb{C})_n$ was a linear map that preserved determinants, then either $T(A) = PAQ$ for all A in $(\mathbb{C})_n$ or $T(A) = PA^+Q$ for all A in $(\mathbb{C})_n$ where P and Q are fixed invertible matrices having $\det(PQ) = 1$ and A^+ denotes the transpose of A.

Many questions concerning linear mappings of matrix rings that preserve specified invariants, including the above determinant preservers, were noted by M. Marcus [10] to be answered once the classification of the linear mappings that preserve rank one matrices was determined. Indeed, we need only know where the elementary matrix units are carried by the linear mapping T in order to determine T since if E_{st} is an elementary matrix unit, then E_{st} has rank one and if $[a_{st}]$ is a matrix, then

$$T([a_{st}]) = T(\sum_{s,t} a_{st}E_{st}) = \sum_{s,t} a_{st}T(E_{st}).$$

Consequently, T is uniquely determined by its action on E_{st}.

Suppose that k is an algebraically closed field of characteristic zero. In 1959 Marcus and Moyls [11] proved that, if $T : (k)_n \to (k)_n$ is a k-linear map with the property that whenever $\text{rank}(A) = 1$ then $\text{rank}(T(A)) = 1$, then T has the form $T(A) = PAQ$ for all A in $(k)_n$ or $T(A) = PA^+Q$ for all A in $(k)_n$ where P and Q are invertible matrices. A careful examination of the proof of Marcus and Moyls shows that if T is an <u>invertible</u> linear transformation, then their theorem is valid for any commutative field of any characteristic. The Marcus-Moyls' result was proven by the use of tensor products of vector spaces and multilinear algebra. An elementary matrix theoretic proof of the same result (over an algebraically closed field) was given by Minc [14] in 1977.

There has been considerable interest in this problem. In addition to the above mentioned survey of Marcus [10], there is an earlier survey by Marcus [9] and, in 1977, a survey in the Ph.D. thesis of Robert Grone [3] which lists 103 related papers on (principally) linear mapping problems over fields. Even here, Grone fails to list the extensive literature concerning the automorphism theory of the classical linear groups which also is relevant to these problems. The approach of these papers is primarily linear and multilinear theoretic in contrast to the geometric-algebraic methods present in the automorphism theory of classical groups or the new group scheme methods introduced by Waterhouse in [16]. Stephen Pierce of the University of Toronto has indicated that he is preparing a survey of the results concerning the invariant preserving linear mappings that have appeared in the literature since 1977.

One should observe that the thrust of these ideas provides a converse of the classical invariant theory. Here an invariant is specified and the structure of the set of invertible mappings preserving that invariant is characterized. In classical invariant theory, a transformation group is given and the hope is to determine an algebraically independent set of invariants that in some fashion generates all the invariants of the group action.

In 1980 D. J. James [7] announced the classification of the linear mappings of $(R)_n$ (where R was an integral domain) that preserved the determinant. It was James' paper that initiated our interest in the classification of invariant preserving linear maps of matrix rings over commutative rings. At the same time, knowing of James' result, W. Waterhouse began work on the problem. In 1983 [13] we classified the rank one and the determinant preserving, invertible linear maps of $(R)_n$ over an arbitrary commutative ring. At the same time, Waterhouse [16] announced the analogous result. Although producing the same conclusion, the two proofs are very different in style and approach. We utilized a linear algebraic approach, extending the original work of Marcus-Moyls, while Waterhouse introduced new and exciting techniques from the theory of group schemes. In the remaining parts of this paper, we discuss briefly both approaches.

Section II The Linear Algebraic Approach of McDonald, Marcus, and Moyls

Extending the work of Marcus and Moyls involved three ingredients: the concept of a rank one matrix, the concept of the transpose, and the concept of the equivalence transformation. These ideas will be discussed below. Greater detail is provided in [13].

If R is a commutative ring, then denote by B(R) the Boolean algebra of idempotents in R. Let e be an idempotent of R. Then e induces a natural ring decomposition of R as $R = R_1 \times R_2$ where $R_1 = Re$ and $R_2 = R(1 - e)$. This decomposition of R induces a natural decomposition of the matrix ring as $(R)_n = (R_1)_n \times (R_2)_n$. If A is a matrix in $(R)_n$, then denote A's decomposition relative to the idempotent e by $A = \langle A_1, A_2\rangle$. Applying the transpose to the second coordinate A_2 of A, determines an invertible R-linear mapping $\Omega_e : (R)_n \to (R)_n$ defined by $\Omega_e(A) = \Omega_e(\langle A_1, A_2\rangle) = \langle A_1, (A_2)^+\rangle$. Denote the set $\{\Omega_e \mid \text{where } e \text{ is an idempotent in } R\}$ by $Inv(R)$. The arithmetic of the Boolean algebra of B(R) induces a group structures on $Inv(R)$. This group is called group of involutions on the spectrum of R. The mapping Ω_e provides the generalization of the transpose. Note that $\Omega_1 = I$ is the identity mapping and that $\Omega_o = (\ )^+$ is the standard transpose mapping.

A finitely generated projective R-module P is said to be a rank one projective module if the localization of P, that is P_w, is a free R_w-module of R_w-dimension one for each prime ideal w of R. For background concerning the theory of rank one projective R-modules see [12]. The rank one projective modules serve as the generalization of the concept of a line where we understand a 'line' to mean a free R-module of R-dimension one. If P is a rank one projective, then let [P] denote the R-isomorphism class of P. Let $Pic(R)$ denote the set of R-isomorphism classes of rank one projective R-modules. The set $Pic(R)$ is a multiplicative abelian group under $[P] \circ [Q] = [P \otimes Q]$. Under this multiplication, induced by the tensor product $\otimes$, the identity $1 = [R]$ and if $P^* = Hom_R(P,R)$ is the dual module, then $[P]^{-1} = [P^*]$.

If A is a matrix over a field k, then A is said to have rank one if its column space has k-dimension one, that is, its column space is a line. Extending this idea, we say that a matrix A over a commutative ring R is of rank one if the R-module generated by the columns of A is a rank one projective R-module. R. Gilmer and R. Heitmann [2] recently characterized rank one matrices. They show the following:

> Let $A = [a_{rs}]$ be an $n \times n$ matrix over R.
>
> Then the following are equivalent:
>
> (a) A has rank one.
>
> (b) (1) The ideal generated by the elements a_{rs} is all of R, and
>
> (2) All 2×2 submatrices of A have determinant equal to 0.

This result can be proven by a localization argument and makes the rank one matrices relatively easy to identify and compute.

The above ideas prove to be correct generalizations of the transpose mapping and the rank one matrix. The generalization of the equivalence transformation is more technical.

Let P and Q be invertible matrices in $(R)_n$. The mapping

$$E_{(P,\ Q)} : (R)_n \to (R)_n$$

given by $E_{(P,\ Q)}(A) = PAQ^{-1}$ for all A in $(R)_n$ is called an <u>equivalence transformation</u> of $(R)_n$. This mapping is an invertible R-linear transformation; further, the set of such transformations, denoted by *Equiv*$(R,\ n)$, forms a multiplicative group whose elements satisfy

(a) $E_{(P,\ Q)}E_{(U,\ V)} = E_{(PU,\ QV)}$.

(b) $[E_{(P,\ Q)}]^{-1} = E_{(U,\ V)}$ where $U = P^{-1}$ and $V = Q^{-1}$.

(c) $E_{(P,Q)} = E_{(U,V)}$ if and only if $P = aU$ and $Q = aV$ where a is a unit of R.

Property (c) indicates that the equivalence transformation $E_{(P,Q)}$ may be identified with the set $\{p(P,Q) \mid p \text{ a unit in } R\}$ in the free R-space $(R)_n \oplus (R)_n$. Thus, actually we may identify the equivalence transformation with the line $\{p(P,Q) \mid p \text{ in } R\}$ generated by the 'unimodular' vector (P,Q). Let $L_{(P,Q)}$ denote this line. Thus, $L_{(P,Q)}$ is a free R-submodule of $(R)_n \oplus (R)_n$ of R-dimension one. Further, since $E_{(P, Q)}E_{(U, V)} = E_{(PU, QV)}$, then there is an natural multiplication of these lines given by $L_{(P, Q)}L_{(U, V)} = L_{(PU, QV)}$. Finally, under this multiplication there is an identity, namely the line $L_{(I, I)}$. What then should serve as a generalization of an equivalence transformation? By the above remarks, we will begin by looking at R-submodules of $(R)_n \oplus (R)_n$ and the natural multiplication induced on these modules by the matrix multiplication. Let N denote the collection of all R-submodules of $(R)_n \oplus (R)_n$. If U and V are in N, define a multiplication (induced by the matrix multiplication) by

$$UV = \{ \Sigma uv \mid u \text{ in } U \text{ and } v \text{ in } V \}$$

where the summation Σ extends over all finite sets of products uv where u is in U and v is in V. Let E denote the R-submodule generated by (I, I), i.e., E is the set of all $a(I, I) = (aI, aI)$ for a in R. An R-submodule U is said to be <u>invertible</u> if there is a submodule V with $UV = VU = E$. The idea of invertibility is not new and was recently discussed with considerable clarity by M. Isaacs for submodules of $(R)_n$ in [5]. Let G denote the set of invertible R-submodules of $(R)_n \oplus (R)_n$. The following results are proven in [13] and summarize the properties of G and its elements:

Let U and V be in G. Then

(a) U is a finitely generated, rank one projective R-module.

(b) If UV = E, then VU = E. The set G is a group. If $U \subseteq V$, then U = V. If UV = E, then V is called the inverse of U and is denoted by U^{-1}.

(c) If t and u are in U and v and w are in U^{-1}, then uv = vu, tu = ut, and vw = wv.

(d) If P and Q are invertible matrices in $(R)_n$, then $L_{(P, Q)}$ is an invertible submodule in G which is free of R-dimension one.

(e) Let X be a projective R-module of rank one. Then X is isomorphic to an element of G if and only if R^n is isomorphic to $X \oplus \ldots \oplus X$ (n summands).

Let G_o denote the subgroup of invertible R-submodules of G of the form $L_{(P, Q)}$ where P and Q are invertible matrices in $(R)_n$. The mapping $E_{(P, Q)} \to L_{(P, Q)}$ determines a group isomorphism between G_o and $Equiv(R, n)$. Further,

(f) The following are equivalent:

(1) $U \cong V$ as R-modules.

(2) U = (P, Q)V where P and Q are invertible matrices in the ring $(R)_n$.

(3) The cosets G_oU and G_oV are equal.

(g) The subgroup G_o is normal in G and the quotient group G/G_o is an abelian group. This group is denoted by $Pic(R, n)$.

(h) $UV \cong U \otimes V$. If U is in G then $U^n = L_{(P, Q)}$ for invertible matrices P and Q.

(i) A submodule W of $(R)_n \oplus (R)_n$ is invertible if and only if its localization is invertible for each prime ideal of R. Further, if R is a local ring, then each invertible submodule has the form $L_{(P, Q)}$.

Suppose that U is an invertible submodule with inverse $V = U^{-1}$. Since $UV = E$, select u_i in U and v_i in V with $\Sigma u_i v_i = (I, I)$. Since the u_i and v_i are pairs of matrices in $(R)_n \oplus (R)_n$, each may be written as $u_i = \langle s_i, t_i \rangle$ and $v_i = \langle y_i, z_i \rangle$. We now define the mapping $E_U : (R)_n \to (R)_n$ by

$$E_U(A) = \sum_i s_i A z_i.$$

Clearly, E_U is an R-linear mapping. This mapping E_U will serve as the appropriate generalization of the equivalence transformation. Denote the set of all E_U for U in G by $EQUIV(R, n)$. The following properties of E_U are given in [12] or [13] :

(a) The localization of E_U at each prime ideal is an equivalence transformation. In particular, this implies that E_U is an R-isomorphism.

(b) E_U depends only on U and not on the choice of elements u_i and v_i.

(c) If $T : (R)_n \to (R)_n$ is an R-linear transformation such that at each localization at a prime w in R, the map T_w is an equivalence transformation, then there is an invertible submodule U with $T = E_U$.

(d) There is a natural exact sequence of groups

$$1 \to \mathit{Equiv}(R, n) \to \mathit{EQUIV}(R, n) \to \mathit{Pic}(R, n) \to 1$$

where $E_{(P, Q)} \to E_U$ for $U = L_{(Q, P)}$ and $E_U \to [U]$.

Let $\mathit{RANK}(R, n)$ denote the group of invertible R-linear transformations of $(R)_n$ that preserve rank one matrices. Let $\mathit{Rank}(R, n)$ denote the semidirect product of the groups $\mathit{Inv}(R)$ and $\mathit{Equiv}(R, n)$, that is the classical rank one preserving invertible linear mappings. The central result of [13] is that the following diagram is commutative and has exact rows and split exact columns:

$$\begin{array}{ccccccccc} & & 1 & & 1 & & 1 & & \\ & & \downarrow & & \downarrow & & \downarrow & & \\ 1 & \to & \mathit{Equiv}(R, n) & \to & \mathit{EQUIV}(R, n) & \to & \mathit{Pic}(R, n) & \to & 1 \\ & & \downarrow & & \downarrow & & \downarrow & & \\ 1 & \to & \mathit{Rank}(R, n) & \to & \mathit{RANK}(R, n) & \to & \mathit{Pic}(R, n) & \to & 1 \\ & & \downarrow & & \downarrow & & \downarrow & & \\ 1 & \to & \mathit{Inv}(R) & \to & \mathit{Inv}(R) & \to & 1 & & \\ & & \downarrow & & \downarrow & & & & \\ & & 1 & & 1 & & & & \end{array}$$

where

$$\mathit{Rank}(R, n) = \{E_U \Omega_e \mid U \text{ is in } G,\ e \text{ is in } B(R)\}.$$

and the maps are natural. In particular, if T is an invertible linear mapping that preserves rank one matrices, then there is an idempotent e and an invertible R-module U with $T = \Omega_e E_U$. In [13] it is shown how the form of the determinant preserving linear maps follows from the result above (a linear map preserving the determinant is always invertible) and, consequently, a modern form of Frobenius's theorem.

The proof of this result requires first a proof of the Marcus-Moyls' result in the case of a local ring. The principal technique is to 'lift' the known field version through the maximal ideal to the local ring. Second, localization and commutative algebra techniques are utilized to obtain a global version. The details may be found in [13] with the necessary background in [12].

Section III The Group Scheme Approach of Waterhouse.

This section is based on the paper [16] by W. Waterhouse. We borrow extensively from his exposition. Additional background may be found in Waterhouse's text Introduction to Affine Group Schemes [15]. Suppose R is a commutative ring and $X = [X_{st}]$ is an n x n matrix commuting of indeterminates. The determinant $\det(X_{st})$ may be viewed as a homogeneous polynomical of degree n in the n^2 variables X_{st} when $n \geq 2$. The question is then, "What linear changes of variables will preserve that form?" Waterhouse views any problem of this type as a question of affine group schemes. By an 'affine group scheme' G, we mean any subgroup of the general linear group $GL(-, n)$ defined by a set of polynomial equations on the matrix entries. In other words, there is a fixed commutative ring R, a fixed set of polynomials in $R[(X_{st})]$, and for each commutative R-algebra A, there is a matrix group G(A) whose elements $[y_{st}]$ are in $GL(A, n)$ and satisfy the polynomials. For each R algebra mapping $A \to B$, there is an induced group morphism $G(A) \to G(B)$, and G is functorial. In many cases, the set of polynomials under consideration have integer coefficients and we may assume that $R = Z$. If this is the case, then G is defined for all commutative rings.

To illustrate this idea of an affine group scheme, suppose that f is a fixed polynomial in $R[X_1, \ldots, X_w]$. We may ask for the linear changes of variables of the form $TX_i = \sum_t a_{it}X_t$ that satisfy $f(T(X)) = f(X)$, i.e. those changes that preserve the polynomial f. This question can be asked for any R-algebra A, thus it defines subsets, denoted by $Aut(A, f)$ and called the 'automorphisms of f' inside of $GL(A, w)$. These changes of variables induce a group action of the general linear group $GL(A, w)$ on the polynomial ring $A[X_1, \ldots, X_w]$. Thus, the 'stabilizer' $Aut(A, f)$ of f is a subgroup of $GL(A, w)$. Further, explicitly calculating $f(TX)$, one produces a polynomial where the coefficients of the powers of X are polynomial expressions involving the matrix entries in T and coefficients of f. Consequently, the stablizer condition $f^{0}T = f$ is equivalent to certain polynomial equations over the matrix entries $[y_{st}]$ of T and fixed coefficients over R. Therefore, $Aut(-, f)$ is an affine group scheme over R.

Instead of preserving a polynomial, one might wish to preserve the condition that one or more polynomials are zero. For example, letting $R = Z$ and using the Gilmer-Heitmann characterization of rank one matrices given in Section II, we might wish to preserve the condition (b)(2) which states that all the determinants of the 2 x 2 submatrices of (X_{st}) be simultaneously zero. Viewing the polynomials as formal expressions, we seek changes of variables T such that the determinant of each 2 x 2 submatrix of TX is a sum of (polynomial) multiples of the determinants of the 2 x 2 submatrices of X. This is equivalent to passing to the ideal I in $Z[X_{11}, X_{12}, \ldots, X_{nn}]$ generated by the determinants of the 2 x 2 submatrices of $[X_{st}]$. The mapping T <u>preserves the ideal I</u> if $f^0 T$ is in I for every f in I. Waterhouse ([16], (1.4.1)) shows that if T preserves an ideal I, then so does T^{-1}. This implies that the set of T preserving the ideal is a group. Unfortunately, in general, such a group may not be an affine group scheme. Waterhouse proves the following result, providing a test as to when a group preserving an ideal is actually an affine group scheme.

Let V be a module of all polynomials in $R[X_1, \ldots, X_w]$ of degree at most r. Let I be an ideal generated by the elements of degree at most r. Suppose there is a basis $f_1, \ldots, f_u$ of V such that among this basis the elements $f_1, \ldots, f_s$ span $I \cap V$. Then the condition of preserving I defines an affine group scheme of R-algebras.

Note that if R is a field, then the condition above is always satisfied.

As Waterhouse notes, a limited amount of knowledge suffices to determine if a set of linear transformations is an affine group scheme. Often it is not necessary to consider arbitrary commutative rings in that many results and proofs need be only stated and verified over an algebraically closed field.

Continuing the example above, let $H_2(R)$ denote the invertible linear maps over R that preserve the ideal generated by all the determinants of the 2 x 2 submatrices of $[X_{st}]$. In [19] Waterhouse shows that $H_2(R)$ equals $Rank(R, n)$ and by the above result one can conclude that $Rank(\cdot, n)$ is an affine group scheme.

To state Waterhouse's main theorem we need some preliminaries on Lie algebras. Suppose that R = k is a field and suppose that G is an affine group scheme over k. Construct the ring $k[\varepsilon]$ where ε satisfies $\varepsilon\varepsilon = 0$. The elements in $G(k[\varepsilon])$ that map to the identity in G(k) under $\varepsilon \to 0$ form a k-vector space which is called the Lie algebra of G and denoted by $Lie(G)$.

The dimension, dim(G), of an affine group scheme G intuitively represents the number independent parameters necessary to represent an element of G -- a precise definition is given in ([15], p. 88). The k-dimension of $Lie(G)$ as a vector space is greater than or equal to the dim(G). Inside G can be found a normal subgroup G^0 of G called the connected component of the identity. The following theorem of Waterhouse describes group schemes of 'finite type'. The subgroups of the general linear group that are under consideration are of finite type. Also, the 'flatness' as described in the theorem is automatic in our case. Finally, if the algebras are restricted to a specified field L, then the restriction will be denoted by G_L.

(Waterhouse [15], Theorem 1.6.1) Let G and H be affine group schemes of finite type over Z with G flat. Let $\phi : G \to H$ be a homomorphism.

Suppose that for all algebraically closed fields L, we can show

(a) $\dim(G_L) \geq \dim_L(Lie(H_L))$,

(b) the mappings $G(L) \to H(L)$ and $G(L[\varepsilon]) \to H(L[\varepsilon])$ given by ϕ are injective, and

(c) all elements inside H(L) normalizing $\phi(G^0(L))$ are in $\phi(G(L))$.

Then ϕ maps G(R) isomorphically onto H(R) for every commutative ring R.

Many group schemes arise naturally as quotients of other group schemes. Waterhouse notes that there is a problem with such a quotient construction. Suppose that N is a normal subgroup scheme of G over Z-algebras. Thus, for each commutative ring R the group N(R) is a normal subgroup of G(R). The group scheme quotient can be defined by utilizing a universal mapping property (See [15], Section 15.4). In particular, any group scheme morphism $G \to H$ having kernel N will give an injection of G/N into H. However, in general, for a commutative ring R, the group (G/N)(R) will be larger than the group G(R)/N(R). This arises because some of the elements of (G/N)(R) may actually come from G(S) where $R \to S$ is a ring extension and not from the elements in G(R). This failure of the surjectivity of the map $G(R) \to (G/N)(R)$ is measured by certain cohomology sets $H^1(R, G)$. These sets are related by an exact sequence of the form

$$1 \to N(R) \to (G/N)(R) \to H^1(R, N) \to H^1(R, G) \to H^1(R, G/N)$$

In the case of $G = GL(-, 1)$, which is the linear automorphism group of the 'line', i.e., the free module of dimension one, then $H^1(R, GL(R, 1))$ is the Picard group $Pic(R)$ of rank one projective

R-modules.

We now examine Waterhouse's description of $Rank(R, n)$. Consider the group scheme $H = GL(-, n) \times GL(-, n)$ over $\mathbf{Z}$. For any commutative ring R there is a copy of $GL(R, 1)$ in H consisting of the set $\{(aI, aI) \mid$ where a is a unit in R$\}$. Thus, there is a group scheme quotient $G^0 = H/GL(-, 1)$. In addition the cyclic group of order 2, namely $\mathbf{Z}/2\mathbf{Z}$, acts on H by interchanging the factors. This action of the cyclic group induces an action on G^0 and permits the construction of a semidirect product. Let G denote the semidirect product of G^0 and $\mathbf{Z}/2\mathbf{Z}$. Waterhouse shows that G is isomorphic to $Rank(-, n)$ (Theorem 3.4 [16]). Since G^0 is a group scheme quotient, the group morphism $GL(R, n) \times GL(R, n) \to G^0(R)$ may not be surjective. The obstruction is measured, as noted above, by a cohomology sequence

$$1 \to GL(R, 1) \to GL(R, n) \times GL(R, n) \to G^0(R) \to$$
$$H^1(R, GL(-, 1)) \to H^1(R, GL(-, n) \times GL(-, n))$$

induced by the group scheme exact sequence

$$1 \to GL(-, 1) \to GL(-, n) \times GL(-, n) \to G^0 \to 1.$$

We have noted that $H^1(R, GL(-, 1))$ is the Picard group of R. Since

$$H^1(R, GL(-, n) \times GL(-, n)) = H^1(R, GL(-, n)) \times H^1(R, GL(-, n)))$$

we need only examine $H^1(R, GL(R, n))$ which classifies the projective R-modules of rank n. Waterhouse describes how this and the above sequence give rise to the projective modules representing the invertible R-submodules of $(R)_n$ described in Section II. In doing this he gives a complete description of $Rank(R, n)$.

Section IV Concluding Remarks.

In [16] Waterhouse describes how his group scheme approach can also be utilized in related questions concerning the automorphisms of Pfaffian ideals and symmetric determinants.

The trace is an natural invariant of $(R)_n$. In 1977 [8] Kovacs described the trace preserving linear transformations over a field. If A is a matrix over a field k, then let $L_A : (k)_n \to (k)_n$ be the linear (left) multiplication mapping given by $L_A(X) = AX$. Let $\sigma_A(X) = [A,X] = AX - XA$ and let I denote the identity transformation. Kovacs proves that any linear transformation T of $(k)_n$ to itself that preserves the trace may be represented as $T = I + \sum_{s,t} \sigma_{A(s)} L_{B(t)}$. This set of trace preserving linear mappings contains maps that are not invertible. Even if one reduces to a reasonably natural subset that is composed of invertible linear mappings (See Kovacs [8]), it may be that T does not have the form $X \to PXQ$ or $X \to PX^tQ$. The subject of the trace preserving invertible linear mappings would seem to deserve additional study, especially due to the role of the trace in classical invariant theory.

In 1948 [4] L. K. Hua determined the bijective (not necessarily linear) coherence preserving mappings of $(k)_n \to (k)_n$. Jacob [6] in 1955 extended this characterization to tensor products of vector spaces over disvision rings. Two matrices A and B are <u>coherent</u> if their difference A - B has rank one. One the surface, 'Coherence' might appear to be an artificial definition; however, it is related to special relativity. Suppose we have a space-time point, i.e., a space-time 'event', consisting of three space coordinates and one time coordinate. Two such events are called coherent if their spacial distance is equal to the product of the time needed to traverse the distance between the points times the speed of light. A space-time event may be represented as a 2 X 2 Hermitian matrix. When this is done, two such events are coherent when the difference of their matrix representations has rank one. Thus, matrix coherence has a natural origin. In concluding, it is worthwhile to make two observations. First, little additional work has been done on bijective coherence preserving linear maps and, to our knowledge, no work on this problem work over commutative rings. Second, Jacob shows the above result may be formulated over division rings. This opens the possibility of the study of coherence and similar related questions over noncommutative scalar rings analogous to the current research concerning the automorphisms of the general linear group over noncommutative scalar rings.

References

[1] G. Frobenius, Über die Darstellung der endlichen Gruppen durch lineare Substitutionen, S.-B. Preuss. Akad. Will. Berlin (1897), 994-1015.

[2] R. Gilmer and R. Heitmann, On Pic (R[X]) for R Seminormal, J. Pure Appl. Alg. 16 (1980), 251-257.

[3] R. Grone, Isometries of Matrix Algebras, Ph.D. Thesis, University of California, Santa Barbara.

[4] L. K. Hua, A theorem on matrices over a field and its applications, J. Chinese Math. Soc. (N. S.) 1 (1951), 110-163

[5] I. M. Isaacs, Automorphisms of matrix algebras over commutative rings, Linear Alg. and Appl. 31 (1980), 215-231

[6] H. G. Jacob, Jr., Coherence invariant mappings on Kronecker products, Amer. J. Math. 77 (1955), 177-189.

[7] D. James, On the automorphisms of det(X_{ij}), Math. Chronicle 9 (1980), 35-40

[8] A. Kovacs, Trace preserving linear transformations on matrix algebras, Lin. Multilin. Alg. 4 (1977), 243-250.

[9] M. Marcus, Linear operations on matrices, Amer. Math. Monthly 69 (1962), 837-847.

[10] ________. Linear operations on matrices, J. Nat. Bureau Standards 75 B (1971), 107-113.

[11] M. Marcus and B. Moyls, Transformations on tensor product spaces, Pacific J. of Math. 9 (1959), 1215-1221.

[12] B. McDonald, Linear Algebra Over Commutative Rings. Marcel Dekker, Inc., New York and Basel (1984).

[13] ________, R-linear endomorphisms of $(R)_n$ preserving invariants, Memoir of the Amer. Math. Soc. 287 (1983). Providence, Rhode Island.

[14] H. Minc, Linear transformations on matrices: Rank 1 preservers and determinant preservers, Lin. Multilin. Alg. 4 (1977), 265-272.

[15] W.C. Waterhouse, Introduction to Affine Group Schemes. Springer-Verlag, New York (1979).

[16] ______________, Automorphisms of $\det(X_{ij})$: The group scheme approach, to appear: Adv. in Math..

[17] ______________, Twisted forms of the determinant, J. of Alg. 86 (1984), 60-75.

[18] ______________, Invertibility of linear maps preserving matrix invariants, Lin. Multilin. Alg. 13 (1983), 105-113.

[19] ______________, On linear transformations preserving rank one matrices over commutative rings, to appear, Lin. Multilin. Alg.

KINEMATIC ALGEBRAS AND THEIR GEOMETRIES

Helmut Karzel and Günter Kist

Technische Universität München

ABSTRACT. In this chapter we will discuss kinematic algebras and their applications in geometry. We begin (§1) with a survey of the development of kinematics. The subjects covered in this paper will be found in §2.

§ 1 MOTIVATION AND HISTORICAL REVIEW

Kinematics is the theory of motions of a metric geometry $(P,\mathfrak{L},\equiv)$. If the metric structure is given by a congruence-relation $\equiv$ then a <u>motion</u> β is a permutation of the point set P such that if $X \in \mathfrak{L}$ is a line then $\beta(X) \in \mathfrak{L}$ is a line and if $a,b \in P$ are any two points, then $(\beta(a),\beta(b)) \equiv (a,b)$. For the study of kinematics one has used additional geometric and algebraic structures which have led to the notions "kinematic space" and "kinematic algebra". The origin of kinematic algebra are already found in the works of L. EULER and C.F. GAUSS.

The metric geometry we know best is euclidean geometry and since GAUSS we have known a nice algebraic representation of the euclidean plane. Let $(\mathbb{C},\mathbb{R})$ be the quadratic field extension consisting of the complex and real numbers $\mathbb{C}$ and $\mathbb{R}$. Then by an <u>euclidean derivation</u> $\mathcal{A}(\mathbb{C},\mathbb{R},^-)$ we obtain the euclidean plane if we consider $\mathbb{C}$ as the set of points, the set $\mathfrak{L} := \{a + \mathbb{R}b \mid a,b \in \mathbb{C}, b \neq 0\}$ as the set of lines, and if we define the congruence relation $\equiv$ by
$(a,b) \equiv (c,d) :\Leftrightarrow (a-b)\overline{(a-b)} = (c-d)\overline{(c-d)}$.

Any motions of the euclidean plane $\mathcal{A}(\mathbb{C},\mathbb{R},^-)$ can be composed of the following elementary maps:

$^- : \mathbb{C} \longrightarrow \mathbb{C}\ ;\ z = x + iy \longrightarrow \overline{z} = x - iy$

$a^+ : \mathbb{C} \longrightarrow \mathbb{C}\ ;\ z \longrightarrow a + z$ where $a \in \mathbb{C}$

$b^\cdot : \mathbb{C} \longrightarrow \mathbb{C}\ ;\ z \longrightarrow b \cdot z$ where $b \in \mathbb{C}_1 := \{z \in \mathbb{C} \mid z\overline{z} = 1\}$

R. Kaya et al. (eds.), Rings and Geometry, 437–509.

and $\mathfrak{B}^+ := \{a^+b^\bullet \mid a \in \mathbb{C}, b \in \mathbb{C}_1\}$ is the group of all proper motions while $\mathfrak{B}^- = \mathfrak{B}^+ \circ \bar{\ }$ is the coset of all improper motions. Thus $\mathfrak{B} = \mathfrak{B}^+ \cup \mathfrak{B}^-$ is the group of all motions of $A(\mathbb{C},\mathbb{R},\bar{\ })$. All rotations fixing a distinct point form a commutative subgroup of M which is isomorphic to the multiplicative group $(\mathbb{C}_1,\cdot)$. Since "$z = x + iy \in \mathbb{C}_1 \Leftrightarrow x^2 + y^2 = 1$", any rotation is determined by its fixed point and a pair (x,y) of real numbers with $x^2 + y^2 = 1$. As a quadratic field extension $(\mathbb{C},\mathbb{R})$ is a vector space of dimension 2 so that the associated projective space $\Pi(\mathbb{C},\mathbb{R}) = (\mathbb{C}^*/\mathbb{R}^*, \{\mathbb{C}^*/\mathbb{R}^*\})$, $\mathbb{C}^* := \mathbb{C} \setminus \{0\}$, $\mathbb{R}^* := \mathbb{R} \setminus \{0\}$ is a line. But $(\mathbb{C}^*/\mathbb{R}^*,\cdot)$ is also a group, namely the factor group and the map

$$\varkappa : (\mathbb{C}_1,\cdot) \longrightarrow \mathbb{C}^*/\mathbb{R}^*;\ e^{i\varphi} \longrightarrow \mathbb{R}^* e^{i\frac{\varphi}{2}} \text{ is an isomorphism.}$$

We will call $\varkappa$ a <u>kinematic map</u> because all rotations fixing a distinct point are mapped bijectively on the points of the projective line.
For $z \in \mathbb{C}_1$, $z \neq -1$ we have also $\varkappa(z) = \mathbb{R}^*(1 + z)$. Therefore these considerations can be extended to any separable quadratic field extension (L,K) if we denote by $\bar{x}$ the image of $x \in L$ by applying the K-automorphism of the field L which is different from the identity.

For the 3-dimensional euclidean space $\mathbb{R}^3$, EULER considered all rotations ρ fixing a distinct point O. He assigned to any quadrupel $(a_0,a_1,a_2,a_3) \in \mathbb{R}^4$ and $a_0^2 + a_1^2 + a_2^2 + a_3^2 = 1$ a rotation ρ. EULER's formula can be written in a short form by using quaternions as CAYLEY showed. For this purpose we have to identify the point set $\mathbb{R}^3$ with the set $J := \{x \in \mathbb{H} \mid x + \bar{x} = 0\} = \{x_1 i + x_2 j + x_3 k \mid x_1,x_2,x_3 \in \mathbb{R}\}$ of pure quaternions. Here for $a = a_0 + a_1 i + a_2 j + a_3 k \in \mathbb{H}$, $\bar{a}$ denotes the conjugate quaternion $\bar{a} = a_0 + (-a_1)i + (-a_2)j + (-a_3)k$.
Now let $a \in \mathbb{H}$ be a quaternion with $1 = a\bar{a} = a_0^2 + a_1^2 + a_2^2 + a_3^2$, then

<u>(1)</u> $a_\iota : J \longrightarrow J;\ x \longrightarrow ax\bar{a} = axa^{-1}$

defines the rotation ρ with the axis $\mathbb{R}(a_1,a_2,a_3) = \mathbb{R}(a_1 i + a_2 j + a_3 k)$ and the angle $2 \arccos a_0$, and we see at once that any rotation fixing O can be described by the formula (1) of EULER, CAYLEY.

As in the case of complex numbers, let $\mathbb{H}_1 := \{x \in \mathbb{H} \mid x\bar{x} = 1\}$ be the set of all quaternions of length 1. Then $\mathbb{H}_1$ is a subgroup of the multiplicative group $(\mathbb{H}^*, \cdot)$ and $a, b \in \mathbb{H}_1$ define the same rotation $a_\iota = b_\iota$ if $a = b$ or $a = -b$. If we denote as usual by O_3^+ the group of all rotations fixing 0, then we see that O_3^+ is isomorphic to the factor group $\mathbb{H}_1 / \{1, -1\}$. Since $\mathbb{R}$ is the center of $\mathbb{H}$, $(\mathbb{R}^*, \cdot)$ is a normal subgroup of $(\mathbb{H}^*, \cdot)$ and we can form the factor group $\mathbb{H}^*/\mathbb{R}^*$. For $a \in \mathbb{H}^* := \mathbb{H} \setminus \{0\}$ let $b = \sqrt{a\bar{a}}^{-1} \cdot a$, then $b \in \mathbb{H}_1$ and $\mathbb{R}^* a = \mathbb{R}^* b$. This shows that the factor group $\mathbb{H}_1 / \{1, -1\}$ and $\mathbb{H}^*/\mathbb{R}^*$ are isomorphic hence

(2) $\quad O_3^+ = \mathbb{H}_1 / \{1, -1\} \cong \mathbb{H}^*/\mathbb{R}^*$

But the pair $(\mathbb{H}, \mathbb{R})$ is also a 4-dimensional vector space, and so the <u>projective derivation</u> $\Pi(\mathbb{H}, \mathbb{R})$ gives us a 3-dimensional projective space. If $\varphi : \mathbb{H}^* \longrightarrow \mathbb{H}^*/\mathbb{R}^*;\ x \longrightarrow \mathbb{R}^* x$ denotes the canonical map and $\mathfrak{S}_2$ the set of all 2-dimensional vector subspaces of $(\mathbb{H}, \mathbb{R})$, then $\Pi(\mathbb{H}, \mathbb{R}) = (\mathbb{H}^*/\mathbb{R}^*, \mathfrak{L})$ is the derived projective space with the point set $\mathbb{H}^*/\mathbb{R}^*$ and the line set $\mathfrak{L} = \{\varphi(S^*) \mid S \in \mathfrak{S}_2\}$ (in this paper, if T is any subset of an additively written group, we always set $T^* := T \setminus \{0\}$).
This consideration leads us to the observation

(3) There is a kinematic map $\varkappa : O_3^+ \longrightarrow \mathbb{H}^*/\mathbb{R}^*;\ a_\iota \longrightarrow \mathbb{R}^* a$ which is a bijection between the group O_3^+ of rotations fixing 0 and the 3-dimensional projective space over the reals.

To sum up one can say that the set $G := \mathbb{H}^*/\mathbb{R}^*$ is provided with an incidence structure $\mathfrak{L}$ as well as with a structure of a group " $\cdot$ ", and one can prove that both structures are compatible according to the following conditions:

(L) For all $a \in G$ the map $a_\ell : G \longrightarrow G;\ x \longrightarrow ax$ is an automorphism of the incidence structure $\mathfrak{L}$ i.e. $a_\ell \in \mathrm{Aut}(G, \mathfrak{L})$

(R) $\forall a \in G,\ a_r : G \longrightarrow G;\ x \longrightarrow xa : a_r \in \mathrm{Aut}(G, \mathfrak{L})$

(F) $\forall X \in \mathfrak{L}$ with $1 \in X$, X is a subgroup of $(G, \cdot)$

In general, we call a triple $(G, \mathfrak{L}, \cdot)$ where $(G, \mathfrak{L})$ is an incidence space (i.e. any two distinct points $a, b \in G$ can be joined by exactly one line $L \in \mathfrak{L}$ and $|X| \geq 2$ for all $X \in \mathfrak{L}$)

and $(G,\cdot)$ a group, an <u>incidence group</u>, if (L) is valid,
a <u>2-sided</u> incidence group, if (L) and (R) are valid,
a <u>fibered</u> incidence group, if (L) and (F) are valid, and
a <u>kinematic space</u>, if (L),(R) and (F) are valid.

Therefore the kinematic map $\varkappa$ maps O_3^+ on the kinematic space $(\mathbb{H}^*/\mathbb{R}^*, \mathfrak{L}, \cdot)$.

<u>Remark 1.</u> In 1881 STEPHANOS gave a geometric definition of the kinematic map $\varkappa$. Let $d \in \mathbb{R}^3$ be a unit vector (hence $d \in J \cap \mathbb{H}_1$) and $\alpha \in [0,\pi]$ then the pair (d,α) determines a rotation $\rho(d,\alpha)$ with the axis $\mathbb{R}d$ and the angle α such that for $x \in \mathbb{R}^3 \setminus \mathbb{R}d$ the orientation of $(x, \rho(x), d)$ is positive provided that $\alpha \neq 0,\pi$. For $\alpha \neq 0,\pi$, $\varkappa(\rho) = \mathrm{tg}\,\frac{\alpha}{2} \cdot d$ is the point of the halfline $\mathbb{R}^+ \cdot d$ which has the distance $\mathrm{tg}\,\frac{\alpha}{2}$ from the origin 0; for $\alpha = 0$, $\varkappa(\rho) = 0$ and for $\alpha = \pi$ STEPHANOS defined $\varkappa(\rho)$ to be the point at infinity of the line $\mathbb{R}d$.

<u>Remark 2.</u> By the kinematic map $\varkappa$ all rotations ρ with the same axis are mapped on the points of a line of the kinematic space. So we have: For the kinematic space $(G,\mathfrak{L},\cdot)$ the set $\mathfrak{F} := \{X \in \mathfrak{L} \mid 1 \in X\}$ is a <u>fibration</u> (= partition) (i.e. $\mathfrak{F}$ is a set of subgroups such that $\cup \mathfrak{F} = G$ and $X \cap Y = \{1\}$ for any $X,Y \in \mathfrak{F}$ with $X \neq Y$) and we have $\mathfrak{L} = \{a \cdot X \mid a \in G, X \in \mathfrak{F}\}$, hence we have a similar situation as in ordinary analytic geometry, where we associate to a vector space (V,K) an affine space by defining V as point set and $\mathfrak{L} := \{a + X \mid a \in V, X \in \mathfrak{S}_1\}$ as line set if $\mathfrak{S}_1$ denotes the set of all 1-dimensional vector subspaces of (V,K). But since in our case the group $(G,\cdot)$ is not commutative we obtain here two parallelisms, a left and a right parallelism $\parallel_\ell$ and $\parallel_r$ by defining for any two lines $X,Y \in \mathfrak{L}$:

$$X \parallel_\ell Y :\Leftrightarrow X^{-1}X = Y^{-1}Y$$
$$X \parallel_r Y :\Leftrightarrow XX^{-1} = YY^{-1}$$

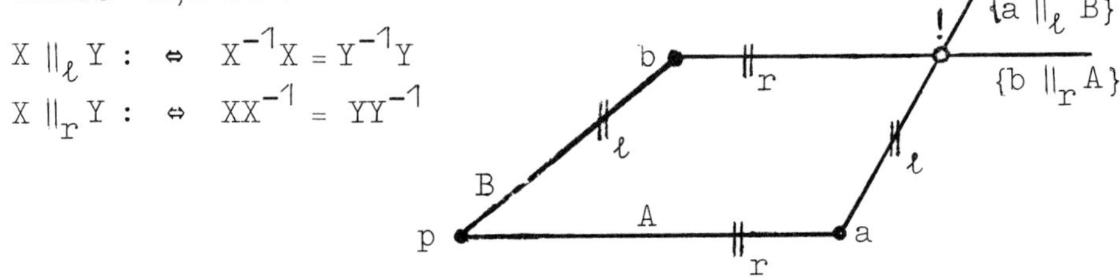

If we consider only one of the parallelisms there are in general no parallelograms. But we always have <u>mixed parallelograms</u>: Let p be a point, A,B two lines with $p \in A,B$ and let $a \in A$, $b \in B$, then $\{a \parallel_\ell B\} \cap \{b \parallel_r A\} \neq \emptyset$ where $\{a \parallel_\ell B\}$ denotes the line through a which is left parallel to B and $\{b \parallel_r A\}$ the right parallel line to A through b.

Remark 3. If one starts from a projective space $(P,\mathfrak{G})$ which is provided with two distinct parallelisms $\parallel_\ell$ and $\parallel_r$ such that mixed parallelograms are closed, then there exists a quaternion division algebra (A,K) such that the structure $(P,\mathfrak{G},\parallel_\ell,\parallel_r)$ can be derived in the same manner as above from the algebra (A,K) instead of $(\mathbb{H},\mathbb{R})$ (cf.[21],[22]).

For both algebras $(\mathbb{C},\mathbb{R})$ and $(\mathbb{H},\mathbb{R})$ the map $q(z) := z\bar{z}$ is a positive definite quadratic form. Therefore, if we apply this time the euclidean derivation on the quaternions $(\mathbb{H},\mathbb{R})$ we obtain again a euclidean geometry $A(\mathbb{H},\mathbb{R},^-) = (\mathbb{H},\mathfrak{L} = \{a + \mathbb{R}b \mid a,b \in \mathbb{H},\ b \neq 0\},\equiv)$ (where "$(a,b) \equiv (c,d) :\Leftrightarrow (a-b)\overline{(a-b)} = (c-d)\overline{(c-d)}$") which is here a 4-dimensional euclidean space. $A(\mathbb{H},\mathbb{R},^-)$ contains the 3-dimensional euclidean space $A(J,\mathbb{R},^-)$. Similarly as in the case of the euclidean plane we now can determine all motions of the 3- and 4-dimensional euclidean space. But since $\mathbb{H}$ is not commutative, any $b \in \mathbb{H}_1$ gives rise to two maps $b_\ell : \mathbb{H} \longrightarrow \mathbb{H};\ x \longrightarrow bx$ and $b_r : \mathbb{H} \longrightarrow \mathbb{H};\ x \longrightarrow xb$ and we have $b_\ell = b_r$ only if $b = 1$ or $b = -1$.

(4) For the group $\mathfrak{B}^+$ of all proper motions of $A(\mathbb{H},\mathbb{R},^-)$ we have $\mathfrak{B}^+ = \{a^+ \circ b_\ell \circ c_r \mid a \in \mathbb{H},\ b,c \in \mathbb{H}_1\}$ and $\mathfrak{B} = \mathfrak{B}^+ \cup \mathfrak{B}^+ \circ ^-$ is the group of all motions of $A(\mathbb{H},\mathbb{R},^-)$. The motion group of the 3-dimensional euclidean space $A(J,\mathbb{R},^-)$ is the subgroup $\{a^+ \circ b_\ell \circ \bar{b}_r \mid a \in J,\ b \in \mathbb{H}_1\} \cup \{a^+ \circ b_\ell \circ \bar{b}_r \circ ^- \mid a \in J,\ b \in \mathbb{H}_1\}$ of $\mathfrak{B}$.

There is a close relationship between euclidean and elliptic geometries. Elliptic geometry was discovered around 1870 by F. KLEIN. Let $(P,\mathfrak{L},\equiv)$ be an euclidean geometry and let $0 \in P$ be a distinct point. From projective geometry we know that the set $P_p := \{L \in \mathfrak{L} \mid 0 \in L\}$ of all lines and the set $\mathfrak{L}_p$ of all planes of the euclidean space passing through 0 form a projective space if we consider P_p as the set of points, and $\mathfrak{L}_p$ as the set of lines, and if we call a point $L \in P_p$ and a line $M \in \mathfrak{L}_p$ incident if $L \subset M$. Furthermore $\dim(P,\mathfrak{L},\equiv) = \dim(P_p,\mathfrak{L}_p) + 1$. Now the projective space can be provided with a metric structure $\equiv_p$ derived from the metric structure of the euclidean space. For $A,B \in P_p$ let $\alpha(A,B)$ be the smaller angle which is determined by the two intersecting lines A,B of the euclidean space. Then we call two pairs of points (A,B), (C,D) of the projective space congruent,

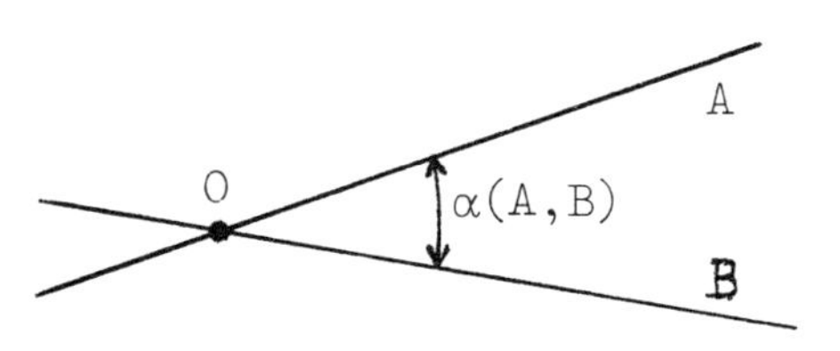

if $\alpha(A,B) = \alpha(C,D)$. We denote the elliptic space defined in this way $\Pi_0(P,\mathfrak{L},\equiv)$ and we call it the elliptic derivation of the euclidean space in the point 0. In case the euclidean space $(P,\mathfrak{L},\equiv)$ is represented by the metric vector space $(V,\mathbb{R},q)$ we have for the two lines $A = \mathbb{R}a$, $B = \mathbb{R}b$, $a,b \in V$, $a,b \neq 0$ the formula

$$\cos^2\alpha(A,B) = \frac{f(a,b)^2}{4q(a)\cdot q(b)} \quad \text{where} \quad f(x,y) = q(x+y) - q(x) - q(y)$$

and hence for the corresponding elliptic space the relation

$$(\mathbb{R}a,\mathbb{R}b) \equiv_p (\mathbb{R}c,\mathbb{R}d) \Leftrightarrow \frac{f(a,b)^2}{4q(a)q(b)} = \frac{f(c,d)^2}{4q(c)q(b)} .$$

Now starting from the quaternions $(\mathbb{H},\mathbb{R})$ we obtain (by the so called <u>elliptic derivation</u> $\Pi(\mathbb{H},\mathbb{R},^-)$ the 3-dimensional elliptic space $(\mathbb{H}^*/\mathbb{R}^*,\mathfrak{L}_p,\equiv_p)$ with $\mathfrak{L} = \{\varphi(S^*) \mid S \in \mathfrak{S}_2\}$ and

$$(\mathbb{R}a,\mathbb{R}b) \equiv_p (\mathbb{R}c,\mathbb{R}d) \Leftrightarrow \frac{(a\bar{b}+b\bar{a})^2}{4a\bar{a}\cdot b\bar{b}} = \frac{(c\bar{d}+d\bar{c})^2}{4c\bar{c}\cdot d\bar{d}}$$

because $f(x,y) = (x+y)\overline{(x+y)} - x\bar{x} - y\bar{y} = x\bar{y} + y\bar{x}$.

Since any motion of an euclidean space maps each angle on a congruent angle, all motions of an euclidean space fixing a distinct point 0 induce motions in the elliptic space corresponding to the point 0 and we have even:

<u>(5)</u> Let $(P,\mathfrak{L},\equiv)$ be an euclidean space, let $0 \in P$, let $\mathfrak{B}_0$ the group of all motions of $(P,\mathfrak{L},\equiv)$ fixing 0 and let $\Pi_0(P,\mathfrak{L},\equiv)$ the derived elliptic space. Then $\mathfrak{B}_0/\{1,-1\}$ is isomorphic to the motion group of the elliptic space $\Pi_0(P,\mathfrak{L},\equiv)$.

Now let us return to the formula (1) of EULER,CAYLEY. Since for any $a \in \mathbb{H}_1$ the map a_ι of (1) is distinct from -1 and since the 3-dimensional euclidean space $A(J,\mathbb{R},^-)$ determines the corresponding elliptic plane $\Pi(J,\mathbb{R},^-)$ all motions of the elliptic plane $\Pi(J,\mathbb{R},^-)$ are given by the formula (1) of EULER,CAYLEY :

<u>(1)'</u> $a_\iota : J^*/\mathbb{R}^* \longrightarrow J^*/\mathbb{R}^*;\ \mathbb{R}^*x \longrightarrow \mathbb{R}^*a x\bar{a} = \mathbb{R}^*axa^{-1}$ for $a \in \mathbb{H}_1$.

Together with (3) we have the result:

<u>(6)</u> Let G be the motion group of the elliptic plane $\Pi(J,\mathbb{R},^-)$. Then there is a kinematic map $\varkappa : G \longrightarrow \mathbb{H}^*/\mathbb{R}^*$ which maps G bijectively on the point set of the 3-dimensional elliptic space, and G is also isomorphic to the group $\mathbb{H}_1/\{1,-1\}$.

By (4) and (5) we obtain:

(7) The motion group Γ of the 3-dimensional elliptic space $\Pi(\mathbb{H},\mathbb{R},^-)$ consists of the maps

$a_\ell \circ b_r : \mathbb{H}^*/\mathbb{R}^* \longrightarrow \mathbb{H}^*/\mathbb{R}^*\,;\ \mathbb{R}^*x \longrightarrow \mathbb{R}^*axb$

$a_\ell \circ b_r \circ {}^- : \mathbb{H}^*/\mathbb{R}^* \longrightarrow \mathbb{H}^*/\mathbb{R}^*\,;\ \mathbb{R}^*x \longrightarrow \mathbb{R}^*a\bar{x}b$

where $a,b \in \mathbb{H}_1$. The subgroup $\Gamma^+ = \{a_\ell \circ b_r \mid a,b \in \mathbb{H}_1\}$ of all proper motions is isomorphic to the direct product $G \times G$ where $G \cong \mathbb{H}^*/\mathbb{R}^*$.

Summarizing we can say:

(8) The following geometric structures correspond to the quaternion field $(\mathbb{H},\mathbb{R})$:

I. A 4-dimensional euclidean space $A(\mathbb{H},\mathbb{R},^-)$ containing the 3-dimensional euclidean space $A(J,\mathbb{R},^-)$. The motion group $\mathfrak{B}_4$ of $A(\mathbb{H},\mathbb{R},^-)$ is given by $\mathfrak{B}_4 = \mathbb{H}^+ \circ \mathbb{H}_{1\ell} \circ \mathbb{H}_{1r} \cup \mathbb{H}^+ \circ \mathbb{H}_{1\ell} \circ \mathbb{H}_{1r} \circ {}^-$ where $\mathbb{H}^+ := \{a^+ : \mathbb{H} \longrightarrow \mathbb{H} \mid a^+(x) = a+x,\ a \in \mathbb{H}\}$,

$\mathbb{H}_{1\ell} := \{b_\ell : \mathbb{H} \longrightarrow \mathbb{H} \mid b_\ell(x) = bx,\ b \in \mathbb{H}_1\}$,

$\mathbb{H}_{1r} := \{c_r : \mathbb{H} \longrightarrow \mathbb{H} \mid c_r(x) = xc,\ c \in \mathbb{H}_1\}$ and

${}^- : \mathbb{H} \longrightarrow \mathbb{H}; x = x_0 + x_1 i + x_2 j + x_3 k \longrightarrow \bar{x} = x_0 - x_1 i - x_2 j - x_3 k$ and the motion group $\mathfrak{B}_3$ of $A(J,\mathbb{R},^-)$ by $\mathfrak{B}_3 = J^+ \circ \mathbb{H}_{1\iota} \cup J^+ \circ \mathbb{H}_{1\iota} \circ {}^-$ where $J^+ := \{a^+ : J \longrightarrow J \mid a \in J\}$ and

$\mathbb{H}_{1\iota} := \{a_\iota :\ J \longrightarrow J;\ x \longrightarrow ax\bar{a} \mid a \in \mathbb{H}_1\}$.

II. A 3-dimensional elliptic space $\Pi(\mathbb{H},\mathbb{R},^-)$ containing the elliptic plane $\Pi(J,\mathbb{R},^-)$ where the point set $G := \mathbb{H}^*/\mathbb{R}^*$ of the elliptic space $\Pi(\mathbb{H},\mathbb{R},^-)$ can be provided with the group structure given by the factor group $\mathbb{H}^*/\mathbb{R}^*$. The set $G = \mathbb{H}^*/\mathbb{R}^*$ is a 3-dimensional projective kinematic space with respect to the incidence structure $\mathfrak{L}$ and the group structure " . ". The group Γ_3^+ of all proper motions of $\Pi(\mathbb{H},\mathbb{R},^-)$ is the direct product $\Gamma_3^+ = G_\ell \circ G_r$ where $G_\ell := \{a_\ell : G \longrightarrow G \mid a_\ell(x) = ax,\ a \in G\}$ and $G_r := \{a_r : G \longrightarrow G \mid a_r(x) = xa,\ a \in G\}$ and hence isomorphic to $G \times G$. The group Γ_2 of all motions of the elliptic plane $\Pi(J,\mathbb{R},^-)$ is isomorphic to the group G and Γ_2 is given by $\Gamma_2 = G_\iota := \{a : E \longrightarrow E\,;\ x \longrightarrow axa^{-1} \mid a \in G\}$ where $E = J^*/\mathbb{R}^*(\subset G)$ denotes the point set of the elliptic plane $\Pi(J,\mathbb{R},^-)$.

The application $\iota : G \longrightarrow \Gamma_2;\ a \longrightarrow a_\iota$ is an isomorphism and $\varkappa = \iota^{-1}$ is a kinematic map which maps any motion of the elliptic plane onto a point of the elliptic space.

Remark 4. Very soon after F. KLEIN's discovery of elliptic geometry, W. CLIFFORD introduced his notion of parallelism in elliptic geometry, which had many properties in common with euclidean parallelism. In euclidean geometry two parallel distinct lines A, B have the properties:

1. $A \cup B$ is contained in a common plane.
2. $A \cap B = \emptyset$
3. For any point $a \in A$ the line $\{a \perp B\}$ through a and perpendicular to B is also perpendicular to A, hence $\{a \perp B\} \perp A$.

3!. There are rectangles (3! is a consequence of 3.).

4. For any two points $a_1, a_2 \in A$ the distances $d(a_i, B)$ of the point a_i from the line B are equal: $d(a_1, B) = d(a_2, B)$.

Furthermore euclidean parallelism $\parallel$ has the following properties:

5. $\parallel$ is an equivalence relation on the line set $\mathfrak{L}$.
6. To any line A and to any point p there is exactly one line $B =: \{p \parallel A\}$ with $p \in B$ and $B \parallel A$.
7. To any two points a, b there is exactly one translation τ (i.e. τ is either the identity or a motion without fixed points such that $\tau(X) \parallel X$ for any line X) with $\tau(a) = b$; if $a \neq b$ and $C \parallel \overline{a,b}$ then $\tau(C) = C$.

If the euclidean geometry is algebraically represented by a vector space $(V, \mathbb{R})$, i.e. V is identified with the point set, and the lines are the cosets of the 1-dimensional vector subspaces hence $\mathfrak{L} = \{a + X \mid a \in V, X \in \mathfrak{J}\}$ where $\mathfrak{J} = \{X = \mathbb{R}b \mid b \in V, b \neq 0\}$, then:

8. For $A, B \in \mathfrak{L}$: $A \parallel B \Leftrightarrow A - A = B - B$.
9. For $a, b \in V$, $(b-a)^+ : V \longrightarrow V;\ x \longrightarrow (b-a)+x$ is the uniquely determined translation mapping a onto b.
10. For $A \in \mathfrak{L}$ and $p \in V$ we have $\{p \parallel A\} = p + (A-A)$.

The property 3! which is a consequence of EUCLID's axiom of parallelism, is one of the hypotheses considered by G. SACCHERI and J.H. LAMBERT: Let (a,b,c,d) be a quadrangle such that the three angles $\alpha := \sphericalangle(d,a,b)$, $\beta := \sphericalangle(a,b,c)$, $\gamma := \sphericalangle(b,c,d)$ are right-angled, then there are the hypotheses $\delta := \sphericalangle(c,d,a) < R$, $\delta = R$ and $\delta > R$ where R denotes the right angle.

SACCHERI and LAMBERT proved that the hypothesis $\delta > R$ is false by assuming all axioms of EUCLID except for the axiom of parallelism, but their efforts to also disprove the hypothesis $\delta < R$ failed. The hypothesis $\delta < R$ is true in hyperbolic geometry and the hypothesis $\delta > R$ in spheric and elliptic geometries. Now one can ask if the property 3! (i.e. hypothesis $\delta = R$) is true in hyperbolic or elliptic geometries if one also admits 3-dimensional quadrangles. So let (a,b,c,d) be a 3-dimensional quadrangle with $\alpha = \beta = \gamma = R$ and let d' be the orthogonal projection of the point d on the plane $\overline{\{a,b,c\}}$ spanned by the points a,b,c hence $d' := \overline{\{a,b,c\}} \cap \{d \perp \overline{\{a,b,c\}}\}$. Then (a,b,c,d') is a plane quadrangle and still $\alpha' := \sphericalangle(d',a,b) = \beta = \gamma' := \sphericalangle(b,c,d') = R$ but $\delta < \delta' := \sphericalangle(c,d',a)$. Therefore, also in hyperbolic geometry there are no 3-dimensional rectangles. In elliptic geometry the situation is different because $\delta \leq \delta'$ and $R < \delta'$. There are in general two points d_1, d_2 on the line $D := \{d \perp \overline{\{a,b,c\}}\}$ such that (a,b,c,d_1) and (a,b,c,d_2) are 3-dimensional rectangles. Let $A := \overline{a,b}$ and $B_i := \overline{c,d_i}$, then the pairs of lines (A,B_1) and (A,B_2) have the properties 2,3,4 but not 1. CLIFFORD showed that for any line A and any point $p \notin A$ of the elliptic space there are exactly two lines B_1, B_2 through p such that the properties 3. and 4. are valid for A,B_1 and A,B_2, if the line A is not contained in the polar plane $p^{\perp} := \{x \in P \mid d(x,p) = R\}$ of the point p. If $A \subset p^{\perp}$ then $B := A^{\perp} \ni p$ is the only line through p having the properties 3. and 4.. In this way CLIFFORD obtained his left and right parallelism in the elliptic geometry. Using the representation (8) of elliptic geometry $\Pi(\mathbb{H},\mathbb{R},^{-}) = (G = \mathbb{H}^*/\mathbb{R}^*, \mathfrak{L}, \equiv)$ by quaternions $(\mathbb{H},\mathbb{R})$, CLIFFORD's parallelism can be described as in Remark 3., hence two lines $A,B \in \mathfrak{L}$ are left parallel denoted by $A \parallel_{\ell} B$, if and only if, $A^{-1}A = B^{-1}B$ and they are right parallel if $AA^{-1} = BB^{-1}$. If we replace the notion translation by left- or right-translation respectively, then each left- and right-parallelism $\parallel_{\ell}$ and $\parallel_r$ has the same properties as euclidean parallelism except for property 1. A translation $\tau \neq id$ is called a left-translation if τ is fixed point free, if for any two points $a,b \in G$, $\overline{a,\tau(a)} \parallel_r \overline{b,\tau(b)}$ and if for each line $L \in \mathfrak{L}$ $\tau(L) \parallel_{\ell} L$.

Our considerations have shown that there is a nice algebraic description of the motion group of the elliptic plane by quaternions using the formula of EULER, CAYLEY. E. STUDY tried to find a similar algebraic representation of the motion group of the euclidean plane in 1904. For this purpose he introduced a new algebraic structure, the so called Study-quaternions $\mathbb{H}_S$. This algebra is also 4-dimensional and has a basis $1,i,\varepsilon,i\varepsilon$ such that $i^2 = -1$, $\varepsilon^2 = 0$, $\varepsilon i = -i\varepsilon$. The Study-quaternions $\mathbb{H}_S$ no longer form a division algebra but rather a local algebra with the maximal ideal $M := \mathbb{F}\varepsilon + \mathbb{F}i\varepsilon$.

If we also set $\bar{a} := a_0 - a_1 i - a_2\varepsilon - a_3 i\varepsilon$ for $a = a_0 + a_1 i + a_2\varepsilon + a_3 i\varepsilon \in \mathbb{H}_S$, $a_0, a_1, a_2, a_3 \in \mathbb{R}$, then $x \rightarrow \bar{x}$ is again an antiautomorphism of $\mathbb{H}_S$, and a is a unit, if and only if, $a \cdot \bar{a} = a_0^2 + a_1^2 \neq 0$. So $U = \mathbb{H}_S \backslash M$ is the set of units of the algebra $\mathbb{H}_S$. By identifying each point (x_1, x_2) of the euclidean plane with the pure Study-quaternion $x := i + x_1\varepsilon + x_2 i\varepsilon$ every proper motion α of the euclidean plane is given by the formula of EULER, CAYLEY

(9) $a_\iota : i + \mathbb{R}\varepsilon + \mathbb{R}i\varepsilon \longrightarrow i + \mathbb{R}\varepsilon + \mathbb{R}i\varepsilon;\ x \longrightarrow ax\bar{a}$

where $a \in \mathbb{H}_S$ is a Study quaternion with $a\bar{a} = 1$. This motion α is the product of the rotation fixing the point i and having the angle $\varphi = \arccos(a_0^2 - a_1^2)$, and the translation $((a_0 a_3 + a_2 a_1)\varepsilon + (a_0 a_2 - a_1 a_3)i\varepsilon)^+$.
If we denote again by $\mathbb{H}_{S1} := \{x \in \mathbb{H}_S \mid x\bar{x} = 1\}$ the set of all Study-quaternions of "length 1", then the proper motion group $\mathfrak{B}_2^+$ of the euclidean plane is isomorphic to the factor group $\mathbb{H}_{S1}/\{1,-1\}$, a similar situation as in the elliptic case.
What other similarities do we have if we study the Study-quaternions $(\mathbb{H}_S, \mathbb{R})$ from the same point of view as we did with the usual quaternions $(\mathbb{H}, \mathbb{R})$?
For Study-quaternions $(\mathbb{H}_S, \mathbb{R})$ the map $q : \mathbb{H}_S \longrightarrow \mathbb{R};\ x \longrightarrow x\bar{x}$ is also a positive quadratic form with the corresponding bilinear form $f(x,y) = x\bar{y} + y\bar{x}$, but this time the radical $\mathrm{rad}\ q = \mathbb{H}_S^{\perp} := \{x \in \mathbb{H}_S \mid \forall y \in \mathbb{H}_S : f(x,y) = 0\}$ is not 0 but $\mathrm{rad}\ q = M = \mathbb{R}\varepsilon + \mathbb{R}i\varepsilon$. For the set U of the units we have $U = \{x \in \mathbb{H}_S \mid q(x) \neq 0\} = \mathbb{H}_S \setminus M$ and for $x \in U$ the element $x^{-1} = (q(x))^{-1}\bar{x}$ is the inverse of x. Since $\mathbb{R}^*$ is the center of U we can form the factor group $G = U/\mathbb{R}^*$. Also here for $a \in U$ we have $b := \sqrt{a\bar{a}}^{\,-1} \cdot a \in \mathbb{H}_{S1}$ and $\mathbb{R}^* a = \mathbb{R}^* b$, hence $\mathbb{H}_{S1}/\{1,-1\}$ and $U/\mathbb{R}^*$ are isomorphic and so:

(10) $\mathfrak{B}_2^+ \cong \mathbb{H}_{S1}/\{1,-1\} \cong U/\mathbb{R}^*$

Applying the projective derivation $\Pi(\mathbb{H}_S, \mathbb{R})$ on the 4-dimensional vector space $(\mathbb{H}_S, \mathbb{R})$ the 2-dimensional subspace $M = \mathbb{R}\varepsilon + \mathbb{R}i\varepsilon$ is mapped on the line $\varphi(M^*) = M^*/\mathbb{R}^*$ and we have $\varphi(U) = U/\mathbb{R}^* = \mathbb{H}_S^*/\mathbb{R}^* \setminus M^*/\mathbb{R}^*$. So the group $G = U/\mathbb{R}^*$ can be also considered as the point set of the 3-dimensional projective space where all points of a distinct line are deleted, a similar situation as we know from projective and affine geometries where the affine space is obtained from the projective space by deleting all points of a plane. The point set $G = U/\mathbb{R}^*$ is called a (3-1)-slit space if we provide G with the trace structure $\mathfrak{L}$: Let $\mathfrak{L}_p$ be the set of lines of the projective space $\Pi(\mathbb{H}_S, \mathbb{R})$, then $\mathfrak{L} := \{X \cap G \mid X \in \mathfrak{L}_p \wedge |X \cap G| \geq 2\}$. Different from the affine space, in the (3-1)-slit space $(G, \mathfrak{L})$ there are two types of lines, the set $\mathfrak{L}_o$ of <u>projective lines</u> defined by

$\mathfrak{L}_o := \{X \in \mathfrak{L}_p \mid X \subset G\} = \{X \in \mathfrak{L}_p \mid X \cap M^*/\mathbb{R}^* = \emptyset\}$ and the set $\mathfrak{L}_1$ of <u>affine lines</u> where $\mathfrak{L}_1 := \{X \cap G \mid X \in \mathfrak{L}_p \wedge |X \cap M^*/\mathbb{R}^*| = 1\}$. So we see that also for the Study-quaternions $(\mathbb{H}_S, \mathbb{R})$ we have in a natural way a set G, which is provided with a group structure "$\cdot$" and an incidence structure $\mathfrak{L}$. This procedure shall be called <u>kinematic derivation</u> and be denoted by $\Pi_\varkappa(\mathbb{H}_S, \mathbb{R})$ (cf. §7), then:

<u>(11)</u> For the Study-quaternions $(\mathbb{H}_S, \mathbb{R})$ the kinematic derivation $\Pi_\varkappa(\mathbb{H}_S, \mathbb{R})$ is a kinematic space and we have the kinematic map $\varkappa : \mathfrak{B}_2^+ \longrightarrow G;\ a_\iota \longrightarrow \mathbb{R}^* a$ where $a_\iota : i + M \longrightarrow i + M;\ x \longrightarrow ax\bar{a}$ and $a \in \mathbb{H}_{S1}$.

<u>Remark 5.</u> A geometric description for this kinematic map was given 1911 independently by W. BLASCHKE [4] and J. GRÜNWALD [9].

As we have seen the motion group of an elliptic plane has a kinematic space which is the 3-dimensional elliptic space and the elliptic plane itself can be considered to be a subplane of the kinematic space. What is the situation in the euclidean case? If we apply the elliptic derivation to the Study-quaternions $(\mathbb{H}_S, \mathbb{R})$ then the kinematic space $\Pi_\varkappa(\mathbb{H}_S, \mathbb{R}) = (G, \mathfrak{L}, \cdot)$ will be provided with a metric structure and becomes the so called <u>quasielliptic space</u>:

"For $\mathbb{R}^*a, \mathbb{R}^*b, \mathbb{R}^*c, \mathbb{R}^*d \in G$ let

$$(\mathbb{R}^*a, \mathbb{R}^*b) \equiv (\mathbb{R}^*c, \mathbb{R}^*d) :\Leftrightarrow \frac{(a\bar{b} + b\bar{a})^2}{4a\bar{a} \cdot b\bar{b}} = \frac{(c\bar{d} + d\bar{c})^2}{4c\bar{c} \cdot d\bar{d}}\text{"}.$$

If $a = a_1 + a_2\varepsilon,\ b = b_1 + b_2\varepsilon \in \mathbb{H}_S = \mathbb{C} + \mathbb{C}\varepsilon$ then

$$\frac{(a\bar{b} + b\bar{a})^2}{4a\bar{a} \cdot b\bar{b}} = \frac{(a_1\bar{b}_1 + b_1\bar{a}_1)^2}{4a_1\bar{a}_1 \cdot b_1\bar{b}_1}.$$

Remark 6. The quasielliptic space was studied by W.BLASCHKE [4], [5].

This quasielliptic space $(G,\mathfrak{L},\equiv)$ is neither a euclidean nor an elliptic space, but still the euclidean plane with respect to the incidence structure can be considered as a subplane of $(G,\mathfrak{L})$. For the point set of the euclidean plane was identified with the set $i+M$, and this set $i+M$ is contained in the set $J := \{x \in \mathbb{H}_S \mid \overline{x} = -x\}$ as well as in U. Applying the projectice derivation we obtain $\varphi(i+M) = \{\mathbb{R}^*(i+x) \mid x \in M\} = \varphi(J \setminus M) =: E$, and $\varphi(i+x) = \varphi(i+y)$ for $x,y \in M$ implies $x = y$. So φ maps $i+M$ bijectively on the plane E of $(G,\mathfrak{L})$ and a line $i+a+\mathbb{R}b$ with $a,b \in M$, $b \neq 0$ of the euclidean plane is mapped on the line $\varphi(i+a+\mathbb{R}b) = \varphi((\mathbb{R}(i+a)+\mathbb{R}b)\setminus M)$ of $\mathfrak{L}$. But the metric of the quasielliptic space $(G,\mathfrak{L},\equiv)$ does not induce the euclidean metric in E because for any two points $\mathbb{R}^*(i+a)$ and $\mathbb{R}^*(i+b)$ of E (with $a,b \in M$) we have

$$\frac{((i+a)\overline{(i+b)} + (i+b)\overline{(i+a)})^2}{4(i+a)\overline{(i+a)} \cdot (i+b)\overline{(i+b)}} = \frac{(1-ib-ai+1-ia-bi)^2}{4(1-ia-ai)(1-ib-bi)} = 1 .$$

As in the elliptic case one can prove:

(12) The Study-quaternions $(\mathbb{H}_S,\mathbb{R})$ determine a corresponding kinematic space $\Pi_\varkappa(\mathbb{H}_S,\mathbb{R}) = (G = U/\mathbb{R}^*, \mathfrak{L}, \cdot)$ which becomes the quasielliptic space $(G,\mathfrak{L},\equiv)$ by applying the elliptic derivation. The motion group Γ_S of the quasielliptic space $(G,\mathfrak{L},\equiv)$ contains the subgroups $G_\iota := \{a_\iota : G \to G;\ x \to axa^{-1} \mid a \in G\}$, $\Gamma_S^+ = G_\ell \circ G_r$ and $\Gamma_S^+ \cup \Gamma_S^+ \circ \bar{\ }$ where Γ_S^+ is the direct product of the group G_ℓ of the "left-translations" and the group G_r of "right-translations". The groups G_ι, G_ℓ, G_r and $\mathfrak{B}_2^+$ are all isomorphic to the group $G = U/\mathbb{R}^*$, and the motion group $\mathfrak{B}_2^+$ of the euclidean plane is obtained by restricting G_ι to the subplane E hence $\mathfrak{B}_2^+ = \{a_\iota|_E \mid a_\iota \in G_\iota\}$.

Also $(G,\mathfrak{L},\cdot,\equiv)$ can be provided with CLIFFORDs left- and right-parallelisms $\parallel_\ell$ and $\parallel_r$, and G_ℓ and G_r are the groups of left- and right-translations respectively. On the subplane E, $\parallel_\ell$ and $\parallel_r$ coincide with the euclidean parallelism, hence if A,B are two lines of the plane $\mathbf{E}$ with $A \neq B$ then $A \parallel_\ell B$ (i.e. $A^{-1}A = B^{-1}B$) $\Leftrightarrow$ $A \parallel_r B$ (i.e. $AA^{-1} = BB^{-1}$) $\Leftrightarrow$ $A \cap B = \emptyset$.

Now let us follow STUDY's train of thought with regard to hyperbolic geometry. Since F. KLEIN we have known that the motion group G of the hyperbolic plane is isomorphic to the group $PGL(2,\mathbb{R})$ consisting of the maps $x \longrightarrow (ax+b)\cdot(cx+d)^{-1}$ where $a,b,c,d \in \mathbb{R}$ with $ad-bc \neq 0$. This group is isomorphic to the factor group $GL(2,\mathbb{R})/\mathbb{R}^*$ where $GL(2,\mathbb{R})$ is the multiplicative group consisting of all 2×2-matrices $\begin{pmatrix} a & b \\ c & d \end{pmatrix}$ whose determinant is $\neq 0$ and where $\mathbb{R}^*$ is identified with the set of matrices $\{\begin{pmatrix} \lambda & 0 \\ 0 & \lambda \end{pmatrix} \mid \lambda \in \mathbb{R}^*\}$. But $GL(2,\mathbb{R})$ is exactly the set U of all units of the 4-dimensional algebra $(\mathfrak{M}_{22},\mathbb{R})$ consisting of all 2×2-matrices with coefficients in $\mathbb{R}$. So also here there is an algebra closely related to the motion group G of the hyperbolic plane. The center of the algebra is the field of real numbers $\mathbb{R}$ and $G \cong U/\mathbb{R}^*$, the same situation as in elliptic and euclidean geometries. $(\mathfrak{M}_{22},\mathbb{R})$ is a simple algebra having the involutorial antiautomorphism

$$-:=\begin{pmatrix} a_0 & a_1 \\ a_2 & a_3 \end{pmatrix} \longrightarrow \bar{a} = \begin{pmatrix} a_3 & -a_1 \\ -a_2 & a_0 \end{pmatrix},$$ the quadratic form

$q: \mathfrak{M}_{22} \longrightarrow \mathbb{R}\,;\ a \longrightarrow a\bar{a} = \det(a)$ and $GL(2,\mathbb{R}) = U = \{a \in \mathfrak{M}_{22} \mid q(a) \neq 0\}$. The quadric $\mathfrak{M}^0_{22} := \{x \in \mathfrak{M}_{22} \mid q(x) = 0\}$ has the index 2 and $U = \mathfrak{M}_{22} \setminus \mathfrak{M}^0_{22}$.

As in the previous cases let us provide the motion group $G = U/\mathbb{R}^*$ of the hyperbolic plane with an incidence structure $\mathfrak{L}$ and a congruence structure $\equiv$ by setting: $\mathfrak{L} := \{X \cap G \mid X \in \mathfrak{L}_p \wedge |X \cap G| \geq 2\}$, where again $\mathfrak{L}_p$ denotes the set of lines of the projective space $\Pi(\mathfrak{M}_{22},\mathbb{R})$ of the 4-dimensional vector space $(\mathfrak{M}_{22},\mathbb{R})$, and

$$(\mathbb{R}^*a,\mathbb{R}^*b) \equiv (\mathbb{R}^*c,\mathbb{R}^*d) :\Leftrightarrow \frac{(a\bar{b}+b\bar{a})^2}{4a\bar{a}\cdot b\bar{b}} = \frac{(c\bar{d}+d\bar{c})^2}{4c\bar{c}\cdot d\bar{d}}$$

for $\mathbb{R}^*a, \mathbb{R}^*b, \mathbb{R}^*c, \mathbb{R}^*d \in G$. Then one obtains

<u>(13)</u> $(G,\mathfrak{L},\cdot,\equiv)$ is a kinematic space provided with a congruence relation $\equiv$. For the motion group Γ_h of the metric space $(G,\mathfrak{L},\equiv)$ we have $\Gamma_h = G_\ell \circ G_r \cup G_\ell \circ G_r \circ -$ and G_ℓ and G_r are the groups of left- and right-translations with regard to the CLIFFORD left- and right-parallelisms of the kinematic space $(G,\mathfrak{L},\cdot)$.

Remark 7. In this case $Q := \varphi(\mathfrak{M}_{22}^{o\,*})$ is a ruled quadric of the 3-dimensional projective space $\Pi(\mathfrak{M}_{22}, \mathbb{R})$. Therefore there are four types of lines: Lines which do not intersect Q, lines which have exactly one point in common with Q (hence tangents), lines which intersect Q in two points and lines which are contained in Q. Accordingly $\mathfrak{L}$ comprises the sets of projective lines $\mathfrak{L}_0$, of affine lines $\mathfrak{L}_1$ and of 2-lines $\mathfrak{L}_2$.

Just as in the elliptic case the hyperbolic plane can be considered in a suitable way as a subplane of its kinematic space $(G, \mathfrak{L}, \cdot, \equiv)$; the difference is only that here $(G, \mathfrak{L}, \equiv)$ is not the hyperbolic space. Let $J := \{x \in \mathfrak{M}_{22} \mid \overline{x} = -x\}$, $\mathfrak{M}_{22}^+ := \{x \in \mathfrak{M}_{22} \mid q(x) > 0\}$ and $\mathfrak{M}_{22}^- := \{x \in \mathfrak{M}_{22} \mid q(x) < 0\}$. Then we get the incidence structure of the hyperbolic plane if we consider the set $H := \varphi(J \cap \mathfrak{M}_{22}^+) = J \cap \mathfrak{M}_{22}^+ / \mathbb{R}^*$ as the set of points and $\mathfrak{H} := \varphi(J \cap \mathfrak{M}_{22}^-) = J \cap \mathfrak{M}_{22}^- / \mathbb{R}^*$ as the set of lines and if we call a point $\mathbb{R}^* a \in H$ and a line $\mathbb{R}^* u \in \mathfrak{H}$ incident if $0 = f(a,u) = q(a+u) - q(a) - q(u) = a\overline{u} + u\overline{a}$. Obviously $H \subset G$ and $(H, \mathfrak{H}, \equiv_H)$ is the hyperbolic plane where $\equiv_H$ denotes the restriction of the congruence relation $\equiv$ on H (cf.[23]). H is an invariant subset of G, and the motion group $\mathfrak{B}_H$ of the hyperbolic plane $(H, \mathfrak{H}, \equiv_H)$ consists of restrictions of all inner automorphisms of the group G, so $\mathfrak{B}_H = \{a_\iota : H \longrightarrow H;\ x \longrightarrow axa^{-1} \mid a \in G\}$. The groups G and $\mathfrak{B}_H$ are isomorphic.

In the elliptic plane one cannot distinguish between proper and improper motions. But as in the euclidean plane one has also in the hyperbolic plane proper and improper motions. If $(\mathfrak{M}_{22})_1 := \{x \in \mathfrak{M}_{22} \mid q(x) = 1\}$, then the formula

$$\begin{array}{ccccccc} a_\iota : & J \cap \mathfrak{M}_{22}^+ & \longrightarrow & J \cap \mathfrak{M}_{22}^+ ; & x & \longrightarrow & ax\overline{a} \\ & \downarrow \varphi & & \downarrow \varphi & \downarrow & & \downarrow \\ & H & \longrightarrow & H ; & \mathbb{R}^* x & \longrightarrow & \mathbb{R}^* ax\overline{a} \end{array} \qquad \text{for } a \in (\mathfrak{M}_{22})_1$$

of EULER, CAYLEY defines exactly the proper motions of the hyperbolic plane, and the factor group $(\mathfrak{M}_{22})_1 / \{1, -1\}$ is isomorphic to the group of all proper motions of the hyperbolic plane.

The result of our reflections is, that for each of the classical geometries, elliptic, euclidean and hyperbolic, there exists an algebra (A,K) with the property that for any $a \in A$, $a^2 = -a\bar{a} + (a+\bar{a})a \in K + Ka$. Such algebras will be called kinematic and shall be studied in this paper.

Remark 8. The pseudo-euclidean plane is another classical metric plane; the metric structure in $\mathbb{R}^2$ is given by $(a,b) \equiv (c,d) :\Leftrightarrow (a_1-b_1)^2 - (a_2-b_2)^2 = (c_1-d_1)^2 - (c_2-d_2)^2$.
Also this plane has a corresponding kinematic algebra as E.M. SCHRÖDER [41] showed.

Remark 9. For every axiomatically determined absolute plane one can construct in an abstract way an associated kinematic space and this kinematic space is a suitable tool for the foundation of the absolute plane (cf.[16]). This was first done in 1934 for the elliptic plane by E. PODEHL and K. REIDEMEISTER in their famous paper [39].

§ 2 PROBLEMS RESULTING FROM CLASSICAL KINEMATICS; A SURVEY OF THE MATERIAL COVERED IN THIS PAPER

From our expositions in § 1 it seems desirable to give a classification of kinematic algebras and furthermore to describe the structure of the different types of kinematic algebras. This will be done in §4 in a more general setting which includes the larger class of alternative kinematic algebras. To any alternative kinematic algebra (A,K) there belongs a quadratic form q such that $U := \{x \in A \mid q(x) \neq 0\}$ is the set of units. The set U is a Moufang-loop with respect to multiplication and even a group if (A,K) is associative. Since K^* is in the center of U, we can form the factor loop $G = U/K^*$ which is also a Moufang-loop. Both loops $(U,\cdot)$ and $(G,\cdot)$ can be provided by so-called affine or projective derivations with an incidence structure $\mathfrak{L}$ obtained as the trace space of the affine or projective space belonging to the vector space (A,K). These incidence spaces $(U,\mathfrak{L})$ and $(G,\mathfrak{L})$ belong to the class of affine porous and projective porous spaces respectively (cf. §5) and $(U,\mathfrak{L},\cdot)$ and $(G,\mathfrak{L},\cdot)$ are incidence loops (cf. §7). A subset Q of the point set of an incidence space is called a 2-set, if any line L is contained in Q, if the intersection $L \cap Q$ consists of more then two points. A porous space is obtained from an affine or projective space by omitting a 2-set Q.
The question whether there are, apart from the kinematic algebras, other algebras, whose affine or projective derivation is a porous space, leads to the larger class of 2-algebras, which are discussed in § 3.

In §6 we study a special class of 2-algebras (A,K) characterized by the property that the set Q of all non-units is a 2-set, consisting of the union $M_1 \cup M_2$ of a hyperplane M_1 and an arbitrary subspace M_2 of (A,K). The projective derivation gives us 2-sided subaffine incidence groups in the sense of E.M. SCHRÖDER. For these incidence groups we give an affine representation.
The projective derivations of kinematic algebras are treated in §7. Here we state the fundamental representation theorem for kinematic Moufang-loops.
From linear algebra we know that the adjoint map is an involutorial antiautomorphism of the matrix ring, which allows one to give an orthogonal decomposition of the matrix ring in the subspaces of Hermitian and skew-Hermitian matrices. In the case of a kinematic algebra (A,K) we understand by an adjoint map $*$ an arbitrary involutorial antiautomorphism. Then the Hermitian and skew-Hermitian subsets $H := \{x \in A \mid x^* = x\}$ and $H^- := \{x \in A \mid x^* = -x\}$ form metric vector spaces over the fixed field $F := \{\lambda \in K \mid \lambda^* = \lambda\}$ with respect to the quadratic form q. The kinematics of the affine and projective derivations of these metric vector spaces (H,F,q) and (H^-,F,q) are studied in §8. We had no time to give a complete classification of the derived geometries. For this purpose it would have been necessary first to know all possible adjoint maps $*$. Under additional assumptions this concept leads to the kinematic model of P. KUSTAANHEIMO for the 3-dimensional hyperbolic space, which will be discussed in §9.

§ 3 2-ALGEBRAS

Let (A,K) be an algebra with identity 1 and let U be the set of all units. By a unit we understand an element $u \in A$ such that the left and right translation u_ℓ and u_r are permutations. Then (A,K) is called a 2-algebra, if the property $(*)$ holds

$(*)$ $\forall a \in A : \ |(a+K) \cap (A \setminus U)| \leq 2$

If we assume that K is not the smallest field Z_2, then $(*)$ implies that for any $a \in A$ there is a $\lambda \in K$ with $a+\lambda \in U$, i.e. $a \in -\lambda + U \subset K + U$; hence

$(**)$ $A = K + U$

A subclass of the 2-algebras are the bilocal algebras (A,K) which are defined by the property:

(B) There are ideals M and N such that $A \setminus U = M \cup N$

A bilocal algebra is called <u>proper</u>, if $M \not\subset N$ and $N \not\subset M$, otherwise <u>local</u>. A special case of local algebras are the <u>division algebras</u> where $M = N = \{0\}$.
Here we will study mainly associative 2-algebras.
Any ring $A = \mathfrak{M}_{22}(D)$ of 2×2-matrices with elements in a skew-field D is an example of an associative 2-algebra (A,K) where K is a subfield of the center of D.

<u>(3.1)</u> Let (A,K) be a 2-algebra. Then
a) Any subalgebra A' of A with $1 \in A'$ is a 2-algebra
b) For any ideal M the factor algebra $(A/M,K)$ is again a 2-algebra.

Proof. b): Obviously $\{u+M \mid u \in U\}$ is a set of units of the algebra $(A/M,K)$, therefore $a + \lambda + M$ is not a unit in A/M if $a + \lambda \notin U$.

As usually we denote by $\text{Rad } A := \{x \in A \mid \forall a \in A \ \exists n \in \mathbb{N} : (ax)^n = 0\}$ the <u>radical</u> of the algebra (A,K) (cf.[1], Theorem 10, p.24). Rad A is the maximal nilpotent ideal of (A,K).

<u>(3.2)</u> (Structure-theorem for finite dimensional 2-algebras)
Any 2-algebra (A,K) with $K \neq \mathbb{Z}_2$ contains at most two maximal ideals and belongs to one of the following classes:
I. If (A,K) is simple, then A is either a skew-field or there is a skew-field D such that K is contained in the center of D and A is the ring $\mathfrak{M}_{22}(D)$ of all 2×2-matrices with elements in D.

II. If (A,K) contains two maximal ideals M_1, M_2, then (A,K) is a proper bilocal algebra.

III. If (A,K) contains one maximal ideal M, then there are the following subclasses:
a) (A,K) is a local algebra, i.e. $U = A \setminus M$; then $\text{Rad } A = M$.
b) $\text{Rad } A = M$ and $U \neq A \setminus M$, then any maximal left-ideal L contains M, and A/M is the matrix ring $\mathfrak{M}_{22}(D)$ over a skew-field D.

Proof. Let us assume (A,K) contains two distinct maximal ideals M_1 and M_2 and let L be a maximal left ideal with $M_1 \not\subset L$. We assume $M_2 \not\subset L$. Then we have $A = M_1 + M_2 = M_1 + L = M_2 + L$ and hence there are $m_1, m_1' \in M_1$, $m_2, m_2' \in M_2$ and $l, l' \in L$ with $1 = m_1 + m_2 = m_1' + l = m_2' + l'$. This gives us $1 = (m_1 m_2' + m_1 l') + (m_2 m_1' + m_2 l) = (m_1 m_2' + m_2 m_1') + m_1 l' + m_2 l$ and $d := m_1 m_2' + m_2 m_1' \in D := M_1 \cap M_2$, $d_1 := m_1 l' \in D_1 := M_1 \cap L$, $d_2 := m_2 l \in D_2 := M_2 \cap L$. Furthermore we have $d \notin L$ because otherwise $1 = d + d_1 + d_2 \in L$, and with the same argument $d_1 \notin M_2$, $d_2 \notin M_1$. Since $|K| \geq 3$ there are $\alpha_1, \alpha_2 \in K^*$ with $\alpha_1 \neq \alpha_2$. For $G := \alpha_1 d_1 + \alpha_2 d_2 + K$ we have

$G \cap M_1 \ni \alpha_1 d_1 + \alpha_2 d_2 - \alpha_2 \cdot 1 = \alpha_1 d_1 + \alpha_2 d_2 - \alpha_2(d + d_1 + d_2) = (\alpha_1 - \alpha_2) d_1 - \alpha_2 d$

$G \cap M_2 \ni (\alpha_2 - \alpha_1) d_2 - \alpha_1 d$, $G \cap L \ni \alpha_1 d_1 + \alpha_2 d_2$.

The elements $(\alpha_1 - \alpha_2) d_1 - \alpha_2 d$, $(\alpha_2 - \alpha_1) d_2 - \alpha_1 d$, $\alpha_1 d_1 + \alpha_2 d_2$ are distinct, because $\alpha_1, \alpha_2, \alpha_1 - \alpha_2 \neq 0$ and $d \notin L$, $d_1 \notin M_2$, $d_2 \notin M_1$. Hence $|G \cap (A \setminus U)| \geq 3$ which contradicts property (*). Therefore $M_2 \subset L$.

Now let us assume that there is another maximal left ideal $L' \neq L$ with $M_2 \subset L'$. Then again $A = L + L'$ and hence $1 = e + e'$ with $e \in L$, $e' \in L'$. Thus $1 = m_1 + m_2 = (m_1 e + m_1 e') + m_2$ and $c := m_1 e \in M_1 \cap L$, $c' := m_1 e' \in M_1 \cap L'$, $m_2 \in M_2 \subset L \cap L'$. For $F := \alpha_1 c + \alpha_2 c' + K$ we obtain as above $|F \cap (A \setminus U)| \geq 3$. Hence L is the only maximal left ideal with $M_2 \subset L$. For $u \in U$ the left ideal Lu is maximal and contains $M_2 = M_2 u$, hence $Lu = L$. Now let $a \in A$. By (**) there are $\lambda \in K$, $u \in U$ with $a = \lambda + u$ and we have $La \subset L\lambda + Lu = L$. So L is 2-sided and therefore $L = M_2$. We have proved that M_1 and M_2 are the only maximal left-ideals, which tells us $A \setminus U = M_1 \cup M_2$. Therefore any 2-algebra (A,K) contains at most two maximal ideals and any 2-algebra which contains two maximal ideals is a proper bilocal algebra. It remains to prove I. and III.

I.: By the Wedderburn theorem (cf.[1],p.39) for any simple algebra (A,K) the ring A is isomorphic to a total matrix ring $\mathfrak{M}_{nn}(D)$ over a skew-field D, where K can be considered as a subfield of the center of D. Since $|K| \geq 3$ there is $\lambda \in K^* \setminus \{1\}$.

If $n \geq 3$ then for $a = \begin{pmatrix} \lambda & & & & 0 \\ & 1 & & & \\ & & 0 & & \\ & & & \ddots & \\ 0 & & & & 0 \end{pmatrix} \in \mathfrak{M}_{nn}(D) = A$ the set

$(a+K) \cap (A \setminus U)$ contains the three elements a, $a-I_n$, $a+(-\lambda)I_n$ where I_n denotes the identity matrix of $\mathfrak{M}_{nn}(D)$.
Therefore $n \leq 2$.

IIIa): Let $x \in M$. If $x = 0$ we have $xM = \{0\} \neq M$, because (A,K) is not simple. If $x \neq 0$, then $x \notin xM$; for $x = xm$ with $m \in M$ we obtain $0 = x(1-m)$. This is a contradiction because $1-m \in U$ and $x \neq 0$. We have proved $xM \neq M$ for all $x \in M$.
In the same way we obtain $x_1 \dots x_n M \neq x_1 \dots x_{n-1} M$ for $x_1, \dots, x_n \in M$, if $x_1 \dots x_{n-1} M \neq 0$. If $d := \dim M$, then $M^d = \{0\}$.
b): Let $m \in M$. Since $\mathrm{Rad}\, A = M$ there is an integer $d \in \mathbb{N}$ with $m^d = 0$ and we have the equation $(1-m)(1+m+m^2+\dots+m^{d-1}) = 1$, hence $1-m \in U$. Now let us assume there is a maximal left-ideal L with $M \not\subset L$. Then $A = M+L$ and hence $1 = m+l$ with $m \in M$, $l \in L$. We obtain the contradiction $l = 1-m \in L \cap U$. By (3.1b) $(A/M,K)$ is a 2-algebra, which is simple. Hence I. gives us the second part of b), because A/M is not a skew-field by our assumption $U \neq A \setminus M$.
To finish the proof of III. we have to show that $\mathrm{Rad}\, A = M$.
First we prove $\mathrm{Rad}\, A \subset L$ for any maximal left ideal L. For $\mathrm{Rad}\, A \not\subset L$ we have $A = L + \mathrm{Rad}\, A$, hence $1 = l+m$ with $l \in L$, $m \in \mathrm{Rad}\, A$. This implies as above $l = 1-m \in L \cap U$ because m is nilpotent. The factor algebra $A' := A/\mathrm{Rad}\, A$ is semisimple and therefore the direct sum $A' = S_1 \oplus \dots \oplus S_r$ of simple ideals S_i (cf.[1],p.39). If $r > 1$, A' contains exactly the r maximal ideals $S_1 \oplus \dots \oplus S_{r-1}, \dots, S_2 \oplus \dots \oplus S_r$.
Since M is the only maximal ideal of A we have $A' = A/\mathrm{Rad}\, A = S_1$, hence $\mathrm{Rad}\, A = M$.

(3.3) Let (A,K) be a proper bilocal algebra such that $(A/\text{Rad}\,A,K)$ is separable and let M_1, M_2 be the two maximal ideals. Then:

a) M_1 contains a maximal skew-field D_1 such that $M_1 = D_1 + \text{Rad}\,A$ and $D_1 \cap \text{Rad}\,A = \{0\}$.

b) If e denotes the identity of the skew-field D_1, then M_2 contains a maximal skew-field D_2 with $1-e \in D_2$ and $M_2 = D_2 + \text{Rad}\,A$, $D_2 \cap \text{Rad}\,A = \{0\}$.

c) If we set $R := \text{Rad}\,A = M_1 \cap M_2$, $N_{11} := eRe$, $N_{22} := (1-e)R(1-e)$, $N_{12} := eA(1-e)$, $N_{21} := (1-e)Ae$, then for $i,j,k,l \in \{1,2\}$ we have: $N_{ij}N_{kl} \subset \delta_{jk}N_{il}$, $D_iN_{jk} = \delta_{ij}N_{jk}$ and $N_{ij}D_k = \delta_{jk}N_{ij}$.

d) $A = D_1 + N_{11} + N_{12} + N_{21} + N_{22} + D_2$ is a direct sum of vector subspaces and $\text{Rad}\,A = N_{11} + N_{12} + N_{21} + N_{22}$.

e) $U = D_1^* + \text{Rad}\,A + D_2^*$.

f) (N_{jk}, D_j) is a left and (N_{jk}, D_k) a right vector space.

g) Let $a = d_1 + n_{11} + n_{12} + n_{21} + n_{22} + d_2$ and $a' = d_1' + \ldots + d_2'$ be the representation of $a, a' \in A$ by d). Then $a \cdot a' =$
$$= d_1d_1' + (d_1n_{11}' + n_{11}d_1' + n_{12}n_{21}') + (d_1n_{12}' + n_{11}n_{12}' + n_{12}n_{22}' + n_{12}d_2')$$
$$+ (d_2n_{21}' + n_{21}n_{11}' + n_{22}n_{21}' + n_{21}d_1') + (d_2n_{22}' + n_{22}d_2' + n_{21}n_{12}') + d_2d_2'.$$

Proof. (i) If $\text{Rad}\,A = \{0\}$, then $A = M_1 \oplus M_2$ is a direct sum. Let $1 = e_1 + e_2$, $e_i \in M_i$, then e_i is an identity of M_i, and $(M_1^*, \cdot)$ is isomorphic to $(M_1^* + e_2, \cdot)$. Since $M_1^* + e_2 \subset U$, M_1 (and also M_2) is a skew-field. In this case we have $D_i = M_i$ and $N_{ij} = \{0\}$ $(i,j = 1,2)$. Hence $A = D_1 \oplus D_2$ is the direct sum of two skew-fields D_1 and D_2 whose centers contain the field K.

(ii) Since $A/\text{Rad}\,A$ is again bilocal, by (i) there are two skew-fields D_1' and D_2' such that $A/\text{Rad}\,A = D_1' \oplus D_2'$. Since $A/\text{Rad}\,A$ is separable, by the pricipal theorem of WEDDERBURN there is a subalgebra S of A, such that $A = S + \text{Rad}\,A$ and

$S \cong A/\mathrm{Rad}\, A$. Now $S = D_1' \oplus D_2'$ gives us that there are two skew-fields D_1 and D_2'' in A with $S = D_1 \oplus D_2''$ and $A = D_1 + D_2'' + \mathrm{Rad}\, A$. Furthermore we have $M_1 = D_1 + \mathrm{Rad}\, A$ and $M_2 = D_2'' + \mathrm{Rad}\, A$.
Let $e \in D_1$ be the identity of the skew-field D_1 . Since $1 \notin M_1$, $1-e \in M_2$ and $1-e \neq 0$. Now we consider the PEIRCE decomposition $A = eAe + N_{12} + N_{21} + (1-e)A(1-e)$ where $eAe = D_1 + N_{11} \subset M_1$ and $N_{12} + N_{21} \subset \mathrm{Rad}\, A$. The algebra $A_{22} := (1-e)A(1-e)$ has the identity element $1-e$ and $\mathrm{Rad}\, A_{22} = N_{22}$. Since $M_2 = A_{22} + \mathrm{Rad}\, A$ and $N_{22} \subset \mathrm{Rad}\, A$, we have $D_2' \cong M_2/\mathrm{Rad}\, A \cong A_{22}/A_{22} \cap \mathrm{Rad}\, A = A_{22}/N_{22}$, hence again by the principal theorem of WEDDERBURN $A_{22} = D_2 + N_{22}$, where the subalgebra D_2 of A_{22} is isomorphic to the skew-field D_2'. Since A_{22} is a local algebra, $1-e \in D_2$. We have proved a)-d), because $e(1-e) = (1-e)e = 0$, $(1-e)^2 = 1-e$, $e^2 = e$. e) follows from a),b),d) and $U = A \setminus (M_1 \cup M_2)$. f) is the consequence of c) and e).

Under (i) we proved the following statement

<u>(3.4)</u> If for a proper bilocal algebra (A,K) the intersection $\mathrm{Rad}\, A = M_1 \cap M_2$ of the two maximal ideals M_1, M_2 is $\{0\}$, then M_1 and M_2 are skew-fields and A is the direct sum $M_1 \oplus M_2$ in the sense of algebras.

<u>(3.5)</u> <u>Corollary</u>. Let (A,K) as in (3.3), then $A_i := D_i + N_{ii}$ $(i = 1,2)$ is a local algebra. If $N_{12} = \{0\} = N_{21}$, then $A = A_1 \oplus A_2$ is the direct sum of two local algebras. If (A_1',K) and (A_2',K) are two local algebras, then the direct sum $A' := A_1' \oplus A_2'$ is a bilocal algebra; if furhtermore $A_i'/\mathrm{Rad}\, A_i'$ $(i = 1,2)$ is separable, then $N_{12}' = N_{21}' = \{0\}$.

(3.6) Let (A,K) as in (3.3) with $N_{11} = N_{22} = \{0\}$. Then $a \cdot a' = d_1 d_1' + (d_1 n_{12}' + n_{12} d_2') + (d_2 n_{21}' + n_{21} d_1') + d_2 d_2'$.

The statements (3.3) and (3.6) tells us how to construct bilocal algebras, where $N_{11} = N_{22} = \{0\}$, but where at least one of the zero algebras N_{12}, N_{21} is $\neq \{0\}$. Let D_1 and D_2 be two skew-fields such that their centers contain a common field K. Now let N_{12} be a left vector space over D_1 and a right vector space over D_2, and N_{21} a left vector space over D_2 and a right vector space over D_1. If we define $A := D_1 + N_{12} + N_{21} + D_2$ as the direct sum of the K vector spaces D_1, N_{12}, N_{21}, D_2 and the multiplication by (3.6) we obtain a bilocal algebra (A,K) with $N_{11} = N_{22} = \{0\}$. A more general situation is given by:

(3.7) For $i,j = 1,2$ let (D_i, K) be an associative division algebra, N_{ij} a left vector space over D_i and a right vector space over D_j and let $A := D_1 + N_{11} + N_{12} + N_{21} + N_{22} + D_2$ be the direct sum of the K-vector-spaces and let in A a multiplication be given by

$$a \cdot a' := d_1 d_1' + (d_1 n_{11}' + n_{11} d_1') + (d_1 n_{12}' + n_{12} d_2') + (d_2 n_{21}' + n_{21} d_1') + (d_2 n_{22}' + n_{22} d_2') + d_2 d_2' .$$

Then (A,K) is a bilocal algebra, where Rad A is the zero algebra. Any bilocal algebra (A,K) where Rad A is a zero algebra, can be constructed in this way.

Remarks. 1. If (L,K) is a local algebra and L is not a skew-field and if $A := \mathfrak{M}_{22}(L)$ is the ring of all 2×2-matrices with coefficients in L, then (A,K) is a 2-algebra of class IIIb).

2. If (A,K) is a 2-algebra of class IIIb) such that $(A/\text{Rad } A, K)$ is separable, then there is a local algebra (L,K) such that L is not a skew-field and $A = \mathfrak{M}_{22}(L)$.

(3.8) Any 2-algebra (A,K) has the following property:
$\forall a,b \in A,\ b \neq 0: \ |(a+Kb)\cap(A\backslash U)| \leq 2$ or $a+Kb \subset A\backslash U$.

Proof. If $b \in U$, then $(a+Kb)\cap(A\backslash U) = [(ab^{-1}+K)\cap(A\backslash U)]b$, hence $|(a+Kb)\cap(A\backslash U)| = |(ab^{-1}+K)\cap(A\backslash U)| \leq 2$ by $(*)$.
Let $b \in A\backslash U$ but $a+Kb \not\subset A\backslash U$. If $(a+Kb)\cap(A\backslash U) \neq \emptyset$, we may assume $a \in A\backslash U$. By theorem (3.2) we have to discuss the following cases:

1. A is a skew-field, then $A\backslash U = \{0\}$ and hence $|(a+Kb)\cap(A\backslash U)| \leq 1$.

2. $A = \mathfrak{M}_{22}(D)$; then $H := A\backslash U = \left\{\begin{pmatrix} x_1 & x_2 \\ x_3 & x_4 \end{pmatrix} \mid x_3x_1^{-1}x_2 = x_4 \text{ if } x_1 \neq 0 \text{ or } x_1 = x_2x_3 = 0\right\}$. First we show:

a) $\forall h,x \in H$ with $h,x \neq 0 \ \exists u,v \in U(=GL(2,D)): uhv = x$.

Proof. We may assume $h = \begin{pmatrix} 1 & 0 \\ 0 & 0 \end{pmatrix}$. Since $x \in H$ we have to consider the following cases:

$x_1 \neq 0$. Then $x_4 = x_3x_1^{-1}x_2$, $u := \begin{pmatrix} 1 & 0 \\ x_3x_1^{-1} & 1 \end{pmatrix}$, $v := \begin{pmatrix} x_1 & x_2 \\ 0 & 1 \end{pmatrix} \in U$ and $x = uhv$.

$x_1 = x_2 = 0$. Then $x_3 \neq 0$ or $x_4 \neq 0$ hence $v := \begin{pmatrix} x_3 & x_4 \\ 0 & 1 \end{pmatrix} \in U$ or $v := \begin{pmatrix} x_3 & x_4 \\ 1 & 0 \end{pmatrix} \in U$ and $x = uhv$ for $u = \begin{pmatrix} 0 & -1 \\ 1 & 0 \end{pmatrix}$.

$x_1 = x_3 = 0$. Then $x_2 \neq 0$ or $x_4 \neq 0$ hence $u := \begin{pmatrix} x_2 & 0 \\ x_4 & 1 \end{pmatrix} \in U$ or $u := \begin{pmatrix} x_2 & 1 \\ x_4 & 0 \end{pmatrix} \in U$ and $x = uhv$ for $v = \begin{pmatrix} 0 & 1 \\ -1 & 0 \end{pmatrix}$.

b) For $u,v \in U$ we have $uHv = H$ and $|(a+Kb)\cap H| = |u[(a+Kb)\cap H]v| = |(uav+Kubv)\cap H|$.

By a) and b) we may now assume $b = \begin{pmatrix} 0 & 1 \\ 0 & 0 \end{pmatrix}$. Let $\lambda \in K$. Since $a \in H$ we have: If $a_1 \neq 0$ hence $a_4 = a_3a_1^{-1}a_2$, then:
$a+\lambda b = \begin{pmatrix} a_1 & a_2+\lambda \\ a_3 & a_4 \end{pmatrix} \in H \Leftrightarrow a_4 = a_3a_1^{-1}(a_2+\lambda) \Leftrightarrow a_3a_1^{-1}\lambda = 0 \Leftrightarrow$
$\Leftrightarrow a+Kb \subset H$ for $a_3 = 0$ and $(a+Kb)\cap H = \{a\}$ for $a_3 \neq 0$.

If $a_1 = 0$ and $a_2 = 0$, then: $\begin{pmatrix} 0 & \lambda \\ a_3 & a_4 \end{pmatrix} \in H \Leftrightarrow \lambda a_3 = 0 \Leftrightarrow$ $a+Kb \subset H$ for $a_3 = 0$ and $(a+Kb)\cap H = \{a\}$ for $a_3 \neq 0$.

If $a_1 = 0$ and $a_3 = 0$, then $\begin{pmatrix} 0 & a_2+\lambda \\ 0 & a_4 \end{pmatrix} \in H$ for all $\lambda \in K$, thus $a + Kb \subset H$.

3. A is bilocal (hence $A \setminus U = M_1 \cup M_2$) or A is local (hence $A \setminus U = M$). If $|(a + Kb) \cap M_i| \geq 2$, then $a + Kb \subset M_i$ because M_i is an ideal.

4. A is of class IIIb). Let $h \in H := A \setminus U$ and let L be a maximal left ideal with $h \in L$. Then by (3.2IIIb) $M \subset L \subset H$ hence $h + M \subset L + M \subset L \subset H$. Thus $H + M \subset H$. Now let $\lambda \in K$. Then $a + \lambda b \in H$ implies $a + \lambda b + M \subset H$. Let us assume $a + Kb \not\subset H$, then $(a + Kb + M)/M \not\subset H/M$ and $|(a + Kb) \cap H| \leq$ $\leq |(a + Kb + M)/M \cap H/M| \leq 2$ by 2. because $A/M = \mathfrak{M}_{22}(D)$.

Remark 3. The structure theorem (3.2) for associative 2-algebras can be extended to the class of alternative 2-algebras. By an alternative algebra the associativity condition is replaced by the weaker alternative laws $a(ab) = a^2b$ and $(ab)b = ab^2$. All alternative algebras with identity belong to the class of all algebras, where the set U of unities forms a loop, i.e. $U \circ U \subset U$. An alternative ring D is called an alternative field, if $U = D \setminus \{0\}$. We obtain the structure theorem for alternative 2-algebras (A,K), if we replace in (3.2) always the word "skew-field" by "alternative field". The proof can be obtained by using the Wedderburn-structure theorems which were extended to alternative algebras by E. ARTIN and M. ZORN. The reader can find these extended theorems with proofs in the book of R.D. SCHAFER [40].

§ 4 ALTERNATIVE KINEMATIC ALGEBRAS

In this section we extend the report on associative kinematic algebras given in [14],[15] to alternative ones. An algebra (A,K) with identity 1 is called <u>kinematic</u>, if the following condition (K) is valid

<u>(K)</u> for any $a \in A$, $a^2 \in K + Ka$

From now on let (A,K) be an alternative kinematic algebra with $K \neq \mathbb{Z}_2$. As in [14],[15] we introduce the sets

$$J := \{x \in A \setminus K^* \mid x^2 \in K\} \quad \text{if char } K \neq 2$$

$$J := \{x \in A \mid x^2 \in K\} \quad \text{if char } K = 2$$

$$N := \{x \in A \mid x^2 = 0\}$$

By (K) to every $x \in A \setminus K$ there are exactly two elements $\alpha_x, \beta_x \in K$ with $x^2 = \alpha_x + \beta_x x$. For $\lambda \in K$ we set $\alpha_\lambda := -\lambda^2$ and $\beta_\lambda := 2\lambda$.

<u>(4.1)</u> For $x,y \in A$, $\tau \in K$ we have $\beta_{\tau x} = \tau\beta_x$ and $\beta_{x+y} = \beta_x + \beta_y$.

Proof. (i) For $x \in K$ we have $\beta_{\tau x} = 2\tau x = \tau 2x = \tau\beta_x$ by definition. Now let $x \notin K$. Then $\alpha_{\tau x} + \beta_{\tau x}\tau x = (\tau x)^2 = \tau^2 x^2 = \tau^2\alpha_x + \tau^2\beta_x x$, $\beta_{\tau x} = \tau\beta_x$.

(ii) For $x \notin K$, $\rho \in K$ we have $\alpha_{\rho+x} + \beta_{\rho+x}(\rho + x) = (\rho + x)^2 = \rho^2 + 2\rho x + x^2 = \rho^2 + 2\rho x + \alpha_x + \beta_x x$ and $\beta_{\rho+x} = 2\rho + \beta_x = \beta_\rho + \beta_x$.

(iii) If $1,x,y$ are linearly dependent, but $1,x$ linearly independent, then $y = \rho + \sigma x$. By (i) and (ii) we obtain $\beta_{x+y} = \beta_{\rho+(1+\sigma)x} = \beta_\rho + \beta_{(1+\sigma)x} = \beta_\rho + (1+\sigma)\beta_x = \beta_x + \beta_\rho + \sigma\beta_x = \beta_x + \beta_\rho + \beta_{\sigma x} = \beta_x + \beta_{\rho+\sigma x} = \beta_x + \beta_y$.

(iv) Let $1,x,y$ linearly independent. Since $K \neq \mathbb{Z}_2$, there is an $\lambda \in K \setminus \{0,1\}$ and from the equation $(x + \lambda y)^2 - \lambda(x + y)^2 = (1 - \lambda)(x^2 - \lambda y^2)$ we obtain

$$\alpha_{x+\lambda y} - \lambda\alpha_{x+y} + (\beta_{x+\lambda y} - \lambda\beta_{x+y})x + \lambda(\beta_{x+\lambda y} - \beta_{x+y})y$$
$$= (1 - \lambda)(\alpha_x - \lambda\alpha_y + \beta_x x - \lambda\beta_y y),$$

and by comparing the coefficients of x and y,
$\beta_{x+\lambda y} - \lambda\beta_{x+y} = (1-\lambda)\beta_x$ and $\lambda(\beta_{x+\lambda y} - \beta_{x+y}) = (1-\lambda)(-\lambda)\beta_y$,
hence $(1-\lambda)(-\lambda)\beta_y = \lambda(\lambda\beta_{x+y}+(1-\lambda)\beta_x-\beta_{x+y}) = \lambda(1-\lambda)(\beta_x - \beta_{x+y})$,
i.e. $\beta_x + \beta_y = \beta_{x+y}$.

(4.2) J is a vector subspace with $\operatorname{codim} J = 1$, if $\operatorname{char} K \neq 2$, and $\operatorname{codim} J \leq 1$, if $\operatorname{char} K = 2$.

Proof. In both cases $\operatorname{char} K \neq 2$ and $\operatorname{char} K = 2$ we have $J = \{x \in A \mid \beta_x = 0\}$ and by (4.1) J is a vector subspace. Let $x \in A\setminus K$. For $\operatorname{char} K \neq 2$ we have $x = \left(x - \frac{1}{2}\beta_x\right) + \frac{1}{2}\beta_x \in J+K$. Since $K \not\subset J$, $\operatorname{codim} J = 1$. For $\operatorname{char} K = 2$ let $J \neq A$. Then there is $a \in A\setminus J$ and $\beta_a \neq 0$. For $\lambda = \beta_x\beta_a^{-1}$ we have $\beta_{x-\lambda a} = 0$, hence $x - \lambda a \in J$, i.e. $x \in \lambda a + J \subset Ka + J$.

(4.3) The map $\bar{\ }: A \longrightarrow A;\ x \longrightarrow \bar{x} := \beta_x - x$ is an involutorial K-antiautomorphism.

Proof. Let $x,y \in A$. Then $\overline{x+y} = \beta_{x+y} - (x+y) = \beta_x + \beta_y - x - y = \bar{x} + \bar{y}$ by (4.1). Furthermore $x y \bar{y} \bar{x} = xy(\beta_y - y)\bar{x} = x(\beta_y y - y^2)\bar{x}$ $= x(\beta_y y - (\alpha_y + \beta_y y))\bar{x} = x(-\alpha_y)\bar{x} = (-\alpha_y)(-\alpha_x) = \alpha_y\alpha_x \in K$ and

$$
\begin{aligned}
xy + \bar{y}\,\bar{x} &= xy + (\beta_y - y)(\beta_x - x)\\
&= xy + yx - \beta_y x - \beta_x y + \beta_x\beta_y\\
&= (x+y)^2 - x^2 - y^2 - \beta_y x - \beta_x y + \beta_x\beta_y\\
&= \alpha_{x+y} - \alpha_x - \alpha_y + \beta_x\beta_y \in K.
\end{aligned}
$$

Since $x y \overline{x y} = -\alpha_{xy} \in K$, $xy + \overline{xy} = \beta_{xy} \in K$, we obtain $\overline{xy} = \bar{y}\,\bar{x}$, if $xy \notin K$. If $xy \in K$ and $x \notin K$, then $x(y+1) = xy + x \notin K$, hence $\overline{xy} + \bar{x} = \overline{xy+x} = \overline{x(y+1)} = \overline{y+1}\cdot\bar{x} = (\bar{y}+1)\bar{x} = \bar{y}\,\bar{x} + \bar{x}$, and so $\overline{xy} = \bar{y}\,\bar{x}$. Finally $\bar{\bar{x}} = \overline{\beta_x - x} = \overline{\beta_x} - \bar{x} = \beta_x - (\beta_x - x) = x$.

(4.4) If char $K \neq 2$, then $\bar{\ }: \begin{cases} A = K+J \longrightarrow K+J \\ a = \lambda+x \longrightarrow \bar{a} = \lambda-x \end{cases}$ and hence $\bar{a} = a$, if and only if $a \in K$.
If char $K = 2$, $A \neq J$, $b \in A \setminus J$, then $\bar{\ }: \begin{cases} A = Kb + J \longrightarrow Kb + J \\ a = \lambda b + x \longrightarrow \lambda b + (\lambda\beta_b + x) \end{cases}$ and hence $a = \bar{a}$, if and only if $a \in J$.
In both cases $J = \{a \in A \mid \bar{a} = -a\}$.

(4.5) Let $u,v,w \in J$. Then
a) $uv + vu \in K$
b) $uv \in J \Rightarrow uv = -vu$ and $vu \in J$
c) $(uv)w \in J \Rightarrow (uv)w = w(vu)$

Proof. $uv + vu = uv + (-v)(-u) = uv + \bar{v}\,\bar{u}$ (by (4.4))
$= uv + \overline{uv} \in K$ (by (4.3)).
$uv \in J$ implies $\overline{uv} = -uv$, hence $uv + vu = 0$. If $(uv)w \in J$, then $-(uv)w = \overline{(uv)w} = \bar{w}(\bar{v}\,\bar{u}) = -w(vu)$.

(4.6) Let $x,y,z \in A$, then
a) $xy - yx \in J$
b) $xy \in J \Leftrightarrow yx \in J$
c) $yx \in K + Kx + Ky + Kxy$
d) $x(yz + zy) = (xy)z + (xz)y$
e) If $yz \in J$, then: $(xy)z = -(x\bar{z})\bar{y}$ and $z(yx) = -\bar{y}(\bar{z}x)$

Proof. a,b): $xy + \overline{xy} = xy + \bar{y}\,\bar{x} = xy + (\beta_y - y)(\beta_x - x)$
$= xy + \beta_y\beta_x - \beta_y x - \beta_x y + yx = (\beta_x - x)(\beta_y - y) + yx$
$= \bar{x}\,\bar{y} + yx$
and so $xy - yx = \overline{yx} - \overline{xy} = -\overline{xy - yx}$, thus $xy - yx \in J$ by (4.4).
a) and (4.2) implies b).
c): $yx = (xy + \overline{xy} - \beta_x\beta_y) + \beta_y x + \beta_x y - xy \in K + Kx + Ky + Kxy$.
d): By applying the right alternative law "$ab^2 = (ab)b$" we obtain $xy^2 + x(yz) + x(zy) + xz^2 = x(y+z)^2 = (x(y+z))(y+z) =$
$= xy^2 + (xy)z + (xz)y + xz^2$
which gives us d).

e): $(xz)y+(xy)z = x(zy+yz)$ by d)

$= x(zy - \overline{z}\,\overline{y})$ because $\overline{yz} = -yz$

$= x(z(y+\overline{y}) - (z+\overline{z})\overline{y})$

$= (xz)(y+\overline{y}) - (x(z+\overline{z}))\overline{y}$ because $y+\overline{y}, z+\overline{z} \in K$

$= (xz)y - (x\overline{z})\overline{y}$

and so $(xy)z = -(x\overline{z})\overline{y}$. Since "$yz \in J \Leftrightarrow zy \in J$" by b) we have also $(xz)y = -(x\overline{y})\overline{z}$, hence $\overline{y}(\overline{z}\,\overline{x}) = \overline{(xz)y} = \overline{-(x\overline{y})\overline{z}} = -z(y\overline{x})$ and so $\overline{y}(\overline{z}x) = -z(yx)$ by replacing $\overline{x}$ by x.

(4.7) Let $a,b,x \in A$ and $c,d,u \in J$.

a) $xu \in J \Leftrightarrow xu = u\overline{x} \Leftrightarrow \overline{x}u = ux \in J$.

b) $au, bu, (ab)u \in J \Rightarrow (ab)u = b(au)$ and $(ba)u \in J$.

c) If $cd, xc, xd, x(cd) \in J$, then $\overline{x}(cd) = (xc)d$.

d) If $cd, uc, ud, u(cd) \in J$, then $-u(cd) = (uc)d$.

e) If $cd \in J$ and $yv \in J$ for $y \in \{a,b,ab\}$, $v \in \{c,d,cd\}$, then:

α) $(ab)(cd) = (ba)(cd) = a(b(cd))$

β) $ab - ba \in N$ or $cd = 0$.

Proof. a): $xu \in J$ implies $\overline{xu} = -xu$, hence $xu = -\overline{u}\,\overline{x} = u\overline{x}$. Since "$xu \in J \Leftrightarrow ux \in J$" by (4.6b), we have also $\overline{x}u = ux$. Conversely, if $u\overline{x} = xu$, then $\overline{xu} = \overline{u}\,\overline{x} = -u\overline{x} = -xu$, because $\overline{u} = -u$, thus $xu \in J$.

b): $b(au) = b(u\overline{a}) = -\overline{u}(\overline{b}\,\overline{a}) = u(\overline{ab}) = (ab)u$ by a) and (4.6e). Since $(ba)u \in Ku + Kau + Kbu + K(ab)u$ by (4.6c) and J is a vector subspace our assumptions imply $(ba)u \in J$.

c): $\overline{x}(cd) = (cd)x = -(c\overline{x})\overline{d} = (xc)d$ by a) and (4.6e)

d) is a consequence of c), because $\overline{u} = -u$.

e): By applying (4.6e) several times we obtain the following equations:

$(ab)(cd) = c((\overline{b}\,\overline{a})d)$ because $c(ab) \in J$ by (4.6b) and $\overline{c} = -c$

$= c((\overline{b}d)a)$ because $\overline{a}d = da \in J$

$= -\overline{(\overline{b}d)}(\overline{c}a)$ because $b(cd) = c(\overline{b}d) \in J$ by (4.6e) and hence $(\overline{b}d)c \in J$ by (4.6b)

$= -(\overline{b}d)(ca)$ because $\overline{b}d = db \in J$ by (4.6b).

Now $ad \in J$ implies $a(dc) = d(\overline{a}c)$ by (4.6e). Since $ac, cd \in J$

we have by a) $\overline{a}c = ca$ and $dc = -cd$, thus $d(ca) = -a(cd)$. So $a(cd) \in J$ implies $d(ca) \in J$. Therefore we have

$$\begin{aligned} -(\overline{b}d)(ca) &= (\overline{b}(ca))d = (c(ba))d && \text{because } bc \in J \\ &= (cd)(\overline{ba}) && \text{since } (ba)d \in J \text{ by b)} \\ &= (ba)(cd) && \text{by a) since } cd, (ab)(cd) \in J \\ &= a(b(cd)) && \text{by b).} \end{aligned}$$

The equation $(ab)(cd) = (ba)(cd)$ gives us $(ab - ba)(cd) = 0$. By (4.6a) $ab - ba \in J$, thus $cd = 0$, if $ab - ba \notin N$.

(4.8) The map $q : A \longrightarrow K;\ a \longrightarrow q(a) := a\overline{a} = \overline{a}a$ is a quadratic form with the bilinear form $f(a,b) = a\overline{b} + b\overline{a} = \overline{a}b + \overline{b}a = f(\overline{a},\overline{b})$, and we have

a) $q(ab) = q(a)\cdot q(b)$, i.e. (A,K,q) is a composition algebra

b) $f(ac,bc) = f(ca,cb) = q(c)\cdot f(a,b)$ for all $a,b,c \in A$

c) $U = \{a \in A \mid q(a) \neq 0\}$, where U denotes the set of units of (A,K)

d) $u^{\perp} := \{x \in A \mid f(u,x) = 0\} = uJ = Ju$ for $u \in U$

Proof. Since $f(a,b) = q(a+b) - q(a) - q(b) = a\overline{b} + b\overline{a}$, q is a quadratic form with $q(ab) = (ab)\overline{ab} = a\,b\,\overline{b}\,\overline{a} = aq(b)\overline{a} = q(a)\cdot q(b)$ by (4.3). Since $f(ac,bc) = q(ac+bc) - q(ac) - q(bc) =$ $= q(c)\cdot(q(a+b) - q(a) - q(b)) = q(c)f(a,b)$, b) is valid by a). For $a \in A$ with $q(a) \neq 0$, $a^{-1} = \frac{1}{q(a)}\,\overline{a}$.

d): Since $1^{\perp} = \{x \in A \mid f(1,x) = \overline{x} + x = 0\} = J$ by (4.4) and $f(u,x) = q(u)\cdot f(1,u^{-1}x) = q(u)\cdot f(1,xu^{-1})$ we have $u^{\perp} = uJ = Ju$.

(4.9) Every kinematic algebra (A,K) is a 2-algebra.

Proof. Since $U = \{x \in A \mid x\overline{x} \neq 0\}$, an element $a+\lambda \in a+K$ does not lie in U, if $0 = (a+\lambda)(\overline{a+\lambda}) = \lambda^2 + (a+\overline{a})\lambda + a\overline{a}$. Since $a+\overline{a},\ a\overline{a} \in K$, this equation has at most two solutions $\lambda \in K$, thus $|(a+K) \cap (A\setminus U)| \leq 2$.

(4.10) Let M be an ideal of (A,K). Then
a) $\overline{M}$ is an ideal.
b) If $M \neq A$, then $M \cap \overline{M} \subset N$.
c) $M^{\perp} := \{a \in A \mid \forall m \in M : f(a,m) = 0\}$ is an ideal.

Proof. b): Let $x \in M \cap \overline{M}$. Then $x + \overline{x}, x\overline{x} \in K \cap (M \cap \overline{M}) = \{0\}$. But $\overline{x} + x = 0 = x\overline{x}$ implies $\overline{x} = -x$, hence $x^2 = 0$ and so $x \in N$.
c): Let $m \in M$, $x \in M^{\perp}$ and $u \in U$, then $f(ux,m) = f(ux,u(u^{-1}m))$ $= q(u)f(x,u^{-1}m) = 0$, because $u^{-1}m \in M$ and $f(xu,m) = 0$.
Now let $a \in A\setminus U$, hence $a\overline{a} = 0$. Then $(\lambda+a)\overline{(\lambda+a)} = \lambda^2+\lambda a+\lambda\overline{a} =$ $= \lambda(\lambda + a + \overline{a})$ for any $\lambda \in K$. Since $|K| \geq 3$ there is a $\lambda \in K^*$ with $\lambda + a + \overline{a} \neq 0$. Thus $\lambda + a \in U$. Then $ax = ((\lambda+a) - \lambda)x =$ $= (\lambda+a)x - \lambda x \in M^{\perp}$ because $\lambda+a, \lambda \in U$, and $xa \in M^{\perp}$.

(4.11) $\mathrm{Rad}\, q := \{x \in A^{\perp} \mid q(x) = 0\} = \mathrm{Rad}\, A \subset N$.

Proof. If char $K \neq 2$, then $\mathrm{Rad}\, q = A^{\perp}$ and $\mathrm{Rad}\, q$ is an ideal by (4.10c). Now let char $K = 2$ and $x,y \in \mathrm{Rad}\, q$. Then $f(x,y) = 0$ and $q(x+y) = q(x) + q(y) + f(x,y) = 0$, hence $x+y \in \mathrm{Rad}\, q$. For $a \in A$, we have $ax, xa \in A^{\perp}$ by (4.10c) and $q(ax) = q(a)q(x)$ $= q(xa) = 0$ by (4.8a). Thus $\mathrm{Rad}\, q$ is an ideal. In both cases $\mathrm{Rad}\, q \neq A$, because $q(1) = 1 \neq 0$. Since $\mathrm{Rad}\, q \subset 1^{\perp} =$ $\{x \in A \mid x+\overline{x} = 0\} = J$ by (4.4), we have $\mathrm{Rad}\, q = \overline{\mathrm{Rad}\, q}$ and by (4.10b) $\mathrm{Rad}\, q \subset N$. This implies $\mathrm{Rad}\, q \subset \mathrm{Rad}\, A$ by definition of $\mathrm{Rad}\, A$.
Now let $x \in \mathrm{Rad}\, A$, hence $x \notin U$ and so $0 = x\overline{x} = x(\beta_x - x) = \beta_x x - x^2$ or $x^2 = \beta_x x$. This gives us $x^n = (\beta_x)^{n-1}x$, thus $\beta_x = 0$ and $x \in N$.
From $\mathrm{Rad}\, A \subset N \subset J$ we obtain for any $y \in A$, $yx \in \mathrm{Rad}\, A$, hence $y\overline{x} = -yx = \overline{yx} = \overline{x}\,\overline{y} = (-x)\overline{y}$ and so $f(x,y) = x\overline{y} + y\overline{x} = 0$. We have proved $\mathrm{Rad}\, A \subset A^{\perp}$ and by $\mathrm{Rad}\, A \subset N$ also $q(x) = 0$ for any $x \in \mathrm{Rad}\, A$. This gives us $\mathrm{Rad}\, A \subset \mathrm{Rad}\, q$.

(4.12) If M is a maximal ideal of (A,K), then $\mathrm{Rad}\, A = M \cap \overline{M}$.

Proof. Suppose $\text{Rad } A \not\subset M$, then $A = M + \text{Rad } A$ and there are $m \in M$, $r \in \text{Rad } A$ with $1 = m+r$, or $m = 1-r \in M$, but $1-r \in U$ because $(1-r)^{-1} = 1+r$. Hence $\text{Rad } A \subset M \cap \overline{M} \subset N$ by (4.10b) and thus $\text{Rad } A = M \cap \overline{M}$.

(4.13) If $\text{Rad } A = \{0\}$, then either $\{0\}$ is the only ideal $\neq A$ (i.e. (A,K) is simple) or $A = K \oplus K$ is the direct sum in sense of algebras (i.e. A is the algebra of the anormal complex numbers over K).

Proof. Suppose there is a maximal ideal $M \neq \{0\}$. Then by (4.12) $M \cap \overline{M} = \text{Rad } A = \{0\}$ and $A = M + \overline{M}$. Consequently there is a unique element $e \in M$ with $1 = e + \overline{e}$; e and $\overline{e}$ are orthogonal idempotents. For any $x \in A$, $x = xe + x\overline{e} = ex + \overline{e}x$, hence $xe - ex = \overline{e}x - x\overline{e} \in M \cap \overline{M} = \{0\}$ and so $xe = ex$ and $\overline{e}x = x\overline{e}$. Therefore $x = \alpha e + \beta\overline{e}$ for $\alpha = xe + \overline{xe}$ and $\beta = x\overline{e} + \overline{x\overline{e}} = x\overline{e} + e\overline{x}$. Since $\alpha, \beta \in K$ we have $A = Ke \oplus K\overline{e}$.

Each kinematic algebra belongs to one of the following classes:

1. J is an ideal
2. J is not an ideal

(4.14) The first class consists of local algebras and comprises the following subclasses:

a) $A = J$; then $\text{char } K = 2$, A is commutative and associative and N is the maximal ideal.

b) $A \neq J$; then $\text{char } K \neq 2$, $J = N = \text{Rad } A$, $K = A/N$ and $N^4 = \{0\}$.
Furthermore: A associative $\Leftrightarrow N^3 = \{0\}$;
A commutative $\Leftrightarrow N^2 = \{0\}$.

Proof. If $J = A$, then $\text{char } K = 2$ by (4.2) and $A = J = \{a \in A \mid \overline{a} = a\}$ by (4.4). Hence for $a,b \in A$, $ab = \overline{a}\,\overline{b} = \overline{ba} = ba$. The associativity follows from (4.7eα), if we substitute $d = 1$, because $A = J$.

Now let $A \neq J$, then J is a maximal ideal. Since for char $K = 2$, $1 \in J$, we have here char $K \neq 2$. By (4.8) for $x \in J$, $q(x) = x\bar{x} = x(-x) = 0$, because x is not a unit. Hence $J = N$.

For any $a,b,c,d \in J$ the conditions of (4.7e) are fulfilled because $J^2 \subset J$. Therefore $(ab)(cd) = a(b(cd)) = (ba)(cd)$ $= -(ab)(cd)$ because $a,b,ab \in J$ implies $ba = -ab$ by (4.5b). Thus $N^4 = 0$, because char $K \neq 2$. Since char $K \neq 2$, $A = K+J$, and for $\alpha+a, \beta+b, \gamma+c \in K+J$ we have

$$(\alpha+a)((\beta+b)(\gamma+c)) = \alpha\beta\gamma + \alpha\beta c + \beta\gamma a + \gamma\alpha b + \alpha bc + \beta ca + \gamma ab + a(bc)$$

and $((\alpha+a)(\beta+b))(\gamma+c) = \quad \ldots \quad +(ab)c$

$= \quad \ldots \quad -a(bc)$

by (4.7d) because $J^2 \subset J$. Thus (A,K) is associative, if and only if $N^3 = \{0\}$. The last part follows from (4.5b) and $J^2 \subset J$.

Remark. The smallest non-associative kinematic algebra (A,K) of class 1. over a given field K we obtain by defining the multiplication on the seven dimensional vector space (N,K). If $B := \{a,b,c,ab,bc,ac,(ab)c\}$ is a base, then the multiplication is given by Table I

	a	b	c	ab	bc	ac	(ab)c
a	0	ab	ac	0	-(ab)c	0	0
b	-ab	0	bc	0	0	(ab)c	0
c	-ac	-bc	0	-(ab)c	0	0	0
ab	0	0	(ab)c	0	0	0	0
bc	(ab)c	0	0	0	0	0	0
ac	0	-(ab)c	0	0	0	0	0
(ab)c	0	0	0	0	0	0	0

From now on let (A,K) be an algebra of class 2., i.e. J is not an ideal.

(4.15) If J is not an ideal, then $N \neq J$ and there is an $s \in U$ with $s - \overline{s} \notin N$.
For any $s \in U$ with $s - \overline{s} \notin N$ we have
a) $K(s) := K + Ks$ is a commutative field or the algebra of the anormal complex numbers over K.

b) If $s \in J$, then $K(s) = K \oplus Ks$

c) $A = K(s) \oplus (J \cap Js)$ is the direct orthogonal sum of vector subspaces.

d) $(\mathrm{Rad}\, A)^3 = \{0\}$

e) If $J \cap Js \subset N$, then $J \cap Js = N = \mathrm{Rad}\, A$,
A associative $\Leftrightarrow$ $(\mathrm{Rad}\, A)^2 = \{0\}$,
A commutative $\Leftrightarrow$ $\mathrm{Rad}\, A = \{0\}$,

and we have

α) If $K(s)$ is a field, then (A,K) is a local algebra and $\dim(A,K)$ is even.

β) If $K(s)$ is not a field, then (A,K) is bilocal and $\dim(A,K)$ can be any number ≥ 2.

Proof. Suppose $J = N$. Then $\mathrm{char}\, K \neq 2$ because otherwise $1 \in K \subset J = N$. Any $\lambda + n$ with $\lambda \in K^*$, $n \in J$ is a unit, because $(\lambda+n)\overline{(\lambda+n)} = \lambda^2 - n^2 = \lambda^2$. Thus $U = A \setminus J$, which implies that J is an ideal.
Let $\mathrm{char}\, K = 2$. Since $J \neq A$ and $|K| \geq 3$, there is an $a \in A \setminus J$ and $\lambda \in K$ with $(a+\lambda)\overline{(a+\lambda)} = \lambda^2 + \lambda(a+\overline{a}) + a\overline{a} \neq 0$. For $s := a + \lambda$ we have $s \in U \setminus J$ and $s - \overline{s} = a + \overline{a} \in K^*$.
Now let $\mathrm{char}\, K \neq 2$. Since $N \neq J$, there is an $s \in J \setminus N$. Then $s \in U$ and $s - \overline{s} = 2s \notin N$.
a): $s - \overline{s} \notin N$ implies $\overline{s} \neq s$ and so $K(s)$ is either a field or the algebra of anormal complex numbers over K.
c): Since $K \perp J$ and $Ks \perp Js$ we have $K(s) \perp (J \cap Js)$.
For $x \in K(s) \cap J \cap Js$ we obtain $x\overline{s} \in Js\overline{s} = J$, hence $-x\overline{s} = \overline{x\overline{s}} = s\overline{x} = -sx = -xs$ because the subalgebra $K(s)$ is commutative. Thus $x(s-\overline{s}) = 0$ and so $x = 0$ because $s-\overline{s} \in J \setminus N \subset U$.
Therefore $K(s) \cap (J \cap Js) = \{0\}$. Since $\dim K(s) = 2 = \mathrm{codim}(J \cap Js)$,

$A = K(s) + (J \cap Js)$.

d): Let $b,c,d \in \text{Rad } A$, then $cd \in \text{Rad } A \subset N \subset J$ and $yv \in J$ for $y \in \{s,b,sb\}$, $v \in \{c,d,cd\}$ because Rad A is an ideal. Hence by (4.7e) $s(b(cd)) = (sb)(cd)$. Since also $sb, s(cd), s(b(cd)) \in \text{Rad } A \subset N \subset J$, $(sb)(cd) = \overline{s}(b(cd))$ by (4.7c). Thus $(s-\overline{s})(b(cd)) = 0$ and so $b(cd) = 0$ since $s-\overline{s} \in U$.

e): Let $x \in K(s)$ and $n \in J \cap Js \subset N$ with $x+n \in N$. Since $N \subset J$ and since J is a vector subspace, $x \in J$, hence $0 = (x+n)\overline{(x+n)} = x\overline{x} = -x^2$ by $x \perp n$.

By a) $x^2 = 0$ implies $x = 0$. Therefore $N = J \cap Js$ is a vector subspace and hence for $n,m \in N$ we have $n+m \in N$, so $0 = (n+m)^2 = nm + mn$. This gives us $nm = -mn$ and so $(nm)^2 = -nm\,mn = 0$, hence $N^2 \subset N$. Now let $y \in K(s)$. Then $y \perp m$, i.e. $0 = y\overline{m} + m\overline{y} = -ym + m\overline{y}$, hence $ym = m\overline{y}$ and $(ym)^2 = ym\,m\overline{y} = 0$. Consequently $(y+n)m = ym + nm \in N + N = N$ and therefore N is an ideal and $N = \text{Rad } A$ by (4.11). For $x \in K(s)$ and $n \in N$, $(x+n)(\overline{x}+\overline{n}) = x\overline{x}$, hence $x+n \in U$ if and only if $x\overline{x} \neq 0$. Therefore (A,K) is a local algebra, if $K(s)$ is a field, and (A,K) is a bilocal algebra with the maximal ideals $M_i = K(s-\lambda_i) + \text{Rad } A$ $(i = 1,2)$, if the minimal polynomial of s has the two roots $\lambda_1, \lambda_2 \in K$.

If A is commutative, then for any $c \in \text{Rad } A$, $cs = sc = c\overline{s}$ by (4.7a), because $sc \in \text{Rad } A \subset J$. Thus $c(s-\overline{s}) = 0$ and so $c = 0$. For $c,d \in \text{Rad } A$ we have $\overline{s}(cd) = (sc)d$ by (4.7c), hence $(\overline{s}-s)(cd) = 0$ and so $cd = 0$, if A is associative. Conversely, if $(\text{Rad } A)^2 = \{0\}$, then A is associative, because $A = K(s) + \text{Rad } A$ and because for any $n \in \text{Rad } A$ the subalgebra $K(s,n)$ is associative.

If $K(s)$ is a field, then $(\text{Rad } A, K(s))$ is a vector space, thus $\dim(A,K) = \dim(K(s) + \text{Rad } A, K) = 2 + \dim(\text{Rad } A, K) =$

$$= 2 + [\text{Rad } A : K(s)]\,[K(s):K] = 2(1 + [\text{Rad } A : K(s)]).$$

β) will be verified later by examples.

(4.16) Under the assumptions of (4.15) let $J \cap Js \not\subset N$. Then there exist $a,b \in J \cap Js$ with $s = ab$ and we have

a) $K(a,b) = K(s) \oplus K(s)\, b = K(s) \oplus (Ka + Kb)$ is a quaternion division algebra or the algebra $\mathfrak{M}_{22}(K)$ of all 2×2-matrices with coefficients in K.

b) $A = K(a,b) \oplus (J \cap Js \cap Ja \cap Jb)$ is the direct orthogonal sum of vector subspaces.

c) $(\mathrm{Rad}\ A)^2 = \{0\}$

d) If $J \cap Js \cap Ja \cap Jb \subset N$, then $J \cap Js \cap Ja \cap Jb = \mathrm{Rad}\ A$, A associative $\Leftrightarrow$ $\mathrm{Rad}\ A = \{0\}$, and we have

α) If $K(a,b)$ is a division algebra, then (A,K) is local, $N = \mathrm{Rad}\ A$ and $4 \mid \dim(\mathrm{Rad}\ A, K)$.

β) If $K(a,b) = \mathfrak{M}_{22}(K)$, then $\dim(\mathrm{Rad}\ A,K)$ can be any even number.

e) If $J \cap Js \cap Ja \cap Jb \not\subset N$, then $\mathrm{Rad}\ A = \{0\}$ and for any $c \in (J \cap Js \cap Ja \cap Jb) \setminus N$ we have $A = K(a,b,c) = K(a,b) \oplus K(a,b) \cdot c$. A is not associative and of degree 8.

α) If $q(x) = x\bar{x} = 0$ only for $x = 0$, then A is an alternative field.

β) If there is an $x \neq 0$ with $q(x) = 0$, then A has the representation $A = M_{22}(K) \oplus M_{22}(K) \cdot d$ with $d^2 = 1$ and the quadratic form q is of index 4.

Proof. Since $J \cap Js \not\subset N$, there are $b,x \in J$ with $b = xs \notin N$. For $a = x^{-1} = (x\bar{x})^{-1} \cdot \bar{x} \in J$ we have $s = ab$ and $a,b \in J \cap Js$.
a): The subalgebra $K(a,b)$ has by (4.6c) the representation $K + Ka + Kb + Kab$ and is not commutative because $ab = s \neq \bar{s} = \bar{b}\,\bar{a} = ba$. Since $Ka + Kb \subset J \cap Js$ and $K(s)b = Kb + Ksb = Kb + Kab^2 = Kb + Ka$ by (4.15c), $K(a,b) = K(s) \oplus K(s)b$.
Now let us assume that $K(a,b)$ is not a division algebra. Then there exists an $x \in K(a,b)$ with $x \neq 0$, but $x\bar{x} = 0$. For $y \in J \cap x^{\perp} \cap K(a,b)$ with $y \neq 0$ we have either $y^2 = 0$, hence

$y \in N^*$, or $xy \neq 0$ and $0 = x\bar{y} + y\bar{x} = -xy + y\bar{x}$, hence $(xy)^2 = (xy)(y\bar{x}) = (x\bar{x})y^2 = 0$ and so $xy \in N^*$. Thus $N \cap K(a,b) \neq \{0\}$. Let $n' \in N^* \cap K(a,b)$. Since $a^\perp \cap b^\perp \cap K(a,b) = K + Kab = K + Ks$ and $(K + Ks) \cap N = \{0\}$ by (4.15a) we have $n' \notin a^\perp$ or $n' \notin b^\perp$. Let us assume $n' \notin a^\perp$, hence $f(a,n') \neq 0$. For $n := -f(a,n')^{-1} \cdot n'$ and $m := a + (a\bar{a})n$ we have $1 = -f(a,n) = -a\bar{n} - n\bar{a} = an + na$ and $n^2 = m^2 = a^2 + (a\bar{a})(an + na) = a^2 - a\bar{a}\, f(a,n) = a^2 + a\bar{a} = 0$. Furthermore $(an)^2 = an(1-na) = an$, $(na)^2 = na(1-an) = na$, $n(an) = n(1-na) = n$, $nm = na$. If we set $e_{11} := an$, $e_{12} := m$, $e_{21} := n$, $e_{22} := na$, then $e_{ij}e_{kl} = \delta_{jk}e_{il}$. This tells us that $K(a,b) = Ke_{11} \oplus Ke_{12} \oplus Ke_{21} \oplus Ke_{22}$ is the algebra of the 2×2-matrices over K.

b) Since $K(a,b) = K + Ks + Ka + Kb$ we have $K(a,b) \perp (J \cap Js \cap Ja \cap Jb)$ by (4.8d). Now let $x \in K(a,b) \cap (J \cap Js \cap Ja \cap Jb)$. Then $x, xa, xb, x\bar{s} \in J$ and so $xa = -ax$, $xb = -bx$, $x\bar{s} = sx$. But $sx = a(bx) = -(ax)b = xab = xs$ and so $x\bar{s} = xs$ or $0 = x(s - \bar{s})$. Since $s - \bar{s} \in U$, $x = 0$. Consequently b) is valid.

c): Let $c,d \in \text{Rad } A$. Since $ab - ba = s - \bar{s} \notin N$, and $\text{Rad } A$ is an ideal with $\text{Rad } A \subset N \subset J$, we have $cd = 0$ by (4.7e).

d): Since $D := J \cap Js \cap Ja \cap Jb$ is a vector subspace with $D \subset N$ we have $0 = (x+y)^2 = x^2 + xy + yx + y^2 = xy + yx$ for any $x,y \in D$, hence $0 = x(-y) + y(-x)$. Therefore $D \subset x^\perp$ and $K(a,b) \subset x^\perp$ for all $x \in D$, hence $A = K(a,b) + D \subset x^\perp$, i.e. $D \subset \text{Rad } A$. Since always $\text{Rad } A \subset D$, $D = \text{Rad } A$. Let $u \in \text{Rad } A$. Then the conditions of (4.7b) are valid and hence $(ab)u = b(au)$. If A is associative, then $su = (ab)u = (ba)u = \bar{s}u$, thus $(s - \bar{s})u = 0$ and so $u = 0$. If $\text{Rad } A = \{0\}$, then $A = K(a,b)$ is either a quaternion field or $A \cong \mathfrak{M}_{22}(K)$ and therefore associative.

α): Here $K(a,b) \cap N = \{0\}$, hence $N = \text{Rad } A$ and N is a left vector space over the skew-field $K(a,b)$, which tells us $\dim(\text{Rad } A, K) \in 4\mathbb{N}$.

β) will be verified later by examples.

e): For $c \in (J \cap Js \cap Ja \cap Jb) \setminus N$ we show first $K(a,b) \perp K(a,b)\cdot c$ i.e. we have to prove $f(x,y) = 0$ for $x \in \{1,a,b,ab\}$, $y \in \{c,ac,bc,(ab)c\}$. By choice of c we have $f(x,c) = 0$. By applying (4.8b) and (4.5b), and $u^{-1} = q(u)^{-1}\cdot\overline{u}$ for $u \in U$, we obtain:

$$
\begin{aligned}
f(1,ac) &= q(a)\cdot f(a^{-1},c) = q(a)\cdot f\left(-\frac{a}{q(a)},c\right) = -f(a,c) = 0,\\
f(1,bc) &= -f(b,c) = 0,\\
f(1,(ab)c) &= f(c^{-1},ab)\cdot q(c) = -f(c,ab) = 0,\\
f(a,ac) &= q(a)\cdot f(1,c) = 0,\\
f(a,bc) &= -f(a,cb) = -q(b)\cdot f(ab^{-1},c) = f(ab,c) = 0,\\
f(a,(ab)c) &= f(ac^{-1},ab)\cdot q(c) = q(a)\cdot f(c^{-1},b)\cdot q(c) = -q(a)\cdot f(c,b) = 0\\
f(b,ac) &= -f(ab,c) = 0,\\
f(b,bc) &= q(b)\cdot f(1,c) = 0,\\
f(b,(ab)c) &= f(-bc,ab) = f(cb,ab) = f(c,a)\cdot q(b) = 0,\\
f(ab,ac) &= q(a)\cdot f(b,c) = 0,\\
f(ab,bc) &= -f(a,c)\cdot q(b) = 0,\\
f(ab,(ab)c) &= q(ab)\cdot f(1,c) = 0
\end{aligned}
$$

Now $K(a,b)\cdot c \subset J \cap Js \cap Ja \cap Jb$ implies together with b) $K(a,b) \cap K(a,b)\cdot c = \emptyset$. Thus the elements $1,a,b,s = ab, c,ac, bc,(ab)c$ are linearly independent and we have by using the formulas of (4.5)-(4.7) the following multiplication <u>Table II</u>:

	a	b	s	c	ac	bc	sc
a	α_a	s	$\alpha_a b$	ac	$\alpha_a c$	$\beta_s c - sc$	$\beta_s ac - \alpha_a bc$
b	$\beta_s - s$	α_b	$\beta_s b - \alpha_b a$	bc	sc	$\alpha_b c$	$\alpha_b ac$
s	$\beta_s a - \alpha_a b$	$\alpha_b a$	$\beta_s s - \alpha_a\alpha_b$	sc	$\alpha_a bc$	$\beta_s bc - \alpha_b ac$	$\beta_s sc - \alpha_a\alpha_b c$
c	$-ac$	$-bc$	$\beta_s c - sc$	α_c	$-\alpha_c a$	$-\alpha_c b$	$\alpha_c\beta_s - \alpha_c s$
ac	$-\alpha_a c$	$-sc$	$\beta_s ac - \alpha_a bc$	$\alpha_c a$	$-\alpha_a\alpha_c$	$\alpha_c s - \alpha_c\beta_s$	$\alpha_a\alpha_c b$
bc	$sc - \beta_s c$	$-\alpha_b c$	$\alpha_b ac$	$\alpha_c b$	$-\alpha_c s$	$-\alpha_b\alpha_c$	$\beta_s\alpha_c b - \alpha_b\alpha_c a$
sc	$\alpha_a bc - \beta_s ac$	$-\alpha_b ac$	$\alpha_a\alpha_b c$	$\alpha_c s$	$-\alpha_a\alpha_c b$	$\alpha_b\alpha_c a - \beta_s\alpha_c b$	$\alpha_a\alpha_b\alpha_c$

As an examplification how to obtain this Table we carry out some of the more complicated calculations:

$$s(bc) = -s(cb) = \overline{c}(\overline{s}b) \quad \text{by (4.6e)}$$
$$= -c((\beta_s - ab)b) = \beta_s bc - \alpha_b ac$$

$$(ac)(bc) = -\overline{b}(\overline{(ac)}c) \quad \text{by (4.6e) because } (ac)b = -(ab)c \in J$$
which is also a consequence of (4.6e)
$$= -b((ac)c) = -b(\alpha_c a) = -\alpha_c ba = -\alpha_c(\beta_s - s)$$

$$(bc)(sc) = (bc)(c\overline{s}) = -c((bc)\overline{s}) \quad \text{by (4.6e) because } (bc)c = \alpha_c b \in J$$
$$= -c((bs)c) \quad \text{by (4.6e) because } sc \in J$$
$$= -c((b(\beta_s - ba))c) = -\beta_s cbc + \alpha_b cac = \beta_s\alpha_c b - \alpha_b\alpha_c a .$$

By the multiplication table we recognize that $K(a,b,c) = K(a,b) \oplus K(a,b)\cdot c$ is a non-associative subalgebra of degree 8. To show $A = K(a,b,c)$ let $d \in K(a,b,c)^{\perp}$. Then $0 = f((ab)c,d) = f(ab,dc^{-1})\cdot q(c) = -f(ab,dc) = f(ab,cd)$ and by (4.7a) $(ab)(cd) \in J$. Hence all assumptions of (4.7e) are fulfilled and we obtain $cd = 0$ because $ab - ba = s - \overline{s} \notin N$ (cf. (4.15)). Since $c \in U$, $d = 0$. But $K(a,b,c)^{\perp} = \{0\}$ implies $K(a,b,c) = A$ and $\text{Rad } A = \text{Rad } q = \{0\}$.

β): By a) we can assume $K(a,b) = \mathfrak{M}_{22}(K)$. For $d := (e_{11} + \alpha_c^{-1} e_{22})c$ we have

$$d^2 = ((e_{11} + \alpha_c^{-1} e_{22})c)(c\overline{(e_{11} + \alpha_c^{-1} e_{22})})$$
$$= \alpha_c(e_{11} + \alpha_c^{-1} e_{22})(\alpha_c^{-1} e_{11} + e_{22}) = \alpha_c\alpha_c^{-1}(e_{11} + e_{22}) = 1 .$$

Remark. There is always a basis such that in Table II, $\beta_s = 0$ for char $K \neq 2$ and $\beta_s = 1$ for char $K = 2$.

(4.17) Duplication of kinematic algebras by CAYLEY-DICKSON. Let (A,K) be an alternative kinematic algebra and $\alpha_d \in K$. Then the direct sum $A_d = A \oplus Ad$ of the vector space A by itself becomes a kinematic algebra of degree $2\cdot[A : K]$, if the multiplication is defined by

$$(x_1 + x_2 d)\cdot(y_1 + y_2 d) := x_1 y_1 + \alpha_d \overline{y_2} x_2 + (x_2\overline{y_1} + y_2 x_1)d .$$

The map $^{-d}: x_1 + x_2 d \longrightarrow \overline{x}_1 - x_2 d$ has the properties:

(i) $q_d(x) := x\overline{x}^d = (x_1 + x_2 d)(\overline{x}_1 - x_2 d) = x_1\overline{x}_1 - \alpha_d \overline{x}_2 x_2 =$
$= q(x_1) - \alpha_d q(x_2)$ is a quadratic form

(ii) $\beta_x = x + \overline{x}^d = x_1 + \overline{x}_1 = \beta_{x_1}$

We have:

A_d is alternative $\Leftrightarrow$ A is associative
A_d is associative $\Leftrightarrow$ A is commutative
A_d is commutative $\Leftrightarrow$ $A = K$ $\Leftrightarrow$ $^-$ is the identity
A_d is a division algebra $\Leftrightarrow$ A is a division algebra and "$q(x_1) - \alpha_d q(x_2) = 0 \Leftrightarrow x_1 = x_2 = 0$".

Classification

By the theorems (4.14), (4.15) and (4.16) we can classify all alternative kinematic algebras (A,K) in the following way:

I. J is an ideal
II. J is not an ideal and codim Rad $A = 2$
III. J is not an ideal and codim Rad $A = 4$
IV. J is not an ideal and codim Rad $A = 8$

Furthermore we have the following subdivision of these classes:

Ia. $A \neq J$ ($\Rightarrow J = N = \text{Rad } A$ and $\text{Char } K \neq 2$)
  α) $N^2 = \{0\}$ ($\Rightarrow (A,K)$ is commutative)
  β) $N^3 = \{0\}$ but $N^2 \neq \{0\}$ ($\Rightarrow (A,K)$ is associative but not commutative)
  γ) $N^4 = \{0\}$ but $N^3 \neq \{0\}$ ($\Rightarrow (A,K)$ is not associative)

Ib. $A = J$ ($\Rightarrow (A,K)$ is associative and commutative and $\text{Char } K = 2$)

IIa. $A/_{\text{Rad } A}$ is a (commutative) field
  α) $\text{Rad } A = \{0\}$ ($\Leftrightarrow (A,K)$ is a quadratic separable field extension)
  β) $\text{Rad } A \neq \{0\}$ and $(\text{Rad } A)^2 = \{0\}$ ($\Rightarrow A$ is associative but not commutative)
  γ) $(\text{Rad } A)^2 \neq \{0\}$ ($\Rightarrow A$ is not associative)

IIb. $A/_{\text{Rad } A} = K \oplus K$

α) Rad $A = \{0\}$ ($\Leftrightarrow A = K \oplus K$)

β) Rad $A \neq \{0\}$ and $(\text{Rad})^2 = \{0\}$ ($\Rightarrow$ A is associative but not commutative)

γ) $(\text{Rad } A)^2 \neq \{0\}$ (A is not associative)

IIIa. $A/_{\text{Rad } A}$ is a quaternion field

α) Rad $A = \{0\}$ ($\Leftrightarrow$ (A,K) is a quaternion field)

β) Rad $A \neq \{0\}$ ($\Rightarrow$ A is not associative)

IIIb. $A/_{\text{Rad } A} = \mathfrak{M}_{22}(K)$

α) Rad $A = \{0\}$ ($\Leftrightarrow A \cong \mathfrak{M}_{22}(K)$)

β) Rad $A \neq \{0\}$ ($\Rightarrow$ A is not associative)

IVa. (A,K) is an octonion-field over K ($\Rightarrow$ A is not associative)

IVb. $(A,K) = \mathfrak{M}_{22}(K) \oplus \mathfrak{M}_{22}(K)\cdot d$, $d^2 = 1$ is the (not associative) split octonion algebra over K.

Since all associative kinematic algebras were discussed in [14] and [15] we restrict ourselves essentially to the non associative cases hence on the classes Iaγ), IIaγ), IIIaβ), IIIbβ), IVa and IVb.

The class Iaγ) is not empty as the examples show given by Table I. Any kinematic algebra (A,K) of class Iaγ) contains elements $a,b,c \in J = N = \text{Rad } A$ such that the eight elements $1,a,b,ab,c,ac,bc,(ab)c$ are linearly independent. Then $K(a,b,c)$ forms a subalgebra of degree 8 and their multiplication is described by Table I.

Now let (A,K) be of type IIaγ). Then by (4.15e) there are $b_1, b_2 \in N = \text{Rad } A$ with $b_1 b_2 \neq 0$ and Rad A is a vector space over the field K(s). Let $x,y,z \in K(s)$ with $xb_1 + yb_2 + z(b_1b_2) = 0$. Then

$$0 = \overline{x}\, b_1^2 + \overline{y}(b_2b_1) + \overline{z}((b_1b_2)b_1) \quad \text{by (4.7c)}$$

$$= \quad -\overline{y}(b_1b_2) \quad \text{by (4.15d)}$$

hence $y = 0$ and in the same way $x = 0$ and so $z = 0$.

Consequently the subalgebra $K(s,b_1,b_2) =$
$= K(s) \oplus \left(K(s)b_1 + K(s)b_2 + K(s)(b_1b_2)\right)$ is of degree 8 and not associative. The multiplication in $K(s,b_1,b_2)$ is determined by the rules (cf. (4.7a,b,c)):
$xb = b\overline{x}$, $x(yb) = (yx)b$, $(xb)c = \overline{x}(bc)$ for $x,y \in K(s)$,
$b,c \in \{b_1,b_2,b_1b_2\}$ and also $b_1^2 = b_2^2 = (b_1b_2)b_1 = b_2(b_1b_2) =$
$= (b_1b_2)^2 = 0$, $b_1b_2 = -b_2b_1$.
If $K(s)$ is a quadratic separable field extension over a field K, and $\{1,b_1,b_2,b_1b_2\}$ is a basis of the vector space $A = K(s)^4$, then (A,K) is a algebra of class IIaγ), if the multiplication is given by the rules above.
An example of class IIbγ) we can obtain in the same way as above, if we start from the algebra $K(s) = K \oplus K$ of anormal complex numbers. But there are also examples with a smaller degree. We can even prove: Any algebra (A,K) of class IIbγ) contains a subalgebra of type IIbγ) of degree 5 (any subalgebra with a degree < 5 is associative). By (4.15e) there are $c,d \in \text{Rad } A$ with $cd \neq 0$. Let $e = (1,0) \in K(s)$, then $\overline{e} = 1-e = (0,1)$ and at least one of the elements $e(cd), \overline{e}(cd) =$ $= cd - e(cd)$ is $\neq 0$. Let us assume $\overline{e}(cd) \neq 0$. For $a := ec$, $b := ed$ we have by (4.7b,c) $ab = (ec)(ed) = \overline{e}(c(ed)) = \overline{e}((ec)d) =$ $= \overline{e}(\overline{e}(cd)) = \overline{e}(cd) \neq 0$ hence $ea = a$, $eb = b$ and $e(ab) = 0$.
The elements a,b,ab are linearly independent and furthermore $ae = \overline{e}a = (1-e)a = a - ea = 0 = be$ and $(ab)e = \overline{e}(ab) =$ $= ab - e(ab) = ab$. Thus the multiplication of the subalgebra $K(e,a,b)$ is given by Table III

	e	a	b	ab
e	e	a	b	0
a	0	0	ab	0
b	0	-ab	0	0
ab	ab	0	0	0

On the other hand for any commutative field K the vector

space $A = K^5$ becomes an algebra of class IIbγ) by defining the multiplication according to the Table III. Any algebra (A,K) of class IIbγ) can be extended to an algebra $(A \oplus Kv, K)$ of class IIbγ) with $[A \oplus Kv : K] = [A : K] + 1$ by defining $ev = v$, $ve = 0$ and $xv = vx = 0$ for $x \in \text{Rad } A$.

If (A,K) is an algebra of type IIIaβ), then A contains a quaternion subfield $K(a,b)$, $\text{Rad } A \neq \{0\}$, and $\text{Rad } A$ is a vector space over $K(a,b)$ (cf. (4.16d)). Hence for $c \in (\text{Rad } A)^*$ the subalgebra $K(a,b,c)$ has the degree 8 and is again of type IIIaβ). The multiplication is given by the duplication process (4.17) for $d = c$ and $\alpha_d = 0$ or by the multiplication Table **II** for $\alpha_c = 0$. Any quaternion field (H,K) can be extended by (4.17) to a algebra of type IIIaβ), if we set $c^2 = 0$. Examples of class IIIbβ) one obtains by applying the duplication process (4.17) on $\mathfrak{M}_{22}(K)$, where K is a commutative field. But every algebra (A,K) of type IIIbβ) contains a subalgebra of type IIIbβ) with degree 6. For $A = \mathfrak{M}_{22}(K) \oplus \text{Rad } A$ and $\text{Rad } A \neq \{0\}$ and for $c \in (\text{Rad } A)^*$ we have $e_{11}d \neq 0$ or $e_{22}d = (1-e_{11})d = d - e_{11}d \neq 0$. Thus we may assume $c := e_{11}d \neq 0$, and we have $c \in (\text{Rad } A)^*$, $e_{11}c = c$, $e_{22}c = (1 - e_{11})c = 0$, $ce_{11} = \overline{e_{11}}\, c = 0$, $ce_{22} = \overline{e_{22}}\, c = c$ and $e_{21}c = e_{21}(e_{11}c) = (e_{11}e_{21})c = 0 = ce_{21}$ by (4.7b). The elements c and $e_{12}c$ are linearly independent because we have the relations $e_{ij}(e_{12}c) = (e_{12}e_{ij})c = \delta_{2i}(e_{1j}c)$ and $(e_{12}c)e_{ij} = \overline{e_{ij}}(e_{12}c) = (e_{12}\overline{e_{ij}})c$ by (4.7b), hence the multiplication Table IV.

These considerations teach us that each algebra of type IIIbβ) can be extended to an algebra of the same type whose degree is greater than 2.

	e_{11}	e_{12}	e_{21}	e_{22}	c	$e_{12}c$
e_{11}	e_{11}	e_{12}	0	0	c	0
e_{12}	0	0	e_{11}	e_{12}	$e_{12}c$	0
e_{21}	e_{21}	e_{22}	0	0	0	0
e_{22}	0	0	e_{21}	e_{22}	0	$e_{12}c$
c	0	$-e_{12}c$	0	c	0	0
$e_{12}c$	$e_{12}c$	0	$-c$	0	0	0

Table IV

By (4.16eβ) to any commutative field K there is exactly one algebra of type IVb), which can be obtained by duplication from $M_{22}(K)$ and $\alpha_d = 1$. Over a given field K of Char $K \neq 2$ there exists an algebra of type IVa), if and only if there are $\alpha_a, \alpha_b, \alpha_c \in K^*$ such that the quadratic form

$$x_o^2 - \alpha_a x_1^2 - \alpha_b x_2^2 + \alpha_a \alpha_b x_3^2 - \alpha_c x_4^2 + \alpha_a \alpha_c x_5^2 + \alpha_b \alpha_c x_6^2 - \alpha_a \alpha_b \alpha_c x_7^2$$

is zero only if $x_o = x_1 = \ldots = x_7 = 0$.

For Char $K = 2$ we have the condition that there are $\alpha_a, \alpha_b, \alpha_c \in K^*$ such that

$$x_o^2 + x_o x_1 + \alpha_a \alpha_b x_1^2 + \alpha_b x_2^2 + \alpha_b x_2 x_3 + \alpha_a \alpha_b^2 x_3^2 + \alpha_c x_4^2 + \alpha_c x_4 x_5 +$$
$$+ \alpha_a \alpha_b \alpha_c x_5^2 + \alpha_b \alpha_c x_6^2 + \alpha_b \alpha_c x_6 x_7 + \alpha_a \alpha_b^2 \alpha_c x_7^2$$

is zero only if $x_o = x_1 = \ldots = x_7 = 0$.

§ 5 GEOMETRIC DERIVATIONS OF 2-ALGEBRAS

Let (A,K) be an associative algebra with identity 1 and let U be the group of all units. Now we can define two geometric derivations of (A,K):

I. The affine derivation $\mathbf{A}_\varkappa(A,K) := (U, \mathfrak{L}, \cdot)$ consists of the set of points U and the set of lines $\mathfrak{L} := \{(a + Kb) \cap U \mid a, b \in A,\ b \neq 0,\ |(a + Kb) \cap U| \geq 2\}$ together with the multiplicative structure of the group U.

II. The projective derivation $\Pi_{\varkappa}(A,K) := (G,\mathfrak{L},\cdot)$ consists of the set of points $G = U/K^* := \{K^*u \mid u \in U\}$ and the set of lines $\mathfrak{L} := \{\varphi(Ku+Kv)^* \cap G \mid u,v \in U \text{ with } Ku \neq Kv\}$ (where φ denotes the canonical map $\varphi : A^* \longrightarrow A^*/K^*$; $a \longrightarrow K^*a$) together with the group structure of the factor group $G = U/K^*$.

As well the affine as the projective derivation of the algebra (A,K) gives us a 2-sided incidence group $(G,\mathfrak{L},\cdot)$ (cf. § 1).

A subset $Q \subset P$, $Q \neq P$ of an affine or a projective space $(P,\mathfrak{L})$ is called an affine 2-set or (projective) 2-set respectively, if $L \subset Q$ or $|L \cap Q| \leq 2$ for any line $L \in \mathfrak{L}$. An incidence space $(P,\mathfrak{L})$ is called an affine porous or (projective) porous space, if there is an affine space $(P_a,\mathfrak{L}_a)$ and an affine 2-set $Q \subset P_a$ with $P = P_a \backslash Q$ and $\mathfrak{L} = \{L \cap P \mid L \in \mathfrak{L}_a, |L \cap P| \geq 2\}$ or a projective space $(P_p,\mathfrak{L}_p)$ and a 2-set $Q \subset P_p$ with $P = P_p \backslash Q$ and $\mathfrak{L} = \{L \cap P \mid L \in \mathfrak{L}_p, |L \cap P| \geq 2\}$ respectively. Every line of a projective porous space belongs to exactly one of the three classes $\mathfrak{L}_i := \{L \cap P \mid L \in \mathfrak{L}_p, |L \cap Q| = i\}$ for $i = 0,1,2$. The lines of $\mathfrak{L}_0$ are called projective lines, of $\mathfrak{L}_1$ affine lines and of $\mathfrak{L}_2$ 2-lines.

Remark. An intrinsic definition of an i-line in an incidence space is given in [14]. G. KIST [25] succeeded to prove that any incidence space $(P,\mathfrak{L})$ with $\mathfrak{L} = \mathfrak{L}_0 \cup \mathfrak{L}_1 \cup \mathfrak{L}_2$ is a projective porous space provided that $\dim(P,\mathfrak{L}) \geq 3$ and $|X| \geq 9$ for $X \in \mathfrak{L}$.

A porous space $(P,\mathfrak{L})$ is called a slit space, if Q is a projective subspace, a 2-slit space, if $Q = M_1 \cup M_2$ is the union of two projective subspaces M_1 and M_2, a subaffine space, if $(P,\mathfrak{L})$ is a 2-slit space where M_1 is a hyperplane. A subaffine space can also be defined as an affine porous space, where Q is an affine subspace.

(5.1) If (A,K) is a 2-algebra with $K \neq \mathbb{Z}_2, \mathbb{Z}_3$, then:

a) The incidence structure $(U,\mathfrak{L})$ of the affine derivation $A_\varkappa(A,K)$ is an affine porous space.

b) The incidence structure $(G = U/K^*, \mathfrak{L})$ of the projective derivation $\Pi_\varkappa(A,K)$ is a porous space.

Proof. a) is a consequence of (3.8). b): Let $K^*a, K^*b \in U/K^*$ with $K^*a \neq K^*b$. Then $K^*a^{-1}b \neq K^*$ and $\varphi(Ka+Kb)^* = \varphi(a\{K^*(\lambda + a^{-1}b) \mid \lambda \in K\} \cup \{K^*a\})$ is a line of the projective space $\Pi(A,K)$. By $(*)$ of § 3 we have $|(K+a^{-1}b) \cap (A \setminus U)| \leq 2$ and therefore $A^*/K^* \setminus U/K^*$ is a 2-set.

(5.2) For a bilocal algebra (A,K) with $K \neq \mathbb{Z}_2, \mathbb{Z}_3$ the projective derivation $\Pi_\varkappa(A,K)$ is a 2-slit incidence group.

§ 6 2-ALGEBRAS WHOSE PROJECTIVE DERIVATION IS AN AFFINE POROUS SPACE

In this section we consider associative 2-algebras (A,K) where $H := A \setminus U$ contains a hyperplane of the vector space (A,K). Then by theorem (3.2) the algebra (A,K) is bilocal and one of the ideals M_1, M_2, say M_1, is a hyperplane of the vector space (A,K). Hence $U = A \setminus (M_1 \cup M_2)$.

(6.1) $(A,+)$ is the direct sum $A = K + M_1$ and for the factor group U/K^* we have $U/K^* = \{K^*(1+x) \mid x \in M_1 \setminus (-1+M_2)\}$.

(6.2) The set $G := M_1 \setminus (-1+M_2)$ is a group with respect to the operation $a \oplus b := a+b+ab$, which is isomorphic to the factor group U/K^*; the map $\psi : U/K^* \longrightarrow (G,\oplus); K^*(1+x) \longrightarrow x$ is an isomorphism.

By definition the set G is the affine space M_1 where the affine subspace $M_1 \cap (-1+M_2)$ is deleted. Therefore G is an affine porous space with the line set
$\mathfrak{L}_G := \{(a+Kb) \cap G \mid a,b \in M_1,\ b \neq 0 \text{ and } |(a+Kb) \cap G| \geq 2\}$.

(6.3) The isomorphism ψ preserves lines.

Proof. Let $1+a,\ 1+b \in U$ with $a,b \in M_1$ and $K^*(1+a) \neq K^*(1+b)$, then $a \neq b$, and $a,b \notin M_1 \cap (-1+M_2)$, hence $a,b \in G$. Now we consider the line L of $\Pi(A,K)$ joining the points $K^*(1+a)$ and $K^*(1+b)$. Then $L = \varphi(\{\lambda(1+a) + \mu(1+b) \mid \lambda,\mu \in K\} \cap U) =$
$= \varphi(\{1 + \lambda(\lambda+\mu)^{-1}a + \mu(\lambda+\mu)^{-1}b \mid \lambda,\mu \in K,\ \lambda+\mu \neq 0\} \cap (A\backslash M_2)) =$
$= \varphi(\{1 + \alpha a + (1-\alpha)b \mid \alpha \in K\} \cap (A\backslash M_2))$ and hence
$\psi(L) = \{\alpha a + (1-\alpha)b \mid \alpha \in K\} \cap G = (a + K(b-a)) \cap G$, which is a line of $\mathfrak{L}_G$.

First we consider the local case, i.e. $M_2 = \{0\}$. Then $G = M_1$ is a nilpotent algebra and $(G, \mathfrak{L}_G, \oplus)$ is an affine 2-sided incidence group.

(6.4) Let (M,K) be a nilpotent algebra. Then:
a) The direct sum $A = K + M$ is a local algebra if the multiplication is defined by $(\lambda+m)(\mu+n) = \lambda\mu + (\lambda n + \mu m + m\cdot n)$; the maximal ideal M is a hyperplane.
b) $(M, \mathfrak{L}, \oplus)$ is an affine 2-sided incidence group, where $\mathfrak{L} := \{a + Kb \mid a,b \in M,\ b \neq 0\}$ and $m \oplus n := m + n + mn$, $m,n \in M$.
c) $\Pi_\varkappa(A,K)$ and $(M, \mathfrak{L}, \oplus)$ are isomorphic incidence groups.

Proof. By our former consideration it is enough to prove a). It is obvious that (A,K) is an algebra and that M is an maximal ideal and a hyperplane. Let $\lambda + m \in A\backslash M$, hence $\lambda \neq 0$, and let $d \in \mathbb{N}$ with $m^{d+1} = 0$. Then $(\lambda+m)^{-1} = \lambda^{-1}(1 + \lambda^{-1}m)^{-1} =$
$= \lambda^{-1}(1 - \lambda^{-1}m + \lambda^{-2}m^2 - \ldots + (-1)^d\lambda^{-d}m^d)$. Thus $U = A\backslash M$.

Now let $M_2 \neq \{0\}$. Then there is an idempotent $e \neq 1$ (cf. [1], p. 23) and $1-e$ is also idempotent. Since $1 = e + (1-e) \notin M_1 \cup M_2$ and $e(1-e) = 0$, either $e \in M_1$ and $1-e \in M_2$ or $e \in M_2$ and $1-e \in M_1$. Let us denote by e the idempotent element of the hyperplane M_1. The pair (M_1, Ke) is a 2-algebra.

(6.5) The set $F := M_1 \setminus M_2 = M_1 \setminus \text{Rad } A$ is a group with respect to the operation $a \odot b := a \cdot b + (a - ae) + (b - eb)$, which is isomorphic to the group $(G,\oplus)$; the map $\chi : (G,\oplus) \longrightarrow (F, \odot);\ x \longrightarrow x + e$ is an isomorphism. We have $\odot = \cdot$ if and only if e is an identity element for M_1, i.e. M_1 is a local algebra.

Proof. $\odot = \cdot$ implies $xe = ex = x$ for all $x \in M_1 \setminus M_2$. Now let $n \in \text{Rad } A$, then $e + n \in M_1 \setminus M_2$ and hence $e + n = (e + n)e = e + ne = e(e + n) = e + en$ which implies $n = ne = en$. But $(F, \odot) = (F, \cdot)$ implies $U = F$ and hence M_1 is a local algebra.

Remarks. 1. The projective derivation $\Pi_\varkappa(A,K)$ is a subaffine incidence group in the sense of E. SCHRÖDER [43]. $\Pi_\varkappa(A,K)$ is isomorphic to the affine derivation $A_\varkappa(M_1,Ke)$ if and only if (M_1,Ke) is a local algebra. SCHRÖDER proved that the left translations a_ℓ of the subaffine incidence group $\Pi_\varkappa(A,K)$ can be uniquely extended to affinities $\overline{a_\ell}$ of the corresponding affine space. Furthermore he showed that (M_1,Ke) is a local algebra if and only if in the affine subspace $Q = \varphi(M_1 \setminus M_2)$ there is a point q such that $\overline{a_\ell}(q) = q$ for all $a \in U/K^*$.

2. SCHRÖDER proved that any 2-sided subaffine incidence group with $Q \neq \emptyset$ (cf. § 5) is isomorphic to $\Pi_\varkappa(A,K)$, where (A,K) is a proper bilocal algebra.

3. For $\dim M_2 = 1$ we have $F = M_1^*$, hence $\odot = \cdot$, and therefore $(F,+,\cdot)$ is a skew-field. Thus $\Pi_\varkappa(A,K) = A_\varkappa(F,Ke)$ is a 2-sided punctured affine incidence group. G. KIST [24] proved that any desarguesian punctured affine incidence group $(G,\mathfrak{L},\cdot)$ is the derivation of a so called Dicksonian nearfield extension (F,K); (F,K) is a division algebra if and only if the incidence group $(G,\mathfrak{L},\cdot)$ is 2-sided.

4. By setting $D_2 = K$ in (3.7) it is easy to construct examples of bilocal algebras where the dimension of Rad A is any

number between 0 and $\dim(A,K)-2$. This implies: Let (D,K) be an associative division algebra, let $d := \dim(D,K)$ and $r \in \mathbb{N}_0$. Then there is always a subaffine incidence group $(G,\mathfrak{L},\cdot)$ such that the dimension of the correspondung affine space $(P_a,\mathfrak{L}_a)$ is equal $d+r$ and $\dim Q = r$.

§ 7 THE KINEMATIC DERIVATION OF AN ALTERNATIVE KINEMATIC ALGEBRA. REPRESENTATION THEOREM.

First we will show that it is possible to extend notions defined in § 1 and § 5. For this purpose we have to recall the following definitions:

1. A set L provided with a binary operation $\cdot: L \times L \longrightarrow L$ is called a loop, if there is an identity element 1 (i.e. $1 \cdot x = x \cdot 1 = x$ for all $x \in L$) and if for any two elements $a,b \in L$ each of the equations $ax = b$ and $ya = b$ has a unique solution in L.

2. A subloop S of a loop L is called normal, if $a(bS) = (ab)S = (aS)b$ for all $a,b \in L$, and even central if furthermore $as = sa$ for all $s \in S$, $a \in L$.

3. If S is a normal subloop of a loop L, then the set $L/S = \{Sa \mid a \in L\}$ becomes a loop with respect to the multiplication $(Sa)(Sb) = S(ab)$, which is called the factor loop.

4. For a loop L, a set $\mathfrak{F}$ of subloops is called a fibration, if $L = \cup \mathfrak{F}$ and if for $X,Y \in \mathfrak{F}$, $X \neq Y$ the conditions $\{1\} \neq X \neq L$ and $X \cap Y = \{1\}$ are valid. The fibration $\mathfrak{F}$ is called kinematic if for any $a \in L$ and $X \in \mathfrak{F}$ there are $b,c \in L$ and $Y,Z \in \mathfrak{F}$ such that $aX = Yb$ and $Xa = cZ$.

5. A loop $(L,\cdot)$ is called a Moufang-loop, if for all $a,b,c \in L$ the so-called Moufang-identity

(M) $(ab)(ca) = (a(bc))a$ is valid. (M) is equivalent to each of the identities (M_ℓ) $a(b(ac)) = ((ab)a)c$;
(M_r) $((ac)b)c = a(c(bc))$ and (M) implies the alternative laws $a(ab) = a^2b$, $(ab)b = ab^2$ and the flexible law $a(ba) = (ab)a$ (cf. BRUCK [8]).

If we allow in the definition of an incidence group $(G,\mathfrak{L},\cdot)$ (cf. § 1) $(G,\cdot)$ to be a loop, then we obtain the notions of an <u>incidence loop</u> and of a <u>2-sided</u> incidence loop. Examples of 2-sided incidence loops we obtain by applying the affine or projective derivation on an algebra (A,K) with identity 1, where the set U of units forms a loop with respect to the multiplication. For such algebras K^* is a central subgroup of the loop U, so that we can form the projective derivation. In the particular case that (A,K) is alternative, the incidence loops are Moufang-loops, and if (A,K) is a 2-algebra, then the incidence structure of the incidence loop is an affine or projective porous space. In the special case that (A,K) is a kinematic algebra we call $\Pi_\varkappa(A,K)$ the <u>kinematic derivation</u>.

Just as for incidence groups we define: An incidence loop $(G,\mathfrak{L},\cdot)$ is called <u>fibered</u>, if every line $X \in \mathfrak{L}$ through 1 is an incidence subloop, and $(G,\mathfrak{L},\cdot)$ is called a <u>kinematic loop</u>, if $(G,\mathfrak{L},\cdot)$ is fibered and 2-sided. For a fibered incidence loop $(G,\mathfrak{L},\cdot)$ the lines $\mathfrak{F} := \{X \in \mathfrak{L} \mid 1 \in X\}$ form a fibration of the loop $(G,\cdot)$. $\mathfrak{F}$ is kinematic if and only if $(G,\mathfrak{L},\cdot)$ is kinematic. On the other hand if $(G,\mathfrak{F})$ is a loop with a fibration and if $\mathfrak{L} := \{aX \mid a \in G, X \in \mathfrak{F}\}$, then $(G,\mathfrak{L},\cdot)$ is a fibered incidence loop.

<u>(7.1)</u> Let (A,K) be an alternative kinematic algebra. Then the kinematic derivation $(G,\mathfrak{L},\cdot) := \Pi_\varkappa(A,K)$ is a kinematic Moufang-loop with the following properties:

(i) $(G,\mathfrak{L})$ is a porous space.

(ii) Any $X \in \mathfrak{F}$ is a commutative subgroup of the loop G.

(iii) For any $X,Y \in \mathfrak{F}$ the subloop $\langle X \cup Y \rangle$ generated by $X \cup Y$ is an incidence group.

(iv) G is a group $\Leftrightarrow$ A is associative.

Proof. By our previous considerations $(G,\mathfrak{L},\cdot)$ is a 2-sided Moufang incidence loop. We know that

$\mathfrak{F} := \{((K + Ka) \cap U)/K^* \mid a \in A \setminus K\}$ is the set of all lines through 1. Since $(K + Ka) \cap U = K(a) \cap U$ is a commutative group, $\mathfrak{F}$ is a fibration consisting of commutative subgroups and hence $(G, \mathfrak{L}, \cdot)$ is a kinematic Moufang loop. (i) is a consequence of (4.9) and (5.1b).

(iii): Let $X = (K(a) \cap U)/K^*$ and $Y = (K(b) \cap U)/K^*$. Then $\langle (K(a) \cap U) \cup (K(b) \cap U) \rangle = K(a,b) \cap U$ and hence $\langle X \cup Y \rangle = (U \cap K(a,b))/K^*$ is a group, because $K(a,b)$ is a associative subalgebra of (A,K). Thus $\langle X \cup Y \rangle$ can be considered as the kinematic derivation of the associative kinematic algebra $(K(a,b),K)$.

(iv): Let G be a group and let $a,b,c \in U$ with $c \notin K$. Then there is a $\lambda \in K^*$, such that $\lambda + c \in U$. If U/K^* is a group, then there exist $\sigma, \tau \in K$ with $a(bc) = \sigma(ab)c$ and $a(b(c+\lambda)) = \tau(ab)(c+\lambda) = \tau(ab)c + \tau\lambda(ab)$, hence $\sigma(ab)c + \lambda(ab) = \tau(ab)c + \tau\lambda(ab)$ or $\sigma c + \lambda = \tau c + \tau\lambda$ which implies $\sigma = \tau = 1$. Hence U is a group. Since any $x \in A \setminus U$ can be expressed in the form $x = \lambda + u$ with $\lambda \in K^*$ and $u \in U$, the algebra (A,K) is associative.

Remark 1. $(G, \mathfrak{L}, \cdot)$ can be provided with a left and right parallelism by defining for any two lines $X, Y \in \mathfrak{L}$:

$$X \parallel_\ell Y :\Leftrightarrow X^{-1}X = Y^{-1}Y \; ; \; X \parallel_r Y :\Leftrightarrow XX^{-1} = YY^{-1} .$$

$(G, \mathfrak{L}, \parallel_\ell, \parallel_r)$ is a double space (i.e. mixed parallelograms are closed (cf. §1, Remark 2.)) only if $(G, \cdot)$ is a group.

(7.2) Any kinematic Moufang loop $(G, \mathfrak{L}, \cdot)$ with $|\mathfrak{L}| \geq 2$ and the properties (i) and (iii) of (7.1) has one of the following representations:

a) $(G, \mathfrak{L}, \cdot)$ is the kinematic derivation of an alternative kinematic algebra (A,K).

b) There is a vector space (V,K) over a non-commutative field K such that $(G, \mathfrak{L}) = \mathbf{A}(V,K)$ and the operation is the vector addition.

c) There is a vector space (V,K) with a congruence $\mathfrak{K}$ (in the sense of ANDRÉ [2]) such that $(G,\cdot) = (V,+)$ and $\mathfrak{L} = \{a + X \mid a \in V, X \in \mathfrak{K}\}$.

d) There is a planar (left-) nearfield F such that $(G,\cdot)$ is the affine group of F consisting of the maps $[c,m] : F \longrightarrow F; x \longrightarrow c + mx$ with $c,m \in F$, $m \neq 0$ and the line set $\mathfrak{L}$ is determined by the fibration $\mathfrak{F} = \{A\} \cup \{G_c \mid c \in F\}$, where $A := \{[a,1] \mid a \in F\}$ and $G_c := \{[c - mc,m] \mid m \in F^*\}$.

Remark 2. A fibration $\mathfrak{K}$ of a vector space (V,K) consisting of vector subspaces is called a congruence, if $V = X + Y$ for any $X,Y \in \mathfrak{K}$ with $X \neq Y$.

The proof of the representation theorem (7.2) follows in the plane case from [18], because then $(G,\cdot)$ is a group by the property (iii): Hence $(G,\mathfrak{L},\cdot)$ is either a translation plane and thus describable by c) or otherweise $(G,\mathfrak{L},\cdot)$ has the representation d). The proof for $\dim(G,\mathfrak{L}) \geq 3$ will be given elsewhere.

Remark 3. In the case that $(G,\cdot)$ is a group fulfilling some further assumptions, a proof of (7.2) is given in L. BRÖCKER [6].

The following theorem (cf.[6],[32]) gives another characterization of kinematic Moufang loops.

(7.3) A Moufang incidence loop $(G,\mathfrak{L},\cdot)$ where $|X| \geq 3$ for $X \in \mathfrak{L}$ is kinematic if and only if the map $\iota : G \longrightarrow G; x \longrightarrow x^{-1}$ is an automorphism of $(G,\mathfrak{L})$.

Proof. Let $(G,\mathfrak{L},\cdot)$ be kinematic and let $aX \in \mathfrak{L}$ with $a \in G$, $X \in \mathfrak{F}$. Then $(aX)^{-1} = X^{-1}a^{-1} = Xa^{-1} \in \mathfrak{L}$. Hence ι is an automorphism of $(G,\mathfrak{L})$. Now let us assume that ι is an automorphism of $(G,\mathfrak{L})$. Then $(G,\mathfrak{L},\cdot)$ is 2-sided, because for $X \in \mathfrak{L}$ and $a \in G$ we have $Xa = (a^{-1}X^{-1})^{-1} \in \mathfrak{L}$. Now let $X \in \mathfrak{L}$ with $1 \in X$ and $x \in X \setminus \{1\}$, then $1,x \in x X^{-1} \in \mathfrak{L}$, hence

$xX^{-1} = X$. Since $|X| \geq 3$ there are $x_1, x_2 \in X \setminus \{1\}$ with $x_1 \neq x_2$, and $1 \in \overline{x_1, x_2}$ implies $x^{-1} \in \overline{x_1 x^{-1}, x_2 x^{-1}} = X$ because $(G, \mathfrak{L}, \cdot)$ is 2-sided and $x_1 x^{-1}, x_2 x^{-1} \in X$. This shows $XX^{-1} = X$, $X^{-1} = X$ and $X^{-1}X = X$. Hence X is a subloop, because in a Moufang loop $y = x_1 x_2^{-1}$ and $z = x_2^{-1} x_1$ are the solutions of the equations $yx_2 = x_1$ and $x_2 z = x_1$.

(7.4) Each fibered Moufang incidence loop $(G, \mathfrak{L}, \cdot)$ where $(G, \mathfrak{L})$ is a porous space, is kinematic, if the property (iii) of (7.1) holds.

Proof. We divide the proof in the following steps:
(1) For any incidence subspace T with $1 \in T$, $T = T^{-1}$, because $T = \cup \{\overline{1, t} \mid t \in T \setminus \{1\}\}$ and $t^{-1} \in \overline{1, t}$.
(2) For any $X \in \mathfrak{F}$, $a \in G \setminus X$ the set Xa is contained in the plane $a^{-1}E$ through 1, where $E := \overline{X \cup \{a\}}$: For $Xa \subset Ea$, and since $1 \in E$, $a^{-1}E$ we have $a^{-1}E = (a^{-1}E)^{-1} = E^{-1}a = Ea$ by (1).
(3) If $xa \in E := \overline{\{1, x, a\}}$, then E is a subgroup.

Proof. By (iii) E is contained in an incidence subgroup and by (1) $E = E^{-1} = \overline{\{1, x^{-1}, a^{-1}\}}$. Since $1 \in xE, aE$ we have by (1) $Ex^{-1} = xE \subset E$, because $xa \in E$, and $Ea^{-1} = aE \subset E$, because $ax^{-1} \in Ex^{-1} \subset E$ and $aE = a\overline{\{1, x^{-1}, a\}} = \overline{\{a, ax^{-1}, a^2\}} \subset E$. Thus for $y \in E$, $yx^{-1}, ya^{-1} \in E$, hence $yE = y\overline{\{1, x^{-1}, a^{-1}\}} \subset E$.
Now let $X \in \mathfrak{F}$, $a \in G \setminus X$ and $E := \overline{X \cup \{a\}}$.
(4) If E is not a subgroup, then $Xa \in \mathfrak{L}$: For $x \in X \setminus \{1\}$ the planes E and xE are distinct by (3). Therefore $X = E \cap xE$ (because $xX = X$) and $Xa = Ea \cap (xE)a = Ea \cap x(Ea) = a^{-1}E \cap (xa^{-1})E$. Here we have used the fact that $\langle E \rangle = \langle \{1, a, x\} \rangle$ is a subgroup by (iii) and statement (1). But $a^{-1}E$ and $(xa^{-1})E$ are distinct planes so that Xa is a line.
It remains to discuss the case that E is a subgroup.

(5) For any $x \in X \setminus \{1\}$ we have $\underline{X \cap \overline{a,xa} = \emptyset}$ and $x^{-1}\overline{a,xa} = \overline{a,xa}$: By (i) $a^{-1}x^{-1}a \in 1, a^{-1}xa$, hence the points $a, xa, x^{-1}a$ are collinear and so $x^{-1}\overline{a,xa} = \overline{a,xa}$. Since $x^{-1}X = X$, $a \notin X$ we have $X \cap \overline{a,xa} = \emptyset$ because the left translation $(x^{-1})_\ell$ does not have fixed points.

(6) If $X \in \mathfrak{L}_1$ is an affine line (cf. §5), then $Xa = aX \in \mathfrak{L}$: Since X is a subgroup, $X \cap aX = \emptyset$ and since E is a subgroup, $aX \subset E$. Therefore aX is the only line in the plane E through a which does not intersect X , because $X \in \mathfrak{L}_1$. By (5) we have $Xa \subset aX$ and in the same way $Xa^{-1} \subset a^{-1}X$. This gives us $Xa = aX$.

(7) If $X \in \mathfrak{L}_2$ is a 2-line (cf. §5) then Xa is contained in one of the two lines Y_1, Y_2 through a in the plane E with $Y_i \cap X = \emptyset$: By (5) $Xa \subset Y_1 \cup Y_2$. Let us assume there are $x_1, x_2 \in X \setminus \{1\}$ with $x_1 a \in Y_1$ and $x_2 a \in Y_2$. Then $x_1^{-1}x_2 a \in Y_1 \cup Y_2$. If $x_1^{-1}x_2 a \in Y_1 = \overline{a, x_1 a}$, then $x_2 a \in x_1 Y_1 = Y_1$ by (5), which contradicts $x_2 a \in Y_2 \setminus \{a\}$. If $x_1^{-1}x_2 a \in Y_2 = \overline{a, x_1^{-1}x_2 a}$, then $x_2^{-1}x_1 a \in Y_2 = \overline{a, x_2 a}$ by (5) and hence $x_1 a \in x_2 Y_2 = Y_2$ again by (5), which contradicts $x_1 a \in Y_1 \setminus \{a\}$.

(8) X is not a projective line, because $X \cap aX = \emptyset$ and $X, aX \subset E$.

(9) From (6),(7),(8) it follows for each line $X \in \mathfrak{J}$ with $X \subset E$, that Xa (and also Xa^{-1}) is contained in a line of the plane E . Hence there are $Y, Z \in \mathfrak{J}$ with $a^{-1}Xa \subset Y$, $aYa^{-1} \subset Z$ and so $X \subset aYa^{-1} \subset Z$. Therefore $X = aYa^{-1}$ i.e. $Xa = aY \in \mathfrak{L}$.

(10) By (4) and (8) we have $Xa \in \mathfrak{L}$ for any $X \in \mathfrak{J}$ and $a \in G$. Now let $bX \in \mathfrak{L}$ with $X \in \mathfrak{J}$ be an arbitrary line and $a \in G$. Then $(ab)(Xa) \in \mathfrak{L}$ and by the Moufang identity (M) $(ab)(Xa) = a(bX)a$ hence $(bX)a \in \mathfrak{L}$. This shows that $(G, \mathfrak{L}, \cdot)$ is kinematic.

Remark 4. This is almost literally the proof which was given in KIST [26] for porous fibered incidence groups.

Under the stronger geometric assumption that $(G,\mathfrak{L})$ is a desarguesian projective space H. WÄHLING [49] proved the following theorem:

(7.5) Let $(G,\mathfrak{L},\cdot)$ be a 2-sided incidence loop such that $(G,\mathfrak{L})$ is a desarguesian projective space. Then there is a (not necessarily associative) division algebra (A,K) such that $(G,\mathfrak{L},\cdot)\cong\Pi_{\varkappa}(A,K)$. For any division algebra (A,K), $\Pi_{\varkappa}(A,K)$ is a 2-sided projective incidence loop. If furthermore $(G,\mathfrak{L},\cdot)$ is kinematic, then the corresponding division algebra is kinematic.

Remark 5. J.M. OSBORN [36] proved that there are kinematic division algebras (A,K) of degree 8 (even with $K=\mathbb{R}$) which are not alternative and R.B. BROWN [7] showed later that there are even such algebras of any degree 2^n with $n\in\mathbb{N}$.

Remark 6. The most far reaching extension of the notion (projective) incidence group was done by H. WÄHLING [49], who studied incidence groupoids.

Problems. 1. Extend the results of WÄHLING (cf.(7.5)) on desarguesian porous incidence loops.

2. Is it possible to obtain the result of KIST (cf.(7.4)) without assuming condition (iii).

3. Is it possible to give an internal characterization of (projective) kinematic loops $(G,\mathfrak{L},\cdot)$ by means of incidence and two parallelisms $\parallel_\ell$ and $\parallel_r$, as it was done for some classes of kinematic spaces (cf.[21],[22],[17],[27],[28]).

§ 8 KINEMATIC ALGEBRAS WITH AN ADJOINT MAP. THE GENERAL NOTATION OF A KINEMATIC MAP.

In this section we consider an alternative kinematic algebra (A,K) together with an involutorial antiautomorphism $*$ of A such that $*(K)=K$ and $u(xu^*)=(ux)u^*$ for all $u,x\in A$. Then the fixed field $F:=\{\lambda\in K \mid \lambda^*=\lambda\}$ is either K, or F is a subfield of K with $[K:F]=2$.

(8.1) $*(J) = J$ and $* \circ \overline{} = \overline{} \circ *$

Proof. For $x \in J$ we have $x^2 \in K$, hence $(x^*)^2 \in *(K) = K$. Thus $x^* \in J$. For Char $K \neq 2$ we have $\overline{(\xi + x)}^* = (\xi - x)^* = \xi^* - x^*$ and $\overline{(\xi + x)^*} = \overline{\xi^*} + \overline{x^*} = \xi^* - x^*$ for $\xi \in K$ and $x \in J$.
Now let Char $K = 2$, $J \neq A$ and $a \in A \setminus J$, hence $A = Ka + J$. Then $\overline{(\xi a + x)}^* = (\xi \overline{a} + x)^* = (\xi \beta_a + \xi a + x)^* = \xi^* \beta_a^* + \xi^* a^* + x^*$ and $\overline{(\xi a + x)^*} = \overline{\xi^* a^* + x^*} = \xi^* \overline{a^*} + x^*$. Since $a^2 = \alpha_a + \beta_a a$ and $\alpha_a^* + \beta_a^* a^* = (a^2)^* = (a^*)^2 = \alpha_{a^*} + \beta_{a^*} \cdot a^*$, we have $\overline{a^*} = \beta_{a^*} + a^* = \beta_a^* + a^*$, so $* \circ \overline{} = \overline{} \circ *$ also in this case.

For the structure $(A, K, *)$ we define the sets of Hermitian and skew-Hermitian elements:
$H := \{a \in A \mid a^* = a\}$ and $H^- := \{a \in A \mid a^* = -a\}$.

Remark 1. For $* = \overline{}$ we have $H = K$ and $H^- = J$, if Char $K \neq 2$
If $Z(A) = K$, then the condition $*(K) = K$ is provable.

For any $u \in U$ let $u_*: x \longrightarrow uxu^*$, then $u_*(H) = H$ and $u_*(H^-) = H^-$. Let $\Gamma^+ := \langle U_\ell \cup U_r \rangle$ be the group generated by all left- and right-translations u_ℓ and v_r with $u, v \in U$ and $\Gamma := \Gamma^+ \cup \Gamma^+ \circ \overline{}$.

Now we are going to determine the subgroups
$\Gamma(H) := \{\gamma \in \Gamma \mid \gamma(H) = H\}$, $\Gamma(H^-) := \{\gamma \in H \mid \gamma(H^-) = H^-\}$, $\Gamma^+(H) := \Gamma^+ \cap \Gamma(H)$ and $\Gamma^+(H^-) := \Gamma^+ \cap \Gamma(H^-)$ in the case that A is associative. If $\gamma = a_\ell b_r \in \Gamma^+(H)$, then for any $x \in H$, $axb = (axb)^* = b^* x a^*$ or $a^{-1} b^* x = xb(a^{-1})^*$. $x = 1$ implies $c := a^{-1} b^* = (a^{-1} b^*)^* \in H$ and we have $cx = xc$ for all $x \in H$. Hence $c \in C := Z(H) \cap H \cap U$ and $\gamma = a_* c_\ell = (aca^{-1})_\ell a_*$. On the other hand if $c \in C$ and $a \in U$, then $a_* c_\ell \in \Gamma^+(H)$.

We have proved the first part of the following theorem:

(8.2) $\Gamma^+(H) = \{a_* c_\ell \mid a \in U, c \in C\}$, C is a normal subgroup of U and for the semidirect product $U \otimes_s C$ of C by U, where the multiplication is given by $(u_1,c_1) \circ (u_2,c_2) := (u_1u_2, u_2^{-1}c_1u_2c_2)$ the map $(u,c) \longrightarrow u_*c_\ell$ is an homomorphism with the kernel $L = \{(u,(u^*u)^{-1}) \mid u \in Z(H) \cap U\}$ which is isomorphic to $Z(H) \cap U$. Hence $\Gamma^+(H) \cong U \otimes_S C/L$. Furthermore $\Gamma(H) = \Gamma^+(H) \cup \Gamma^+(H) \circ -$, $\Gamma^+(H) \cap \Gamma^+(H^-) = \{a_*c_\ell \mid a \in U, c \in C \cap Z(H^-)\}$ and $\Gamma(H) \cap \Gamma(H^-) = \Gamma^+(H) \cap \Gamma^+(H^-) \cup (\Gamma^+(H) \cap \Gamma^+(H^-)) \circ -$. If Char $K \neq 2$, then $C \cap Z(H^-) = Z(A) \cap H \cap U$, and if furthermore $Z(H) \cap U = K^*$, then $U \otimes_S C = U \otimes F^*$ is a direct product and $\Gamma^+(H) \cong U \otimes F^*/K^* = U/K^* \otimes F^*$.

Proof. Let $c_1, c_2 \in C = Z(H) \cap H \cap U$, then $c_1c_2 \in U \cap Z(H)$ and $(c_1^{-1}c_2)^* = c_2^*c_1^{*-1} = c_2c_1^{-1} = c_1^{-1}c_2$, hence $c_1^{-1}c_2 \in H$.
Now let $c \in C$ and $u \in U$, then $u^*u \in H$, $ucu^{-1} \in U$, $(ucu^{-1})^* = u^{-1*}c(u^*u)u^{-1} = u^{*-1}(u^*u)cu^{-1} = ucu^{-1}$ (because $c \in Z(H)$), hence $ucu^{-1} \in H$. For $h \in H$ we have $h' := u^*hu \in H$, hence $ucu^{-1}h = uc(u^{-1}u^{*-1})h'u^{-1} = u(u^{-1}u^{*-1})ch'u^{-1} = u^{*-1}h'cu^{-1} = hucu^{-1}$ and so $ucu^{-1} \in C$. Let $(u,c) \in L$, then for any $x \in H$, $ucxu^* = x$. Since $1 \in H$, $c = u^{-1}u^{*-1}$ and $u^{*-1}xu^* = x$ or $ux = xu$, thus $u \in Z(H) \cap U$.
By (8.1) we have $(\overline{h})^* = \overline{h^*} = \overline{h}$ for $h \in H$ and so $\overline{H} = H$ and also $\overline{H^-} = H^-$. For $a \in U$ and $k \in H^-$ we have $(a_*(k))^* = (aka^*)^* = ak^*a^* = -aka^*$, hence $a_*(H^-) = H^-$. Now $c \in C$, then $c_\ell(k) = ck \in H^-$ implies $-ck = (ck)^* = k^*c^* = -kc$, and so $c \in Z(H^-)$, if $c_\ell(H^-) = H^-$.
For Char $K \neq 2$ we have for any $x \in A$ the equation $x = \frac{1}{2}(x + x^*) + \frac{1}{2}(x - x^*)$ with $\frac{1}{2}(x + x^*) \in H$ and $\frac{1}{2}(x - x^*) \in H^-$. This gives us the remaining statements.

<u>(8.3)</u> The map $f_*: A \times A \longrightarrow H$; $(a,b) \longrightarrow ab^* + ba^*$ is symmetric and bilinear and the map $[\ ,\ ]: A \times A \longrightarrow H^-$; $(a,b) \longrightarrow [a,b] := ab^* - ba^*$ skew-symmetric and bilinear. If $a,b,c \in H^-$, then:

a) $f_*(a,[a,b]) = a^2b - ba^2$

b) $[a,[b,c]] + [b,[c,a]] + [c,[a,b]] = 0$

c) $[a,[b,c]] = f_*(a,c)\cdot b + b\cdot f_*(a,c) - f_*(a,b)\cdot c - c\cdot f_*(a,b)$

d) $f_*([a,b],[a,b]) = 2(f_*(a,a)f_*(b,b) - f_*(a,b)^2) +$
$$+ 4(ab^2a + ba^2b - 2a^2b^2)$$

Proof. a): $f_*(a,[a,b]) = a(-(ab^* - ba^*)) + (ab^* - ba^*)a^*$
$$= a^2b - aba + aba - ba^2 .$$

c): $(ac^* + ca^*)b + b(ac^* + ca^*) - (ab^* + ba^*)c - c(ab^* + ba^*) =$
$= -acb - bca + abc + cba = a(bc - cb) - (bc - cb)a = [a,[b,c]]$

d): $f_*([a,b],[a,b]) = -2\,[a,b]^2 = -2(ab - ba)^2 =$
$= -2((ab)^2 + (ba)^2 - ab^2a - ba^2b) = -2f_*(a,b)^2 + 4(ab^2a + ba^2b) =$
$= 2(f_*(a,a)f_*(b,b) - f_*(a,b)^2) + 4(ab^2a + ba^2b - 2a^2b^2)$.

In the particular case $* = {}^-$ we have $f := f_*$ and we set $a \times b := [a,b]$. Since then $J = H^-$, $x^2 \in K$ for $x \in J$ and $f(A,A) \subset K$. We obtain from (8.3):

(8.4) If $a,b,c \in J$, then:

a) $f(a,a\times b) = f(b,a\times b) = 0$

b) $a\times(b\times c) = 2(f(a,c)\cdot b - f(a,b)\cdot c)$

c) $f(a\times b,a\times b) = 2(f(a,a)f(b,b) - f(a,b)^2)$

(8.5) For $(A,K,*)$ we have the further formulas:

a) $a\times b \in J$ for all $a,b \in A$ and $a\times b = 0 \Rightarrow a,b$ are linearly dependent or $q(a) = q(b) = f(a,b) = 0$.

b) $(a\times b)^* = \overline{b}^* \times \overline{a}^*$, $\overline{a\times b} = b\times a$ for $a,b \in A$

c) $(a\times b)^* = b^* \times a^*$, if $a,b \in J$

d) $(a\times b)^* = \overline{b}\times\overline{a}$, if $a,b \in H$ or $a,b \in H^-$

e) $(a\times b)^* = b\times a$, if $a,b \in H\cap J$ or $a,b \in H^-\cap J$

f) Let $a,b \in H$ or $a,b \in H^-$, $x \in Ka + Kb$, $y \in a^\perp \cap b^\perp$, then: $(a\times b)x = x(a\times b)^*$ and $(a\times b)y = -y(a\times b)^*$.

Proof. a): $a\times b = a\overline{b} - b\overline{a} = -\overline{a\times b}$, hence $a\times b \in J$ by (4.4). $a\times b = 0$ implies $\lambda := a\overline{b} = b\overline{a} \in K$, hence $q(b)\cdot a = \lambda b$,

$q(a)\cdot b = \lambda a$ and $f(a,b) = 2\lambda$.
e): $(a\times b)a = (a\overline{b} - b\overline{a})a = a(\overline{b}a - \overline{a}b) = a(\overline{b}\times\overline{a}) = a(a\times b)^*$ by c) and $(a\times b)b = b(a\times b)^*$ because $a\times b = -b\times a$. $y \in a^\perp \cap b^\perp$ implies $a\overline{y} = -y\overline{a}$, $\overline{a}y = -\overline{y}a$ and so $(a\times b)y = (a\overline{b} - b\overline{a})y =$
$= y(\overline{a}b - \overline{b}a) = y(\overline{a}\times\overline{b}) = -y(a\times b)^*$.

In the following let E be one of the sets H or H^-. For $a,b \in A$ we have $c := a\times b \in J$ by (8.5a) and by (8.5e) the elements of $Ka + Kb$ are solutions of the equation $cx = xc^*$ provided that $a,b \in E$. Now let $c \in J\setminus\{0\}$ be arbitrary and $\gamma \in K$, then also $c^* \in J\setminus\{0\}$. We are going to consider the sets $c(\gamma,A)$ and $c(\gamma,E)$ of solutions of the equation

(*) $cx = \gamma xc^*$ in A and in E respectively.

Then $c(\gamma,A)$ and $c(\gamma,E)$ are vector subspaces of (A,K) and (E,F) respectively.
First we assume $c(\gamma,A) \neq \{0\}$. We set $\alpha := \alpha_c = c^2 \in K$. Then $x \in c(\gamma,A)$, $x \neq 0$ implies $\alpha x = c^2x = c\gamma xc^* = \gamma^2 x(c^*)^2 = \gamma^2\alpha^* x$, hence $\alpha = \gamma^2\alpha^*$. Applying $*$ we obtain $\alpha^* = (\gamma^*)^2\alpha =$
$= (\gamma^*)^2\gamma^2\alpha^* = (\gamma\gamma^*)^2\alpha^*$, hence $\gamma\gamma^* = \pm 1$ if $\alpha \neq 0$. So $c(\gamma,A) \neq \{0\}$ implies $\alpha = 0$ or $\gamma^2 = \alpha(\alpha^*)^{-1}$.
Conversely, if $\alpha = 0$, then $\alpha^* = 0$ and for any $\gamma \in K$ we have $c(c + \gamma c^*) = \gamma cc^* = \gamma(c + \gamma c^*)c^*$, hence $c(\gamma,A) \neq \{0\}$ if $\gamma c^* \neq -c$. If $\gamma c^* = -c$, then $-c^* = \gamma^* c = -\gamma^*\gamma c^*$ and $\gamma\gamma^* = 1$. But $cc = 0$ and $\gamma cc^* = -cc = 0$ implies $c \in c(\gamma,A)$ with $c \neq 0$.
In the case $\alpha \neq 0$ and $\gamma^2 = \alpha(\alpha^*)^{-1}$, for any $y \in A$ the element $x := cy + \gamma yc^*$ is a solution of (*): $cx = \alpha y + \gamma cyc^* =$
$= \gamma^2 y\alpha^* + \gamma cyc^* = \gamma(\gamma yc^* + cy)c^* = \gamma xc^*$. Thus $c(\gamma,A) \neq \{0\}$ or $cy = -\gamma yc^*$ for all $y \in A$. But $cy = -\gamma yc^*$ for all $y \in A$ implies (for $y = 1$) $c = -\gamma c^*$ and $cy = yc$. Hence $c \in Z(A)$ and (*) has the form $cx = -xc$, i.e. $c(\gamma,A) = \{0\}$ for $\mathrm{Char}\, K \neq 2$ and $c(\gamma,A) = A$ for $\mathrm{Char}\, K = 2$. We have proved

(8.6) Let $c \in J\setminus\{0\}$, $\gamma \in K$ and let $c(\gamma,A)$ be the set of all solutions of (*). Then $c(\gamma,A)$ is a K-vector space and we have:

a) If $c(\gamma,A) \neq \{0\}$, then $\alpha = \gamma^2\alpha^*$ (for $\alpha = c^2$) and $(\gamma\gamma^*)^2 = 1$ if $\alpha \neq 0$

b) If $\alpha = \gamma^2\alpha^*$, then $c(\gamma,A) \supset \{cy + \gamma y c^* \mid y \in A\}$.

c) If $\alpha = 0$, then $c(\gamma,A) \neq \{0\}$ for any $\gamma \in K$.

d) If $\alpha = \gamma^2\alpha^*$ and $\alpha \neq 0$, then either $c(\gamma,A) \neq \{0\}$ or $c \in Z(A)$, $c^* = -\gamma^* c$ and Char $K \neq 2$.

e) If $\alpha = \gamma^2\alpha^* \neq 0$ and Char $K \neq 2$, then $c(\gamma,A) = \{cy + \gamma y c^* \mid y \in A\}$ and for $x \in c(\gamma,A)$, $x = c(2^{-1}\alpha^{-1}cx) + \gamma(2^{-1}\alpha^{-1}cx)c^*$.

Next we assume $c(\gamma,E) \neq \{0\}$ and $\alpha \neq 0$. Let $x \in c(\gamma,E)$, $x \neq 0$, hence $cx = \gamma x c^*$, then applying $*$ we obtain $x^*c^* = \gamma^* c x^*$ and since $x \in E$ implies $x^* = x$ or $x^* = -x$ we have $xc^* = \gamma^* cx$. Thus $xc^* = \gamma^*\gamma x c^*$ and so $\gamma\gamma^* = 1$ if $xc^* \neq 0$. But $c^2 = \alpha \neq 0$ implies $(c^*)^2 = \alpha^* \neq 0$ and so $xc^* \neq 0$. Therefore $\gamma^*\gamma = 1$.

Now let us assume $\alpha = \gamma^2\alpha^* \neq 0$ and $\gamma\gamma^* = 1$. Then for any $x \in c(\gamma,A)$, i.e. $cx = \gamma x c^*$, we have $x^*c^* = \gamma^* c x^*$, hence $\gamma x^* c^* = \gamma\gamma^* c x^* = cx^*$ and so $x^* \in c(\gamma,A)$.

If $\gamma \neq -1$ let $d := (1+\gamma^*)c$. Then $dx = (1+\gamma^*)cx = (1+\gamma^*)\gamma x c^* = (\gamma+1)xc^* = xd^*$ and hence $c(\gamma,A) = d(1,A)$. Further $d^2 = (1 + 2\gamma^* + \gamma^{*2})\alpha = (\gamma + 2 + \gamma^*)\gamma^*\alpha \in F$, because $\gamma\gamma^* = 1$ and $\gamma^*\alpha = \gamma\alpha^*$. If $\gamma = -1$, let $\lambda \in K\setminus F$ and $d := (\lambda - \lambda^*)c$. Then $dx = (\lambda - \lambda^*)cx = -(\lambda - \lambda^*)xc^* = xd^*$ and so $c(-1,A) = d(1,A)$. Also $d^2 = (\lambda - \lambda^*)^2\alpha \in F$ because $-\alpha = \gamma^*\alpha = \gamma\alpha^* = -\alpha^*$. So we have proved the parts a) and b) of (8.7).

<u>(8.7)</u> Let $c \in J\setminus\{0\}$ with $\alpha = c^2 \neq 0$ and $\gamma \in K$. Then:

a) If $c(\gamma,E) \neq \{0\}$, then $\gamma\gamma^* = 1$ and $\alpha\gamma^* = \gamma\alpha^*$.

b) Let $\gamma\gamma^* = 1$ and $\alpha\gamma^* = \gamma\alpha^*$. Then

α) $c(\gamma,A) = c(\gamma,A)^*$.

β) If $\gamma \neq -1$ and $d := (1+\gamma^*)c$, then $c(\gamma,A) = d(1,A)$ and $d^2 \in F$

γ) If $\gamma = -1$, $K \neq F$, $\lambda \in K\setminus F$, $\varepsilon := \lambda - \lambda^*$ and $d := \varepsilon c$, then $c(\gamma, A) = d(1, A)$ and $c^2, d^2 \in F$.

c) Let $\gamma\gamma^* = 1$, $\alpha\gamma^* = \gamma\alpha^*$ and Char $K \neq 2$. Then $c(\gamma, A) = c(\gamma, H) \oplus c(\gamma, H^-)$.

α) For $\gamma \neq -1$: $c(\gamma, H) = \{0\} \Leftrightarrow -\gamma c^* = c \in Z(H)$

β) For $\gamma = -1$: $c(\gamma, H) = \{0\} \Leftrightarrow -\gamma c^* = c^* = c \in Z(H^-)$
$\Leftrightarrow c(\gamma, A) = \{0\}$

γ) For $\gamma \neq 1, -1$: $c(\gamma, H^-) = \{0\} \Leftrightarrow -\gamma c^* = c \in Z(H^-)$

δ) For $\gamma = -1$: $c(-1, H^-) = \{0\} \Leftrightarrow c(-1, H) = \{0\}$

ε) For $\gamma = 1$: $c(1, H^-) = \{0\} \Leftrightarrow \forall x \in H^-: \ cx = -xc^*$

If $c \in H^-$, then $c(1, H^-) = \{0\} \nLeftrightarrow c \in Z(H^-)$;

if $c \notin H^-$, then $c(1, H^-) = \{0\} \Rightarrow H^- \subset J$

Proof. c): For $x \in c(\gamma, A)$ we have $x = \frac{1}{2}(x + x^*) + \frac{1}{2}(x - x^*)$, hence by b) $\frac{1}{2}(x + x^*) \in c(\gamma, A) \cap H = c(\gamma, H)$ and $\frac{1}{2}(x - x^*) \in c(\gamma, A) \cap H^- = c(\gamma, H^-)$. Since $H \cap H^- = \{0\}$, $c(\gamma, A) = c(\gamma, H) \oplus c(\gamma, H^-)$.

Let $c(\gamma, H) = \{0\}$. Then $x^* = -x$ for any $x \in c(\gamma, A) = \{cy + \gamma y c^* \mid y \in A\}$ (cf.(8.6e)). For $y = c^*$ we have $(cc^* + \gamma\alpha^*)^* = cc^* + \gamma^*\alpha = -(cc^* + \gamma\alpha^*)$, thus $(c + \gamma c^*)c^* = 0$, i.e. $c + \gamma c^* = 0$. The element $(1 + \gamma^*)(cy - yc) = (1 + \gamma^*)(cy + \gamma y c^*) = (1 + \gamma^*)cy + (1 + \gamma)yc^*$ lies in $c(\gamma, H)$ for all $y \in H$, hence $\gamma^* = -1$ or $c \in Z(H)$ by our assumption $c(\gamma, H) = \{0\}$. If $\gamma^* = -1$ and $-\gamma c^* = c \in Z(H)$, then (∗) has the form $cx = -xc$ and hence $c(\gamma, H) = \{0\}$. This gives us α).

Now let $\gamma^* = -1$, then $c = -\gamma c^* = c^*$, and for all $y \in H^-$ we have $cy + \gamma y c^* = cy - yc \in c(\gamma, H)$ because $(cy - yc)^* = -yc + cy$, thus $c(\gamma, H) = \{0\}$ implies $c \in Z(H^-)$. If $\gamma^* = -1$ and $c^* = c \in Z(H^-)$, then (∗) has the form $cx = -xc$ and hence $c(\gamma, H^-) = \{0\}$. Since $c(\gamma, H) = \{cy - yc \mid y \in H^-\}$ our assumption $c^* = c \in Z(H^-)$ implies $c(\gamma, H) = \{0\}$ and so $c(\gamma, A) = c(\gamma, H) \oplus c(\gamma, H^-) = \{0\}$.

Now let $c(\gamma, H^-) = \{0\}$. Then $x^* = x$ for any $x \in c(\gamma, A) = \{cy + \gamma y c^* \mid y \in A\}$. For $y = 1$ we have

$(c+\gamma c^*)^* = c^* + \gamma^* c = c + \gamma c^* = \gamma(c^* + \gamma^* c)$, hence $\gamma = 1$ or $-\gamma c^* = c$. For $\gamma \neq 1$ the element $(1+\gamma^*)(cy - yc) = (1+\gamma^*)(cy + \gamma y c^*) = (1+\gamma^*)cy + (1+\gamma)yc^*$ lies in $c(\gamma,H^-)$ for all $y \in H^-$, hence $\gamma^* = -1$ or $c \in Z(H^-)$. If $\gamma^* \neq -1$ and $-\gamma c^* = c \in Z(H^-)$, then (∗) has the form $cx = -xc$ and hence $c(\gamma,H^-) = \{0\}$. Thus γ) is proved. Now let $\gamma^* = -1$, then $c = -\gamma c^* = c^*$, and for all $y \in H$ we have $cy + \gamma y c^* = cy - yc \in c(\gamma,H^-)$ because $(cy - yc)^* = yc - cy$. Thus $c(\gamma,H^-) = \{0\}$ implies $c \in Z(H)$ and furthermore $c(\gamma,H) = \{0\}$ because (∗) has again the form $cx = -xc$.

ε): If $\gamma = 1$, then (∗) has the form $cx = xc^*$. Then by (8.6e) $c(1,H^-) = \{cx + xc^* \mid x \in H^-\}$, because $\alpha = \alpha^*$. Hence $c(1,H^-) = \{0\} \Leftrightarrow cx = -xc^*$ for all $x \in H^-$. Since with $x \in H^-$ also $\bar{x} \in H^-$, $c(1,H^-) = \{0\}$ implies $c(x+\bar{x}) = -(x+\bar{x})c^* = -c^*(x+\bar{x})$ for all $x \in H^-$, hence $H^- \subset J$ if $c \notin H^-$.

In the same way one obtains the following statements:

<u>(8.8)</u> Let $c \in J\setminus\{0\}$, $\alpha = c^2$ and Char $K \neq 2$. Then:

a) $c(1,A) \perp c(-1,A)$ and $c(1,A) \cap c(-1,A) = \{x \in A \mid cx = xc^* = 0\}$

b) If $\alpha \neq 0$, then $A = c(1,A) \oplus c(-1,A)$ and $x = x_1 + x_2$ with
$x_1 = c(2^{-1}\alpha^{-1}cx) + (2^{-1}\alpha^{-1}cx)c^* \in c(1,A)$,
$x_2 = c(2^{-1}\alpha^{-1}cx) - (2^{-1}\alpha^{-1}cx)c^* \in c(-1,A)$.

c) $c(1,A) = A \Leftrightarrow c \in H \cap Z(A)$; $c(1,H) = H \Leftrightarrow c \in H \cap Z(H)$;
$c(-1,A) = A \Leftrightarrow c \in H^- \cap Z(A)$; $c(-1,H) = H \Leftrightarrow c \in H^- \cap Z(H)$.

d) For $c \in H$: $c(1,H^-) = H^- \Leftrightarrow c \in Z(H^-)$
For $c \notin H$: $c(1,H^-) = H^-$ implies $H^- \subset J$ and $c^2 = 0$

e) If $\alpha \neq 0$, then the map $c_*: x \longrightarrow cxc^*$ has eigenvalues in K if $\alpha\alpha^* \in K^{(2)}$, in F if $\alpha\alpha^* \in F^{(2)}$ and the eigenvalue 1 (and then also -1), if $\alpha\alpha^* = 1$.

Proof. a): Let $x \in c(1,A)$ and $y \in c(-1,A)$. Then $cx = xc^*$, $cy = -yc^*$ and by applying $^-$ also $\bar{x}c = c^*\bar{x}$ and $\bar{y}c = c^*\bar{y}$.

Hence $f(x,y)\cdot c = (x\overline{y} + y\overline{x})c = -xc^*\overline{y} + yc^*\overline{x} = -cx\overline{y} - cy\overline{x} =$
$= -c\cdot f(x,y) = -f(x,y)\cdot c$, thus $f(x,y) = 0$.

If $c(1,H) \neq \{0\}$, then $\alpha = c^2 \in F$ by (8.7a). Therefore let $J_F := \{x \in J \mid x^2 \in F\}$. The set $J_E := \{a \times b \mid a,b \in E\}$ is a subset of J_F , because $(a \times b)^2 = f(a,b)^2 - 4q(a)q(b)$ and $f(a,b)^* =$
$= f(\overline{b}^*,\overline{a}^*) = f(b^*,a^*) = f(a^*,b^*)$, $q(a)^* = q(a^*)$ by (4.8).

<u>(8.9)</u> Let $c \in J_F\setminus\{0\}$ and $\alpha = c^2$. Then:

a) $\overline{c(1,E)} = c^*(1,E)$

b) $a,b \in c(1,E) \Rightarrow a\overline{b}a \in c(1,E)$ and $a\overline{b}$, $a \times b \in Z(c)$

c) If $\alpha \neq 0$, then $E = c(1,E) \oplus c(-1,E)$

d) If $c(1,E) \cap U \neq \emptyset$, then $c \in J_E$, $cu \in c(1,E)$, $cu \perp u$ and $c = (2u\overline{u})^{-1}\cdot(cu) \times u$ for any $u \in c(1,E) \cap U$.

e) $c(1,E) \cap U = \emptyset \Rightarrow cc^* = -c^*c \notin U$

f) For $\alpha = 0$: $c + c^* \in U \Leftrightarrow c - c^* \in U \Leftrightarrow cc^* + c^*c \neq 0 \Rightarrow c \in J_E$

<u>(8.10)</u> Let $c = a \times b \in J_E\setminus\{0\}$ and $\alpha = c^2$. Then:

a) $c(1,A) \supset Ka + Kb$ and $c(1,E) \supset Fa + Fb$;
$c(-1,A) = a^\perp \cap b^\perp$ and $c(-1,E) = a^\perp \cap b^\perp \cap E$. If $\alpha \neq 0$, then $c(1,A) = Ka + Kb$ and $c(1,E) = Fa + Fb$

b) There are $u,v \in c(1,E)$ with $u \perp v$ and $c = u \times v = 2u\overline{v}$.

c) If $\alpha = 0$ and $c(1,E) \cap U = \emptyset$, then $q(a) = q(b) = f(a,b) = 0$

d) If $F \neq K$ then there is an $i \in K\setminus F$ with $i^2 \in F$ and for $d := ic$ we have $c(-1,A) = d(1,A)$.

In the sequence we will consider algebras $(A,K,*)$ fulfilling the additional axioms:

<u>(A1)</u> For any $c \in J_E\setminus\{0\}$: $Z(c) = K + Kc$

<u>(A2)</u> For any $c \in J_E\setminus\{0\}$: $c(1,E) \cap U \neq \emptyset$

(8.11) Let $c \in J_E \setminus \{0\}$ with $\alpha = c^2 = 0$ and let $a, b \in c(1,E) \setminus \{0\}$ with $Fa \neq Fb$. Then:

a) If (A1) is valid, then $a \times b \in Kc$

b) If (A1) and (A2) are valid, then $c(1,E) = Fa + Fb$.

Proof. a): By (8.5a),(8.9b) and (A1) we have $a \times b \in Z(c) \cap J = (K + Kc) \cap J = Kc$.

b): By (A2) there is an $u \in c(1,E) \cap U$. We show that $c(1,E) = Fu + Fcu$. Let $x \in c(1,E)$ with $Fx \not\subset Fu$ and let $y \in (Fu + Fx) \cap u^{\perp}$, $y \neq 0$. Then $Fu + Fx = Fu + Fy$ and $y \times u = 2y\bar{u} \in Kc$ by a), hence $y \in Kcu$ and thus $c(1,E) = Fu + Fcu$.

In the remaining part of this section we are going to associate to our algebraic structure $(A,K,*)$ some geometric ones and to define the notion of the general kinematic map. For this purpose A is not necessarily associative, but we assume for the set E the following condition

(Z) $u \in U$ and $x^{-1}ux = y^{-1}uy$ for all $x,y \in E \cap U$ implies $u \in K$

Then $\mathbf{C} := Z(E) \cap H \cap U = F^*$ and by (8.2) $\Gamma^+(E) = \{u^*\lambda_\ell \mid u \in U, \lambda \in F\}$

Since (E,F) is a F-vector space, the triple (E,F,q) is a metric vector space. Let $O(E,F,q)$ be the orthogonal group consisting of all isometries and let $O_R(E,F,q)$ be the subgroup generated by the set R of reflections

$$R := \{a^{\square} : x \longrightarrow -x + \frac{q(a+x) - q(a) - q(x)}{q(a)} a \mid a \in E \cap U\}.$$

In our case we have $a^{\square}(x) = -x + \dfrac{a\bar{x} + x\bar{a}}{a\bar{a}} a = \dfrac{a\bar{x}a}{a\bar{a}}$.

Now we are looking for the maps $\gamma \in \Gamma(E)$ with $\gamma \in O(E,F,q)$. Since $\overline{E} = E$, $^{-} \in O(E,F,q)$. Hence let $\gamma = u_*\lambda_\ell$ with $u \in U$, $\lambda \in F^*$. Then $q(x) = q(u_*\lambda_\ell(x)) = q(u\lambda x u^*) = \lambda^2 q(uu^*)q(x)$ for all $x \in E$ and so $1 = \lambda^2 q(u)q(u)^*$, i.e. $q(u)q(u)^* \in F^{(2)}$.

If we denote $U_F := \{u \in U \mid q(uu^*) \in F^{(2)}\}$, $u_\square := u_* \circ \left(\sqrt{q(uu^*)}^{-1}\right)_\ell$ for $u \in U_F$ and $U_\square := \{u_\square \mid u \in U_F\}$, then:

(8.12) a) $E \cap U \subset U_F$ and for $a,b \in E \cap U$ we have:
$a^{\square} = a_{\square} \circ ^{-}$ if $E = H$, $a^{\square} = (-1)_{\ell} \circ a_{\square} \circ ^{-}$ if $E = H^{-}$ and $a^{\square}b^{\square} = a_{\square}\overline{b}_{\square}$.
b) $u_{\square}v_{\square} = (uv)_{\square}$ for $u,v \in U_F$ if A is associative.
c) $u_{\square} = Id_E \Leftrightarrow u \in K^*$
d) $u_{\square}$ is an involution, if and only if $u \in (J\backslash K) \cap U_F$.
e) $O(E,F,q) \cap \langle \Gamma(E) \rangle = \langle U_{\square} \rangle \cup (-1)_{\ell} \circ \langle U_{\square} \rangle \cup \langle U_{\square} \rangle \circ ^{-} \cup (-1)_{\ell} \circ \langle U_{\square} \rangle \circ ^{-}$
f) U_F is a subgroup of U with $K^* \subset U_F$
g) For $u,v \in U_F$ let $u_{\square} \,\square\, v_{\square} := (uv)_{\square}$, then $(U_{\square}, \square)$ is a loop and the map $\square : U_F \longrightarrow (U_{\square}, \square)\,;\; u \longrightarrow u_{\square}$ is a homomorphism of the loop U_F in the loop $(U_{\square}, \square)$ with the kernel K^* and $(U_{\square}, \square) \cong U_F / K^*$. If A is associative, then $\square$ is the composition of maps.

Proof. c): Let $u_{\square}$ be the identity on E and let $x,y \in E$. Then $x = u_{\square}(x) = \sqrt{q(uu^*)}^{\,-1} \cdot uxu^*$ implies $u^* = \sqrt{q(uu^*)} \cdot x^{-1}ux$ and so $x^{-1}ux = y^{-1}uy$ for all $x,y \in E$, hence $u \in K$ by (Z).
d): $Id_E = (u_{\square})^2 = (u^2)_{\square} \Leftrightarrow u^2 \in K^*$ by c). But $u^2 \in K^*$ implies $u \in J$ or $u \in K^*$.

Now we are able to define the general kinematic map for the euclidean derivation $\mathbf{A}(E,F,^{-}) := (E, \{a+Fb \mid a,b \in E,\, b \neq 0\}, \equiv)$, where $(a,b) \equiv (c,d) :\Leftrightarrow q(a-b) = q(c-d)$.

(8.13) For the group of all proper isometries (proper motions with a distinct point as fixed point) $O_R^+(E,F,q) := \langle R^2 \rangle$ of the metric affine space $\mathbf{A}(E,F,^{-})$ we have $O_R^+(E,F,q) \subset \langle U_{\square} \rangle$. The map $\varkappa$ defined by the following commutative diagram is called the general kinematic map for $\mathbf{A}(E,F,^{-})$:

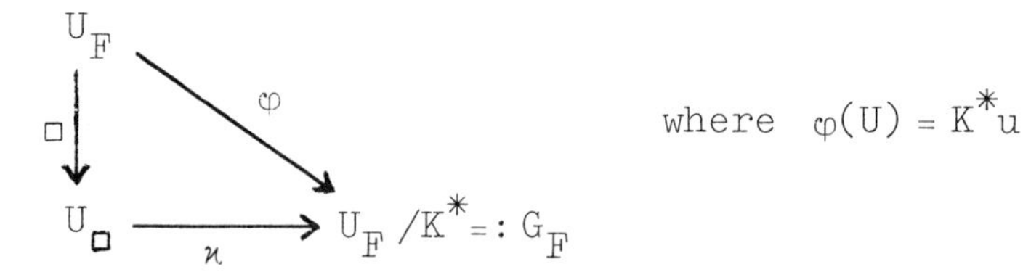

G_F is a subloop of the kinematic Moufang loop $(G := U/K^*, \mathfrak{L}, \cdot) := \Pi_\varkappa(A,K)$.

For the metric vector space (E,F,q) the projective derivation $\Pi(E,F,^-) := (E^*/F^*, \mathfrak{L}, \equiv)$ gives us the projective space with the point set $E^*/F^* = \{F^*a \mid a \in E^*\}$ and the line set $\mathfrak{L} = \{(Fa + Fb)^*/F^* \mid a,b \in E^* \text{ with } Fa \neq Fb\}$ together with the congruence $(F^*a, F^*b) \equiv (F^*c, F^*d) :\Leftrightarrow \frac{f(a,b)^2}{q(a)\cdot q(b)} = \frac{f(c,d)^2}{q(c)\cdot q(d)}$ which is only defined on the subset $(U \cap E)/F^*$. By the motion group $\mathfrak{B}$ of this <u>metric projective space</u> $\Pi(E,F,^-)$ we understand all collineations γ of $(E^*/F^*, \mathfrak{L})$ such that $(\gamma(F^*a), \gamma(F^*b)) \equiv (F^*a, F^*b)$ for all $a,b \in U \cap E$. Since all maps $\gamma \in \Gamma(E)$ are F-linear (even K-linear) permutations, they induce collineations $\hat{\gamma} : F^*x \longrightarrow F^*\gamma(x)$ of $\Pi(E,F)$. Again we are asking for maps $\gamma \in \Gamma(E)$, such that $\hat{\gamma} \in \mathfrak{B}$. It is enough to consider the case $\gamma = u_* \lambda_\ell$ with $u \in U$, $\lambda \in F^*$, because $^-$ induces a collineation of $\mathfrak{B}$. Let $a,b \in E \cap U$. Then

$$\frac{f(\gamma(a),\gamma(b))^2}{q(\gamma(a))\cdot q(\gamma(b))} = \frac{f(u\lambda a u^*, u\lambda b u^*)^2}{q(u\lambda a u^*)\cdot q(u\lambda b u^*)} = \frac{q(u)^2\cdot f(a,b)^2\cdot q(u^*)^2}{q(u)^2\cdot q(a)\cdot q(b)\cdot q(u^*)^2} =$$

$= \frac{f(a,b)^2}{q(a)\cdot q(b)}$. Thus $\hat{\Gamma}(E) \subset \mathfrak{B}$. Let us assume $\hat{\gamma} = \mathrm{Id}$. Then $\hat{u}_* = \mathrm{Id}$ and hence $u_* = u_\ell$ for $\mu \in F^*$. But $u_*(x) = uxu^* = \mu x$ implies $u^* = \mu x^{-1} u^{-1} x$ for all $x \in E \cap U$, hence $u \in K^*$ by (Z). Here we obtain the following theorem:

<u>(8.14)</u> a) Let $a \in E \cap U$. Then $\hat{a}^{\square}$ is a reflection fixing the point F^*a and the polar plane $\varphi(a^\perp)$ pointwise.

b) For $u \in U$: $\hat{u}_\square = \hat{u}_* = \mathrm{Id} \Leftrightarrow u \in K^*$

c) For $u \in U$: $\hat{u}_*$ is involutorial $\Leftrightarrow u \in J \setminus K$

d) $^-$ induces the identity $\Leftrightarrow E \subset J$

e) If $\hat{U} := \{\hat{u}_* \mid u \in U\}$ and if $\iota : F^*x \longrightarrow F^*\overline{x}$, then $\langle \hat{\Gamma}(E) \rangle = \langle \hat{U} \rangle \cup \langle \hat{U} \rangle \circ \iota \subset \mathfrak{B}$, $PO_R(E,F,q) := \{\hat{\sigma} \mid \sigma \in O_R(E,F,q)\} \subset \langle \hat{\Gamma}(E) \rangle$ and $PO_R^+(E,F,q) \subset \langle \hat{U} \rangle$.

f) For $u,v \in U$ let $\hat{u}_* \otimes \hat{v}_* := \widehat{(u,v)}_*$, then $(\hat{U}, \otimes)$ is a loop and the map $\wedge: U \longrightarrow (U, \otimes);\ u \longrightarrow \hat{u}_*$ is a homomorphism with the kernel K^* and $(\hat{U}, \otimes) = U/K^*$.

g) The map $\varkappa : \hat{U} \longrightarrow U/K^* = G$ defined by the following commutative diagram is called the <u>kinematic map</u> for the metric projective space $\Pi(E,F,^-)$:

$\Pi_\varkappa(A,K) = (G,\Omega,\cdot)$ is the kinematic Moufang loop of the metric projective space $\Pi(E,F,^-)$.

h) The map $\beta : E^*/F^* \longrightarrow A^*/K^*;\ F^*x \longrightarrow K^*x$ is injective.

§ 9 KUSTAANHEIMO'S KINEMATIC MODEL OF THE HYPERBOLIC SPACE

We mentioned in §1 that the hyperbolic plane can be embedded in the kinematic space belonging to the kinematic algebra $(\mathfrak{M}_{22}(\mathbb{R}),\mathbb{R})$. If we replace $\mathbb{R}$ by the complex numbers $\mathbb{C}$, then it is possible to give a representation of the 3-dimensional hyperbolic geometry inside the kinematic space $\Pi_\varkappa(\mathfrak{M}_{22}(\mathbb{C}),\mathbb{C})$. This will be then the model of KUSTAANHEIMO [29],[31].

Let $(A,K,*)$ be an associative kinematic algebra with an adjoint map and let $E = H$ or $E = H^-$ such that for E (A1) and the following is valid (cf. §8):

<u>(A2')</u> For any $c \in J_F \setminus \{0\}$: $c(1,E) \cap U \neq \emptyset$

Since $J_E \subset J_F$, (A2') implies (A2) and by (8.9d) $J_E = J_F$. Furthermore we assume $K \neq F$ and we denote by i a distinct element of $K \setminus F$ with $i^2 \in F$.

<u>(9.1)</u> Let $J_E \neq \{0\}$, then:

a) $J_E \cap U \neq \emptyset$

b) $\dim(E,F) = 4 = \dim(A,K)$

c) $\mathrm{Rad}(E,F,q) = \{0\} = \mathrm{Rad}(A,K,q)$

d) A is either a quaternion field (with center K) or the matrix algebra $\mathfrak{M}_{22}(K)$.

Proof. a): Since $J_E \neq \{0\}$, there are $a,b \in E$ with $c := a \times b \neq 0$. By (A2) and (8.9d) we may assume $a \in U$. By (8.10a,d), $a^\perp \cap b^\perp = c(-1,E) = d(1,E)$ and $d(1,E) \cap U \neq \emptyset$ by (A2'). Let $u \in d(1,E) \cap U$, then $a \times u = 2a\bar{u} \in J_E \cap U$.
b) and c) are consequences of a), (8.10a) and (8.9c), and d) follows from c) and (4.16).

In the 3-dimensional projective space $\Pi(E,F)$ the bilinear form $f(x,y) = x\bar{y} + y\bar{x}$ defines a bijective polarity by (9.1c). Therefore any $x \in E\setminus\{0\}$ represents at the same time a point Fx and a plane $(Fx)^\perp := \varphi(x^\perp)$, and a point Fx and a plane $\varphi(y^\perp)$ are incident if and only if $x \in y^\perp$; i.e. $f(x,y) = 0$. The lines of our projective space are described by elements of $J_E\setminus\{0\}$: For $c \in J_E\setminus\{0\}$ the corresponding line $(Fc)^\ell$ consists of the point set $\varphi(c(1,E))$, hence if Fx is a point, then the incidence is given by $cx = xc^*$. The line $(Fc)^\ell$ is contained in the plane $(Fy)^\perp$, if and only if $y \in c(-1,E)$, hence if $cy = -yc^*$. Two elements $c_1, c_2 \in J_E\setminus\{0\}$ determine the same line, if and only if c_1 and c_2 are linearly F-dependent. For two distinct points Fa and Fb the joining line is given by $(F(a\times b))^\ell$, and for two distinct planes $(Fu)^\perp$, $(Fv)^\perp$ the line of intersection is determined by $(Fi(u\times v))^\ell$

(9.2) In the 3-dimensional metric projective space $\Pi(E,F,{}^-)$ with the set $\mathbb{P} := \{Fx \mid x \in E\setminus\{0\}\}$ of points, the set $\mathfrak{E} := \{(Fx)^\perp \mid x \in E\setminus\{0\}\}$ of planes and the set $\mathfrak{L} := \{(Fc)^\ell \mid c \in J_F\setminus\{0\}\}$ of lines, the incidence I can be expressed by the equations: Let $Fa \in \mathbb{P}$, $(Fb)^\perp \in \mathfrak{E}$ and $(Fc)^\ell \in \mathfrak{L}$, then:

$Fa \quad I\,(Fb)^{\perp} \Leftrightarrow f(a,b) = a\bar{b} + b\bar{a} = 0$

$Fa \quad I\,(Fc)^{\ell} \Leftrightarrow ca - ac^{*} = 0$

$(Fc)^{\ell}\, I\,(Fb)^{\perp} \Leftrightarrow cb + bc^{*} = 0$

We recall that in a 3-dimensional projective space with a polarity two planes A,B are called orthogonal, if B is incident with the pole of A, a plane A is orthogonal to a line L, if L is incident with the pole of A and two lines L and M are orthogonal, if M intersects both lines, L and the polar line of L.

(9.3) In $\Pi(E,F,\bar{\ })$ the orthogonality can be expressed by the following equations: Let

$(Fa)^{\perp},(Fb)^{\perp} \in \mathfrak{E}$ and $(Fc)^{\ell},(Fd)^{\ell} \in \mathfrak{L}$, then:

$(Fa)^{\perp} \perp (Fb)^{\perp} \Leftrightarrow f(a,b) = a\bar{b} + b\bar{a} = 0$

$(Fa)^{\perp} \perp (Fc)^{\ell} \Leftrightarrow ca - ac^{*} = 0$

$(Fc)^{\ell}$ and $(Fd)^{\ell}$ are polar $\Leftrightarrow d \in Fic$

$(Fc)^{\ell} \perp (Fd)^{\ell} \Leftrightarrow f(c,d) = c\bar{d} + d\bar{c} = -cd - dc = 0$

In constructive geometry one considers the following fundamental incidence and orthogonal operations (L1),...,(M1),... . For our 3-dimensional metric projective space $\Pi(E,F,\bar{\ })$ these operations have algebraic descriptions:

(L1) Joining line of two distinct points $Fa, Fb \in \mathbb{P}$: $(F(a \times b))^{\ell}$

(L2) Plane determined by three non collinear points $Fu, Fv, Fw \in \mathbb{P}$: $(F((u \times v) \times (u \times w))u)^{\perp}$

(L3) Plane determined by a point Fa and a line $(Fc)^{\ell}$ with $Fa \not I (Fc)^{\ell}$: $(F(c \times (b \times a))b)^{\perp}$ where $Fb\, I\,(Fc)^{\ell}$

(L4) Deciding whether two distinct lines $(Fc)^{\ell},(Fd)^{\ell}$ intersect: $(Fc)^{\ell} \cap (Fd)^{\ell} \neq \emptyset \Leftrightarrow f(c,d) \in F$

(L5) Intersection of a line $(Fc)^{\ell}$ and a plane $(Fa)^{\perp}$, if $(Fc)^{\ell} \not I (Fa)^{\perp}$: $F(ca + ac^{*})$

(L6) Intersection of two distinct planes $(Fa)^{\perp}, (Fb)^{\perp}$: $(Fi(a \times b))^{\ell}$

(M1) Construction of a line through a point Fa and perpendicular to a plane $(Fb)^{\perp}$ with $Fa \neq Fb$: $(F(a \times b))^{\ell}$

(M2) Construction of a plane through a given point Fa and perpendicular to a line $(Fc)^{\ell}$, if $ca + ac^* \neq 0$: $(F(ca + ac^*))^{\perp}$.

(M3) Orthogonal projection on a given plane $(Fa)^{\perp}$: $\mathbb{P} \setminus \{Fa\} \longrightarrow (Fa)^{\perp}$; $Fu \longrightarrow F(u \times a)a$

(M4) Orthogonal projection on a line $(Fc)^{\ell}$: $\mathbb{P} \setminus (Fic)^{\ell} \longrightarrow (Fc)^{\ell}$; $Fu \longrightarrow F(c^2 u + cuc^*)$

If the multiplicative group $F \setminus \{0\}$ of the subfield F contains a subgroup F_+ of index 2, then we can define the following hyperbolic derivation:
The subset $\mathbb{P}_+ := \{Fx \in \mathbb{P} \mid q(x) \in F_+\}$ is the set of points of the associated hyperbolic space, the subset $\mathbb{P}_0 := \{Fx \in \mathbb{P} \mid q(x) = 0\}$ is the set of ends and $\mathfrak{E}_- := \{(Fx)^{\perp} \in \mathfrak{E} \mid q(x) \notin F_+ \cup \{0\}\}$ is the set of hyperbolic planes. The lines of the hyperbolic space are given by $\mathfrak{L}_- := \{(Fc)^{\ell} \in \mathfrak{L} \mid q(c) \notin F_+ \cup \{0\}\}$.

For the particular case, $A = \mathfrak{M}_{22}(\mathbb{C})$, $K = \mathbb{C}$ and $\begin{pmatrix} z_0 & z_1 \\ z_2 & z_3 \end{pmatrix}^* = \begin{pmatrix} z_0^* & z_2^* \\ z_1^* & z_3^* \end{pmatrix}$, where z_i^* denotes the conjugate complex number of z_i, we have $F = \mathbb{R}$ and for $E = H$ the hyperbolic derivation gives us the model of P. KUSTAANHEIMO of the classical 3-dimensional hyperbolic space.

Remark. The development of kinematics is covered in the book "Geschichte der Geometrie seit Hilbert" by H. KARZEL and H.-J. KROLL which will be published by "Wissenschaftliche Buchgesellschaft Darmstadt".

BIBLIOGRAPHY

[1] ALBERT,A.A.: Structure of algebras. (Colloq. Publ. 24) Amer. Math. Soc., Providence 1939.

[2] ANDRE,J.: Über nicht-Desarguessche Ebenen mit transitiver Translationsgruppe. Math. Z. 60 (1954) 156-186.

[3] BAER,R.: The group of motions of a two dimensional elliptic geometry. Compositio math. 9 (1951) 244-288.

[4] BLASCHKE,W.: Euklidische Kinematik und nichteuklidische Geometrie. Zschr. f. Math. u. Phys. 60 (1911) 61-91 u. 203-204.

[5] ——, and MÜLLER,H.R.: Ebene Kinematik. München 1956.

[6] BRÖCKER,L.: Kinematische Räume. Geometriae Dedicata 1 (1973) 241-278

[7] BROWN,R.B.: On generalized Cayley-Dickson algebras. Pacific J. of Math. 20 (1967) 415-422

[8] BRUCK,R.H.: A survey of binary systems. Berlin-Göttingen Heidelberg 1958

[9] GRÜNWALD,J.: Ein Abbildungsprinzip, welches die ebene Geometrie und Kinematik mit der räumlichen Geometrie verknüpft. S.B. Akad. Wien, math.-nat. Kl. IIa, 80 (1911) 677-741

[10] HERZER,A.: Endliche nicht kommutative Gruppen mit Partition π und fixpunktfreiem π-Automorphismus. Archiv d. Math. 34 (1980) 385-392

[11] HOTJE,H.: Beziehungen zwischen einbettbaren Berührstrukturen und kinematischen Räumen. In: Beiträge zur geometrischen Algebra, hrsg. v. H.J. Arnold ... Basel, Stuttgart (1977) 153-156

[12] ——: Die Algebren einbettbarer Berührstrukturen. Geometriae Dedicata 7 (1978) 355-362

[13] KARZEL,H.: Bericht über projektive Inzidenzgruppen. Jber. Deutsch. Math.-Verein. 67 (1965) 58-92. Also in: Wandel von Begriffsbildungen in der Mathematik, hrsg. v. H. Karzel u. K. Sörensen. Darmstadt 1984

[14] ——: Kinematic spaces. Symposia Matematica, Ist. Naz. di Alta Matematica 11, (1973) 413-439

[15] KARZEL,H.: Kinematische Algebren und ihre geometrischen Ableitungen. Abh. Math. Sem. Univ. Hamburg 41, (1974) 158-171.

[16] ——: Gruppentheoretische Begründung metrischer Geometrien. Vorlesungsausarbeitung der von H. Karzel gehaltenen Vorlesung an der Univ. Hamburg (1963), ausgearbeitet von G. GRAUMANN.

[17] ——, and KROLL,H.-J.: Eine inzidenzgeometrische Kennzeichnung projektiver kinematischer Räume. Arch. Math. 26 (1975) 107-112.

[18] ——, and MARCHI,M.: Plane fibered incidence groups. J. of Geometry 20 (1983) 192-201.

[19] ——, and MAXSON,C.J.: Fibered groups with non-trivial centers. Res. d. Math. 7 (1984) 1-17.

[20] ——, and PIEPER,I.: Bericht über geschlitzte Inzidenzgruppen. Jber. Deutsch. Math.-Verein. 72 (1970) 70-114.

[21] ——, KROLL,H.-J. and SÖRENSEN,K.: Invariante Gruppenpartitionen und Doppelräume. J. reine angew. Math. 262/263 (1973) 153-157.

[22] ——, KROLL,H.-J. and SÖRENSEN,K.: Projektive Doppelräume. Arch. Math. 25 (1974) 206-209.

[23] ——, SÖRENSEN,K. and WINDELBERG,D.: Einführung in die Geometrie. Göttingen 1974.

[24] KIST,G.P.: Punktiert-affine Inzidenzgruppen und Fastkörpererweiterungen. Abh. Math. Sem. Univ. Hamburg 44 (1975) 233-248.

[25] ——: Projektiver Abschluß 2-gelochter Räume. Res. d. Math. 3 (1980) 192-211.

[26] ——: Theorie der verallgemeinerten kinematischen Räume. Habilitationsschrift TU München 1980.

[27] KROLL,H.-J.: Bestimmung aller projektiven Doppelräume. Abh. Math. Sem. Univ. Hamburg 44 (1975) 139-142.

[28] ——: Zur Struktur geschlitzter Doppelräume. J. of Geometry 5 (1974) 27-38

[29] KUSTAANHEIMO,P.: Darstellung des klassischen hyperbolischen Raumes durch relativistische Motoren. Abh. Math. Sem. Univ. Hamburg 32 (1968) 89-96

[30] KUSTAANHEIMO,P.: Relativistic spinor linearization of the Kepler motion. Astron. Nachr. 296 (1975) 163-

[31] ——: Über die Geometrie der relativistischen Motoren. Abh. Math. Sem. Univ. Hamburg 44 (1975) 110-121.

[32] LÜNEBURG,H.: Einige methodische Bemerkungen zur Theorie der elliptischen Ebenen. Abh. Math. Sem. Univ. Hamburg 34 (1969) 59-72.

[33] MARCHI,M.: Fibered incidence groups which are not kinematic. J. of Geometry 20 (1983) 95-100.

[34] MISFELD,J.: Zur Struktur stetiger Inzidenzgruppen. Mitt. Math. Ges. Hamburg 10 (1971) 56-69

[35] ——, and SIGMON,K.: Completion of topological incidence groups. J. of Geometry 11 (1978) 150-160

[36] OSBORN,J.M.: Quadratic division algebras. Trans. Amer. Math. Soc. 105 (1962) 202-221.

[37] PIEPER,I.: Darstellung zweiseitiger geschlitzter Inzidenzgruppen. Abh. Math. Sem. Univ. Hamburg 32 (1968) 97-126.

[38] ——: Zur Darstellung zweiseitiger affiner Inzidenzgruppen. Abh. Math. Sem. Univ. Hamburg 35 (1970) 121-130

[39] PODEHL,E. and REIDEMEISTER,K.: Eine Begründung der ebenen elliptischen Geometrie. Abh. Math. Sem. Univ. Hamburg 10 (1934) 231-255.

[40] SCHAFER,R.D.: An introduction to nonassociative algebras. New York 1966

[41] SCHRÖDER,E.M.: Darstellung der Gruppenräume Minkowskischer Ebenen. Archiv d. Math. 21 (1970) 308-316

[42] ——: Kennzeichnung und Darstellung kinematischer Räume metrischer Ebenen. Abh. Math. Sem. Univ. Hamburg 39 (1973) 184-229

[43] ——: Zur Theorie subaffiner Inzidenzgruppen. J. of Geometry 3 (1973) 31-69.

[44] ——: Kreisgeometrische Darstellung metrischer Ebenen und verallgemeinerte Winkel- und Distanzfunktionen. Abh. Math. Sem. Univ. Hamburg 42 (1974) 154-186

[45] SCHÜTTE,K.: Der projektiv erweiterte Gruppenraum der ebenen Bewegungen. Math. Ann. 134 (1957) 62-92.

[46] SÖRENSEN,K.: Eine Beschreibung des kinematischen Raumes von Blaschke und Grünwald durch topologische Inzidenzgruppen. Abh. Math. Sem. Univ. Hamburg 35 (1970) 89-91.

[47] WÄHLING,H.: Darstellung zweiseitiger Inzidenzgruppen durch Divisionsalgebren. Abh. Math. Sem. Univ. Hamburg 30 (1967) 197-240.

[48] ——: Kongruenzerhaltende Permutationen von Kompositionsalgebren. In: Beiträge zur geometrischen Algebra, hrsg. v. H.J. Arnold ... Basel,Stuttgart (1977) 327-335.

[49] ——: Projektive Inzidenzgruppoide und Fastalgebren. J. of Geometry 9 (1977) 109-126.

COORDINATIZATION OF LATTICES *

Ulrich Brehm
Technische Universität Berlin
Fachbereich Mathematik
Straße des 17. Juni 135
D 1000 Berlin 12

ABSTRACT. We characterize the lattices which are isomorphic to submodule lattices of torsion free modules of Goldie dimension at least three over left Ore domains. The characterizing lattice-theoretic properties (axioms) are simple, natural, independent and of a geometric flavor.

In order to get uniqueness of the coordinatizing module (and ring) we start with a lattice L together with a distinguished subset P of "points" which shall correspond exactly to the non-zero cyclic submodules. In the proof of the coordinatization theorem we first construct a factor lattice $L/\sim$ to which we can apply a lattice theoretic version of the classical coordinatization theorem of projective geometry. Thus we get a skew field K and a K-vectorspace V such that $L/\sim$ is isomorphic to the lattice of linear subspaces of V. Then we construct a subring $R \subseteq K$, an R-submodule $M \subseteq K$ and a lattice isomorphism f between L and the lattice of R-submodules of M with $f[P] = \{Rx \mid x \in M \setminus \{0\}\}$.

1. INTRODUCTION

To coordinatize a lattice means in a very general sense to construct an algebraic structure such that a lattice of subalgebras or such that a lattice of congruences is isomorphic to the given lattice. The most natural algebraic structures for the coordinatization of lattices are modules.

---

* The author wants to thank the Deutsche Forschungsgemeinschaft for their support during the formulation of the present paper.

R. Kaya et al. (eds.), Rings and Geometry, 511–550.

Since not every lattice can be represented as the submodule lattice of a module we have to find lattice theoretic properties (called axioms) characterizing the lattices which are isomorphic to submodule lattices of modules of some given class of modules.

It is interesting for lattice theory, as well as for module theory to see how properties of modules are reflected in their submodule lattices and vice versa.

The coordinatization of Desarguesian projective planes and projective spaces of dimension at least three by vector spaces of dimension at least three over skew fields is the fundamental classical result for which many generalizations have been found.

One of these is von Neumann's famous result that every complemented modular lattice with a homogeneous basis of order $n \geqslant 4$ is isomorphic to the lattice of finitely generated submodules of R^n where R is a (von Neumann) regular ring. (cf [vN]).

Another well known generalization of the classical coordinatization theorem of projective geometry is the coordinatization of primary lattices which has been obtained by Baer [Ba 2], Inaba [In] and Jónsson / Monk [JM] under different assumptions on the geometric dimension with the final result: Every primary Arguesian lattice of geometric dimension $\geqslant 3$ is isomorphic to the submodule lattice of a finitely generated module over a completely primary uniserial ring. These papers and further papers and books on coordinatization theorems for lattices are listed in the references.

In the present paper we characterize those lattices which are isomorphic to submodule lattices of torsion free modules of Goldie dimension at least three over left Ore domains. The choice of this class of modules is motivated by the fact that such modules can be regarded as R-submodules of K-vector spaces over the skew field K of left quotients of R.

Most of the results of the present paper are contained in the second chapter of the author's thesis. (cf [Br 1]).

It would be desirable to have the coordinatizing ring and module determined uniquely up to (semilinear) isomorphism, but in general a module is not determined up to isomorphism by its submodule lattice but only up to Morita equivalence (at most), since for every equivalence $F: R\text{-Mod} \to S\text{-Mod}$ between the categories of R-(resp. S-)modules the submodule lattices of M and of F(M) are isomorphic for every R-module M. (cf.[AF], p. 256 f).

If ${}_RM$ and ${}_SN$ are torsion free modules of Goldie dimension at least three over left Ore domains and if f is a lattice isomorphism between the submodule lattices of M and of N with $f(Rx) = Sy$ for some $x \in M\setminus\{0\}$, $y \in N\setminus\{0\}$ then f is induced by a semilinear isomorphism between M and N. (cf [Br 1] for this result, cf [Br 2] for the representation of sum-preserving lattice homomorphisms and of isomorphisms between submodule lattices of much more general modules).

So in order to get uniqueness of the representation of lattices it is natural to start with a lattice together with a distinguished set of "points" which shall correspond to the cyclic submodules.

Alternatively we can start with a lattice together with just one distinguished "point". This will be done in theorem 2.

In theorem 3 we investigate the case that the set of "points" is the set of all compact elements of rank 1.

Finally we show by examples that each one of the axioms is independent of the others.

2. BASIC DEFINITIONS AND NOTATIONS

Most of the definitions and notations are standard and recalled for the conveniance of the reader.

Convention. All rings have a unit and all modules are unitary left modules.

If M is an R-module then $L({}_RM)$ denotes the lattice of R-submodules of M.

Definition. A ring with 1 is called a left Ore domain if $R \neq 0$ and R has no non-zero zerodivisors and if for all $a, b \in R \setminus \{0\}$ holds $Ra \cap Rb \neq 0$.

Definition. A module M over a left Ore domain R is called torsion free if $\forall r \in R\ \forall x \in M: (rx = 0 \Rightarrow r = 0 \text{ or } x = 0)$.

We mention some well known basic facts about left Ore domains and torsion free modules over such rings. (cf [Co])

If R is a left Ore domain then there exists a skew field $K \supseteq R$ with $K = \{a^{-1}b \mid a, b \in R, a \neq 0\}$. K is called the skew field of left quotients of R.

If M is a torsion free module over a left Ore domain R and K is the skew field of left quotients of R then the canonical mapping $i: M \longrightarrow K\otimes_R M$ is injective. Thus we can regard M as an R-submodule of a K-vectorspace.

Especially this last mentioned fact makes the torsion free modules over left Ore domains an interesting class of modules. For the submodule lattices we can hope for axioms which are still of a geometric nature. Our class of modules is quite large and contains for example all torsion free abelian groups and all vector spaces.

Now let L be a lattice with smallest element O.

Definition. L is called upper continuous if L is complete and for all $a\in L$, $U\subseteq L$ holds
$a \cap \Sigma U = \Sigma\{a \cap \Sigma E \mid E \subseteq U,\ E \text{ finite}\}$.

Definition. Let L be complete. Then $a\in L$ is called compact if for all $U\subseteq L$ with $a\leqslant\Sigma U$ there is a finite subset $E\subseteq U$ with $a\leqslant\Sigma U$.

Notation. $[a,b]: = \{c\in L \mid a\leqslant c\leqslant b\}$ where $a,b\in L$.

Definition: A finite subset $A\subseteq L$ is called independent if $O\notin A$ and $a\cap\Sigma(A\setminus\{a\}) = O$ for all $a\in A$.

rank L: $= n$ if $n\in\mathbb{N}\cup\{O\}$ is the maximal cardinality of independent subsets of L.

rank L: $= \infty$ if no such n exists.

For $a\in L$ we define rank a: $=$ rank $[O,a]$.

Definition. For an R-module M we define
rank M: $=$ rank $L({}_R M)$.
rank M is called the Goldie dimension (or the rank) of M.

M is called uniform if rank M $= 1$.

Definition. $a,b,c\in L$ are called strictly collinear if $a+b = a+c = b+c$ and $a\cap b = a\cap c = b\cap c = 0$.
We write $[a,b,c]$ for the statement 'a,b,c are strictly collinear'.

Definition: Let be $a,b\in L$. a is called perspective to b if there exists a $c\in L$ with $a+c = b+c$ and $a\cap c = b\cap c = 0$.
a is called projective to b if there exist $n\in\mathbb{N}, c_1,\dots,c_n\in L$ with $c_1 = a$, $c_n = b$ such that c_i is perspective to c_{i+1} for $i = 1,\dots,n-1$.

Definition. L is called Arguesian if for all $a_1,a_2,a_3,b_1,b_2,b_3\in L$ holds

$$(a_1+b_1)\cap(a_2+b_2) \leq a_3+b_3 \Rightarrow$$
$$(a_1+a_2)\cap(b_1+b_2) \leq ((a_1+a_3)\cap(b_1+b_3))+((a_2+a_3)\cap(b_2+b_3)).$$

Remark. It is known (cf [GLJ]) that a lattice is Arguesian if and only if the following inequality holds for all $a_1,a_2,a_3,b_1,b_2,b_3 \in L$:

$$(a_1+b_1)\cap(a_2+b_2)\cap(a_3+b_3) \leq (a_1\cap(a_2+x))+(b_1\cap(b_2+x)) \text{ where}$$
$$x := (a_1+a_2)\cap(b_1+b_2)\cap(((a_2+a_3)\cap(b_2+b_3))+((a_1+a_3)\cap(b_1+b_3))).$$

This condition is obviously equivalent to an equation. Thus the Arguesian lattices form a variety in the sense of universal algebra.

3. THE AXIOMS AND FORMULATION OF THE COORDINATIZATION THEOREM

Let L be a complete modular lattice and $P\subseteq L$ a subset.

axiom I : For all $a,b\in L$ with $a<b$ there exists a $p\in P$ with $p\leq b$, $p\not\leq a$.

axiom II : All $p\in P$ are compact in L.

axiom III : For all $a,b,c\in L$ with $a\cap c = 0$ and $(a+b)\cap c \neq 0$ there exists a $c'\in L$ with $a\cap c' = 0$ and $b\cap c' \neq 0$.

axiom IV : For all $p\in P$ holds rank $p = 1$ in L.

axiom V : For all $p_1,p_2\in P$ with $p_1\cap p_2 = 0$ there exists a $p_3\in P$ with $p_1+p_2 = p_1+p_3 = p_2+p_3$.

axiom VI: rank $L\geqslant 3$ and L is Arguesian or rank $L\geqslant 4$.

axiom VII: For all $p, q\in P$, $a\in L\setminus\{0\}$ with $p\leqslant q+a$ there exists a $q'\in P$ with $q'\leqslant a$ and $p\leqslant q+q'$.

Theorem 1. Let L be a complete modular lattice and $P\subseteq L$ a subset. Then L, P satisfy the axioms I - VII if and only if there exists a left Ore domain R, a torsion-free R-module M with rank $M\geqslant 3$ and a lattice isomorphism $f:L\longrightarrow L({}_RM)$ with $f[P] = \{Rx \mid x\in M\setminus\{0\}\}$.

Remark: We give two simple conditions each of which can replace the two axioms III and IV.
Let L be a complete modular lattice, $P\subseteq L$ such that the axioms I, II and VII hold. Then are equivalent:

i) axioms III and IV hold for L, P,

ii) for all $p, q\in P$, $a\in L$ holds:
$p\leqslant q+a$ and $p\cap a = 0 \Rightarrow q\cap a = 0$,

iii) for all $p\in P$, $a\in L$ holds:
rank $(p+a)+$ rank $(p\cap a) = 1+$ rank (a).

Proof. The equivalence of these three conditions has been shown in [Br 1]. Since we don't use this result we will give only a proof of i) $\Rightarrow$ iii) in lemma 12.

We will first investigate in lemma 1-3 for each axiom what it means for R and M if $L = L({}_RM)$ and $P = \{Rx \mid x \in M \setminus \{0\}\}$. Axiom III means that the module M is non-singular, axiom IV means that every cyclic submodule of M is uniform, axiom VI means that rank $M \geqslant 3$ and all other axioms hold in every submodule lattice. In lemma 4 we show that the axioms I-VII hold for $L({}_RM)$, $P = \{Rx \mid x \in M \setminus \{0\}\}$ if R is a left Ore domain and M is torsion free.

In lemma 6 we show that axiom III holds in a lattice if and only if a certain equivalence relation $\sim$ is a congruence relation. In lemma 7 and 8 we investigate the factor lattice $L/\sim$.

In lemma 9 we apply a lattice theoretic version of the classical coordinatization theorem of projective geometry to the factor lattice $L/\sim$. Thus we get a skew field K, a K-vector space V and a lattice isomorphism $i: L/\sim \longrightarrow L({}_KV)$.

We will then construct a subring $R \subseteq K$, an R-submodule $M \subseteq V$ and a lattice isomorphism $f: L \longrightarrow L({}_RM)$ such that the following diagram commutes

$$\begin{array}{ccc} L & \xrightarrow{f} & L({}_RM) \\ {\scriptstyle nat}\downarrow & & \downarrow{\scriptstyle K\otimes_R -} \\ L/\sim & \xrightarrow{i} & L({}_KV) \end{array}$$

where nat denotes the natural mapping and $K \otimes_R -$ denotes the mapping with $K \otimes_R -(U) := K \otimes_R U$. Note that $K \otimes_R U = K\cdot U$ since R will be a left Ore domain.

4. LEMMATA

Lemma 1. Let ${}_RM$ be an R-module, $L := L({}_RM)$ and $P := \{Rx \mid x \in M \setminus \{0\}\}$. Then:

a) For L, P the axioms I, II, V and VII hold
b) $L({}_RM)$ is Arguesian
c) For L, P the axiom IV holds if and only if every non-zero cyclic submodule of M is uniform.
d) For L, P axiom VI holds if and only if rank $M \geqslant 3$.

Proof. a) Axioms I and II hold obviously, more generally the compact elements of $L({}_RM)$ are exactly the finitely generated submodules of ${}_RM$. Axiom V holds since $Rx+Ry = Rx+R(x+y) = Ry+R(x+y)$ for all $x,y \in M$. Axiom VII holds since $Rx \subseteq Ry+U$ with $U \in L({}_RM)$, $U \neq 0$ implies that there is a $z \in U$ and an $r \in R$ with $x = ry+z$, thus $Rx \subseteq Ry+Rz'$ with $z' := z$ if $z \neq 0$, $z' \in U \setminus \{0\}$ arbitrary if $z = 0$.

b) This is well known and can be shown directly by a short computation.

c) obvious.
d) obvious. □

We recall that a left ideal $I \subseteq R$ is called essential if there is no left ideal $J \neq 0$ with $I \cap J = 0$ and that an R-module M is called non-singular if

$$\{x \in M \mid \text{ann } x \text{ is an essential left ideal in } R\} = \{0\},$$

where $\text{ann } x := \{r \in R \mid rx = 0\}$.

Note that a ring R is a left Ore domain if and only if rank ${}_RR = 1$ and ${}_RR$ is non-singular, where ${}_RR$ denotes the left R-module R. If R is a left Ore domain then an R-module M is torsion free if and only if M is non-singular.

Lemma 2. If ${}_RM$ is a non-singular module then axiom III holds for $L({}_RM)$.

Proof. If $C\cap(A+B) \neq 0$, $C\cap A = 0$ choose $x\in C\cap(A+B)\setminus\{0\}$. Then there are $y\in A$, $z\in B$ with $x = y+z$. Since ann x is not essential there is an $r\in R\setminus\{0\}$ with $Rr \cap \text{ann } x = 0$, thus $rx \neq 0$, thus $rz = 0$ (otherwise we would get the contradiction $0 \neq rx = ry \in C\cap A = 0$). Choose $C' := Rrz$. Then $C' \subseteq B$ and $C' \neq 0$. It remains to show $Rrz\cap A = 0$. Let be $u \in Rrz\cap A$. Then there exists an $s\in R$ with $srz = u$, thus $srx = sr(y+z) \in C\cap A = 0$, thus $sr \in Rr\cap \text{ann } x = 0$, thus $u = 0$. □

To show the converse we have to assume a weak additional property of the module (since there are modules which are not non-singular but have a chain as submodule lattice).

Lemma 3. If ${}_RM$ is a module such that axiom III holds for $L({}_RM)$ and if for all $x\in M\setminus\{0\}$ there exist $r\in R$ and $y\in M$ such that $rx \neq 0$, $Rrx\cap Ry = 0$ and ann $y \subseteq$ ann r then ${}_RM$ is non-singular.

Proof. Let be $x\in M\setminus\{0\}$, $r\in R$, $y\in M$ such that $rx \neq 0$, $Rrx\cap Ry = 0$ and ann $y \subseteq$ ann r.
Then $Rrx\cap(Ry+R(rx-y)) = Rrx \neq 0$. Now axiom III implies that there is a $z\in R(rx-y)\setminus\{0\}$ such that $Rz\cap Ry = 0$. Thus there is an $s\in R$ such that $z = s(rx-y)$. $srx \neq 0$ since $z \neq 0$ and $Rx\cap Ry = 0$.

We show that $Rsr \cap \text{ann } x = 0$. Let be $t\in R$ with $tsrx = 0$. Then $tz = -tsy \in Rz\cap Ry = 0$, thus $ts\in$ ann $y \subseteq$ ann r, thus $tsr = 0$.

Thus ann x is not essential, thus M is non-singular. □

Lemma 4. If R is a left Ore domain and ${}_RM$ a torsion free R-module with dim $M \geqslant 3$ then the axioms I-VII hold for $L({}_RM)$, $P := \{Rx \mid x\in M\setminus\{0\}\}$.

Proof. Since ann x = 0 for all $x \in M \setminus \{0\}$, M is non-singular. Thus axiom III holds (using lemma 2).

Since R is a left Ore domain and ann x = 0 if $x \neq 0$ we have rank Rx = rank $_R R = 1$ for all $x \in M$, thus axiom IV holds. Now lemma 4 follows from lemma 1. □

Lemma 5. If for L, P the axioms I and II hold then L is upper continuous.

Proof. Axioms I and II imply that L is compactly generated. It is well known (cf [CD], p. 15) that this implies that L is upper continuous. □

Let L be a lattice with 0. Then we define an equivalence relation $\sim$ on L by

$$a \sim b :\iff \forall c \in L : (a \cap c = 0 \iff b \cap c = 0).$$

$\sim$ is obviously an equivalence relation and

$$a \sim b \Rightarrow a \cap c \sim b \cap c \quad \text{for all } a,b,c \in L.$$

$a \not\sim b$ means that $a \sim b$ does not hold.

Lemma 6. Let L be a lattice with 0. Then $\sim$ is a congruence relation if and only if axiom III holds.

Proof: " $\Leftarrow$ ": We have to show $a \sim b \Rightarrow a+c \sim b+c$ for all $a,b,c \in L$. If $a+c \not\sim b+c$ then there exists without loss of generality a $d \in L$ with $(a+c) \cap d \neq 0$, $(b+c) \cap d = 0$, thus by axiom III there is a $d' \in L$ with $a \cap d' \neq 0$, $(b+c) \cap d' = 0$, thus $a \not\sim b$.

" $\Rightarrow$ ": Let $\sim$ be a congruence relation and $a,b,c \in L$ with $a \cap c = 0$ and $(a+b) \cap c \neq 0$. Then $a \not\sim (a+b)$, thus $a \cap b \not\sim b$, thus there is a $d \in L$ with $a \cap b \cap d = 0$ and $b \cap d \neq 0$, thus for $c' := b \cap d$ holds $a \cap c' = 0$ and $b \cap c' \neq 0$.

Lemma 7. Let L be a modular upper continuous lattice in which axiom III holds. Let $f: L \longrightarrow L/\sim$ denote the natural homomorphism. Then

$L/\sim$ is a complemented modular upper continuous lattice,

f preserves arbitrary sums,

$f^{-1}[\{0\}] = \{0\}$,

rank a = rank f(a) for all $a \in L$.

Proof. $f^{-1}[\{0\}] = \{0\}$ is obvious. We show that $L/\sim$ is complete and that f preserves arbitrary sums.
Let be $A \subseteq L$, $A \neq \emptyset$. We show that $f(\Sigma A)$ is the least upper bound of $f[A]$. Obviously $f(\Sigma A)$ is an upper bound of $f[A]$. Now assume $f(b) \geq f(a)$ for all $a \in M$. Then $f(b) \geq f(\Sigma E)$ for all $E \subseteq A$ finite, thus for all $c \in L$ with $c \cap b = 0$ holds

$$c \cap \Sigma A \overset{1}{=} \Sigma\{c \cap \Sigma E \mid E \subseteq A \text{ finite}\} \overset{2}{=} 0.$$

1 since L is upper continuous
2 since $0 = f(c \cap b) \geq f(c \cap \Sigma E)$ and $f^{-1}[\{0\}] = \{0\}$.
Thus $f(b) \geq f(\Sigma A)$, since f = nat. Thus $\Sigma f[A]$ exists and f preserves arbitrary sums. This implies that $L/\sim$ is upper continuous because L is upper continuous.
Since f is a surjective homomorphism with $f^{-1}[\{0\}] = \{0\}$ a subset $A \subseteq L$ is independent if and only if $f[A]$ is independent in $L/\sim$. Thus rank a = rank f(a) for all $a \in L$.

We show that $L/\sim$ is complemented.

Since L is a modular upper continuous lattice, every element a has a pseudo complement a' (i.e. a maximal element a' with $a' \cap a = 0$). This well known implication (cf [St], p. 74) can be seen as follows: Define $A := \{c \in L \mid c \cap a = 0\}$. Using the upper continuity of L one sees that every totally ordered subset of A has an upper bound in A. Hence Zorn's lemma can be applied to give a maximal element in A, which will be a pseudocomplement of a.

Now let be a' a pseudocomplement of a. Assume $a+a' \not\sim 1$. Then there exists a $b \in L$ with $(a+a') \cap b = 0$, $b \neq 0$, thus $a \cap (a'+b) = a \cap (a'+(b \cap (a+a'))) = a \cap a' = 0$ which is a contradiction to the maximality of a'.
Thus f(a') is a complement of f(a). □

Lemma 8. Let L be a lattice with 0, L'a complemented modular lattice and $f: L \to L'$ a surjective lattice homomorphism with $f^{-1}[\{0\}] = \{0\}$. Then axiom III holds in L and there is a lattice isomorphism $g: L/\sim \to L'$ with $f = g \circ nat$, where $nat : L \to L/\sim$ denotes the natural homomorphism.

Proof. We show $f(a) = f(b) \Leftrightarrow a \sim b$. Then lemma 8 follows from lemma 6. Let be $f(a) = f(b)$. Then for all $c \in L$ holds $f(a \cap c) = f(b \cap c)$, thus using $f^{-1}[\{0\}] = \{0\}$ we get $a \cap c = 0$ if and only if $b \cap c = 0$, thus $a \sim b$.

Now let be $f(a) \neq f(b)$. We can assume w.l.o.g. $f(a) \nleq f(b)$. Let be $c \in L$ such that f(c) is a complement of $f(a \cap b)$. Then $f(a \cap b \cap c) = f(a \cap b) \cap f(c) = 0$ and $f(a \cap c) = f(a) \cap f(c) \neq 0$ since $f(a) \nleq f(b)$ and L' is modular. Thus $a \cap b \cap c = 0$ and $a \cap c \neq 0$, thus $a \not\sim b$. □

Lemma 9. Let L be a complete modular lattice and $P \subseteq L$ such that the axioms I-VI hold. Then there exists a skewfield K, a K-vectorspace V and a lattice isomorphism $i: L/\sim \to L({}_K V)$

with $i \circ nat\,[P] = \{Kx \mid x \in V \setminus \{O\}\}$.

Proof. Axioms I-II imply that L is upper continuous (lemma 5).

Axioms I-III imply that $\sim$ is a congruence relation and $L/\sim$ is a complemented modular upper continuous lattice (lemma 7).

Axioms I-IV imply that $nat[P] = \{a \mid a \text{ atom in } L/\sim\}$ and that each element in $L/\sim$ is the sum of atoms.

Proof. nat(p) is an atom in $L/\sim$ for all $p \in P$ since rank (nat(p)) = rank p = 1 (lemma 7) and since $L/\sim$ is complemented. If a is an atom in $L/\sim$ then axiom I implies that there is a $p \in P$ with $nat(p) \leq a$, thus nat(p) = a. Since nat preserves arbitrary sums axiom I now implies that each element in $L/\sim$ is the sum of atoms.

Axioms I-V imply that for any two atoms $a_1, a_2 \in L/\sim$ with $a_1 \cap a_2 = O$ there is an atom $a_3 \in L/\sim$ with $a_1+a_2=a_1+a_3=a_2+a_3$.

Axioms I-III and VI imply that rank $(L/\sim) \geq 3$ and that $L/\sim$ is Arguesian or that rank $(L/\sim) \geq 4$ (lemma 7 and since the homomorphism nat preserves the Arguesian law).

Note that any upper continuous modular lattice in which the greatest element is the sum of atoms, is compactly generated and complemented (for a proof cf [CD], p. 32).

We have shown that axioms I-VI imply that $L/\sim$ has the properties which are required in a lattice theoretic version of the classical coordinatization theorem of projective geometry (cf. [CD], thm. 13.4 combined with 13.2 and 13.3

or cf. [Ba 1],VII.6 and verify that Postulate IX follows from the Arguesian law.)
Applying the classical coordinatization theorem we get a skew field K, a K-vectorspace V and an isomorphism $i: L/\sim \longrightarrow L({}_K V)$. Furthermore $i \circ nat[P] = \{Kx \mid x \in V \setminus \{0\}\}$, since this is the set of atoms in $L({}_K V)$. □

Remark: Lemma 9 is the only place in the proof of the coordinatization theorem where we use the Arguesian law for L (in case rank L = 3). Thus we could replace axiom VI by axiom VI': rank $L \geqslant 3$ and $L/\sim$ is Arguesian or rank $L \geqslant 4$.

Remark: In a modular lattice L holds for all $a,b,c \in L$
$a \cap (b+c) = a \cap (b+c) \cap (b+a) = a \cap (b+(c \cap (a+b)))$ and
$a+(b \cap c) = a+(b \cap c)+(b \cap a) = a+(b \cap (c+(a \cap b)))$.
These equalities will be used in the proofs.
Furthermore if L has a 0 then $a \cap (b+c) = b \cap c = 0$ implies
$c \cap (a+b) = c \cap (b+(a \cap (b+c))) = c \cap b = 0$.

Lemma 10. Let L be a lattice with 0. Then for all $a,b,c,d,e \in L$ holds

a) If $[a,b,c]$ and $[a,d,e]$ and $(b+d) \cap (c+e) = 0$
then $b = d$ and $c = e$.

b) If $[a,b,c]$ and $[c,d,e]$ and $a \cap (d+e) = 0$, then
$[a,d,(a+d) \cap (b+e)]$.

Proof. a) $b = b+((b+d) \cap (c+e)) = (b+d) \cap (b+c+e) =$
$= (b+d) \cap (a+b+e) = b+d$, thus $d \leqslant b$.

$b \leqslant d$ and $c = e$ follow similarly.

b) $a+((a+d) \cap (b+e)) = (a+d) \cap (a+b+e) =$
$= (a+d) \cap (a+c+e) = a+d$.
$d+((a+d) \cap (b+e)) = a+d$ similarly.
$e \cap (a+c) = e \cap (c+(a \cap (c+e))) = e \cap c = 0$, since
$0 = a \cap (d+e) = a \cap (c+e)$ by assumption.

In the same way we can deduce from $e\cap(a+c) = 0$ and $[a,b,c]$ that $a\cap(b+e) = 0$ and $b\cap(c+e) = 0$.
From $b\cap(c+e) = 0$ and $[c,d,e]$ we get $d\cap(b+e) = 0$. Thus $a\cap(b+e) = d\cap(b+e) = 0$ and finally $[a,d,(a+d)\cap(b+e)]$. □

Lemma 11. Let L be a complete lattice, $P\subseteq L$, such that the axioms I, II and VII hold. Then holds
(VII') For all $p\in P$, $a,a'\in L\setminus\{0\}$ with $p\leq a+a'$ there exist $q,q'\in P$ with $q\leq a, q'\leq a'$ and $p\leq q+q'$.

Proof. Since $a = \Sigma\{q\in P | q\leq a\}$ and p compact there are $n\in\mathbb{N}$, $q_1,\ldots,q_n\in P$ with $q_i\leq a$ and $p\leq q_1+q_2+\ldots+q_n+a'$. Applying axiom VII (n+1) times we get (VII'). □

Lemma 12. Let L be a complete modular lattice, $P\subseteq L$ such that the axioms I-IV hold. Then for all $p\in P$, $a\in L$

$$\text{rank } (p+a)+\text{rank } (p\cap a) = 1+\text{rank } a.$$

Proof. From the proof of lemma 9 we know that $L/\sim$ is a complemented atomic modular lattice. It is easy to show that this implies rank a = dim u for all $u\in L/\sim$, where u denotes the lattice theoretic dimension of $[0,u]$ (i.e.every maximal chain in $[0,u]$ has length 1+dim u). It is well known that in modular lattices holds $\dim(u+v)+\dim(u\cap v) = \dim u + \dim v$ (cf. [CD], p. 27). Thus rank $(p+a)$+rank $(p\cap a)$= rank p + rank a = 1 + rank a for all $a\in L$, $p\in P$, noting that rank a = rank nat(a) for all $a\in L$ by lemma 7. □

Lemma 13. Let L be a complete modular lattice, $P\subseteq L$, such that the axioms I-V hold. Then for all $p_1,p_2\in P$ with $p_1\cap p_2=0$ there exists a $p_3\in P$ with $[p_1,p_2,p_3]$.

Proof. Axiom V implies that there exists a $p_3\in P$ with $p_1+p_2 = p_1+p_3 = p_2+p_3$. Since $p_1\cap p_2 = 0$ we get with lemma 12 that rank $(p_1+p_2) = 2$ since $p_1 \cap p_2 = 0$ and we get rank $(p_1 \cap p_3) = $ rank $(p_2 \cap p_3) = 0$, thus $[p_1,p_2,p_3]$. □

Lemma 14. Let L be a complete modular lattice, $P \subseteq L$ such that the axioms I, II, V and VII hold for L, P. Let be $p \in P$. $a,b \in L$ with $[p,a,b]$. Then $a \in P$.

Proof. Lemma 11 implies that there are $q, q' \in P$ with $q \leq a$, $q' \leq b$, $p \leq q+q'$. Thus $a = a \cap (p+b) \leq a \cap (q+q'+b) = a \cap (q+b) = $ $= q+(a \cap b) = q$, thus $a = q \in P$. □

5. PROOF OF THE COORDINATIZATION THEOREM

In lemma 9 we have already constructed a skew field K and a K-vectorspace V. In the next step we construct a subring $R \subseteq V$. Let $\varphi: L \longrightarrow L(_K V)$ be $\varphi := i \circ nat$, where i is the isomorphism which we constructed in lemma 9.

We start by constructing a 3-frame in L, which will be fixed until the end of the proof.

Let be $p_1,p_2,p_3 \in P$ with $p_1 \cap p_2 = p_3 \cap (p_1+p_2) = 0$. Such p_1,p_2,p_3 exist because rank $L \geq 3$ and using axiom I. According to lemma 13 there exist $p_{1j} \in P$ for j=2,3 with $[p_1,p_j,p_{ij}]$. This implies $[\varphi(p_1),\varphi(p_j),\varphi(p_{1j})]$. Now it is easy to see that there exist vectors $x_1,x_j \in V \setminus \{0\}$ such that for j = 2,3

$$Kx_1 = \varphi(p_1),\ Kx_j = \varphi(p_j),\ K(x_1+x_j) = \varphi(p_{1j}).$$

$\{x_1,x_2,x_3\}$ is linearly independent since $p_1 \cap p_2 = p_3 \cap (p_1+p_2) = 0$ and $\varphi(0) = 0$.

Now we define $p_{123} := (p_{12}+p_3) \cap (P_{13}+p_2)$ and $p_{23} := (p_1+p_{123}) \cap (p_2+p_3)$.

A short computation using the linear independence of x_1,x_2,x_3 yields $\varphi(p_{123}) = K(x_1+x_2+x_3)$, $\varphi(p_{23}) = K(x_2+x_3)$.

Applying lemma 10 b) twice we get first $[p_2,p_{123},p_{13}]$ and then $[p_2,p_3,p_{23}]$. Note that $\varphi(a\cap b) = 0$ if and only if $a\cap b=0$.

We define a subset $R\subseteq K$ as

$$R:=\{r\in K \mid \exists a\in L: a+p_2=p_1+p_2, \varphi(a)=K(x_1+rx_2)\}.$$

We show that R is a unitary subring of K.

$1\in R$ holds (choose $a:=p_{12}$).

Let be $r_1,r_2\in R$. We have to show that $r_1-r_2\in R$ and $r_1\cdot r_2\in R$.

Let be $r_1,r_2\in R$. Then there exist $a_i\in L$ $(i=1,2)$ with

$$a_i+p_2=p_1+p_2 \text{ and } \varphi(a_i)=K(x_1+r_ix_2).$$

We define (see fig. 1)

$$a_3:=(((((a_1+p_{13})\cap(p_2+p_3))+a_2)\cap(p_2+p_{13}))+p_3)\cap(p_1+p_2).$$

$$a_4:=(((((p_{12}+p_3)\cap(p_2+p_{13}))+a_1)\cap(p_1+p_3))+((a_2+p_3)\cap(p_2+p_{13})))\cap$$
$$\cap(p_1+p_2).$$

Our construction of a_3 (resp. a_4) corresponds exactly to the well known construction of r_2-r_1 (resp. $r_1\cdot r_2$) in projective geometry (arithmetic on the projective line). Note that ${}_K V$ is a left-vectorspace.

Using that x_1,x_2,x_3 are linearly independent and that φ is a lattice homomorphism with $\varphi^{-1}[0] = \{0\}$, an easy computation yields

$$\varphi(a_3)=K(x_1+(r_2-r_1)x_2),\quad \varphi(a_4)=K(x_1+r_1r_2x_2).$$

See also fig. 1.

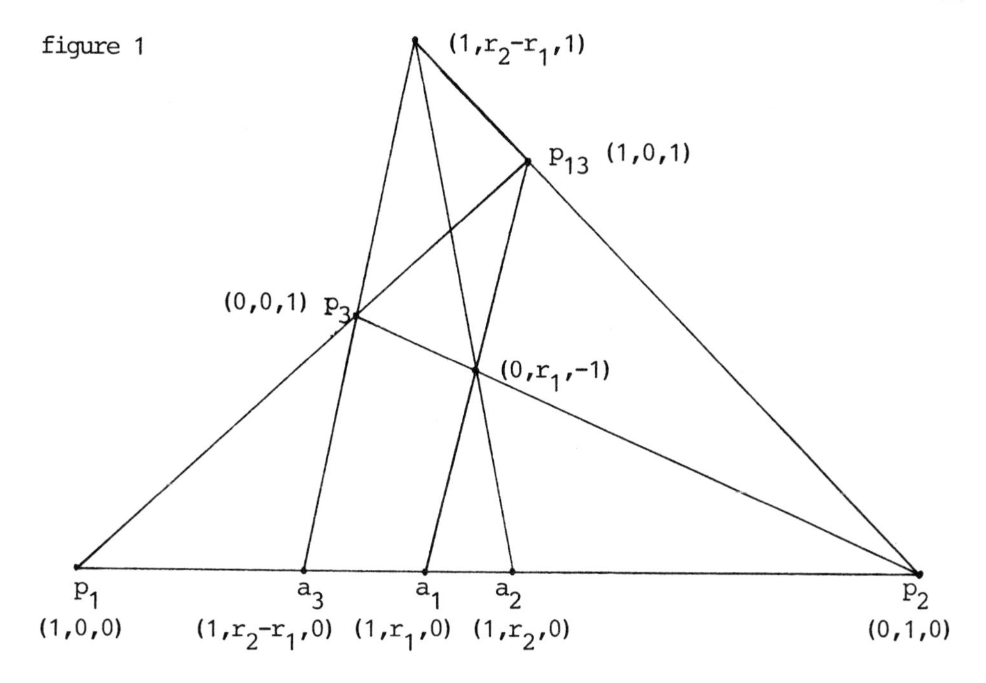
figure 1
(1,r2-r1,1)
p13 (1,0,1)
(0,0,1) p3
(0,r1,-1)
p1
(1,0,0)
a3
(1,r2-r1,0)
a1
(1,r1,0)
a2
(1,r2,0)
p2
(0,1,0)

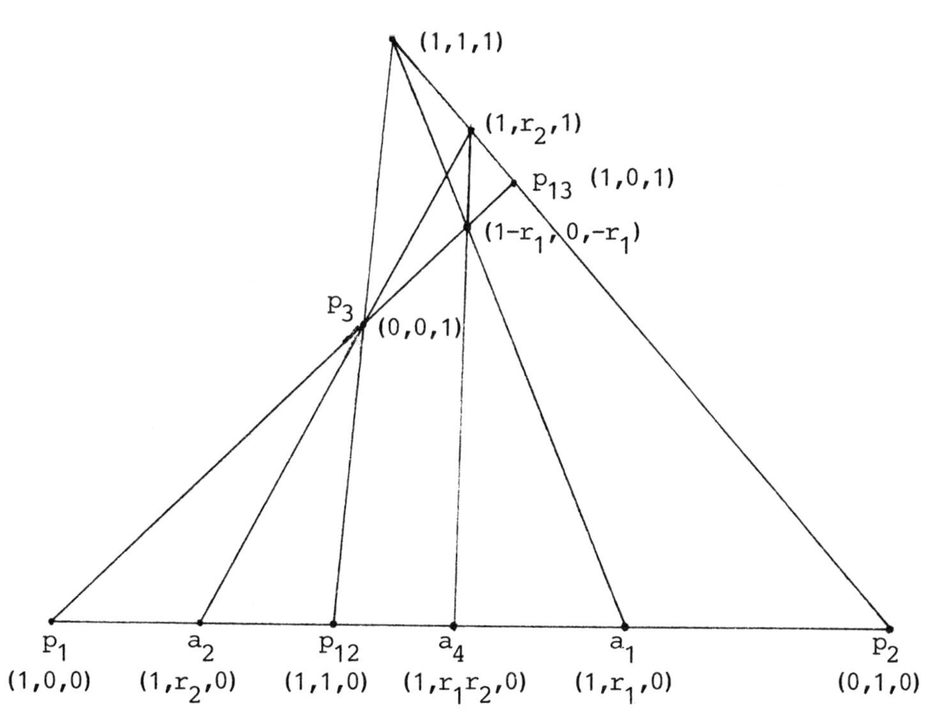
(1,1,1)
(1,r2,1)
p13 (1,0,1)
(1-r1,0,-r1)
p3
(0,0,1)
p1
(1,0,0)
a2
(1,r2,0)
p12
(1,1,0)
a4
(1,r1r2,0)
a1
(1,r1,0)
p2
(0,1,0)

Furthermore a short computation (which we carry out in the appendix) yields

$$a_3+p_2 = p_1+p_2 \quad \text{and} \quad a_4+p_2 = p_1+p_2.$$

Thus $r_2-r_1\in R$ and $r_1r_2\in R$.

Construction of $f : L_1 \longrightarrow L(_R V)$.

We define for i=1,2,3

$L_i:=\{a\in L \mid a\cap p_i = 0\}$

We define for i=1,2,3 mappings f_i from L_i to the powerset $\mathcal{P}(V)$ of V by

$$f_i(a):=\{z\in V \mid \exists a_1,a_2\in L: a_1\leq a, [p_i,a_1,a_2], \varphi(a_1)=Kz, \varphi(a_2)=K(z+x_i)\} \cup \{0\}.$$

$f := f_1$.

We show next ((1)-(13)) that for all $a\in L_1$ holds $f(a)\in L(_R V)$ and $Kf(a) = \varphi(a)$. Let be $a\in L_1$.

(1) $Kf(a) = \varphi(a)$.

$Kf(a)\subseteq\varphi(a)$ is obvious. Let be $z\in\varphi(a)\setminus\{0\}$. Then there exists $q_1\in P$ with $q_1\leq a$ and $\varphi(q_1) = Kz$. Since $p_1\cap q_1\leq p_1\cap a = 0$ there exists $q_2\in P$ with $[p_1,q_1,q_2]$. Thus there exists $t\in K\setminus\{0\}$ with $\varphi(q_2)=K(x_1+tz)$. Thus $tz\in f(a)$, thus $z\in Kf(a)$.

(2) f(a) is closed under sums of linearly independent vectors.

Let be $y,z\in f(a)$ linearly independent. Then there are $a_1,a_2,b_1,b_2\in L$ with $a_1\leq a$, $b_1\leq a$, $[p_1,a_1,a_2]$, $[p_1,b_1,b_2]$, $\varphi(a_1) = Ky$, $\varphi(b_1) = Kz$, $\varphi(a_2) = K(y+x_1)$, $\varphi(b_2) = K(z+x_1)$.

We define

$c_2 := (a_1+b_2) \cap (a_2+b_1)$

$c_1 := (p_1+c_2) \cap (a_1+b_1)$.

Then hold $\varphi(c_2)=K(x_1+y+z)$, $\varphi(c_1)=K(y+z)$,

$c_1+p_1=(c_2+p_1)\cap(a_1+b_1+p_1)=c_2+p_1$, since $a_1+b_1+p_1 = a_1+b_2+p_1 \geq c_2+p_1$

$c_1+c_2=(c_2+p_1)\cap(a_1+b_1+c_2)=(c_2+p_1)\cap(a_1+b_1+((a_2+b_1)\cap(a_1+b_2))$
$\qquad =(c_2+p_1)\cap(a_2+b_1+a_1)\cap(a_1+b_2+b_1)\geq p_1$,

thus $[p_1,c_1,c_2]$ (note that $c_1 \cap c_2=0$ etc., since $\varphi(c_1)\cap\varphi(c_2)= =0$ etc.).

Thus $y+z\in f(a)$.

(3) If $z\notin Kx_1+Kx_2$ and $z\in f(a)$ then $Rz\subseteq f(a)$.

Let be $z\in f(a)\setminus(Kx_1+Kx_2)$ and $r\in R\setminus\{0\}$.

Then there is an $a\in L$ with $a+p_2=p_1+p_2$ and $\varphi(a)=K(x_1+rx_2)$. Since $z\in f(a)$ there are $a_1,a_2\in L$ with $a_1\leq a$, $\varphi(a_1)=Kz$, $\varphi(a_2) = K(x_1+z)$, $[p_1,a_1,a_2]$. We define

$b_2 := (((p_2+a_1)\cap(p_{12}+a_2))+a)\cap(p_1+a_1)$

$b_1 := a_1\cap(p_1+b_2)$.

Then a short computation using the linear independence of x_1,x_2,z and $r \neq 0$ yields

$\varphi(b_2)=K(x_1+rz)$, $\varphi(b_1)=Kz=Krz$.

Thus $p_1\cap b_1=p_1\cap b_2=b_1\cap b_2 = 0$. A short computation (which we carry out in the appendix) yields $[p_1,b_1,b_2]$.

Thus $rz\in f(a)$.

(4) If rank $a \geqslant 2$ and $z \in f(a)$ then $Rz \subseteq f(a)$.

Let be rank $a \geqslant 2$, thus $\dim \varphi(a) = \dim K \cdot f(a) \geqslant 2$.
If $r \in R \setminus \{0\}$, $z \in Kx_1 + Kx_2$ (otherwise use (3)), then there exists $u \in f(a) \setminus (Kx_1 + Kx_2)$ since $Kx_1 \cap f(a) = 0$.

Thus by (2) and (3) we get $u+z \in f(a)$, $-ru \in f(a)$, $r(u+z) \in f(a)$, $rz = r(z+u) - ru \in f(a)$.

(5) If rank $a \geqslant 2$ then $f(a) \in L({}_R V)$

If $z_1, z_2 \in f(a)$, $z_1 \neq z_2$, $Kz_1 = Kz_2$ then there exists $u \in f(a)$ which is linearly independent to z_1, thus $u+z_1$, $-(u+z_2)$ are linearly independent. Thus by (2) we get $u+z_1 \in f(a)$, $-(u+z_2) \in f(a)$, $z_1 - z_2 = u + z_1 - (u+z_2) \in f(a)$. Combining this with (2) and (4) we get $f(a) \in L({}_R V)$.

(6) $f(a \cap b) = f(a) \cap f(b)$ for all $a, b \in L_1$

$f(a \cap b) \subseteq f(a) \cap f(b)$ is obvious.
Let be $z \in f(a) \cap f(b)$, $z \neq 0$. Then there are $a_1, a_2, b_1, b_2 \in L$ with $a_1 \leqslant a$, $b_1 \leqslant b$, $[p_1, a_1, a_2]$, $[p_1, b_1, b_2]$, $\varphi(a_1) = Kz = \varphi(b_1)$, $\varphi(a_2) = K(x_1 + z) = \varphi(b_2)$.
Lemma 10a) implies $a_1 = b_1$, thus $a_1 = b_1 \leqslant a \cap b$, thus $z \in f(a \cap b)$.

(7) $f(a) \in L({}_R V)$ for all $a \in L_1$.

If rank $a=1$ then there is a $q_1 \in P$ with $q_1 \cap (p_1 + a) = 0$ since rank $L \geqslant 3$. Axiom V implies that there is a $q_2 \in P$ with $[p_1, q_1, q_2]$. Now rank $(a+q_1) = $ rank $(a+q_2) = 2$ and $a = (a+q_1) \cap (a+b)$.
Thus $f(a) = f(a+q_1) \cap f(a+q_2)$ is an R-module by (6) and (5).

From now on we regard f as a mapping to $L({}_RV)$ which is possible by (7).

$$f : L_1 \longrightarrow L({}_RV).$$

(8) $f: L_1 \longrightarrow L({}_RV)$ is injective.

Let be $a,b \in L_1$, $f(a)=f(b)$, $q_1 \in P$ with $q_1 \leq a$. Then there is a $q_2 \in L$ with $[p_1,q_1,q_2]$ and a $z \in V$ with $\varphi(q)=Kz$, $\varphi(c)=K(x_1+z)$, thus $z \in f(a)=f(b)$. Thus there are $q_1', q_2' \in L$ with $q_1' \leq b$, $\varphi(q_1') = Kz$, $\varphi(q_2') = K(x_1+z)$, $[p_1,q_1',q_2']$.
Lemma 10 a) implies that $q_1=q_1'$, thus $q_1 \leq b$. Thus $a \leq b$, (use axiom I). Exchanging the role of a and b we get $b \leq a$, thus $a = b$.

(9) For all $a,b \in L$ with $a+b \in L_1$ and $a \cap b=0$ holds $f(a+b) = f(a)+f(b)$.

$f(a+b) \supseteq f(a)+f(b)$, since $f: L_1 \longrightarrow L({}_RV)$ is order-preserving.
First let be $z \in f(a+b)$, but $z \notin \varphi(a)$ and $z \notin \varphi(b)$.
Then there are $c_1 \leq a+b, c_2 \in L$ with $\varphi(c_1)=Kz$, $\varphi(c_2)=K(x_1+z)$, $[p_1,c_1,c_2]$. Using $a \cap b = c_1 \cap a = c_1 \cap b = 0$ and $c_1 \leq a+b$, a short computation yields $[c_1,a',b']$ with $a':=a \cap (b+c_1)$, $b':=b \cap (a+c_1)$; (for example $c_1 = c_1 \cap (a'+b) = c_1 \cap (a'+(b \cap (c_1+a'))) \leq a'+b'$). Since $p_1 \cap (a'+b') = 0$, lemma 10 b) implies $[p_1,a',a'']$ and $[p_1,b',b'']$ with $a'':=(p_1+a') \cap (c_2+b')$, $b'':=(p_1+b') \cap (c_2+a')$.

Since $[c_1,a',b']$ there exists a $u \in V$ with $\varphi(a')=Ku$, $\varphi(b')=K(u+z)$. A short computation yields

$$\varphi(a'')=K(x_1-u) \quad , \quad \varphi(b'')=K(x_1+u+z).$$

Thus $-u \in f(a)$, $u+z \in f(b)$, thus $z \in f(a)+f(b)$.

Now let be $z \in f(a+b)$ and $z \in \varphi(a) \setminus \{0\}$ and $b \neq 0$.
Let be $z_1 \in f(b)$. Then $z+z_1 \in f(a+b)$ and $z+z_1 \notin \varphi(a)$ and $z+z_1 \notin \varphi(b)$, thus we can apply what we have just shown and get $z+z_1 \in f(a)+f(b)$, thus $z \in f(a)+f(b)$. In the same way we get $z \in f(a)+f(b)$ in the case that $z \in f(a+b)$ and $z \in \varphi(b) \setminus \{0\}$.

Remark: Till now we have not yet used axiom VII at all. In the next step we will use axiom VII for the first time.

(10) For all $a,b \in L$ with $a+b \in L_1$ holds $f(a+b)=f(a)+f(b)$.

$f(a+b) \supseteq f(a)+f(b)$ is obvious.

Let be $z \in f(a+b) \setminus \{0\}$. Then there are $c_1 \leq a+b, c_2 \in L$ with $\varphi(c_1)=Kz$, $\varphi(c_2)=K(x_1+z)$, $[p_1,c_1,c_2]$. Lemma 14 implies $c_1,c_2 \in P$.

According to lemma 11 there are $a_1,b_1 \in P$ with $a_1 \leq a$, $b_1 \leq b$, $c_1 \leq a_1+b_1$.

If $a_1 \cap b_1 = 0$ then $z \in f(c_1) \subseteq f(a_1)+f(b_1) \subseteq f(a)+f(b)$ because of (9).

Now assume $a_1 \cap b_1 \neq 0$ and w.l.o.g. assume $a_1 \neq b_1$. Since $a_1,b_1,c_1 \in P$ we have $\varphi(a_1)=\varphi(b_1)=\varphi(c_1)$.

Let be $q \in P$ with $q \cap (p_1+a_1)=0$.

Lemma 13 implies that there are $a_2,b_2 \in P$ with $[q,a_1,a_2]$ and $[q,b_1,b_2]$.

Since $\varphi(a_1)=\varphi(b_1)$ and $a_1 \neq b_1$, lemma 10 a) implies that $a_2 \cap b_2 = 0$.

Since $c_1 \leq q+a_2+b_2$, axiom VII implies that there is a $d \in P$ with $c_1 \leq q+d$ and $d \leq a_2+b_2$. $q \cap d = 0$ since $q \cap c_1 = 0$ and $c_1 \leq q+d$.
Now we apply (9) and get $f(c_1) \subseteq f(q+d) = f(q)+f(d)$, $f(d) \subseteq f(a_2)+f(b_2)$, $f(a_2) \subseteq f(q)+f(a_1)$, $f(b_2) \subseteq f(q)+f(b_1)$, thus $f(c_1) \subseteq f(a_1)+f(b_1)+f(q)$.

This together with $Kf(c_1)=Kf(a_1)=Kf(b_1)$ and $Kf(q)\cap Kf(a_1)=0$ implies that $f(c_1)\subseteq f(a_1)+f(b_1)$, thus $z\in f(c_1)\subseteq f(a)+f(b)$.

(11) $f(p)\in\{Rz \mid z\in V\setminus\{0\}\}$ for all $p\in P\cap L_1$.

We show first $f(p_2)=Rx_2$.

$x_2\in f(p_2)$, thus $Rx_2\subseteq f(p_2)$.
Let be $z\in f(p_2)\setminus\{0\}$, then there is a $k\in K\setminus\{0\}$ with $z=kx_2$. Furthermore there are $a_1\leq p_2, a_2\in L$ with $\varphi(a_1)=Kz=Kx_2$ and $\varphi(a_2)=K(x_1+z)$ and $[p_1,a_1,a_2]$. Thus $p_2+a_2=p_2+a_1+a_2 = p_1+p_2$, thus $k\in R$, thus $z\in Rx_2$.

Now let be $q\in P$ with $q\cap(P_1+P_2)=0$, then there is a $q'\in P$ with $[P_2,q,q']$, thus $[f(p_2),f(q), f(q')]$.
Using lemma 4 and lemma 14 we then get $f(q)=Rz$ for some $z\in V\setminus\{0\}$.
Repeating the same reasoning with q in place of p_2 (and p in place of q) we get

$$f(p)\in\{Rz \mid z\in V\setminus\{0\}\} \quad \text{for all } p\in P\cap L_1.$$

(12) For all $a\in L_1$ and $z\in f(a)\setminus\{0\}$ there is a $p\in P$ with $p\leq a$ and $f(p)=Rz$.

We assume without loss of generality $z\notin Kx_1+Kx_2$ (otherwise $z\notin Kx_1+Kx_3$ and we replace x_2 by x_3, p_2 by p_3, p_{12} by p_{13}). Then there are $a_1,a_2\in L$ with $a_1\leq a$, $[p_1,a_1,a_2]$, $\varphi(a_1)=Kz$, $\varphi(a_2)=K(x_1+z)$. Lemma 10b) yields $[p_2,a_1,b]$, where $b:=(p_{12}+a_2)\cap(p_2+a_1)$. Thus $[f(p_2),f(a_1), f(b)]$ and $\varphi(b) = K(x_2-z)$. Now $f(p_2)=Rx_2$, $[Rx_2,Rz,R(x_2-z)]$ together with lemma 10 a) implies $f(a_1)=Rz, f(b)=R(x_2-z)$.

Now $a_1\in P$ according to lemma 14 , so $p:=a_1$ is the wanted element.

(13) R is a left Ore domain and
K is a skewfield of left quotients of R.

Let be $k \in K \setminus \{0\}$, $q_1 \in P$ with $q_1 \cap (p_1+p_2)=0$ and $z \in V$ with $f(q_1)=Rz$. Then there is a $q_2 \in P$ with $q_2 \leqslant p_2+q_1$ and $\varphi(q_2) = K(x_2+kz)$ and a $y \in f(q_2) \setminus \{0\}$.
Thus $y \in f(q) \subseteq K(x_2+kz) \cap (f(p_2)+f(q_1)) = K(x_2+kz) \cap (Rx_2+Rz)$, thus there are $r_1, r_2 \in R$ with $y=r_1x_2+r_2z \in K(x_2+kz)$, which implies $k=r_1^{-1} \cdot r_2$. Thus K is a skewfield of left quotients of R, which implies that R is a left Ore domain.

Extension of f to $\bar{f}: L \longrightarrow L({}_RV)$.

We are going to extend $f=f_1$ by combining the three mappings $f_i: L \longrightarrow \mathcal{P}(V)$.

First of all we note that all the properties (1) - (13) hold not only for $f_1=f$ but also for f_i $(i=2,3)$ by replacing L_1 by L_i, p_1 by p_i and x_1 by x_i and R by R_i where the rings R_i are defined like R with p_1 and x_1 being replaced by p_i and x_i, and p_2, p_{12}, x_2 being replaced by p_1, p_{1i}, x_1 respectively.

(14) $f_i(a) = f_j(a)$ for all $a \in L$ with $a \cap (p_i+p_j) = 0$,
$(i,j \in \{1,2,3\},\ i \neq j)$.

Let be $z \in f_i(a) \setminus \{0\}$, then there are $a_1 \leqslant a$, $a_2 \in L$ with $\varphi(a_1) = Kz$, $\varphi(a_2) = K(x_i+z)$, $[p_i, a_1, a_2]$. Now we define $b := (a_1+p_{ij}) \cap (a_2+p_{ij})$ and $a_2' := (p_i+b) \cap (p_j+a_1)$. Applying lemma 10 b) twice we get first $[p_i, b, a_2]$ and then $[p_j, a_1, a_2']$. A short computation using the linear independence of $\{x_i, x_j, z\}$ yields $\varphi(b) = K(x_i+x_j+z)$ and then $\varphi(a_2') = K(x_i+z)$. Thus $z \in f_j(a)$. Thus $f_i(a) \subseteq f_j(a)$. Similarly we get $f_j(a) \subseteq f_i(a)$, thus $f_i(a) = f_j(a)$.

(15) $R_i = R$ for $i=2,3$.

Let be $p \in P$ with $p \cap (p_1+p_i)=0$. Then (11) and (14) imply $f_1(p)=Rz=f_2(p)=R_i z'$ for some $z,z' \in V \setminus \{0\}$.
Since R, R_i are unitary subrings of K there is an $r \in R$ with $z'=rz$, thus $R=R_i r$. Now there is an $r' \in R_i$ with $r' \cdot r=1$. Since $r^2 \in R$ there is an $s \in R_i$ with $r^2=sr$, thus $r=s \in R_i$. Thus r is a unit in R_i thus $R=R_i$.

From now on we regard the f_i's as mappings to $L({}_R V)$ which is possible by (7) and (15).

$$f_i : L_i \longrightarrow L({}_R V).$$

(16) For all $p \in P \cap L_i \cap L_j$ holds $f_i(p)=f_j(p)$, $(i,j \in \{1,2,3\}, i \neq j)$.

If $p \cap (p_i+p_j)=0$ then $f_i(p)=f_j(p)$ by (14), otherwise $f_i(p) = f_k(p)=f_j(p)$ by (14), where $k \in \{1,2,3\} \setminus \{i,j\}$ since in this case $\varphi(p) \subseteq (Kx_i+Kx_j) \setminus (Kx_i \cup Kx_j)$.

We define $\overline{f} : L \longrightarrow L({}_R V)$ by

$$\overline{f}(a) := \bigcup \{ f_i(p) \mid i \in \{1,2,3\}, p \in P \cap L_i, p \leq a \},$$

where $\bigcup \emptyset := 0$.

(17) $\overline{f}(a) \in L({}_R V)$ for all $a \in L$.

$\overline{f}(a)$ is closed under multiplication with elements of the ring R since this is true for f_1, f_2 and f_3.
Now let be $z,z' \in \overline{f}(a) \setminus \{0\}$. Then there are $p,p' \in P$ with $p \leq a, p' \leq a, z \in \overline{f}(p), z' \in \overline{f}(p')$. There is an $i \in \{1,2,3\}$ with $(p+p') \cap p_i=0$. Using (16) we get $z \in f_i(p)$, $z' \in f_i(p')$, thus $z+z' \in f_i(p)+f_i(p')=f_i(p+p')$, thus $z+z' \in f_i(q)$ for some $q \in P$ with $q \leq p+p' \leq a$ (using (12)), thus $z+z' \in \overline{f}(a)$.

(18) $\overline{f}(a\cap b)=\overline{f}(a)\cap\overline{f}(b)$.

$\overline{f}(a\cap b)\subseteq\overline{f}(a)\cap\overline{f}(b)$ is obvious.
Let be $z\in\overline{f}(a)\cap\overline{f}(b)\setminus\{O\}$. Since $z\in\overline{f}(a)$ there is an $i\in\{1,2,3\}$ and a $p\in P\cap L_i$ with $z\in f_i(p)$. Since $z\in\overline{f}(b)$ there is a $p'\in P$ with $p'\leqslant b$ and $z\in\overline{f}(p')$. Since $\varphi(p')=Kz=\varphi(p)$ we have $p'\cap p_i=O$.
Now (16) and (6) imply $z\in f_i(p')\cap f_i(p)=f_i(p\cap p')$.
Now (12) implies that there is a $p''\in P$ with $p''\leqslant p\cap p'$ and $z\in f_i(p'')$, thus $z\in\overline{f}(a\cap b)$.

(19) $\overline{f}(a+b)=\overline{f}(a)+\overline{f}(b)$.

$\overline{f}(a+b)\supseteq\overline{f}(a)+\overline{f}(b)$ is obvious (using (17)).
Let be $z\in\overline{f}(a+b)$. Then there is a $p\in P$ with $p\leqslant a+b$ and $z\in\overline{f}(p)$.
According to lemma 11 there are $a',b'\in P$ with $a'\leqslant a, b'\leqslant b, p\leqslant a'+b'$. Now there is an $i\in\{1,2,3\}$ with $(a'+b')\cap p_i=O$. Thus $z\in f_i(p)\subseteq f_i(a'+b')=f_i(a')+f_i(b')$, thus $z\in\overline{f}(a)+\overline{f}(b)$.

(20) $\overline{f}$ preserves arbitrary sums.

Let be $M\subseteq L$. $\overline{f}(\Sigma M)\supseteq\Sigma\overline{f}[M]$ is obvious.
Let be $z\in\overline{f}(\Sigma M)$. Then there is a $p\in P$ with $p\leqslant\Sigma M$ and an $i\in\{1,2,3\}$ with $p\cap p_i=O$ and $z\in f_i(p)$. Since p is compact, there is a finite subset $E\subseteq M$ with $p\leqslant\Sigma E$, thus $z\in\overline{f}(\Sigma E)=\Sigma\overline{f}[E]\subseteq\Sigma\overline{f}[M]$.

(21) $\overline{f}$ is injective.

Let be $\overline{f}(a)\subseteq\overline{f}(b)$. Let be $p\in P$ with $p\leqslant a$. Then by (11) there is a $z\in V$ and $i\in\{1,2,3\}$ with $p\cap p_i=O$ and $f_i(p)=Rz$, thus $z\in\overline{f}(a)\subseteq\overline{f}(b)$. Thus there is a $p'\in P$ with $p'\leqslant b$ and a $j\in\{1,2,3\}$ with $p'\cap q_j=O$ and $z\in f_j(p')=f_i(p')$, thus $f_i(p)\subseteq f_i(p')$. Since f_i is injective by (8) we get $p\leqslant p'\leqslant b$. Since this holds for all $p\in P$ with $p\leqslant a$ we get $a\leqslant b$. Thus $\overline{f}(a)=\overline{f}(b)$ implies $a=b$.

We define $M:=\overline{f}(1)$. We restrict the range of $\overline{f}$ to $L({}_RM)$.

(22) $\overline{f}:L \longrightarrow L({}_RM)$ is surjective.

Let be $U \in L({}_RM)$. We define $a:=\Sigma\{p\in P \mid \overline{f}(p)\subseteq U\}$. Since $\overline{f}$ preserves arbitrary sums we get $\overline{f}(a)\subseteq U$.

Let be $z \in U\setminus\{0\}$. Then there is a $p\in P$ and an $i\in\{1,2,3\}$ with $p\cap p_i=0$ and $z\in f_i(p)$. By (12) there is a $p'\in P$ with $p'\leqslant p$ and $f_i(p')=Rz$, thus $\overline{f}(p')=Rz\subseteq U$, thus $p'\leqslant a$, thus $z\in\overline{f}(p')\subseteq\overline{f}(a)$. Thus $\overline{f}(a)=U$.

We have shown that $\overline{f}:L \longrightarrow L({}_RM)$ is a lattice isomorphism, where R is a left Ore domain and M a torsion free R-module since $M\subseteq V$. Since $\varphi:L \longrightarrow L({}_KV)$ is surjective and $K\overline{f}(p)=\varphi(p)$ for all $p\in P$ we have $K\cdot M=V$, thus rank M = dim V = rank L $\geqslant$ 3.

$\overline{f}[P] = \{Rz \mid z\in M\setminus\{0\}\}$ follows from (11) and (12). □

6. A DIFFERENT APPROACH

Instead of starting with a lattice L together with a given subset P we can start with L together with a given compact element $p \in L$ with rank $p = 1$ and then construct a subset $P \subseteq L$ such that the axioms hold for L and P. To do this we need a lemma which may be useful also for other questions.

Lemma 15. Let L be a modular lattice and R the relation which is defined by $aRb \iff \exists c\in L : [a,b,c]$. Let be $p_0\in L$ with rank $p_0 = 1$ and $p_1,p_2\in L$ with $p_0Rp_1, p_0Rp_2, (p_0+p_1)\cap p_2=0$. Let $\overline{R}$ be the equivalence relation generated by R.

Then for all $p,q \in L$ with $p\overline{R}p_0$, $p\cap q = 0$ and $p\overline{R}q$ holds pRq.

Furthermore if axiom III holds in L and rank $p = 1$ then pRq holds if and only if $p \nparallel q$ and p is perspective to q, i.e. there is a $c \in L$ with $p \cap c = q \cap c = 0$ and $p+c = q+c$.

Proof. We define $P := \{p \in L \mid p\bar{R}p_0\}$.

Obviously pRq implies that p and q are perspective and $p \neq q$. It is well known that for perspective elements p,q in a modular lattice with 0 the intervals $[0,p]$ and $[0,q]$ are isomorphic. Thus rank p = rank $p_0 = 1$ for all $p \in P$.

By assumption there are p_1, p_1', p_2, p_2' with $p_2 \cap (p_0+p_1) = 0$ and $[p_0, p_1, p_1']$, $[p_0, p_2, p_2']$. Now let be $p_3, p_3' \in L$ with $[p_0, p_3, p_3']$. Then $p_3 \cap (p_0+p_1) = 0$ or $p_3 \cap (p_0+p_2) = 0$ since rank $p_3 = 1$ and $(p_0+p_1) \cap (p_0+p_2) \cap p_3 = (p_0+((p_0+p_1) \cap p_2)) \cap p_3 =$ $= p_0+p_3 = 0$. Let be $p_3 \cap (p_0+p_1) = 0$ w.l.o.g..
Then lemma 10 b) yields $[p_1, p_3, (p_1+p_3) \cap (p_1'+p_3')]$, thus p_1Rp_3. Furthermore $(p_3+p_1) \cap p_0 = (p_1+(p_3 \cap (p_0+p_1))) \cap p_0 =$ $= p_1 \cap p_0 = 0$.

Iterating this procedure we get that
(1) for all $p,q \in P$ with pRq there is a $q' \in P$ with pRq' and $(p+q) \cap q' = 0$.

Next we show
(2) If $q \in P$, qRq_1, q_1Rq_2 and $q \cap q_2 = 0$ then qRq_2.

Case 1: $q_2 \cap (q+q_1) = 0$. Then lemma 10 b) yields (2).

Case 2: $q_2 \cap (q+q_1) \neq 0$. Then (1) implies that there is a $p \in P$ with $p \cap (q+q_1) = 0$ and pRq_1. Since rank $q_2 = 1$ and $q_2 \cap (q+q_1) \neq 0$ and $q_2 \cap (p+q) \cap (q+q_1) = q_2 \cap q = 0$ we have $q_2 \cap (p+q) = 0$ and $q_2 \cap (p+q_1) = 0$ similarly.

Now we can apply lemma 10 b) twice and get qRp and pRq_2. Applying lemma 10 b) again we finally get qRq_2.

(3) Assume $q \in P$, $[q,q_1',q_1]$, q_1Rq_2, q_2Rq_3, $q \cap q_2 \neq 0$ and $q_1 \cap q_3 \neq 0$. Then $0 = q_1' \cap q_2 = q_1' \cap q_3 = q \cap q_3$ since rank $q =$ = rank $q_1' = 1$. Applying (2) twice we get $q_1'Rq_3$ and applying (2) again we get qRq_3.

Now for all $p,q \in P$ with $p\bar{R}p_0$, $p \cap q = 0$ and $p\bar{R}q$ we get pRq by iterated applications of (2) and (3).

Now assume axiom III and let be $p,q \in L$ with $p \neq q$, rank $p = 1$ and p perspective to q. Then there is an $a \in L$ with $a \cap p = a \cap q = 0$ and $a+p = a+q$. Let be $a' := a \cap (p+q)$. Then $a'+p = a'+q = p+q$ and $a' \cap p = a' \cap q = 0$ and $a' \neq 0$ since $p \neq q$. Because of axiom III there is a $p' \in L$ with $p' \cap p \neq 0$ and $p' \cap q = 0$. Now rank $p = 1$ implies $p \cap q = 0$, thus $[p,q,a']$, thus pRq. □

Theorem 2. Let L be an upper continuous modular lattice and $p_0 \in L$ a compact element with rank $p_0 = 1$.
Let P denote the set of all elements in L which are projective to p_0. If the axioms I, III, VI and VII hold for L,P then there exists a left Ore domain R and a torsion free R-module M with rank $M \geq 3$ and a lattice isomorphism $f: L \longrightarrow L({}_RM)$ with $f[P] = \{Rx \mid x \in M \setminus \{0\}\}$.

Proof. Since L is upper continuous, $p \in L$ is compact if and only if for all $A \subseteq L$ with $\Sigma A = p$ there exists a finite subset $E \subseteq A$ with $p = \Sigma E$. Since all $p \in P$ are projective to p_0 the intervals $[0,p]$ and $[0,p_0]$ are isomorphic lattices. Since p_0 is compact and rank $p_0 = 1$, all $p \in P$ are compact with rank $p = 1$. Thus the axioms II and IV hold for L,P.

Next we show that the assumption on p_0 of lemma 15 holds for some element of P.

Since rank $L \geq 3$ and since axiom I holds for L,P, there

are $q_1, q_2, q_3 \in P$ with $q_1 \cap (q_2+q_3) = q_1 \cap q_3 = 0$. There are $b_i \in P$, $i = 1, \ldots n$ with $c_1 = q_1$, $c_i R c_{i+1}$, $c_n = q_3$, $c_m = q_2$ for some $m \leqslant n-1$ and all $i=1, \ldots n-1$.
From this we get with the help of lemma 12 that there exists a $k \in \{1, \ldots n-2\}$ with $c_k \cap (c_{k+1} + c_{k+2}) = 0$.

Now we can apply lemma 15 with $p_0 := c_{k+1}$, which shows that axiom V holds for L,P.

Now theorem 2 follows from theorem 1. □

For a torsion free module M over a left Ore domain R we have $\{Rx \mid x \in M \setminus \{0\}\} \subseteq \{p \in L({}_R M) \mid \text{rank } p = 1 \text{ and } p \text{ is compact}\}$. We now characterize when equality holds.

Lemma 16. Let R be a left Ore domain and M a torsion free R-module. Let be $L := L({}_R M)$, $P := \{Rx \mid x \in M \setminus \{0\}\}$. Then are equivalent:

(i) $P = \{p \in L \mid \text{rank } p = 1 \text{ and } p \text{ is compact}\}$

(ii) $Rx+Ry$ is free for all $x,y \in M$.

Proof. (i) ⇒ (ii). Let be $x,y \in M \setminus \{0\}$. If $Rx \cap Ry = 0$ then $Rx+Ry$ is free. If $Rx \cap Ry \neq 0$ then rank $(Rx+Ry) = 1$ since M is torsion free. Furthermore $Rx+Ry$ is compact, thus by assumption $Rx+Ry \in P$, thus there is a $z \in M \setminus \{0\}$ with $Rx+Ry = Rz$ and Rz is free.

(ii) ⇒ (i). If $Rx+Ry$ is free for all $x,y \in M$ and $Rx \cap Ry \neq 0$ then $Rx+Ry$ is free and rank $(Rx+Ry) = 1$, thus there is a $z \in M \setminus \{0\}$ with $Rx+Ry = Rz$. By induction on the number of generators we get that every finitely generated submodule $U \subseteq M$ with rank $(U) = 1$ is of the form $U = Ru$ for some $u \in M \setminus \{0\}$. Thus (i) holds. □

Theorem 3. Let L be a complete modular lattice and $P:=\{p \in L \mid \text{rank } p = 1 \text{ and } p \text{ is compact}\}$. Then the axioms I, III, V and VII are satisfied for L,P if and only if there is a left Ore domain R and an R-module M with rank $M \geq 3$ such that Rx+Ry is free for all $x,y \in M$ and a lattice isomorphism $f:L \to L({}_R M)$ with $f[P] = \{Rx \mid x \in M \setminus \{0\}\}$.

Proof. Theorem 3 follows from theorem 1 and lemma 16. Note that M is torsion free if every cyclic submodule is free. □

Remark. If R is a left Ore domain and $M \neq 0$ an R-module such that Rx+Ry is free for all $x,y \in M$, then every finitely generated left ideal of R is principal, i.e. R is a left Bezout domain. The converse is not true in general as the following example shows.

Example. Let R be the noncommutative ring of polynomials over the field $\mathbb{Q}(x)$ which is defined by

$$R:=\{\sum_{i=0}^{n} r_i(x)y^i \mid n \in \mathbb{N},\ r_i(x) \in \mathbb{Q}(x)\} \text{ with } y\cdot r(x):=r(x^3)\cdot y.$$

R is a left Ore domain and every left ideal is principal(cf[Co]). Let K be the skew field of left quotients of R and $M:={}_R K^3$. Then $R(1,0,0) + R(y^{-1}xy,0,0)$ is not free.
For a proof cf [Br 1].

Example. Let R be the left Ore domain of the previous example and $M:=R^3 + R(y^{-1}xy,\ y^{-1}x^2y,0) \subseteq K^3$. Then Rx+Ry is free for all $x,y \in M$ but
$R(1,0,0) + R(0,1,0) + R(y^{-1}xy,\ y^{-1}x^2y,0)$ is not free. A proof of these facts can be found in [Br 1].

Remark. Instead of working with $P:=\{Rx \mid x \in M \setminus \{0\}\}$ we can also work with $\{Rx \mid x \in M\}$. To do this we have to change the axioms as follows:

axiom IV$^\bullet$: For all $p \in P$ holds rank $p \leq 1$ in L.

axiom VII$^\bullet$. For all $p, q \in P$, $a \in L$ with $p \leq q+a$ there exists a $q' \in P$ with $q' \leq a$ and $p \leq q+q'$.

Then theorem 1 holds with axioms IV and VII being replaced by axioms IV$^\bullet$ and VII$^\bullet$ and $\{Rx \mid x \in M \setminus \{0\}\}$ being replaced by $\{Rx \mid x \in M\}$.

To see this, note that the axioms I-VII are satisfied for L, $P \setminus \{0\}$ if and only if the axioms I, II, III, IV$^\bullet$, V, VI, VII$^\bullet$ are satisfied for L, $P \cup \{0\}$. Furthermore axiom VII$^\bullet$ implies that $0 \in P$ (choose $p = q$ and $a = 0$).

Theorem 2 holds with axiom VII being replaced by axiom VII$^\bullet$ and $\{Rx \mid x \in M \setminus \{0\}\}$ being replaced by $\{Rx \mid x \in M\}$.

Theorem 3 holds with axiom VII being replaced by axiom VII$^\bullet$, $\{p \in L \mid$ rank $p = 1$ and p is compact$\}$ being replaced by $\{p \in L \mid$ rank $p \leq 1$ and p is compact$\}$ and $\{Rx \mid x \in M \setminus \{0\}\}$ being replaced by $\{Rx \mid x \in M\}$.

7. THE INDEPENDENCE OF THE AXIOMS

The following 7 examples of complete modular lattices together with subsets of "points" show that each of the axioms I - VII is independent to the combination of all the other axioms (of I - VII).

Example 1: $L := L({}_{\mathbb{Z}}\mathbb{Z}^4) \times L({}_{\mathbb{Z}}\mathbb{Z}^4)$,

$P := \{(\mathbb{Z}x, 0) \mid x \in \mathbb{Z}^4 \setminus \{0\} \cdot \}$.

All the axioms except axiom I are satisfied.

Example 2: $L := L({}_{\mathbb{Z}}\mathbb{Z}^4) \times L({}_{\mathbb{Z}}\mathbb{Z}^4)$,

$P := \{(\mathbb{Z}x, 0) \mid x \in \mathbb{Z}^4 \setminus \{0\}\} \cup \{(0, \mathbb{Z}x) \mid x \in \mathbb{Z}^4 \setminus \{0\}\}$

All the axioms except axiom V are satisfied.

Example 3: If L is the lattice of subspaces of a non-Desarguesian projective plane and P the set of points, then rank L = 3 and all the axioms except axiom VI are satisfied.

Example 4: $R := \mathbb{Z}_4$, $L := L({}_R R^3)$, $P := \{Rx \mid x \in R^3 \setminus \{0\}\}$.

All the axioms except axiom III are satisfied, since R is uniform, but ${}_R R^3$ is not non-singular.

Example 5: $R := \mathbb{Z}_2 \times \mathbb{Z}_2$, $L := L({}_R R^3)$, $P := \{Rx \mid x \in R^3 \setminus \{0\}\}$.

All the axioms except axiom IV are satisfied since ${}_R R^3$ is non-singular but R is not uniform.

Example 6: Let $L_e({}_{\mathbb{Z}}\mathbb{Q}^3)$ denote the lattice of finitely generated subgroups of $\mathbb{Q}^3$ and let be $L := L_e({}_{\mathbb{Z}}\mathbb{Q}^3) \cup \{1\}$ the lattice which arises from adding a largest element 1 to $L_e({}_{\mathbb{Z}}\mathbb{Q}^3)$. Let be $P := \{\mathbb{Z}x \mid x \in \mathbb{Q}^3 \setminus \{0\}\}$.

Then all the axioms except Axiom II are satisfied. In fact 0 is the only compact element in L.

Example 7: $L := L({}_{\mathbb{Z}}\mathbb{Z}^3)$,

$$P := \{\mathbb{Z}x \mid x \in \mathbb{Z}^3 \setminus \{0\}\} \setminus \{\mathbb{Z}(1,0,0)\}.$$

Then all the axioms except axiom VII are satisfied.

Axiom V is satisfied since for all linear independent $x, y \in \mathbb{Z}^3$ holds $[\mathbb{Z}x, \mathbb{Z}y, \mathbb{Z}(x+y)]$ and $[\mathbb{Z}x, \mathbb{Z}y, \mathbb{Z}(x-y)]$ and $\mathbb{Z}(x+y) \neq \mathbb{Z}(1,0,0)$ or $\mathbb{Z}(x-y) \neq \mathbb{Z}(1,0,0)$.

Axiom VII is not satisfied since $\mathbb{Z}(1,1,0) \subseteq \mathbb{Z}(0,1,0) + \mathbb{Z}(1,0,0)$ but for no proper submodule $M \subseteq \mathbb{Z}(1,0,0)$ holds $\mathbb{Z}(1,1,0) \subseteq \mathbb{Z}(0,1,0) + M$.

APPENDIX

Computations for the proof that R is closed under + and $\cdot$ (after figure 1). $a_3+p_2 = p_1+p_2$: After using the modular law three times and thus getting p_2 in the place indicated by $\downarrow$ we get

$$a_3+p_2 = (((((\underbrace{a_1+p_{13}+\overset{\downarrow}{p_2}}_{=p_1+p_{13}+p_2\ \geq\ p_3+p_2}) \cap (p_2+p_3))+a_2) \cap (p_2+p_{13}))+p_3) \cap (p_1+p_2)$$

$$= p_3+p_2+a_2 = p_3+p_2+p_1 \geq p_2+p_{13}$$

$$= p_2+p_{13}+p_3 \geq p_1+p_2$$

thus $a_3+p_2 = p_1+p_2$.

$a_4+p_2 = p_1+p_2$: After using the modular law twice and thus getting p_2 in the place indicated by $\downarrow$ we get

$$a_4+p_2 = (((((p_{12}+p_3) \cap (p_2+p_{13}))+a_1) \cap (p_1+p_3)) +$$
$$+ ((\underbrace{a_2+p_3+\overset{\downarrow}{p_2}}_{=p_1+p_3+p_2\ \geq\ p_2+p_{13}}) \cap (p_2+p_{13}))) \cap (p_1+p_2)$$
$$= p_2+p_{13}$$

Now we apply the modular law twice and get p_{13} in the place indicated by $\downarrow$, thus

$$a_4+p_2 = (((((\underbrace{p_{12}+p_3+\overset{\downarrow}{p_{13}}}_{=p_{12}+p_1+p_{13}\ \geq\ p_2+p_{13}}) \cap (p_2+p_{13}))+a_1) \cap (p_1+p_3))+p_2) \cap (p_1+p_2)$$

$$= p_2+p_{13}+a_1 = p_2+p_1+p_{13} \geq p_1+p_3$$

$$= p_1+p_3+p_2$$

thus $a_4+p_2 = p_1+p_2$.

Computation for the proof of (3) for f.

$p_1+b_2 = (p_1+a_1)\cap(p_1+b_2) = p_1+b_1$ is obvious.

Applying the modular law twice we get

$$b_2+a_1 = (\underbrace{((p_2+a_1)\cap\underbrace{(p_{12}+a_2+\overset{\downarrow}{a}_1)}_{\substack{=p_{12}+p_1+a_1 \\ \geqslant p_2+a_1}})+a}_{=p_2+a_1+a\ =\ p_1+p_2+a_1})\cap(p_1+a_1) = p_1+a_1$$

thus $b_1+b_2 = (a_1+b_2)\cap(p_1+b_2) \geqslant p_1$.

Thus $[p_1,b_1,b_2]$.

REFERENCES

Besides the quoted publications we give some additional references (marked by '). For further references cf. [Day 2].

[Am]' Amemiya, I., 'On the representation of complemented modular lattices', J. Math. Soc. Japan 9 (1957), 263-279.

[AF] Anderson, W., Fuller, K., Rings and Categories of Modules, Springer-Verl., New York (1974).

[Ar]' Artmann, B., 'On coordinates in modular lattices', Illinois J. Math. 12 (1968), 626-648.

[Ba 1] Baer, R., Linear algebra and projective geometry, Academic Press, New York (1952).

[Ba 2] Baer, R., 'A unified theory of projective spaces and finite Abelian groups', Trans. Amer. Math. Soc. 52 (1942).

[Bi]' Birkhoff, G., Lattice theory, third ed., AMS Colloq. Publ. (1967).

[Br 1] Brehm, U., 'Untermodulverbände torsionsfreier Moduln', Dissertation, Freiburg i. Br. (1983).

[Br 2] Brehm, U., 'Representation of sum-preserving mappings between submodule lattices by R-balanced mappings', to appear.

[Co] Cohn, P., Free rings and their relations, Academic Press, London (1971).

[CD] Crawley, P., Dilworth, R., Algebraic theory of lattices , Prentice Hall, Englewood Cliffs (1973).

[Day 1]' Day, A., 'Equational theories of projective geometries', Math. Report 3-81, Dept. Math. Sci. Lakehead Univ. (1981).

[Day 2]' Day, A., 'Geometrical applications in modular lattices', Universal Algebra and Lattice Theory, Lect. Notes in Math. 1004, Springer-Verl., Proceedings, Puebla (1982), 111-141.

[Day 3]' Day, A., 'A lemma on projective geometries as modular and / or arguesian lattices', Canad. Math. Bull. 26 (1983), 283-290.

[DP]' Day, A., Pickering, D., 'The coordinatization of Arguesian lattices', Trans. Amer. Math. Soc. 278 (1983), 507-522.

[Fa]' Faltings, G., 'Modulare Verbände mit Punktsystem', Geometriae Dedicata 4 (1975), 105-137.

[Fr]' Freese, R., 'Projective geometries as projective

modular lattices', Trans. Amer. Math. Soc. 251 (1979), 329-342.

[Fri]' Frink, O., 'Complemented lattices and projective spaces of infinite dimension', Trans. Amer. Math. Soc. 60 (1946), 452-467.

[FH]' Fryer, K., Halperin, I., 'The von Neumann coordinatization theorem for complemented modular lattices', Acta Sci. Math. Szeged 17 (1956), 203-249.

[GJL] Grätzer, G., Jónsson, B., Lasker, H., 'The amalgamation property in equational classes of modular lattices', Pacific J. Math. 45 (1973), 507-524.

[HH]' Herrmann, C., Huhn, A., 'Lattices of normal subgroups generated by frames', Colloq. Sci. Janos Bolyai 14 (1975), 97-136.

[Hu 1]' Hutchinson, G., 'On the representation of lattices by modules', Trans. Amer. Math. Soc. 209 (1975), 311-351.

[Hu 2]' Hutchinson, G., 'On classes of lattices representable by modules', Proc. Univ. of Houston, Lattice Theory Conf., Houston (1973), 69-94.

[In] Inaba, E., 'On primary lattices', J. Fac. Sci. Hokkaido Univ. 11 (1948), 39-107.

[Jo 1]' Jónsson, B., 'Representations of complemented modular lattices', Trans. Amer. Math. Soc. 97 (1960), 64-94.

[Jo 2]' Jónsson, B., 'Modular lattices and Desargues theorem', Math. Scand. 2 (1954), 295-314.

[JM]' Jónssen, B., Monk, G., 'Representation of primary Arguesian lattices', Pacific J. Math. **30** (1969), 95-139.

[Kl]' Klingenberg, W., 'Projektive Geometrien mit Homomorphismen', Math. Annalen **132** (1956), 180-200.

[Mä]' Maeda, F., Kontinuierliche Geometrien, Springer-Verl., Berlin (1958).

[vN] von Neumann, J., Continuous Geometry, Princeton Univ. Press, Princeton (1960).

[Sc]' Scoppola, C., 'Sul reticolo dei sottogruppi di un gruppo abeliano senza torsione di rango diverso da 1: una caratterizzazione reticolare', Rend. Sem. Mat. Univ. Padova **65** (1981), 205-221.

[Sk]' Skornyakov, L., Complemented modular lattices and regular rings, Oliver and Boyd, Edinburgh (1964).

[St] Stenström, B., Rings of quotients, Springer-Verl., Berlin (1975).

Epilog

THE ADVANTAGE OF GEOMETRIC CONCEPTS IN MATHEMATICS

C. Arf
Research Institute for Basic Sciences
P.O.Box 74 Gebze-Kocaeli, TURKEY

As an outsider to the subject of this meeting my talk will be necessarily restricted to the expression of my views concerning the importance of geometric concepts in the evolution of mathematics as a creatian of the human mind.

If we try to review the historical evolution of the human mind, we notice a close parallelism with our individual intellectual evolution. We can even say that the programmation of our brain in the today's human community is a highly speeded reproduction of the evolution the human mind during the prehistorical and historical several millenia. That process of programation seems to begin at our babyhood, with the passage from a short stage of not permanently registered sets of chaotic sensations to that of almost permanently registered perceptions. The very first of such perceptions seems to occur by the isolation of a particular sensation in a chaotic set of almost simultaneous sensations. That isolation seems to be result of the reaction of the baby's self preservation instinct which seems to be inherent to every organized matter. The baby seems to register almost permanently such isolated sensations in the form of a duality versus the chaotic multitude within which it makes that selection. The baby's mind identifies such almost permanently registered sensations with other that may occur at different times by the location of the registration in its brain, and this process of identification makes out of a registered sensation a concept. In the process of identification of different registered and therefore remembered sensations, it is necessary that the baby's mind is able to recognize, even if vaguely, the chaotic multitude of sensations within which it isolates the registered one. This necessity of recognition leads the baby's mind to isolate other sensations included in that chaotic multitudes. Thus the baby gets the concept of food and that of its mother as two of the firsts, by isolation of some smelling, tasting, optical, acoustical etc... sensations. This process of registering of some sets of sensations as concepts, repeats itself in different forms and this leads to increase the degree of abstraction of the acquired concepts. Thus for example a baby that is fed at different times by its mother, or by another person uses a milk bottle or a spoon, becomes aware of the more abstract concept of a source of food. Such abstractions

R. Kaya et al. (eds.), Rings and Geometry, 553–556.

create, of course only vaguely, the concepts of units and that of equivalence of units, that can yet be differentiated by preferences and therefore by orderings. That vague concepts of units, of ordering and of almost equivalent units create in the baby's mind, of course very vaguely the process of recursive counting and ordering. These experiences prepare the baby to the abstraction of the recursive concept of natural numbers and to the abstraction of the cause and effect relations. However those concepts seems to occur at the stage of childhood.

By the time the human baby reaches the childhood, the concepts of equivalence and of recursive ordering are already realized with enough precision to enable the child to represent its perceptions by words and pictures. Thus the child as well as the early historical human being is ready to discover or to learn some arithmetic such as addition and multiplication of numbers and the formation of fractions. In all these the behaviour of the human mind is more or less explicitly recursive. In a more advanced childhood or in a more advanced historical period the human being becomes aware of the recursive totality of the natural numbers and notices the need to imagine the Cartesian products of the set of natural numbers. Thus for example he describes a tribe as consisting of 100 males, 110 females, 80 children, 30 horses, 40 camels, 30 cows and 200 sheep. Furthermore he is led to concentrate on some, still recursively characterized subsets of such Cartesian product. Thus for example a fraction is actually the subset of the Cartesian product $\mathbb{N} \times \mathbb{N}$ consisting of the pairs $((a_i \varepsilon \mathbb{N}, b_i \varepsilon \mathbb{N}: a_i b_j \; a_j b_i, \; i,j \varepsilon \mathbb{N})$. From that stage on he feels the need of getting busy with mathematics, He notices the concept of divisibily, the distinction of some first prime numbers, formations of some sets of natural numbers and making some mathematical statements in the form of finding the common part of several recursively defined sets of natural numbers. This was what we mean by solving some sets of algebraic equations. As once stated by Kronecker, the human being was thus equipped with all that was necessary to formulate and to answer mathematical questions. However, with this alone he could not go very much farther than the mathematics of the mesopotamiens.The explanation of this, that sounds reasonable to me, is the following:

As I have just stated the problems that are naturally presented to human beings lead them to form Cartesian products of several copies of the set $\mathbb{N}$ of natural numbers and consider some subsets or some sets of classes of members of such products. Some of those sets consist of very small numbers of members, in which case they can be defined by explicit enumerations of their members. But most of the times they are either infinite or consist of very large numbers of members. In the last case the recursive processes are the only means of characterization of those sets. Of course the recursions do not in general occur explicitly in the formulation and in the solutions of the problems. They do occur through properties previously derived from recursive processes. Otherwise nobody would be patient enough to follow the large multitudes of recursions involved in those problems. The methods that are used in the design of the set of implicit or explicit recursions that leads to the solutions, consist of elimination some factors of the Cartesian products until the problem is reduced to a problem in a small product of copies of $\mathbb{N}$, where the solution becomes trivial; and of finding a way of formulating the problem in other even larger Cartesian products where it offers enough symmetries to reduce it automatically to a problem in a small enough Cartesian product where it becomes almost trivial. The use of these methods as well as of their combinations requires some virtuosity on the part of the mathematician in making appropriate guesses in the choices of the factors, of the Cartesian products, to be eliminated and of the other Cartesian products into which the problem should be transfered in order to get it endowed with simplifying symmetries. Thus a mathematician needs next to a patient logic, a large amount of what one calls intuition. That is just where the use of geometric concepts comes in. Intuition seems to be the mysterious capacity of the human brain to make, statistical and as such imprecise judgements. It is seemingly based on the simultaneous contemplation of large amounts of comparable perceptions, in opposition to the very nature of the formation of the concept of the set of natural numbers. The nature endoves the human mind with the concept of a three dimensional space as whole consisting of a multitude, where we do not feel the need of isolating units in order to get rid of a sensation of chaos. Unlike the concepts of the set of natural numbers which we bild up step by step, the concept of space is realized in one single step. The human baby is aware of a sensation of space as the location of certains of units that are created in the above described process of isolation of sensations. It distinguishes directions and distances of the locations of its unit, but it cannot isolate those location without those units as landmarks. I think that this is just what leads the human mind to concieve the space as a homogeneous continuum. Our first idea of real numbers is based on that direct perception of space as a whole. This direct perception of a multitude enables the human mind to notice directly a multitude of relations between elements of that global concepts of space. Then the mathematician can use this ability to guess some kind of a scheme of the solution of a problem interpreting it as a geometric problem.

The early settled human communities were agricultural communities. They were making their livings on the soil, and needed to count properly by the process of exchanging their goods. So they did develop some arithmetical rules. They did not need much geometry. They could use landmarks to fix distances and directions and notice experimentally a few rules to evaluate areas and volumes in terms of their dimensions. Unlike the Egyptians and Mesopotamians, the Hellenic Communities were living on Egean shores and islands, and the navigation was an important part of their everyday's lives There are no landmarks on the sea except for the stars in the sky and distant silhouettes of the shores. So they did need to workout some deductive axiomatic geometry. Their axiomatic geometry was the beginning of the mathematics proper as a science. They did not make much use of the above mentioned geometric intuition to formulate and solve problems concerning numbers more sophisticated than simple arithmetics. However, they did prove that the set of primes is not finite, and they did almost create the concept of real numbers with their theory of propositions. They also did almost reach the concept of limit in their efforts to evaluate the ratio of the length of the circumference of a circle to its radius and to evaluate the area of domains bounded by a curve.

Thus in my opinion the main advantage of the geometric concepts is the fact that we can intuitively perceive simultaneously a large number of conceptual objects and their interrelations by locating them in our space. This permits us to make intuitive choices of the processes which have chances to lead to the solutions of our problems. In addition we get the impression that we have grasped in almost one single step the structure of the solution and this is in my opinion a big source of happines for us.

Of course opinions cannot be proved. They can only be justified by facts. The facts that one uses tables, diagrams, geometric terminology in all branches of mathematics is my justification. The following simple example illustrates my point:

Let us consider the following statement: If

$$B,C\varepsilon a, \quad C,A\varepsilon b, \quad A,B\varepsilon c, \quad B',C'\varepsilon a', \quad C',A'\varepsilon b', \quad A',B'\varepsilon c'$$

$$d\cap a=d\cap a', \quad d\cap b=d\cap b', \quad d\cap c=d\cap c'$$

then

$$D\varepsilon AA', \quad D\varepsilon BB', \quad D\varepsilon CC' .$$

An experienced person recognizes immediately here the Desargues Configuration. But a less experienced person sees the statement more easily in the drawing below, that suggests immediately the proof in the 3-space.

I wish for the mathematician to be not deprived of this wonderful tool of discovery and source of happiness.

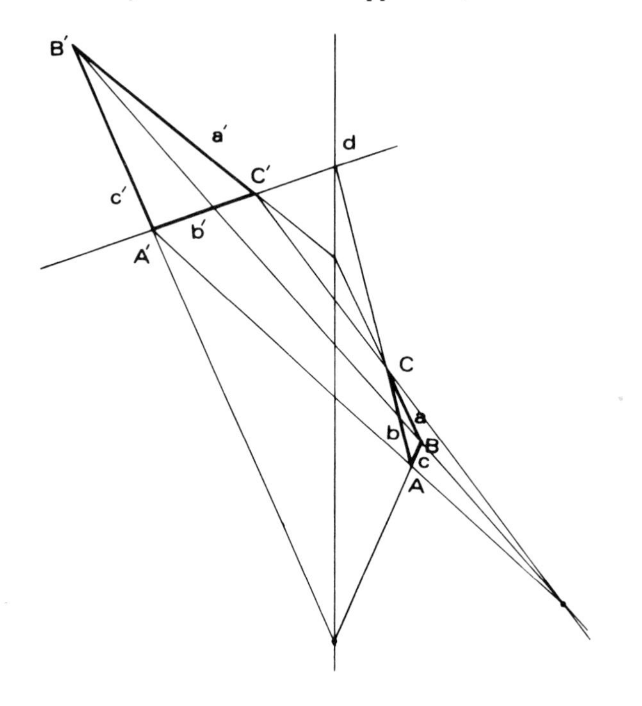

INDEX OF SUBJECTS